T0201277

INVASION GENETICS

Participants at the symposium *Invasion Genetics: The Baker and Stebbins Legacy* held during August 13–15, 2014, Asilomar, California, USA, organized by Spencer C. H. Barrett, Robert I. Colautti, Katrina M. Dlugosch, and Loren H. Rieseberg.

INVASION GENETICS

THE BAKER AND STEBBINS LEGACY

Edited by

Spencer C. H. Barrett

Department of Ecology and Evolutionary Biology, University of Toronto,
25 Willcocks Street, Toronto, ON M5S 3B2, Canada

Robert I. Colautti

Department of Biology, Queen's University, Kingston, ON K7L 3N6, Canada

Katrina M. Dlugosch

Department of Ecology and Evolutionary Biology, University of Arizona,
PO Box 210088, Tucson, AZ 85721, USA

Loren H. Rieseberg

Department of Botany, University of British Columbia, 1316-6270 University Blvd.,
Vancouver, BC V6T 1Z4, Canada; Department of Biology, Indiana University,
Bloomington, IN 47405, USA

WILEY

Registered Office
John Wiley & Sons, Ltd, The Atrium, Southern Gate, Chichester, West Sussex, PO19 8SQ, UK

Editorial Offices
9600 Garsington Road, Oxford, OX4 2DQ, UK
The Atrium, Southern Gate, Chichester, West Sussex, PO19 8SQ, UK
111 River Street, Hoboken, NJ 07030-5774, USA

For details of our global editorial offices, for customer services and for information about how to apply for permission to reuse the copyright material in this book please see our website at www.wiley.com/wiley-blackwell.

Library of Congress Cataloging-in-Publication Data

Names: Barrett, Spencer Charles Hilton, editor. | Colautti, Robert I., editor. |
    Dlugosch, Katrina M., editor. | Rieseberg, Loren H., editor.
Title: Invasion genetics : the Baker and Stebbins legacy / edited by Spencer C. H.
    Barrett, Robert I. Colautti, Katrina M. Dlugosch, Loren H. Rieseberg.
Description: Chichester, West Sussex, UK ; Hoboken, NJ : John Wiley & Sons, Inc., 2016. |
    Includes bibliographical references and index.
Identifiers: LCCN 2016011611 | ISBN 9781118922163 (hardback)
Subjects: | MESH: Baker, Herbert G. | Stebbins, G. Ledyard (George Ledyard), 1906–2000. |
    Genetics of colonizing species. | Genetic Phenomena | Biological Evolution | Introduced Species
Classification: LCC QH431 | NLM QU 500 | DDC 576.5–dc23 LC record available at http://lccn.loc.gov/2016011611

A catalogue record for this book is available from the British Library.

Cover credit: Starling: Professor Tim Blackburn; Cane toad: Dr Matthew Greenlees

Wiley also publishes its books in a variety of electronic formats. Some content that appears in print may not be available in electronic books.

Set in 9/11pt Photina by SPi Global, Pondicherry, India
Printed in Singapore by C.O.S. Printers Pte Ltd

1   2017

CONTENTS

CONTRIBUTORS

SAMANTHA R. ANDERSON, *Department of Ecology and Evolutionary Biology, University of Arizona, PO Box 210088, Tucson, AZ 85721, USA*

BRIAN BARRETT, *Department of Integrative Biology, University of Texas, Austin, TX 78712, USA*

SPENCER C. H. BARRETT, *Department of Ecology and Evolutionary Biology, University of Toronto, 25 Willcocks Street, Toronto, ON M5S 3B2, Canada*

JOCELYN E. BEHM, *Department of Ecological Sciences, Section of Animal Ecology, Vrije Universiteit, 1081 HV Amsterdam, the Netherlands; Center for Biodiversity, Department of Biology, Temple University, Philadelphia, PA 19122, USA*

TIM M. BLACKBURN, *Department of Genetics, Evolution and Environment, Centre for Biodiversity and Environment Research, University College London, Gower Street, London, WC1E 6BT, UK; Institute of Zoology, ZSL, Regent's Park, London, NW1 4RY, UK*

MARK W. BLOWS, *School of Biological Sciences, University of Queensland, St Lucia, QLD 4072, Australia*

DAN G. BOCK, *Department of Botany, University of British Columbia, 1316-6270 University Blvd., Vancouver, BC V6T 1Z4, Canada*

OLIVER BOSSDORF, *Institute of Plant Sciences, University of Bern, Altenbergrain 21, CH-3013 Bern, Switzerland; Institute of Evolution and Ecology, University of Tübingen, Auf der Morgenstelle 5, D-72076 Tübingen, Germany*

JOSEPH BRAASCH, *Department of Ecology and Evolutionary Biology, University of Arizona, PO Box 210088, Tucson, AZ 85721, USA*

F. ALICE CANG, *Department of Ecology and Evolutionary Biology, University of Arizona, PO Box 210088, Tucson, AZ 85721, USA*

CELINE CASEYS, *Department of Botany, University of British Columbia, 1316-6270 University Blvd., Vancouver, BC V6T 1Z4, Canada*

PHILLIP CASSEY, *School of Biological Sciences, University of Adelaide, Adelaide, SA 5005, Australia*

JOANNE CLAVEL, *Conservation des Espèces, Restauration et Suivi des Populations – CRBPO, Muséum National d'Histoire Naturelle-CNRS-Université Pierre et Marie Curie, 55 rue Buffon, 75005, Paris, France*

ROBERT I. COLAUTTI, *Department of Biology, Queen's University, Kingston, ON K7L 3N6, Canada*

ROGER D. COUSENS, *School of BioSciences, The University of Melbourne, Melbourne, VIC 3010, Australia*

MELANIA E. CRISTESCU, *Department of Biology, McGill University, Montreal, QC H3A 1B1, Canada*

WAYNE DAWSON, *School of Biological and Biomedical Sciences, Durham University, South Road, Durham, DH1 3LE, UK*

TROY DAY, *Department of Mathematics and Statistics, Jeffery Hall, Queen's University, Kingston, ON K7L 3N6, Canada; Department of Biology, Queen's University, Kingston, ON K7L 3N6, Canada*

KATRINA M. DLUGOSCH, *Department of Ecology and Evolutionary Biology, University of Arizona, PO Box 210088, Tucson, AZ 85721, USA*

CYRIL DUTECH, *BIOGECO, INRA, Université Bordeaux, 33610, Cestas, France*

JACINTHA ELLERS, *Department of Ecological Sciences, Section of Animal Ecology, Vrije Universiteit, 1081 HV Amsterdam, the Netherlands*

LAURENT EXCOFFIER, *Institute of Ecology and Evolution, University of Berne, Berne 3012, Switzerland; Swiss Institute of Bioinformatics, Lausanne 1015, Switzerland*

ALICE FEURTEY, *Ecologie Systématique Evolution, Univ. Paris-Sud, CNRS, AgroParisTech, Université Paris-Saclay, 91400 Orsay, France*

MARKUS FISCHER, *Institute of Plant Sciences, University of Bern, Altenbergrain 21, CH-3013 Bern, Switzerland*

HEATHER D. GILLETTE, *Department of Ecology and Evolutionary Biology, University of Arizona, PO Box 210088, Tucson, AZ 85721, USA*

TATIANA GIRAUD, *Ecologie Systématique Evolution, Univ. Paris-Sud, CNRS, AgroParisTech, Université Paris-Saclay, 91400 Orsay, France*

PIERRE GLADIEUX, *Ecologie Systématique Evolution, Univ. Paris-Sud, CNRS, AgroParisTech, Université Paris-Saclay, 91400 Orsay, France*

MIN A. HAHN, *Department of Botany, University of British Columbia, 1316-6270 University Blvd., Vancouver, BC V6T 1Z4, Canada*

JILL A. HAMILTON, *Department of Evolution and Ecology, University of California, Davis, CA 95616, USA; Department of Biological Sciences, North Dakota State University, Fargo, ND 58102, USA*

MATTHEW R. HELMUS, *Department of Ecological Sciences, Section of Animal Ecology, Vrije Universiteit, 1081 HV Amsterdam, the Netherlands; Center for Biodiversity, Department of Biology, Temple University, Philadelphia, PA 19122, USA*

SYLVIA M. HEREDIA, *Department of Botany, University of British Columbia, 1316-6270 University Blvd., Vancouver, BC V6T 1Z4, Canada*

KATHRYN A. HODGINS, *School of Biological Sciences, Monash University, Clayton, VIC 3800, Australia*

MICHAEL E. HOOD, *Department of Biology, Amherst College, Amherst, MA 01002, USA*

SARIEL HÜBNER, *Department of Botany, University of British Columbia, 1316-6270 University Blvd., Vancouver, BC V6T 1Z4, Canada*

WENDY A.M. JESSE, *Department of Ecological Sciences, Section of Animal Ecology, Vrije Universiteit, 1081 HV Amsterdam, the Netherlands*

MARK KIRKPATRICK, *Department of Integrative Biology, University of Texas, Austin, TX 78712, USA*

MARK VAN KLEUNEN, *Ecology, Department of Biology, University of Konstanz, Universitätsstrasse 10, Konstanz D-78457, Germany*

JASON J. KOLBE, *Department of Biological Sciences, University of Rhode Island, Kingston, RI 02881, USA*

TONIA KORVES, *Data Analytics Department, The MITRE Corporation, 202 Burlington Rd., Bedford, MA 01730, USA*

RUSSELL LANDE, *Division of Biology, Imperial College London, Silwood Park Campus, Ascot, Berkshire SL5 7PY, UK*

JENNIFER A. LAU, *Kellogg Biological Station and Department of Plant Biology, Michigan State University, 3700 E. Gull Lake Dr., Hickory Corners, MI 49060, USA*

CAROL EUNMI LEE, *Center of Rapid Evolution (CORE), Department of Zoology, University of Wisconsin, 430 Lincoln Drive, Birge Hall, Madison, WI 53706, USA*

JULIE L. LOCKWOOD, *Ecology, Evolution, and Natural Resources, Rutgers University, New Brunswick, NJ 08901, USA*

JONATHAN B. LOSOS, *Department of Organismic and Evolutionary Biology and Museum of Comparative Zoology, Harvard University, Cambridge, MA 02138, USA*

NOËLIE MAUREL, *Ecology, Department of Biology, University of Konstanz, Universitätsstrasse 10, Konstanz D-78457, Germany*

KATRINA McGUIGAN, *School of Biological Sciences, University of Queensland, St Lucia, QLD 4072, Australia*

MIKI OKADA, *Department of Evolution and Ecology, University of California, Davis, CA 95616, USA*

JOHN R. PANNELL, *Department of Ecology and Evolution, University of Lausanne, Biophore Building, 1015 Lausanne, Switzerland*

MADALIN PAREPA, *Institute of Plant Sciences, University of Bern, Altenbergrain 21, CH-3013 Bern, Switzerland; Institute of Evolution and Ecology, University of Tübingen, Auf der Morgenstelle 5, D-72076 Tübingen, Germany*

INGRID M. PARKER, *Department of Ecology and Evolutionary Biology, University of California, Santa Cruz, CA 95064, USA*

STEPHAN PEISCHL, *Institute of Ecology and Evolution, University of Berne, Berne 3012, Switzerland; Swiss Institute of Bioinformatics, Lausanne 1015, Switzerland*

MARK F. RICHARDSON, *Centre for Integrative Ecology, School of Life and Environmental Sciences, Deakin University, Pigdons Road, Geelong, VIC 3217, Australia*

LOREN H. RIESEBERG, *Department of Botany, University of British Columbia, 1316-6270 University Blvd., Vancouver, BC V6T 1Z4, Canada; Department of Biology, Indiana University, Bloomington, IN 47405, USA*

LEE A. ROLLINS, *Centre for Integrative Ecology, School of Life and Environmental Sciences, Deakin University, Pigdons Road, Geelong, VIC 3217, Australia*

MÉLANIE ROY, *Evolution et Diversité Biologique, Université Toulouse Paul Sabatier-Ecole National de Formation Agronomique-CNRS, 118 route de Narbonne, 31062, Toulouse, France*

JOHANNA SCHMITT, *Department of Evolution and Ecology, University of California, Davis, CA 95616, USA*

RICHARD SHINE, *School of Biological Sciences A08, University of Sydney, Sydney, NSW 2006, Australia*

ALODIE SNIRC, *Ecologie Systématique Evolution, Univ. Paris-Sud, CNRS, AgroParisTech, Université Paris-Saclay, 91400 Orsay, France*

CASEY P. terHORST, *Department of Biology, California State University, Northridge, 18111 Nordhoff Street, Northridge, CA 91330-8303, USA*

KATHRYN G. TURNER, *Department of Botany, University of British Columbia, 1316-6270 University Blvd., Vancouver, BC V6T 1Z4, Canada*

KENNETH D. WHITNEY, *Department of Biology, University of New Mexico, Albuquerque, NM 87131-0001, USA*

YUAN-YE ZHANG, *Institute of Plant Sciences, University of Bern, Altenbergrain 21, CH-3013 Bern, Switzerland*

PREFACE

Invasion biology is concerned with the introduction and spread of non-native species and their environmental, human health and economic impacts. Although a relatively young discipline, it now represents a major growth area in applied biology and conservation science with a dedicated journal – *Biological Invasions* – and numerous meetings and annual symposia. For most of its short history, the study of invasions has largely addressed ecological and environmental questions, many originating from Charles Elton's influential book *The Ecology of Invasions by Plant and Animals*, published in 1958. The impressive progress made in this area, especially at the population and community levels, was recently reviewed in the volume *Fifty Years of Invasion Ecology: The Legacy of Charles Elton*, edited by David Richardson in 2011.

Since its inception, the growth of invasion biology has not been accompanied by equivalent attention to the genetics and evolution of invasive species. Fundamental evolutionary principles such as natural selection, genetic drift and the evolution of local adaptation were rarely considered in the early literature on biological invasions. Moreover, until recently investigations of the patterns of genetic diversity in invasive species were relatively sparse. This difference in attention to studies on the ecology versus the genetics of invading species is particularly intriguing because, as Daniel Simberloff has pointed out (*Annual Review of Ecology and Systematics*, 1988, **19**: 473–511), genetic studies of rare and endangered species played a major role in stimulating the development of conservation science in the 1970s and 1980s. Similar questions on the genetic causes and consequences of rarity and abundance can inform a variety of general topics in ecology and evolutionary biology.

The early disassociation between ecological and genetic studies of biological invasions is much less evident today. The rapid development of diverse molecular approaches for assaying genetic diversity in populations, combined with advances in evolutionary theory, have given rise to a burgeoning interest in the genetics and evolution of biological invasions during the past two decades. This work is often interpreted in the context of information on the life histories, demography and ecology of populations. Because of these exciting developments, we believe it is therefore timely to provide a synthesis of the new field of invasion genetics – *the study of processes shaping genetic diversity and contemporary evolution in introduced species and their influence on biological invasions*. When fully integrated with invasion ecology these fields should provide deeper insights into the causes and consequences of biological invasions, as well as providing a unique framework for studies of contemporary evolution.

Another important motivation for assembling this volume on invasion genetics is to celebrate the 50th anniversary of one of the most important books in evolutionary biology: *The Genetics of Colonizing Species* (1965), edited by Herbert G. Baker and G. Ledyard Stebbins. The book was based on a symposium held in 1964 at Asilomar, California, and was remarkable for a variety of reasons, not least because the contributors were many of the leading evolutionary biologists and geneticists of the time including Theodosius Dobzhansky, Richard C. Lewontin, Ernst Mayr, Edward O. Wilson, Conrad H. Waddington, Hampton L. Carson, Charles B. Heiser Jr. and Robert W. Allard, to name just a few. The Asilomar symposium was based on the idea that successful colonizing species – including weeds, pests and diseases – represent natural experiments for investigating key questions concerning the ecological and evolutionary genetics of introduced populations and can therefore inform more generally evolutionary biology. Many of the topics that were discussed by participants at the Asilomar meeting are being actively studied today using diverse invasive species. The Baker and Stebbins volume can therefore be rightly considered as the foundational document for invasion genetics, despite the 'lag phase' in its influence on the early development of invasion biology.

To commemorate the 50th anniversary of the Baker and Stebbins meeting, the editors of this volume organized a symposium – *Invasion Genetics: The Baker and Stebbins Legacy* – held at Asilomar during August 13–15, 2014. Seventy-four participants from 12 countries met to hear 19 talks, contribute towards three panel discussions and view 37 posters in evening sessions. The meeting was an outstanding success and was characterized by a high level of collegiality, spirited exchange and enthusiasm about the future prospects for invasion genetics. The talks and a sample of selected posters at Asilomar were recently published as a special issue of the journal *Molecular Ecology* (Volume **24**, Number 9, 2015). In addition, to provide a more comprehensive treatment of the meeting, and as a fitting companion to the Baker and Stebbins volume, the editors planned this book to include articles from the *Molecular Ecology* volume as well as additional content from the symposium.

One of the most popular features of the Baker and Stebbins volume was the inclusion of verbal exchanges between the participants at the meeting, which were often insightful, and also helped to guide future work on the genetics of colonizing species. In planning the meeting we decided to continue with this tradition by recording questions to speakers and their answers. In addition, we organized three panel discussions at the end of each section of the meeting and recorded all exchanges. In this volume we include a selection of these questions and answers to provide a historical record of the intellectual exchanges that were a feature of the meeting, and to provide in-depth coverage of some of the key issues in invasion genetics. The volume also includes two chapters by authors that were invited to the meeting but were unable to attend, as well as introductory summaries to each section of the book written by the editors and guest authors. *Invasion Genetics: The Baker and Stebbins Legacy* contains 20 chapters grouped into three sections: 1 – Evolutionary Ecology, 2 – Evolutionary Genetics and 3 – Invasion Genomics. These parts reflect the three primary themes in which considerable progress has been made in research on the evolution and genetics of invasive species over the past two decades. These include studies of a wide range of plants, animals and microorganisms, as well as theoretical analyses on some of the key problems in evolutionary genetics relevant to invasive populations.

The organization of the 2015 Asilomar meeting, the compilation of the special issue of *Molecular Ecology* and the production of this book have involved the cooperation and hard work of many people. Each chapter in this volume was reviewed by at least two reviewers, and we thank the following individuals for their efforts in helping authors improve the quality of chapters: Jake Alexander, Jill Anderson, Allan Baker, Regina Baucom, Mark Blows, Dan Bock, Benjamin Brachi, Jutta Burger, Jeremiah Busch, Mark Chapman, David Chapple, Melania Cristescu, Angela Dale, Troy Day, Jennifer Dechaine, Charles Fenster, Lila Fishman, Jannice Friedman, Richard Glor, Ruth Hufbauer, David Houle, Boris Igic, Mark Kirkpatrick, Beth Leger, Julie Lockwood, Andrew Lowe, Luke Mahler, Lynn Martin, Ayub Oduor, John Pannell, Diana Pilson, Lisa Pope, Peter Prentis, Marcel Rejmánek, Christina Richards, David Richardson, Fabrice Roux, Howard Rundle, Frank Shaw, Jessica Stapley, Neal Stewart, John Stinchcombe, Michael Whitlock, Kenneth Whitney and Amber Wright. We also thank Tim Vines and Jennifer Gow of the *Molecular Ecology* editorial office for their assistance with expediting manuscript reviews and prompt responses to many of our questions, and Tobias Mankis for help in transcribing the audiotapes at Asilomar. At Wiley-Blackwell, we very much appreciated the assistance of Alice Wood (Journal Publishing Manager), Leah Webster (Journal Publishing Assistant) and Vicci Parr (Senior Marketing Manager) who helped to make the meeting at Asilomar run smoothly and also provided organizational and logistical support; Ward Cooper (Former Senior Commissioning Editor, Ecology, Conservation and Evolution) and Liz Ferguson (Vice President, Editorial Development) for their early advice and enthusiasm for the project; and David McDade (Executive Editor, *Natural Sciences*), Kelvin Matthews (Senior Project Editor) and Emma Strickland (Assistant Editor, *Natural Sciences*) for their assistance with the final production of the book. Finally, we would like to thank Wiley-Blackwell for their generous financial support of the meeting in Asilomar, without which this volume could not have been produced.

Spencer C. H. Barrett
Toronto
Robert I. Colautti
Kingston
Katrina M. Dlugosch
Tucson
Loren H. Rieseberg
Vancouver
September, 2015

Chapter 1
———————

FOUNDATIONS OF INVASION GENETICS: THE BAKER AND STEBBINS LEGACY

Spencer C. H. Barrett

Department of Ecology and Evolutionary Biology, University of Toronto,
25 Willcocks Street, Toronto, ON, M5S 3B2, Canada

Abstract

Invasion genetics is a relatively new discipline that investigates patterns of genetic variation in populations of invasive species and their ecological and evolutionary consequences. Evolutionary biologists have a long-standing interest in colonizing species, owing to their short life cycles and widespread distributions, but not until publication of *The Genetics of Colonizing Species* (1965), edited by H.G. Baker and G.L. Stebbins, was a synthesis on the genetics and evolution of colonizers available. Here, I make the case that the Baker and Stebbins volume is the foundational document for invasion genetics, and in conjunction with the increased use of genetic markers and development of invasion biology, resulted in the birth of this new field over the past two decades. I consider the historical origins and legacy of the Baker and Stebbins volume and review some of the key issues that were addressed. I provide biographical sketches of the two editors, emphasizing their contrasting backgrounds and personalities. I review examples from my own work on plant invasions that are relevant to issues discussed by contributors to the volume. These include the following: determinants of invasion success, life history trade-offs, generalist vs. specialist strategies, general-purpose genotypes, adaptive phenotypic plasticity, mating systems and the influence of bottlenecks on genetic variation. I conclude by posing several key questions in invasion genetics and argue that one of the main challenges that the area faces is to integrate experimental field studies of the ecology and demography of populations with the largely descriptive approaches that have tended to dominate most research to date.

Previously published as an article in *Molecular Ecology* (2015) 24, 1927–1941, doi: 10.1111/mec.13014

———————

INTRODUCTION

Colonization is the establishment of a species at a site that it does not currently occupy and is necessarily a feature of the population biology of all organisms. Territorial expansion occurs at a range of spatial and temporal scales, from intercontinental migration to the local patch, and over geological epochs to the transport of species by humans in more recent times. The ecological and evolutionary consequences of colonization are therefore highly scale dependent and species vary in the extent to which recurrent colonizing episodes have shaped their ecology, life histories and genetic systems. The scale of colonization also has important genetic consequences as the amounts and kinds of genetic variation transferred from one place to another can influence the likelihood of successful establishment, future spread and evolutionary potential.

Long-distance dispersal can expose colonizing populations to novel selective forces because of different abiotic and biotic conditions in the introduced compared to the native range. Most species introductions fail owing to maladaptation or chance, but those that are successful represent 'experiments in evolution', particularly when adaptive responses occur over short timescales. Among introduced species, some are successful at confronting the many challenges presented by novel environments, as a result they can become highly invasive and exhibit rapid range expansion. A key question of importance to biologists interested in biological invasions concerns the extent to which *in situ* evolutionary changes occur during the invasion process. Here, I trace the foundations of the fledgling field of invasion genetics and identify the publication of the edited volume *The Genetics of Colonizing Species* (Baker & Stebbins 1965) as being particularly influential because it considered for the first time in detail the evolutionary processes that occur in species particularly adept at colonization.

Invasion biology is an applied scientific discipline concerned with the introduction and spread of introduced (non-native) species throughout the world, along with their environmental, health and economic impacts. Although invasion biology is multidisciplinary, addressing diverse basic and applied questions, a dominant paradigm focuses on determining the factors that cause species to become invasive and trying to predict which features of organisms and their new environments promote invasion success. Although several

early naturalists reported on species introductions (e.g. Darwin discussed the invasiveness of thistles and cardoon in Argentina during his voyage on the Beagle, reviewed in Chew 2011), it was not until Charles Elton (1958) published *The Ecology of Invasions by Plant and Animals* that a synthetic treatment of numerous case studies was attempted. Elton's monograph is often considered the foundation for the scientific study of biological invasions (Richardson & Pyšek 2008), but significantly, it was not associated with a surge of interest on the topic. Simberloff (2011) has persuasively argued that the real impetus for the birth of invasion biology came later in the 1980s, from the volumes published on the ecology of invasions by the Scientific Committee on Problems of the Environment (SCOPE) beginning in 1982 (e.g. Mooney & Drake 1986). Subsequently, the field of invasion biology experienced exponential growth, and by the end of the 1990s, the journal *Biological Invasions* appeared, devoted to publications on species introductions.

For most of its short history, invasive biology has focused primarily on ecological questions, and until recently, there has been a striking disassociation between studies on the ecology of invasions from those concerned with their genetics and evolution. Although Elton (1958) briefly mentioned the possible role of genetics in the decline of Canadian Pondweed (*Elodea canadensis*) in the United Kingdom, the evolution of resistance in insect pests and fungi and the occurrence of hybridization and polyploidy in *Spartina* invasions, he did not consider in any detail the possibility that many invasive populations may have the capacity to respond adaptively to novel ecological conditions. Elton was an ecologist not an evolutionist, and because of this, his perspective was mainly on species interactions and community ecology. It is noteworthy that his book was barely cited in the Baker & Stebbins (1965) volume, despite the fact that both works are concerned with species invasions (Simberloff 2011). Similarly, with few exceptions (e.g. Baker 1986; Barrett & Richardson 1986), the SCOPE volumes, following the mandate of the committee, were largely restricted to ecological studies of invasive species, with little consideration of whether the genetic characteristics of invasive populations might have relevance to their spread and management.

During the 1970–80s, evolutionary biologists, following the lead provided by the Baker & Stebbins (1965) volume, began to investigate a variety of questions in ecological and evolutionary genetics using invasive species as study systems (e.g. Allard *et al.* 1972; Selander & Kaufman 1973; Richardson *et al.* 1980). Significantly, this work had little influence on the early development of invasion biology (Callaway & Maron 2006), but over time several volumes (Parsons 1983; Williamson 1996; Cox 2004; Sax *et al.* 2005) dealt with genetic issues in invasion biology, and this helped to integrate ecological and evolutionary approaches to the study of species introductions. In concert with advances in molecular techniques for assaying genetic variation and the development of a growing body of evolutionary theory relevant to evolutionary processes in colonizing populations, this led to the birth of invasion genetics.

In this introductory chapter, I provide a historical background to the *The Genetics of Colonizing Species* and consider its scientific legacy. My treatment does not attempt to be comprehensive and instead involves selected examples, particularly on plant invasions as I know these best. It is written from a personal perspective. I was a former PhD student of H.G. Baker who went to California in the early 1970s after being 'turned on' by reading the Baker and Stebbins volume at Reading University, U.K., where I was taking a degree in the Department of Agricultural Botany. The book had an enormous influence on my thinking and initiated a lifelong interest in the ecology and genetics of plant invasions.

I begin this article by considering the goals of the 1964 Asilomar symposium that gave rise to the volume edited by Baker and Stebbins the following year. I provide short biographical sketches of the two editors based in part on my own interactions with them. I consider how their backgrounds and research interests may have influenced the choice of contributors and the main themes of the meeting. I make the case that the Baker and Stebbins volume helped to initiate research on a range of fundamental problems concerned with the ecological and evolutionary genetics of colonization, which now form the conceptual foundations of invasion genetics. I briefly review selected topics in invasion genetics, evaluating progress made since Asilomar, and conclude by considering key questions and challenges for the fledgling field. *The Genetics of Colonizing Species* can lay claim to being the foundational document for invasion genetics and its historical legacy was celebrated by a 50th anniversary symposium held at Asilomar in August 2014, which forms the basis of this volume.

HISTORICAL BACKGROUND

The Genetics of Colonizing Species is a collection of studies and discussions that arose from a symposium held from 12 to 16 February 1964 at Asilomar, a charming seaside retreat on the Pacific coast of California, near Monterey. The initial idea for the symposium came from the influential British geneticist Cyril H. Waddington, who at the time was the President of the International Union of Biological Sciences (IUBS), a nongovernmental organization for advancing knowledge of biology in the service of human improvement. Waddington, Baker and Stebbins selected the speakers for the meeting and their contributions resulted in 27 articles, including the introduction to the symposium by Waddington and a summary chapter by Ernst Mayr.

The objective of the Asilomar meeting was to bring together geneticists, ecologists, taxonomists and applied scientists (e.g. workers in weed control, biological control of insects pests and wildlife biologists) to exchange ideas about the types of evolutionary change that would be likely to occur when organisms are introduced to regions of the world to which they are not native. The meeting lasted 5 days and was attended by approximately 30 participants, a relatively small gathering by today's standards. The contributors represented an international selection coming from USA (12), UK (4), Australia (3), New Zealand (2), Israel (2), Austria (1), Canada (1), Japan (1) and the West Indies (1), and all were male, an unfortunate sign of the times.

Several features of the 1965 volume are particularly noteworthy and have made it a classic and an attractive read for those interested in the history of evolutionary biology. First, the contributors included many individuals who were either leaders in their field or were to become so in later years. These included the following: R.W. Allard, L.C. Birch, Hampton L. Carson, Theodosius Dobzhansky, Friedrich Ehrendorfer, John L. Harper, Charles B. Heiser Jr., R. C. Lewontin, Ernst Mayr, Edward O. Wilson and Daniel Zohary, among others. The contributors also included less well-known scientists whose careers were given an important boost from being invited to speak (e.g. Gerald Mulligan; see Mulligan 2014). Second, the organizers decided to publish after each contribution the verbal exchanges among the participants over questions that arose. Conducted with grace and wit, and reflecting the personalities of the participants, some of these exchanges provide valuable insights into the thinking at the time, sometimes prescient, in other instances flawed. A fine example is the exchange between Lewontin and Mayr (p. 481) on the influence of founder events on genetic variation. Lewontin gives a lesson in population genetics to Mayr concerning the founder principle that Mayr had earlier made famous.

An unusual feature of *The Genetics of Colonizing Species* is the significant number of botanists among the authors; a striking contrast to today where they are often sparsely represented in symposium volumes on general topics, in part, owing to the slow attrition of plant organismal biology faculty positions at many academic institutions. Twelve of the 27 contributions involved plants, and they featured prominently in the published exchanges. The significant number of plant scientists represented in the volume was undoubtedly a reflection of the fact that the two editors—Baker and Stebbins—were both established botanists and therefore well informed about leading researchers and work being conducted on colonizing plants. Weed biology was a thriving discipline during the 1960s, and many plant ecologists and biosystematists were investigating weedy taxa because of their experimental tractability and interesting variation patterns. Another reason for the significant botanical representation at Asilomar may have been because plants display a greater diversity of genetic and reproductive systems than occurs in most animal groups, a point emphasized by Lewontin (p. 77) in the volume. This diversity lends itself to comparative studies, and such approaches were a prominent feature of many of the botanical contributions including those by Baker and Stebbins. Finally, it is probably not an accident that many of the botanists invited to the meeting had worked in California, which was and still is today a centre for evolutionary research, particularly on plants because of the amazing diversity of the California Floristic Province.

THE EDITORS

Herbert G. Baker—Renaissance botanist and incurable holist

Baker was an outstanding natural historian and field botanist with a broad knowledge of plant diversity, especially crops and weeds. He can probably be considered one of the first genuine plant evolutionary ecologists and was insistent that ecology and evolution were inseparable disciplines. He published extensively on the breeding systems and pollination biology of flowering plants but is perhaps best known for 'Baker's Law', coined by Stebbins (1957), which refers to the benefits of self-compatible hermaphroditism in establishment following long-distance dispersal, especially to islands where mates or pollinators may be in short supply or absent (Baker 1955, 1967). This topic, which generally concerns the constraints imposed by low-density conditions on colonization and reproduction (e.g. 'allee effects'), continues to stimulate new work (Pannell & Barrett 1998; Dornier *et al.* 2008; Cheptou 2012). Further details of Baker's scientific contributions, which included around 175 publications and a book on plant domestication, can be found in Barrett (2001).

Born in Brighton, England in 1920, Baker received his PhD in 1945 from the University of London. His thesis topic on the consequences of invasion for hybridization and species replacement in *Silene* initiated a lifelong interest in plant invasions. Significantly, Baker explicitly used the term 'invasion' in the title of his classic thesis study (Baker 1948) published in the *Journal of Ecology*. Using this military metaphor may have been associated with the times, Baker had experienced living and working through the Second World War. By sampling variation in populations of *Silene dioica* (then *Melandrium dioicum*) and *Silene latifola* (*M. album*) from selected regions of the United Kingdom, Baker recognized different stages in the invasion process leading to the replacement of one species by another. He documented extensive hybridization between the two *Silene* species, especially in populations occurring in disturbed habitats.

Baker's first university position was as Lecturer at the University of Leeds (1945–54), where he came under the influence of the distinguished cytologist Irene Manton and through her developed cytological skills and a strong appreciation for chromosomal variation and the evolution of genetic systems. After a short spell as Professor of Botany at the University of Ghana (1954–57), where his long-term interest in the reproductive biology of tropical plants first began, he moved permanently to the USA to take up a position at the University of California (U.C.) until his retirement, where he was Director of the U.C. Botanical Garden and later Professor of Botany. His wife Irene Baker provided both technical and emotional support throughout his career and the two published numerous studies together. The 'Baker laboratory' at Berkeley was always a welcoming place for students, and Baker rarely turned anyone away who wanted to talk about plants.

He supervised 49 PhD students during his career, although many of the theses remained on his shelf unpublished because he was uncomfortable putting pressure on his students and was always occupied by numerous projects of his own.

It was in California that Baker developed his long friendship with G. Ledyard Stebbins. Their shared interest in the Californian flora resulted in many field trips together (Fig. 1) and an appreciation of each other's expertise. In temperament, the two were polar opposites and perhaps this enabled them to get along with one another so well. Baker was gentle, sweet, retiring and hated confrontation. He had few interests outside of research (except track and field sports) and routinely worked on campus during weekends when he was not in the field. Despite his diffident manner, Baker was competitive, ambitious and quite capable of subtle criticism when it was merited. But this was always delivered politely with a minimum of histrionics. This made Baker the perfect foil for the mercurial Stebbins.

Fig. 1 Herbert G. Baker and G. Ledyard Stebbins in the field, Napa County, California 1973, on an excursion organized by the Bay Area Biosystematists. (*See insert for color representation of the figure.*)

G. Ledyard Stebbins—Botanical architect of the evolutionary synthesis

Stebbins is generally considered the botanical architect of the evolutionary synthesis and his monumental work *Plant Variation and Evolution* (Stebbins 1950), in which he synthesized existing knowledge of the genetics and evolution of plants, provided the foundation for the emerging field of plant evolutionary biology. Stebbins was the only botanist included in the group of scientists responsible for the modern evolutionary synthesis—Theodosius Dobzhansky, Ernst Mayr, George Gaylord Simpson and Julian Huxley—and was a dominant intellectual figure in mid-20th century evolutionary biology. He authored several other books and monographs, which covered a remarkable range of topics including local floras, chromosome evolution, macroevolution and developmental biology, as well as several general texts on evolution. Stebbins was especially adept at synthesis and among the ~260 articles that he wrote, his reviews were especially notable. His biographer Vassiliki Betty Smocovitis has written extensively on Stebbins' life and scientific accomplishments (Smocovitis 2001, 2006; Crawford & Vassiliki 2004), and I therefore provide only a brief summary, primarily on aspects of his career relevant to the Asilomar meeting and his relationship with Baker.

Born in 1906 in Lawrence, New York, USA, Stebbins obtained his PhD in 1931 from Harvard University where he worked on geographical variation and evolution in *Antennaria*, focusing in particular on chromosomal variation, hybridization and apomixis. At Harvard, he was strongly influenced by the geneticist Karl Sax, much to the chagrin of his supervisor, morphologist E.C. Jeffrey, who was not a fan of the 'new genetics' being promoted by Thomas Hunt Morgan and Sax (Smocovitis 2001). In 1935, Stebbins moved to U.C. Berkeley to work with E.B. Babcock on an ambitious project to understand the genetic mechanisms governing variation and evolution in *Crepis*, an herbaceous genus composed of polyploid and apomictic forms, and in which there were several introduced weedy taxa. Their monograph on *Crepis* (Babcock & Stebbins 1938) foreshadowed many of the themes that were later to become major components of his 1950 book—geographical variation, hybridization, polyploidy, speciation and variation in reproductive systems, all themes that also appear in *The Genetic of Colonizing Species*. In 1936, Stebbins met Dobzhansky for the first time. 'Dobie' became the single most important influence on Stebbins and was largely responsible for Stebbins' transformation from a plant geneticist to an evolutionary biologist (Smocovitis 2006).

It is unclear when Stebbins and Baker first met. By the time Baker arrived at Berkeley in 1957, Stebbins had left to help organize a new genetics department at U.C. Davis, where he widened his interests to include studies of crop plants and developmental genetics. However, Stebbins continued to make regular visits to the Berkeley campus where he continued to teach well into the 1960s. Therefore, it seems quite likely that the two developed their friendship through these visits and during field excursions and regular meetings of the 'Bay Area Biosystematists', a group of like-minded evolutionists and systematists who met regularly at various locations in the San Francisco Bay area to hear invited lectures and discuss the latest efforts to integrate ecology and genetics into systematics. This was a select group to which graduate students were not generally invited and nor were women until the 1970s (V.B. Smocovitis, personal communication).

Stebbins had boundless energy, was full of ideas and had a deep passion for plants and conservation. He was an engaging and charismatic undergraduate lecturer and could be warm, generous and funny, especially when he broke into songs from Gilbert and Sullivan, which he often did. He had an eccentric streak and in class would occasionally step into waste paper baskets by accident and appeared to be completely oblivious about combing his hair and zipping up his fly in front of the class. Those who had the experience of driving with Stebbins never forgot the experience, as he talked constantly at the same time as scanning the passing countryside for interesting plants. Stebbins also had a reputation for being a difficult person, and he was prone to losing his temper and being overly domineering. In conversation, it was often nearly impossible to be on an equal footing because of his impatient, quick mind and his tendency to constantly interrupt. Stebbins was also not especially open to having his ideas questioned, as occurred when the late David G. Lloyd (University of Canterbury) pointed out to him publically in Christchurch that some of his interpretations on the evolution of genetic systems involved group selection. Stebbins blew up and had a temper tantrum!

In contrast to Baker, Stebbins was the primary supervisor to very few graduate students and while at Davis, he fell out with several prominent faculty members including Robert Allard and Leslie Gottlieb, who were often not on speaking terms with him. Yet, Stebbins also played important mentorship roles in the careers of Verne Grant, Charles Heiser Jr. and Peter Raven, and he could be gracious and generous with his ideas. Baker was fully aware of Stebbins' volatile personality, and perhaps because Baker was deferential and full of admiration for Stebbins, he was happy to live in his colleague's shadow, chirping up politely when Stebbins monologues had ended. Others were less tolerant. Nevertheless, Stebbins was admired by many senior figures in evolutionary biology, and the inclusion of so many of them in *The Genetics of Colonizing Species* was undoubtedly a result of his influence.

SEVERAL TOPICS IDENTIFIED IN *THE GENETICS OF COLONIZING SPECIES*

The 27 contributions that make up *The Genetics of Colonizing Species* can be grouped into three loosely connected themes—concepts related to colonization, case histories of particular taxonomic groups and the management of invasive species, especially through biological control. Here, I highlight several topics that emerged in the volume that have subsequently been the focus of increased attention, and some of which stimulated work in my own laboratory.

Conceptual beginnings

The book begins with a series of contributions on general concepts relevant to colonizers, including species interactions, island colonization, mating systems, selection, population differentiation and genetic drift. Among these contributions was the only theoretical chapter in the volume, by Lewontin on selection for colonizing ability (p. 77), including the influences of interdemic selection and changing environments. His analysis highlights the classic trade-off between development rate and fecundity. Our recent work indicates the importance of life history trade-offs in colonizing species. Local adaptation to growing season length with northern migration in eastern N. American populations of the wetland invader *Lythrum salicaria* involves a trade-off between flowering time and size (which determines reproductive output) and suggests a genetic constraint to further northward migration for populations at the current range margin (Colautti *et al.* 2010a; Colautti & Barrett 2013). Lewontin's chapter represents one of the earliest efforts to use theory to predict optimal strategies for colonizing species and presaged the subsequent development of a rich theoretical literature concerned with many different aspects of biological invasions (e.g. Andow *et al.* 1990; Shigesada & Kawasaki 1997; Higgins & Richardson 1999; García-Ramos & Rodríguez 2002). As is evident from this volume, theory is now an integral component of invasion genetics.

Determinants of invasion success—the comparative approach

The second series of contributions in *The Genetics of Colonizing Species* largely focused on case histories and analyses of successful colonizers in particular geographical regions. A recurrent theme was the effort to predict which traits characterize successful colonizers. Baker's chapter (p. 147) on the mode of origin of weeds exemplifies this approach and spurred much subsequent work and some controversy. Through comparative experimental studies of several taxa, he identified a suite of traits that distinguished closely related weeds and nonweeds. These included self-compatibility, high phenotypic plasticity, short life cycles and rapid flowering. Based on these comparisons and his wide knowledge of common weeds, Baker drew up a list of 14 characteristics that might be expected in the 'ideal weed' (p. 166). This was obviously a heuristic exercise, and Baker was clear that it was unlikely that any species possessed all of the features he listed. Nevertheless, subsequent studies (Perrins *et al.* 1992; Williamson & Fitter 1996; Moles *et al.* 2008) questioned the value of Baker's ideal weed list and argued that environmental conditions in the introduced range, particularly biotic challenges, will play a crucial role in whether an introduced species becomes invasive. They also argued that invasive plants as a group were simply too heterogeneous to draw the kinds of generalizations implied by Baker's list.

Despite these valid concerns, efforts to identify the determinants of invasiveness have burgeoned over the past few decades. Considerable progress has been made in identifying traits of invaders using phylogenetic and experimental approaches and taking into account a variety of other influences including historical, biogeographical and habitat factors (e.g. Rejmánek &

Richardson 1996; Gravuer *et al.* 2008; Ahern *et al.* 2010; van Kleunen *et al.* 2010a,b; Kuester *et al.* 2014). By today's standards, Baker's pairwise congeneric comparisons of weeds and nonweeds seem rudimentary. The comparisons were limited to a few taxa and did not appear to involve sister taxa. Congeneric species pairs were not grown in experimental mixtures, as suggested by Harper in his chapter as the most appropriate way to detect differences in ecology (p. 262). However, despite these shortcomings, Baker's work on weeds did identify an important question and pointed the way forward to the use of more robust comparative approaches.

Both generalists and specialists

One tension that emerged in *The Genetics of Colonizing Species* concerned the extent to which invaders are commonly generalists or specialists. In his chapter, Harper repeatedly stressed the specialized character of many plant invaders (p. 244), whereas by contrast, Baker viewed weeds as commonly exhibiting a 'jack-of-all-trades-master-of-none' strategy arising from 'general-purpose genotypes' (p. 158). He suggested that such genotypes provide colonizers with wide environmental tolerance and an ability to grow in a multitude of climates and edaphic conditions through phenotypic plasticity. Today, we recognize that both Harper and Baker were partially right. Because of the wide range of strategies that are evident among invasive species, both generalists and specialists occur. Comparative studies of members of the barnyard grass complex (*Echinochloa crus-galli* and relatives) illustrate the diversity of strategies in invasive plants (Fig. 2; reviewed in Barrett 1983, 1988, 1992). *Echinochloa crus-galli* is cosmopolitan in distribution and considered one of the world's worst weeds (Holm *et al.* 1977). Native to the Old Word, it ranges in distribution from 50°N to 40°S, occurring in a wide range of disturbed environments, and is recorded from 36 crops in 61 countries. It is a generalist par excellence. In contrast, *Echinochloa phyllopogon* and *Echinochloa oryzoides* (*E. crus-galli* var. *oryzicola*) are specialized mimics of rice restricted in distribution to cultivated rice fields. The generalist and specialist species differ in a suite of life history traits reflecting contrasting ecological preferences: the generalist flowers faster and produces larger numbers of smaller, dormant seeds. In contrast, flowering in the mimics is delayed, coinciding with that

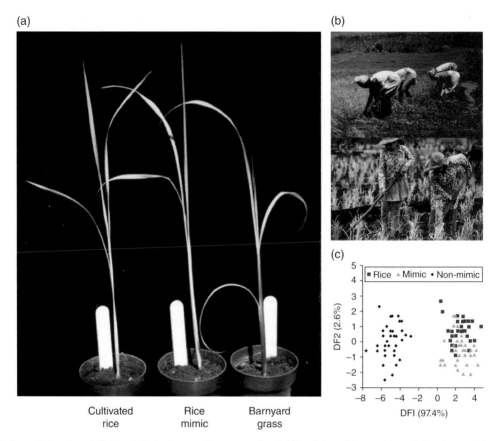

Fig. 2 Generalist and specialist weeds in the barnyard grass complex; (a) from left to right—cultivated rice, the specialist rice mimic *Echinochloa phyllopogon*, and the generalist *Echinochloa crus-galli*; (b) weeding practices in rice exert selection pressures on the morphology of weed populations favouring variants of barnyard grass that resemble rice; (c) phenotypic resemblance between the generalist, the rice mimic and rice based on a discriminant functions analysis of nine quantitative characters. For further details, see Barrett (1983). (*See insert for color representation of the figure.*)

of rice, and plants produce fewer, larger seeds that lack dormancy. In addition, the generalist is more plastic in its growth and development and maintains more genetic variation in populations than the two rice weed specialists. Thus, even among this closely related complex of annual selfing weeds, both generalist and specialist strategies have evolved.

General-purpose genotypes and adaptive plasticity

In his chapter, Baker used a variety of examples to illustrate his 'general-purpose genotype' concept, but in all cases, the species he identified possess uniparental reproduction. These included species with prolific clonal reproduction, such as the sterile pentaploid *Oxalis pes-caprae* and the floating aquatic *Eichhornia crassipes* (Fig. 3a,b), as well as apomictic *Poa pratensis* and autogamous *Eupatorium microstemon*. He proposed that outbreeders of undisturbed natural communities would be unlikely to possess general-purpose genotypes and instead would evolve finely adapted ecotypes specialized to local conditions, and at least in their own habitats, they would be superior to generalist weeds. Following Baker, the concept of general-purpose genotypes has been extended to several animal groups, particularly obligately asexual polyploid species that exhibit geographical parthenogenesis (reviewed in Lynch 1984). In principle, there is no reason why

(a)

(b)

Fig. 3 Two invasive weeds identified by Baker (1965) as possessing general-purpose genotypes. (a) The sterile pentaploid short-styled morph of tristylous *Oxalis pes-caprae*, Tel Aviv, Israel (2013); (b) The clonal aquatic *Eichhornia crassipes* at Bacon Island Slough near Stockton, California (2014). The population of *E. crassipes* is composed of a single clone of the mid-styled morph and has persisted at this site for 40 years and during this time has dramatically increased in size as a result of clonal growth. Sexual reproduction is prevented at the site despite seed production because of unsuitable conditions for seed germination and seedling establishment (see Barrett 1980). (*See insert for color representation of the figure.*)

generalist genotypes could not develop in outbreeding populations, depending on the 'grain of the environment' (see Levins 1968), but as yet most cases where the general-purpose genotype concept has been applied involve species with uniparental reproduction.

Baker's embryonic ideas on general-purpose genotypes are directly relevant to the evolution of reaction norms and adaptive phenotypic plasticity (Sultan 1987; Pigliucci 2001). A recent meta-analysis reported that invasive species possess higher phenotypic plasticity than native species (Davidson *et al.* 2011), a finding that supports Baker's ideas on the benefits of plasticity in generalist weeds. However, other studies comparing related invasive and noninvasive native species have failed to show any consistent pattern (Palacio-López & Gianoli 2011), perhaps because plasticity evolves and changes during the invasion process. Although many successful weeds do indeed display extraordinarily high phenotypic plasticity, many nonweeds of heterogeneous environments show similar behaviour, leaving open the relative importance of plasticity and local adaptation for invasion success. Addressing this issue and finding support for the general-purpose genotype concept are now a focus of current research (e.g. Parker *et al.* 2003; Dybdahl & Kane 2005; Richards *et al.* 2006; Hulme 2008). Experimental studies of the annual selfing herb *Polygonum cespitosum*, introduced to North America from eastern Asia, have revealed individuals that resemble the kind of general-purpose genotype envisioned by Baker (Matesanz & Sultan 2013). Eight of 14 invasive populations investigated by these authors contained varying (3–21%) proportions of 'high-performance genotypes' that maintained high reproductive output across a range of moisture and light levels. Additional evidence from a 'resurrection study' over an 11-year period demonstrated rapid postintroduction evolutionary change in adaptive plasticity (Sultan *et al.* 2012). In this species, it will be interesting to determine the extent to which plasticity might be gradually replaced by local adaptation as the invasion matures or whether ongoing population turnover and colonizing episodes maintain a high degree of plasticity among most genotypes. Of course, both plasticity and local adaptation are likely to play important roles in most sexual invaders; the main future challenge will be to determine their relative contribution to fitness and invasive spread, as well as the role of pre-adaptation.

Mating systems in invasive populations

Several contributors to *The Genetics of Colonizing Species* considered the extent to which the mating system was important for colonizing success. Allard (p. 49) pointed out that among the world's most successful plant colonizers, the vast majority were predominantly selfing, and Baker (p. 147) and Mulligan (p. 127) emphasized the importance of self-compatibility in the evolution of weediness. However, Stebbins found no evidence for an overrepresentation of selfing species in his survey of native Californian weeds, and Heiser (p. 391) pointed out that among annual sunflowers (*Helianthus*), the majority are self-incompatible, including the widespread and weedy *Helianthus annuus*. This raises two questions for workers today: Is there an optimal mating system for an invasive species, and is there evidence for the selection of selfing during the invasion process?

Currently, there are no definitive answers to either of these questions. Some progress has been made in clarifying why reproductive systems are of importance for invasion success (reviewed in Barrett 2011), and several phylogenetically controlled analyses show that the facility for autonomous self-pollination is overrepresented among invasive species (van Kleunen & Johnson 2007; van Kleunen *et al.* 2008; Burns *et al.* 2011), a pattern consistent with Baker's Law. However, many perennials and even some annual colonizers are self-incompatible, and there is still scant empirical evidence that selection for reproductive assurance during the invasion process drives evolutionary transitions from outcrossing to selfing. Support for Baker's Law comes largely from comparative evidence or case studies of island colonization by weedy species (e.g. Barrett & Shore 1987; Barrett *et al.* 1989), but at more restricted spatial scales, it has proven more difficult to find evidence for predictable patterns of mating-system variation, such as the breakdown of self-incompatibility to self-compatibility along gradients of colonization or succession (Colautti *et al.* 2010b; but see Cheptou *et al.* 2002). Selection for selfing in colonizing populations depends on several factors including the spatial scale of colonization, gene flow, inbreeding depression and the availability of standing genetic variation in mating-system modifiers.

Models investigating the benefits of reproductive assurance in a metapopulation suggest that an optimal mating system for an invader should include the ability to modify selfing rates according to the patch density (Pannell & Barrett 1998). During colonizing episodes when populations are small or are at low density, plants should self to maximize fecundity. However, when populations become larger and mates or pollinators are less likely to be limiting, outcrossing should become more beneficial, promoting recombination

and adaptive evolution. Future work on invasive populations might usefully investigate the extent to which the mating systems of self-compatible colonizers are indeed flexible and that patterns of mating are context dependent. In addition, experimental field studies demonstrating selection for traits providing reproductive assurance in small populations, similar to those recently conducted on noninvasive *Clarkia* by Moeller & Geber (2005), would be most valuable.

Evolutionary history, bottlenecks and genetic diversity

Many of the contributors to *The Genetics of Colonizing Species* discussed the importance of understanding the evolutionary history of colonization and the extent to which demography may influence the amount of genetic variation in populations. Whereas some authors (e.g. Mayr, Carson) placed considerable emphasis on the role of small population size in reducing diversity, others (e.g. Lewontin, Fraser) were less convinced that bottlenecks were likely to be important in limiting evolutionary potential (see for example, pp. 123–125, 481). Several of the exchanges were in essence versions of the classic Fisher–Wright debate on the significance of stochastic forces in evolution.

At the time of the Asilomar meeting, most inferences about the evolutionary history of colonizing species were based on guesswork, or less often, records from herbaria and museum collections. Today, through the use of genetic markers, we are in a much better position to reconstruct the migratory history of invasions and assess the magnitude of genetic bottlenecks and founder events. There is now evidence from neutral loci that many populations of introduced species have less genetic variation than populations in the native range. However, a survey of 80 species of plants, animals and fungi revealed that the overall average loss in allelic richness was only 15.5% (Dlugosch & Parker 2008), much less than might have been predicted by several of the participants at Asilomar. It is now recognized that assessing the genetic and evolutionary consequences of bottlenecks depends on a variety of biological and historical factors including the types of genes examined (e.g. Mendelian loci vs. quantitative variation; Lewontin 1984), the reproductive systems of species (biparental vs. uniparental; Novak & Mack 2005), the frequency of bottlenecks (single vs. repeated; Nei *et al.* 1975), the occurrence of multiple introductions and admixture (Keller *et al.* 2014), and the extent of interspecific hybridization (Ellstrand & Schierenbeck 2000). These influences, in addition to knowledge of the ecology and demography of populations, should be taken into account when interpreting patterns of genetic diversity in invasive species.

Founder events and bottlenecks are not unexpected in introduced species with uniparental reproduction. In selfers, inbreeding preserves multilocus associations established through founder events and genetic drift (Golding & Strobeck 1980; Brown 1983), and the lack of sexual reproduction in many asexual populations freezes standing variation following a bottleneck and prevents opportunities to regain diversity through recombination, although somatic mutations may play some role in increasing diversity as reported in some clonal plants (Ally *et al.* 2008; Bobiwash *et al.* 2013). Bottlenecks of varying severity have been commonly detected in selfing and clonal species using neutral genetic markers (e.g. Husband & Barrett 1991; Kliber & Eckert 2005; Zhang *et al.* 2010). However, even highly selfing populations are able to maintain considerable amounts of quantitative genetic variation because of the high mutability of polygenic characters and the fact that many genes contribute to the expression of these traits. Theoretical studies by Lande (1976, 1977) suggest that if populations expand after a bottleneck, as occurs during many invasions, sufficient genetic variability at quantitative trait loci can be generated for rapid adaptive evolution. Unfortunately few, if any, studies of invasive species have compared additive genetic variation and evolvability of ecologically relevant traits in native and introduced populations using appropriate breeding designs. So, it is too early to say if introduced populations generally have less quantitative genetic variation than native populations. However, based on the spate of recent examples of rapid evolutionary change in invasive species over the past decade (reviewed in Cox 2004; Whitney & Gabler 2008; Suarez & Tsutsui 2008), it seems probable that most invasive populations of both outbreeding and inbreeding species have sufficient standing genetic variation to respond adaptively to local ecological conditions. In contrast, clonal species with limited or no sexual reproduction occurring in invasive populations (Fig. 3) meet the challenges of novel environments through a different strategy—phenotypic plasticity.

Native or alien invasions?

Occasionally, in the study of invasive species, it is unclear whether populations in a particular region are native or alien. For example, this occurred with the originally described Californian endemic *Bacopa nobsiana*, which on further study turned out to be the introduced *B. rotundifolia* (Barrett & Strother 1978). Although most biological invasions involve introduced species, this is not necessarily the case if human disturbance opens up novel environments and native species are provided an opportunity to multiply and spread (e.g. many weeds of Californian rice fields; Barrett & Seaman 1980). Indeed, in his contribution to *The Genetics of Colonizing Species*, Stebbins (p. 173) reviewed many other examples of colonizing species of the native Californian flora that, following European settlement, spread rapidly to become successful weeds. Harper (p. 244) also discussed native species that have invaded agricultural land from native plant communities in Britain. As is often the case for plant invasions, anthropogenic disturbance is usually the key ecological factor promoting spread.

Our molecular studies of the annual aquatic *Eichhornia paniculata* illustrate how information on demographic history can be used to determine whether a species is native or introduced to a particular region. Populations of *E. paniculata* are native to N.E. Brazil where they are largely outcrossing, inhabit temporary pools and ditches and are pollinated by specialist long-tongued bees (Fig. 4). However, populations also occur in Cuba and Jamaica where they are predominantly selfing and infest cultivated rice fields and other disturbed habitats associated with agricultural land. Phylogeographical studies and comparisons of nucleotide diversity indicate a moderate bottleneck associated with long-distance dispersal from Brazil to the Caribbean (Husband & Barrett 1991; Ness *et al.* 2010). Caribbean populations are considered native to the islands but an alternative possibility is that they were introduced in historic times, perhaps associated with agriculture. We investigated these alternative hypotheses using coalescent simulations of the demographic history of populations. The results clearly indicate that *E. paniculata* was not introduced to the Caribbean in historic times. Rather, natural colonization probably mediated by long-distance dispersal by migratory birds, occurred ~125 000 years before present, well before the origins of agriculture (Ness *et al.* 2010). Here, a species with a markedly disjunct neotropical

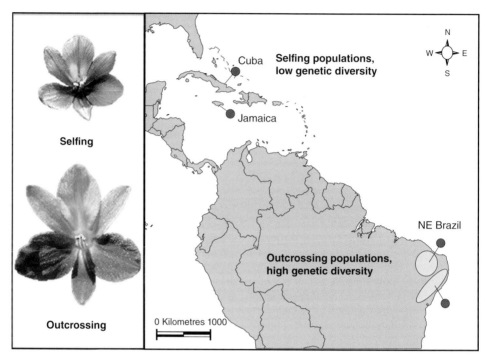

Fig. 4 The geographical distribution of *Eichhornia paniculata* illustrating part of its disjunct distribution; populations in N.E. Brazil are large flowered, outcrossing, genetically diverse and noninvasive, in Cuba and Jamaica, the species has smaller flowers, is highly selfing, has much less genetic diversity and has invaded rice fields on both islands. The establishment of selfing populations in the Caribbean is an example of Baker's Law.

distribution has become invasive through colonization of a novel niche (rice fields) not available at the centre of its range in N.E. Brazil.

WHAT IS INVASION GENETICS?

History and definition

In contrast to the limited influence that Elton's book had on the early development of invasion ecology in the first few decades after it appeared (Simberloff 2011), *The Genetics of Colonizing Species* stimulated considerable research activity soon after its publication, leading to a steady increase in citations to the book (Fig. 5). The growing interest in the genetics of colonizing species was undoubtedly also associated with the 'electrophoresis revolution', following landmark studies by Lewontin & Hubby (1966) and Harris (1966) reporting the utility of electrophoretic techniques for measuring genetic diversity in populations. Within a short period, many laboratories adopted these

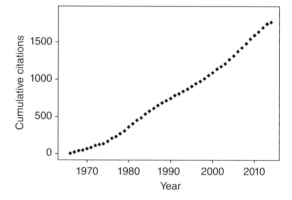

Fig. 5 Cumulative citations of the Baker and Stebbins volume *The Genetics of Colonizing Species* and chapters therein from its publication to the present; data obtained from Web of Science July 2014.

approaches, resulting in a flood of data on allozyme variation in plant and animal populations, including many colonizers (reviewed in Nevo 1978; Brown 1979; Hamrick *et al.* 1979; Barrett & Shore 1989).

In his contribution to *The Genetics of Colonizing Species* on the genetic systems of selfing plants, Allard (p. 49) reported estimates of outcrossing and the amounts of quantitative genetic variation in populations using morphological markers and phenotypic traits, respectively. But shortly after, it became possible using enzyme polymorphisms to obtain more precise estimates of mating-system parameters and to survey numerous natural populations to measure heterozygosity, allelic richness and population genetic structure. Anthony Brown (Fig. 6), then a new graduate student with Allard from Australia, was dispatched to Lewontin's laboratory at Chicago in the late 1960s to learn electrophoretic techniques. Soon, the Allard laboratory was leading the way in studies of the population genetics of plant populations and attracting many trainees, several of whom (e.g. S.K. Jain, A.H.D. Brown, M.T. Clegg, J.L. Hamrick) would go on to make valuable contributions to our understanding of the genetics of plant colonization. In particular, Brown returned to CSIRO in Australia where his laboratory pioneered investigations of the mating systems and population genetic structure of diverse colonizing species (e.g. Brown & Burdon 1983) and co-authored an influential review on evolutionary change during invasion (Brown & Marshall 1981). Significantly, the extensive data that accumulated in the 1970s and 80s from allozyme studies of genetic variation in colonizing species while enriching evolutionary biology had relatively little influence on the early development of invasion biology.

The range of genetic markers diversified during the 1980s and 90s to includes RAPDs, AFLPs, cpDNA, microsatellites and finally DNA sequences, giving rise to the birth of molecular ecology, in which questions in ecology and evolution were addressed using a diversity of molecular genetic techniques. In 1991, the journal *Molecular Ecology* appeared for the first time and was to provide an important forum for studies concerned with genetic variation in invasive species. The awakening of public awareness of the 'invasive species problem' in the 1990s, and the increasing availability of funding sources to investigate nuisance species, resulted in a greater number of workers from the long-standing sub-disciplines of ecology and evolution becoming interested in invasion biology. Because of the threat posed by invasive species to biodiversity and ecosystem function, invasion biology became incorporated into the broader field of conservation biology, and *Molecular Ecology* widened its scope to include articles relevant to conservation, some of which dealt with genetic aspects of biological invasions. The first study explicitly using 'invasion genetics' in the title appeared at the end of the decade in a study of the Mediterranean fruit fly (Villablanca *et al.* 1998), and the term has subsequently been used frequently in the titles of articles reporting work on a wide range of organisms (e.g. spiny water flea—Colautti *et al.* 2005; freshwater mussel—Therriault *et al.* 2005; Eurasian round goby—Brown & Stepien 2009; vase tunicate *Ciona*—Zhan *et al.* 2010; black rat—Konečný *et al.* 2013). A Web of Science search conducted during the preparation of this article using the key words 'invasion genetics' revealed numerous articles, and, although the field is relatively young, it is developing rapidly.

The first definition of invasion genetics in the literature appears to be by Colautti *et al.* (2005), who described the field as 'the application of genetic techniques to investigate biological invasions'. While this is straightforward and sufficient, I offer an alternative definition that attempts to place more emphasis on the biological questions commonly addressed. In my view invasion genetics is *the study of the historical, ecological and demographic processes responsible for the patterns of genetic diversity in populations and their influence on invasion success and contemporary evolution during biological invasion*'. As this definition makes clear, invasion genetics is not only an integral part of invasion biology, serving as an equal partner to invasion ecology, it is also very much a contributor to the broader area of

Fig. 6 Anthony H.D. Brown an early pioneer in the electrophoretic study of enzyme polymorphisms in plant populations, including many colonizing species. Brown is seen here scoring starch gels at CSIRO, Canberra, Australia, 1984.

contemporary evolution (Stockwell *et al.* 2003; Carroll *et al.* 2007; Westley 2011), in which evolutionary change over timescales of a few hundred years is the focus of attention.

Future challenges and key questions

What are the challenges and opportunities ahead for invasion genetics? I conclude by providing a series of questions that collectively form a solid foundation for the field and consider how invasion ecology and invasion genetics could become more fully integrated. The genesis of many of these questions was the Baker & Stebbins (1965) chapters, including the published dialogue between authors. At the time, the toolbox of techniques available to researchers was much more limited and thus opportunities to definitively answer these questions were difficult. Additional questions have arisen more recently as a result of new theory in evolutionary genetics and techniques in molecular genetics.

1 What are the source populations for biological invasions and how many times, when and where have immigrants been introduced to the alien range?

2 How important is pre-adaptation vs. postinvasion adaptation to invasive spread?

3 How important is genetic variation to colony establishment and as a contributory factor to the lag-phase that characterizes many biological invasions?

4 Is population genetic structure different between native and introduced populations and is diversity reduced compared to the native range as a result of bottlenecks, or increased because of gene flow and hybridization?

5 What is the relative importance of new mutations vs. standing genetic variation to evolution in invasive populations?

6 What are the agents of natural selection resulting in the evolution of local adaptation in introduced populations? How quickly does adaptation occur and what types of constraints limit selection response?

7 Does the evolution of local adaptation increase invasion success resulting in further range expansion?

8 Are particular reproductive and genetic systems favoured in invasive species and does the invasion process itself promote evolutionary changes in them?

9 What are the relative roles of phenotypic plasticity and local adaptation to fitness during invasion and how might this be quantified?

10 Do some species possess general-purpose genotypes and how do they originate and spread?

Perhaps the biggest future challenge that invasion genetics faces is to fully integrate existing approaches with ecological and demographic studies of invasive species. Much work in invasion genetics to date can be characterized as an extension of the 'find them and grind them' approach that characterized the early stages of the electrophoresis revolution but using more powerful markers and sophisticated population genetics software programs. In future, it will be important to complement such descriptive work with manipulative field experiments that go beyond the common garden studies that are commonly employed today (reviewed in Colautti *et al.* 2009). Early pioneering work by Martins & Jain (1979) tracked artificially established colonies of *Trifolium hirtum* with known genetic inputs to investigate the adaptive role of genetic variation to colonizing ability. Surprisingly, there has been little subsequent work of this type on invasive species despite the diversity of markers that are now available and the popularity of large-scale field experiments in ecology. Colonizing species generally possess short life cycles and offer tractable systems for field studies of experimental evolution, so long as quarantine regulations and management concerns can be thoroughly satisfied. When combined with genomic analysis and field manipulations of environmental and demographic variables, they offer exciting opportunities to provide new insights into the genetics of biological invasions and on contemporary evolution.

ACKNOWLEDGEMENTS

I thank Barbara Pickersgill, Hugh Bunting, Roy Snaydon, Tony Brown, Subodh Jain, Dick Mack, Chris Eckert, Brian Husband and Rob Colautti for discussions over the years on plant colonization and invasive species; Ron and Nancy Dengler, Kermit Ritland and, particularly, Betty Smocovitis for anecdotes about Stebbins; Andrew Hendry and Ken Whitney for providing references; Chris Balogh for conducting the Web of Science search, Bill Cole for preparing the figures; and the Natural Sciences and Engineering Research Council of Canada and Canada Research Chair funds for providing financial support for my research on invasive plant species.

REFERENCES

Ahern RG, Landis DA, Reznicek AA, Schemske DW (2010) Spread of exotic plants in the landscape: the role of time, growth habit, and history of invasiveness. *Biological Invasions*, **12**, 3157–3169.

Allard RW, Babbel GR, Clegg MT, Kahler AL (1972) Evidence for coadaptation in *Avena barbata*. *Proceedings of the National Academy of Sciences of the United States of America*, **69**, 3043–3048.

Ally D, Ritland K, Otto SP (2008) Can clone size serve as a proxy for clone age? An exploration using microsatellite divergence in *Populus tremuloides*. *Molecular Ecology*, **17**, 4897–4911.

Andow DA, Kareiva PM, Levin SA, Okubo A (1990) Spread of invading organisms. *Landscape Ecology*, **4**, 177–188.

Babcock EB, Stebbins GL Jr (1938) The American species of *Crepis*: their inter-relations and distribution as affected by polyploidy. Carnegie Institute of Washington Publication No 504. Washington, District of Columbia.

Baker HG (1948) Stages in invasion and replacement by species of *Melandrium*. *Journal of Ecology*, **36**, 96–119.

Baker HG (1955) Self-compatibility and establishment after "long-distance" dispersal. *Evolution*, **9**, 347–349.

Baker HG (1965) Characteristics and modes of origin of weeds. In: *The Genetics of Colonizing Species* (eds Baker HG, Stebbins GL), pp. 147–168. Academic Press, New York.

Baker HG (1967) Support for Baker's Law – As a rule. *Evolution*, **21**, 853–856.

Baker HG (1986) Patterns of plant invasion in North America. In: *Ecology of Biological Invasions of North America Ecological Studies 58* (eds Mooney HA, Drake JA), pp. 44–57. Springer-Verlag, New York.

Baker HG, Stebbins GL (eds) (1965) *The Genetics of Colonizing Species*. Academic Press, New York.

Barrett SCH (1980) Sexual reproduction in *Eichhornia crassipes* (water hyacinth) II. Seed production in natural populations. *Journal of Applied Ecology*, **17**, 113–124.

Barrett SCH (1983) Crop mimicry in weeds. *Economic Botany*, **37**, 255–282.

Barrett SCH (1988) Genetics and evolution of agricultural weeds. In: *Weed Management in Agroecosystems: Ecological Approaches* (eds Altieri M, Liebman MZ), pp. 57–75. CRC Press Inc., Boca Raton, Florida.

Barrett SCH (1992) Genetics of weed invasions. In: *Applied Population Biology* (eds Jain SK, Botsford L), pp. 91–119. Wolfgang Junk, Netherlands.

Barrett SCH (2001) The Bakers and Stebbins era comes to a close. *Evolution*, **55**, 2371–2374.

Barrett SCH (2011) Why reproductive systems matter for the invasion biology of plants. In: *Fifty Years of Invasion Ecology The Legacy of Charles Elton* (ed. Richardson D), pp. 195–210. Wiley-Blackwell, Oxford.

Barrett SCH, Richardson BJ (1986) Genetic attributes of invading species. In: *Ecology of Biological Invasions: An Australian Perspective* (eds Groves RH, Burdon JJ), pp. 21–23. Australian Academy of Science, Canberra.

Barrett SCH, Seaman DE (1980) The weed flora of Californian rice fields. *Aquatic Botany*, **9**, 351–376.

Barrett SCH, Shore JS (1987) Variation and evolution of breeding systems in the *Turnera ulmifolia* L. complex (Turneraceae). *Evolution*, **41**, 340–354.

Barrett SCH, Shore JS (1989) Isozyme variation in colonizing plants. In: *Isozymes in Plant Biology* (eds Soltis D, Soltis P), pp. 106–126. Dioscorides Press, Portland, Oregon.

Barrett SCH, Strother JL (1978) The taxonomy and natural history of *Bacopa* (Scrophulariaceae) in California. *Systematic Botany*, **3**, 408–419.

Barrett SCH, Morgan MT, Husband BC (1989) The dissolution of a complex genetic polymorphism: the evolution of self-fertilization in tristylous *Eichhornia paniculata* (Pontederiaceae). *Evolution*, **41**, 1398–1416.

Bobiwash K, Schultz ST, Schoen DJ (2013) Somatic deleterious mutation rate in a woody plant: estimation from phenotypic data. *Heredity*, **111**, 338–344.

Brown AHD (1979) Enzyme polymorphism in plant populations. *Theoretical Population Biology*, **15**, 1–42.

Brown AHD (1983) Multilocus organization of plant populations. In: *Population Biology and Evolution* (eds Wohrmann K, Loescheke V), pp. 159–169. Springer-Verlag, Berlin.

Brown AHD, Burdon JJ (1983) Multilocus diversity in an outbreeding weed, *Echium plantagineum* L. *Australian Journal of Biological Sciences*, **36**, 503–509.

Brown AHD, Marshall DR (1981) Evolutionary changes accompanying colonization in plants. In: *Evolution Today: Proceedings of the Second International Congress of Systematic and Evolutionary Biology* (eds Scudder GGT, Reveal JL), pp. 351–363. Carnegie-Mellon University, Pittsburgh, Pennsylvania.

Brown JE, Stepien CA (2009) Invasion genetics of the Eurasian round goby in North America: tracing sources and spread patterns. *Molecular Ecology*, **18**, 64–79.

Burns JN, Ashman T-L, Steets JA, Harmon-Threatt A, Knight TM (2011) A phylogenetically controlled analysis of the roles of reproductive traits in plant invasions. *Oecologia*, **166**, 1009–1017.

Callaway RM, Maron JL (2006) What have exotic plant invasions taught us over the past 20 years? *Trends in Ecology and Evolution*, **21**, 369–374.

Carroll SP, Hendry AP, Reznick DN, Fox CW (2007) Evolution on ecological time scales. *Functional Ecology*, **21**, 387–393.

Cheptou P-O (2012) Clarifying Baker's Law. *Annals of Botany*, **109**, 633–641.

Cheptou P-O, Lepart J, Escarre J (2002) Mating system variation along a successional gradient in the allogamous colonizing plant *Crepis sancta* (Asteraceae). *Journal of Evolutionary Biology*, **15**, 753–762.

Chew MK (2011) Invasion biology: historical precedents. In: *Encyclopaedia of Biological Invasions* (eds Simberloff D & Rejmánek M), pp. 369–375. University of California Press, Berkeley, California.

Colautti RI, Barrett SCH (2013) Rapid adaptation to climate facilitates range expansion of an invasive plant. *Science*, **342**, 364–366.

Colautti RI, Manca M, Viljanen M *et al.* (2005) Invasion genetics of the Eurasian spiny waterflea: evidence for bottlenecks and gene flow using microsatellites. *Molecular Ecology*, **14**, 1869–1879.

Colautti RI, Maron JL, Barrett SCH (2009) Common garden comparisons of native and introduced plant populations: latitudinal clines can obscure evolutionary inferences. *Evolutionary Applications*, **3**, 187–189.

Colautti RI, Eckert CG, Barrett SCH (2010a) Evolutionary constraints on adaptive evolution during range expansion in an invasive plant. *Proceedings of the Royal Society of London Series B*, **277**, 1799–1806.

Colautti RI, White NA, Barrett SCH (2010b) Variation of self-incompatibility within invasive populations of purple loosestrife (*Lythrum salicaria* L.) from eastern North America. *International Journal of Plants Sciences*, **171**, 158–166.

Cox GW (2004) *Alien Species and Evolution: The Evolutionary Ecology of Exotic Plants, Animals, Microbes, and Interacting Native Species*. Island Press, Washington, District of Columbia.

Crawford DJ, Vassiliki VB (2004) *The Scientific Papers of G. Ledyard Stebbins (1929–2000)*. ARG Gantner Verlag, Ruggell, Liechtenstein.

Davidson AM, Jennions M, Nicotra AB (2011) Do invasive species show higher phenotypic plasticity than native species and, if so, is it adaptive? A meta-analysis. *Ecology Letters*, **14**, 419–431.

Dlugosch KLM, Parker IM (2008) Founding events in species invasions: genetic variation, adaptive evolution, and the role of multiple introductions. *Molecular Ecology*, **17**, 431–449.

Dornier A, Munoz F, Cheptou P-O (2008) Allee effect and self-fertilization in hermaphrodites: reproductive assurance in a structured metapopulation. *Evolution*, **62**, 2558–2569.

Dybdahl MF, Kane SL (2005) Adaptation vs. phenotypic plasticity in the success of a clonal invader. *Ecology*, **86**, 1592–1601.

Ellstrand NC, Schierenbeck KA (2000) Hybridization as a stimulus for the evolution of invasiveness in plants? *Proceedings of the National Academy of Sciences of the United States of America*, **97**, 7043–7050.

Elton CS (1958) *The Ecology of Invasions by Animals and Plants*. Methuen, London.

García-Ramos G, Rodríguez D (2002) Evolutionary speed of species invasions. *Evolution*, **56**, 661–668.

Golding GB, Strobeck C (1980) Linkage disequilibrium in a finite population that is partially selfing. *Genetics*, **94**, 777–789.

Gravuer K, Sullivan JJ, Williams PA, Duncan RP (2008) Strong human association with plant invasion success for *Trifolium* introductions to New Zealand. *Proceedings of the National Academy of Sciences of the United States of America*, **105**, 6344–6349.

Hamrick JL, Linhart YB, Mitton JB (1979) Relationships between life history characteristics and electrophoretically detectable genetic variation in plants. *Annual Review of Ecology and Systematics*, **10**, 173–2000.

Harris H (1966) Enzyme polymorphism in man. *Proceedings of the Royal Society of London Series B*, **164**, 298–310.

Higgins SI, Richardson DM (1999) Predicting plant migration rates in a changing world: the role of long-distance dispersal. *American Naturalist*, **153**, 464–475.

Holm LG, Plucknett DL, Pancho JV, Herberger JP (1977) *The World's Worst Weeds: Distribution and Biology*. University of Hawaii, Press, Honolulu.

Hulme PE (2008) Phenotypic plasticity and plant invasions: is it all Jack? *Functional Ecology*, **22**, 3–7.

Husband BC, Barrett SCH (1991) Colonization history and population genetic structure of *Eichhornia paniculata* in Jamaica. *Heredity*, **66**, 287–296.

Keller SR, Fields PD, Berardi Taylor DR (2014) Recent admixture generates heterozygosity-fitness correlations during the range expansion of an invading species. *Journal of Evolutionary Biology*, **27**, 616–627.

van Kleunen M, Johnson SD (2007) Effects of self-compatibility on the distribution range of invasive European plants in North America. *Conservation Biology*, **21**, 1537–1544.

van Kleunen M, Manning JC, Pasqualetto V, Johnson SD (2008) Phylogenetically independent associations between autonomous self-fertilization and plant invasiveness. *American Naturalist*, **171**, 195–201.

van Kleunen M, Weber E, Fischer M (2010a) A meta-analysis of trait differences between invasive and non-invasive plant species. *Ecology Letters*, **13**, 235–245.

van Kleunen M, Dawson W, Schlaepfer D, Jeschke JM, Fischer M (2010b) Are invaders different? A conceptual framework of comparative approaches for assessing determinants of invasiveness. *Ecology Letters*, **13**, 947–958.

Kliber A, Eckert CG (2005) Interaction between founder effect and selection during biological invasion in an aquatic plant. *Evolution*, **59**, 1900–1913.

Konečný A, Estoup A, Duplantier J-M *et al.* (2013) Invasion genetics of the introduced black rat (*Rattus rattus*) in Senegal, West Africa. *Molecular Ecology*, **22**, 286–3000.

Kuester A, Conner JK, Culley T, Baucom RS (2014) How weeds emerge: a taxonomic and trait-based examination using Unites States data. *New Phytologist*, **202**, 1055–1068.

Lande R (1976) The maintenance of genetic variability by mutation in a polygenic character with linked loci. *Genetical Research*, **26**, 221–235.

Lande R (1977) The influence of the mating system on the maintenance of genetic variability in polygenic characters. *Genetics*, **86**, 485–498.

Levins R (1968) *Evolution in Changing Environments*. Princeton University Press, Princeton, New Jersey.

Lewontin RC (1984) Detecting population differences in quantitative characters as opposed to gene frequencies. *American Naturalist*, **123**, 115–124.

Lewontin RC, Hubby JL (1966) A molecular approach to the study of genic heterozygosity in natural populations II. Amount of variation and degree of heterozygosity in natural populations of *Drosophila pseudoobscura*. *Genetics*, **54**, 595–609.

Lynch M (1984) Destabilizing hybridization, general-purpose genotypes and geographic parthenogenesis. *Quarterly Review of Biology*, **59**, 257–290.

Martins PS, Jain SK (1979) Role of genetic variation in the colonizing ability of rose clover (*Trifolium hirtum* All.). *American Naturalist*, **114**, 591–595.

Matesanz S, Sultan SE (2013) High-performance genotypes in an introduced plant: insights into future invasiveness. *Ecology*, **94**, 2464–2474.

Moeller DA, Geber MA (2005) Ecological context of the evolution of self-pollination between incipient *Clarkia* species. *Evolution*, **66**, 1210–1225.

Moles AT, Gruber MA, Bonser SP (2008) A new framework for predicting invasive species. *Journal of Ecology*, **96**, 13–17.

Mooney HA, Drake JA (eds) (1986) *Ecology of Biological Invasions of North America*. Ecological Studies 58. Springer-Verlag, New York.

Mulligan J (2014) *The Real Weed Man. Portrait of a Canadian Botanist. Gerald A. Mulligan*. Privately Published. Library and Archives Canada Cataloguing in Publication, ISBN 978-0-9937698-0-1.

Nei M, Maruyama T, Chakraborty R (1975) The bottleneck effect and genetic variability in populations. *Evolution*, **29**, 1–20.

Ness RW, Wright SI, Barrett SCH (2010) Mating-system variation, demographic history and patterns of nucleotide diversity in the tristylous plant *Eichhornia paniculata*. *Genetics*, **184**, 381–392.

Nevo E (1978) Genetic variation in natural populations: patterns and theory. *Theoretical Population Biology*, **13**, 121–177.

Novak SJ, Mack RN (2005) Genetic bottlenecks in alien plant species. Influence of mating system and introduction dynamics. In: *Species Invasions Insights into Ecology, Evolution, and Biogeography* (eds Sax DF, Stachowicz JJ, Gaines DD), pp. 201–228. Sinauer & Associates, Sunderland, Massachusetts.

Palacio-López K, Gianoli E (2011) Invasive plants do not display greater phenotypic plasticity than native or non-invasive counterparts: a meta-analysis. *Oikos*, **120**, 1393–1401.

Pannell JR, Barrett SCH (1998) Baker's Law revisited: reproductive assurance in a metapopulation. *Evolution*, **52**, 657–668.

Parker IM, Rodriguez J, Loik M (2003) An evolutionary approach to understanding the biology of invasions: local adaptation and general-purpose genotypes in the weed *Verbascum thapsus*. *Conservation Biology*, **17**, 59–72.

Parsons PA (1983) *The Evolutionary Biology of Colonizing Species*. Cambridge University Press, Cambridge.

Perrins J, Williamson M, Fitter A (1992) Do annual weeds have predictable characters? *Acta Oecologica*, **13**, 517–533.

Pigliucci M (2001) *Phenotypic Plasticity Beyond Nature and Nurture*. John Hopkins University Press, Baltimore, Maryland.

Rejmánek M, Richardson DM (1996) What attributes make some plant species more invasive? *Ecology*, **77**, 1655–1661.

Richards CL, Bossdorf O, Muth NZ, Gurevitch J, Pigliucci M (2006) Jack of all trades, master of some? On the role of phenotypic plasticity in plant invasions. *Ecology Letters*, **9**, 981–993.

Richardson DM, Pyšek P (2008) Fifty years of invasion ecology – the legacy of Charles Elton. *Diversity and Distributions*, **14**, 161–168.

Richardson BJ, Rogers PM, Hewitt GM (1980) Ecological genetics of the wild rabbit in Australia. II Protein variation in British, French and Australian rabbits and the geographical distribution of variation in Australia. *Australian Journal of Biological Sciences*, **33**, 371–383.

Sax DF, Stachowicz JJ, Gaines SD (eds) (2005) *Species Invasions: Insights into Ecology, Evolution and Biogeography*. Sinauer Associates, Sunderland, Massachusetts.

Selander RK, Kaufman DW (1973) Self-fertilization and genic population structure in a colonizing land snail. *Proceedings of the National Academy of Sciences of the United States of America*, **70**, 1186–1190.

Shigesada N, Kawasaki K (1997) *Biological Invasions: Theory and Practice*. Oxford University Press, Oxford.

Simberloff D (2011) Charles Elton: neither founder nor siren. In: *Fifty Years of Invasion Ecology The Legacy of Charles Elton* (ed. Richardson D), pp. 11–24. Wiley-Blackwell, Oxford.

Smocovitis VB (2001) G. Ledyard Stebbins and the evolutionary synthesis. *Annual Review of Genetics*, **35**, 803–814.

Smocovitis VB (2006) Keeping up with Dobzhansky: G. Ledyard Stebbins, Jr., plant evolution, and the evolutionary synthesis. *History and Philosophy of the Life Sciences*, **28**, 9–47.

Stebbins GL (1950) *Variation and Evolution in Plants*. Columbia University Press, New York.

Stebbins GL (1957) Self-fertilization and population variability in the higher plants. *American Naturalist*, **91**, 337–354.

Stockwell CA, Hendry AP, Kinnison MT (2003) Contemporary evolution meets conservation biology. *Trends in Ecology and Evolution*, **18**, 94–101.

Suarez AV, Tsutsui ND (2008) The evolutionary consequences of biological invasions. *Molecular Ecology*, **17**, 351–360.

Sultan SE (1987) Evolutionary implications of phenotypic plasticity in plants. *Evolutionary Biology*, **21**, 127–178.

Sultan SE, Horgan-Kobelski T, Nichols LM, Riggs C, Waples RK (2012) A resurrection study reveals rapid adaptive evolution within populations of an invasive plant. *Evolutionary Applications*, **6**, 266–278.

Therriault TW, Orlova MI, Docker MF, MacIsaac HJ, Heath DD (2005) Invasion genetics of a freshwater mussel (*Dreissena rostriformis bugensis*) in eastern Europe: high gene flow and multiple introductions. *Heredity*, **95**, 16–23.

Villablanca FX, Roderick GK, Palumbi SR (1998) Invasion genetics of the Mediterranean fruit fly: variation in multiple nuclear introns. *Molecular Ecology*, **7**, 547–560.

Westley PAH (2011) What invasive species reveal about the rate and form of contemporary phenotypic change in nature. *American Naturalist*, **177**, 496–509.

Whitney KD, Gabler CA (2008) Rapid evolution in introduced species, 'invasive traits' and recipient communities: challenges for predicting invasive potential. *Diversity and Distributions*, **14**, 569–580.

Williamson M (1996) *Biological Invasions*. Chapman and Hall, London.

Williamson MH, Fitter A (1996) The characters of successful invaders. *Biological Conservation*, **78**, 163–170.

Zhan A, MacIsaac HJ, Cristescu ME (2010) Invasion genetics of the *Ciona intestinalis* species complex: from regional endemism to global homogeneity. *Molecular Ecology*, **19**, 4678–4694.

Zhang Y-Y, Zhang D-Y, Barrett SCH (2010) Genetic uniformity characterizes the invasive spread of water hyacinth (*Eichhornia crassipes*), a clonal aquatic plant. *Molecular Ecology*, **19**, 1774–1786.

Part 1

Evolutionary Ecology

INTRODUCTION

Katrina M. Dlugosch and Ingrid M. Parker*[†]

* Department of Ecology and Evolutionary Biology, University of Arizona, PO Box 210088, Tucson, AZ 85721, USA
[†] Department of Ecology and Evolutionary Biology, University of California, Santa Cruz, CA 95064, USA

One of the most recognizable legacies of *The Genetics of Colonizing Species* (Baker & Stebbins 1965) was the bringing together of evolutionary biologists and ecologists to jointly consider how the ecology of colonizing and invasive species might evolve. The authors debated the traits that would facilitate or inhibit colonization success, the environments that would select for such traits and the essential features of post-colonization evolutionary ecology. While many general predictions were proposed at the time and in the decades since, invasion ecology and evolution have often appeared idiosyncratic and resistant to generalization (Richardson & Pyšek 2008). Despite these challenges, the study of the evolutionary ecology of invaders has continued to thrive and inform our general understanding of how ecology can evolve (Cadotte *et al.* 2005; Sax *et al.* 2005) (Chapter 20). The chapters in this section highlight a variety of ways in which the field of invader evolutionary ecology is being both broadened and refined to better address the fundamental questions posed in 1965.

Blackburn and colleagues (Chapter 2) set the stage by reviewing one of the most important areas where our understanding of invasion ecology has broadened: the pervasive influence of population size on all stages of the invasion process. Evidence from well-studied groups clearly indicates that invasion must first and foremost involve enough individuals to proceed through establishment and spread. This 'numbers game' may both influence and be influenced by genetic diversity, for example by reducing inbreeding depression, providing evolutionary rescue or increasing resource use complementarity. While there has been particular interest in whether low genetic diversity in founding populations can inhibit invasion, Blackburn *et al.* argue that numbers appear to limit early establishment mainly through effects on demographic stochasticity, with little role for genetics. This conclusion fits with mounting evidence that large losses of genetic diversity are rare and manifest only in extremely small founder populations (Chapter 14). In contrast, Blackburn *et al.* note that positive associations between population size and post-establishment invasiveness suggest that density and range expansion are promoted by increased genetic variation associated with larger introductions. This variation might counter deleterious founder effects and facilitate adaptation over longer timescales (Dlugosch & Parker 2008) (Chapter 13). Indeed, the integral role of population size in both demography and evolution has established it as being among the most important characteristics in the study of species' invasions.

Unquestionably, the most well-known treatment of invader traits to date is Baker's list of the characteristics of 'ideal' weeds, which first appeared in his contribution to the 1965 volume (Baker 1965). It would be hard to overestimate the influence this paper has had on invasion biology and weed science, and Baker's predictions are still being tested today (Chapter 1). Support for Baker's list has been mixed, but van

---

Invasion Genetics: The Baker and Stebbins Legacy, First Edition. Edited by Spencer C. H. Barrett, Robert I. Colautti, Katrina M. Dlugosch, and Loren H. Rieseberg.

Kleunen and colleagues (Chapter 3) revisit these ideas and discuss several areas where we are refining our thinking to better identify the traits that enhance invasiveness. Most critically, it is now recognized that there are fundamentally different paths to invasion success, and these paths lead to different expectations for invader traits. For example, invaders can fill unique ecological roles in a community (i.e. Darwin's naturalization hypothesis; Mack 1996), or they can outcompete resident species with similar ecological roles. As a result, van Kleunen *et al.* caution that we must alter our expectations to avoid problematic common assumptions of past studies, specifically recognizing that (i) invaders may succeed by having traits similar to, rather than different from, native species; (ii) invaders may benefit from having intermediate, rather than extreme, phenotypes; and (iii) suites of traits are likely to work together in different combinations to form invader syndromes (the latter having been foreshadowed by pioneering work on introduction decision trees; e.g. Reichard & Hamilton 1997). Van Kleunen *et al.* note that Baker's list and subsequent hypotheses may in fact be correct about many favourable invader traits such as high dispersal ability, but that optima for individual traits may be better understood by looking beyond past assumptions.

In Chapter 4, Pannell tackles another long-standing debate regarding invader traits, disentangling a history of confusion regarding the evolution of mating systems and its relation to dispersal and establishment in colonizing species. Baker's law posits that successful long-distance colonizers are most often those capable of self-fertilization (Baker 1955), and this idea was extended to predict an association between selfing and invasion (Baker 1965). However, Pannell discusses the importance of distinguishing between the capacity to self-fertilize (self-compatibility) and selection for an autogamous mating system, and from there he delineates several distinct hypotheses that explain why populations on the edge of species' ranges should be more likely to be self-fertilizing. Re-examining Lewontin's (1965) insightful chapter on 'Selection for colonizing ability' in the Baker and Stebbins' volume, Pannell explores how 'colonizing episodes' of invasive species, metapopulations and demes within populations are all contexts in which ideas about the evolutionary dynamics of mating systems have general relevance. Pannell argues that two important areas for future research include more experimental tests of the importance of reproductive assurance for colonizing species, and

theoretical analyses that include the evolution of inbreeding depression.

These first chapters in the section highlight that it is reasonable to expect successful invaders to share some common features, as all of these species must overcome the same set of barriers: introduction, establishment, invasion and spread. Comparative analyses among taxa can help identify where these commonalities exist, and evaluate how genetics and trait evolution can influence the probability of overcoming invasion barriers. Recently, molecular ecological approaches have revolutionized the study of microbes, as we have gained the ability to detect and delimit the ranges of species that are nearly invisible and yet can have major ecological effects, such as decomposers and mycorrhizal fungi. Gladieux and colleagues (Chapter 5) review the rapidly emerging field of fungal invasion ecology and evolution, and strikingly identify many of the classic plant and animal invader traits (e.g. dispersal ability and mating system) as being important for this very divergent group. In addition, certain aspects of fungi and fungal invasion offer new opportunities for investigating evolutionary processes. For example, invasive fungi often abandon sexual reproduction, suggesting they could be valuable study systems for testing Baker's law. The close link between pathogenic fungi and their hosts provides a particularly good opportunity to study the importance of major niche shifts in the invasion process and the relative contributions of plasticity and evolution in ecologically important traits such as pathogenicity. These questions are experimentally tractable in rapidly evolving fungal systems, and there are likely to be significant new insights gained from further research in this area.

In Chapter 6, Colautti and Lau review what we have learned thus far from patterns of rapid evolution in invading species, and what these patterns can and cannot tell us about natural selection on invader traits and the likelihood that trait evolution enhances invasiveness. In recent decades, there has been a growing body of research documenting rapid evolution both among and within invaded regions. Yet Colautti and Lau point out that there is still little rigorous evidence that these evolutionary changes are adaptive and/or have contributed to invasion success. They demonstrate a number of ways in which apparent patterns of adaptation during invasion can instead be generated by genetic drift, inferred from improper experimental design, or misinterpreted in terms of the geographic source of divergence. They note that to critically improve our understanding of invader evolution, we need to embrace classical approaches to quantifying adaptation, such as reciprocal transplants and selection analyses. In a novel comparison of selection analyses in invading and native species, Colautti and Lau find that estimates of the strength of selection (selection gradient) were similar to that experienced by native species, but the predicted response to selection (selection differential) tended to be greater for invaders. These results suggest that most invaders might not be encountering especially novel selective environments in their new ranges, but rather might experience reduced constraint on their response to selection – potentially from simplified species interactions. These provocative ideas suggest that there is much more to learn about the nature of selection on invading species.

In addition to studies of evolution in single species, invasion ecology has also been expanding to examine the evolutionary consequences of invasion for entire communities. In Chapter 7, Helmus and colleagues use a phyloecological perspective of *Anolis* lizard distributions in the Caribbean to consider how species introductions affect the evolutionary diversity of communities. They point out that biogeographic processes strongly limit the phylogenetic diversity of most assemblages, and that human-assisted invasions will not only increase species richness but also the phylogenetic 'evolutionary history' of communities. This increase in community evolutionary history and phylogenetic diversity may have ecological implications for total resource use and for the success of future invasions. These patterns are likely to be general, given the accumulating evidence that invaders are often particularly phylogenetically (e.g. Strauss *et al.* 2006) and/or functionally (e.g. Ordonez 2014) distinct from invaded communities. Helmus *et al.* further posit that for Caribbean anoles, the possibility of future extinctions (extinction debt) could mean that local phylogenetic diversity is currently at a temporary maximum. Their results highlight that the evolution of ecological communities will be particularly dynamic during invasion, and potentially over longer periods of time as extinction debts play out.

In the final chapter of this section, Lau and Terhorst (Chapter 8) explore the potential evolutionary consequences of invasion for the ecology of native species. It is intuitive to think that invasions would drive adaptive evolution of native species' traits, given the ecological dominance of an invader. Instead, Lau and Terhorst present case studies to the contrary and clarify reasons

why our expectations may not be met. First, the potential ecological uniqueness of invaders is highlighted here again, because invaders that fill novel niches in recipient communities might not interact strongly with native species. Weak and/or novel ecological interactions could generate relatively little selection on native traits, or result in negative ecological pleiotropy (wherein a trait affects multiple different interactions in an antagonist manner for the native species) and no net directional selection by the invader. Second, where invaders do generate strong patterns of directional selection, these will act in the context of selection regimes that vary in time and space for native species. Spatial and/or temporal variation in the impact of invaders should often result in adaptive evolution that is driven by favourable (high fitness) environments for native species, where invaders are relatively less common and impose little selection. Lau and Terhorst make a clear case for the complexity of the selective environment for native species, again demonstrating the value of pursuing classical experimental studies of selection and adaptation in invaded communities in their natural field environment (Chapter 6).

Across the chapters in this section, two themes emerge regarding important future directions in the study of the evolutionary ecology of invaders. First, the manner and extent to which invaders are ecologically unique in their recipient communities should shape our predictions for 'invasiveness' traits, natural selection on invading and native species, and the likelihood of ecological constraints on trait evolution. The burgeoning fields of community phylogenetic ecology and trait-based ecology are already providing new avenues for quantifying ecological similarity among invasive and native species (e.g. Ordonez 2014; Parker *et al.* 2015), offering tremendous opportunities to refine and test long-standing hypotheses about invader traits and impacts.

Second, the time has come to quantify the contribution of adaptation to the success and severity of species invasions. We need more experiments that disentangle population size and genetic variation, that quantify the fitness effects of trait evolution, and that scale fitness differences among genotypes up to their population growth and spread in the field. Microbial systems with fast evolutionary timescales may be especially tractable for these purposes and appear to share many important features with larger scale systems. Studies of ecologically and socially important invaders in combination with studies of model systems should reveal in what ways adaptation is contributing directly to the success and impact of invasions. Such experiments will allow us to address fundamental questions about how often the ecology of invasive species is shaped by adaptation to local environments, adaptation for colonizing ability itself, or the absence of adaptation. In these ways, we should expect to see a vibrant stream of insights into invader evolutionary ecology – sparked in 1965 – continue for decades to come.

REFERENCES

Baker HG (1955) Self-compatibility and establishment after 'long-distance' dispersal. *Evolution*, **9**, 347–349.

Baker HG (1965) Characteristics and modes of origin of weeds. In: *The Genetics of Colonizing Species* (eds. Baker HG, Stebbins GL), pp. 147–172. Academic Press, New York.

Baker HG, Stebbins GL (Eds.) (1965) *The Genetics of Colonizing Species*. Academic Press, New York.

Cadotte MW, McMahon SM, Fukami T (Eds.) (2005) *Conceptual Ecology and Invasion Biology: Reciprocal Approaches to Nature*. Kluwer Academic Publishers, Dordrecht.

Dlugosch KM, Parker IM (2008) Founding events in species invasions: genetic variation, adaptive evolution, and the role of multiple introductions. *Molecular Ecology*, **17**, 431–449.

Mack RN (1996) Biotic barriers to plant naturalization. In: *Proceedings of the IX International Symposium on Biological Control of Weeds* (eds. Moran VC, Hoffman JH), pp. 39–46. University of Cape Town, Stellenbosch.

Ordonez A (2014) Functional and phylogenetic similarity of alien plants to co-occurring natives. *Ecology*, **95**, 1191–1202.

Parker IM, Saunders M, Bontrager M *et al.* (2015) Phylogenetic structure and host abundance drive disease pressure in communities. *Nature*, **520**, 542–544.

Reichard SH, Hamilton CW (1997) Predicting invasions of woody plants introduced into North America. *Conservation Biology*, **11**, 193–203.

Richardson DM, Pyšek P (2008) Fifty years of invasion ecology – the legacy of Charles Elton. *Diversity and Distributions*, **14**, 161–168.

Sax D, Stachowicz J, Gaines S (Eds.) (2005) *Species Invasions: Insights into Ecology, Evolution, and Biogeography*. Sinauer Associates Inc., Sunderland.

Strauss SY, Webb CO, Salamin N (2006) Exotic taxa less related to native species are more invasive. *Proceedings of the National Academy of Sciences USA*, **103**, 5841–5845.

Chapter 2

THE INFLUENCE OF NUMBERS ON INVASION SUCCESS

Tim M. Blackburn,[*][†] *Julie L. Lockwood,*[‡] *and Phillip Cassey*[§]

[*]Department of Genetics, Evolution and Environment, Centre for Biodiversity and Environment Research, University College London, Gower Street, London, WC1E 6BT, UK
[†]Institute of Zoology, ZSL, Regent's Park, London, NW1 4RY, UK
[‡]Ecology, Evolution, and Natural Resources, Rutgers University, New Brunswick, NJ 08901, USA
[§]School of Biological Sciences, University of Adelaide, Adelaide, SA 5005, Australia

Abstract

The process by which a species becomes a biological invader, at a location where it does not naturally occur, can be divided into a series of sequential stages (transport, introduction, establishment and spread). A species' success at passing through each of these stages depends, in a large part, on the number of individuals available to assist making each transition. Here, we review the evidence that numbers determine success at each stage of the invasion process and then discuss the likely mechanisms by which numbers affect success. We conclude that numbers of individuals affect transport and introduction by moderating the likelihood that abundant (and widespread) species are deliberately or accidentally translocated; affect establishment success by moderating the stochastic processes (demographic, environmental, genetic or Allee) to which small, introduced populations will be vulnerable; and affect invasive spread most likely because of persistent genetic effects determined by the numbers of individuals involved in the establishment phase. We finish by suggesting some further steps to advance our understanding of the influence of numbers on invasion success, particularly as they relate to the genetics of the process.

Previously published as an article in *Molecular Ecology* (2015) 24, 1942–1953, doi: 10.1111/mec.13075

INTRODUCTION

Hindsight is a wonderful thing. It is easy to look back with a critical eye on the best research produced by previous generations and feel good about oneself by observing only the shortcomings. A case in point is Baker & Stebbins' (1965) edited Proceedings of the First International Union of Biological Sciences Symposia on General Biology. It is titled *The Genetics of Colonizing Species*, but the then state of the art means that it contains little focus on molecular genetics, and mainly considers ecological and quantitative genetics, and phenotypes assumed to have a genetic basis. It is also not always obvious what is meant by a 'colonizing species', and indeed, three different types of such species are identified (Mayr 1965). The bulk of the discussion in Baker & Stebbins (1965) relates to species we would now term non-native or alien (i.e. species whose presence in a region is attributable to human actions, which have enabled them to overcome fundamental biogeographical boundaries; Richardson *et al.* 2011), and their volume is recognized as a classic text of invasion biology. Yet, to the eye of the modern invasion biologist, there are some glaring omissions that serve to highlight how far the field has come over the last half century. Perhaps the most important of these is the lack of appreciation that invasion biology is primarily a succession of numbers games.

The process by which an alien species invades can be divided into a series of consecutive stages: transport (beyond native range limits), introduction (into the wild in a new environment), establishment (of a viable alien population) and finally (invasive) spread (Blackburn *et al.* 2011). Recognition that the number of individuals matters greatly to this process arose from developments in conservation biology, where it had become apparent that the persistence of small populations depends fundamentally on population size (Caughley 1994). Initially, the importance of numbers was largely considered in terms of establishment success, but it was quickly realized that the number of individuals matter at all invasion stages. Thus, abundant native species are more likely to be entrained in a transportation mechanism and later released into a new location (Blackburn & Duncan 2001). Species more abundant in captivity are also more likely to be released (Cassey *et al.* 2004a; Chang *et al.* 2009). Introduced populations are more likely to establish if more individuals are released (higher 'propagule pressure', which is the sum over all release events of the

number of individuals released to form a population, sometimes also termed 'introduction effort': Cassey *et al.* 2004b, 2005; Lockwood *et al.* 2005; Hayes & Barry 2008; Blackburn *et al.* 2009; Simberloff 2009), and populations introduced with higher propagule pressure, or that produce more offspring in the new environment, are more likely to spread (Duncan *et al.* 1999, 2001; Caswell *et al.* 2003; Signorile *et al.* 2014). Several chapters in Baker & Stebbins mention numbers in the context of colonization by alien species, but primarily as a consequence of invasion, not as a cause (e.g. Birch 1965; Fenner 1965; Fraser 1965; Harper 1965; Sakai 1965). None presage the prominence this issue has now achieved. Today, the key questions are not about whether numbers influence invasion success, but how.

The invasion stage that has received most attention in terms of the influence of numbers is establishment. Most alien populations start out at very small numbers (Blackburn *et al.* 2009; fig. 3.1). Population dynamic theory and conservation practice both demonstrate that small populations are more likely to go extinct, on average, than larger populations (see any ecology or conservation text book). We would expect extinction risk to vary with population size for alien as well as for native populations, and so it is no surprise to find that propagule pressure is generally strongly positively correlated with establishment success (Lockwood *et al.* 2005; Colautti *et al.* 2006; Hayes & Barry 2008; Blackburn *et al.* 2009; Simberloff 2009). Nevertheless, small populations are vulnerable to a variety of processes, including demographic stochasticity, environmental heterogeneity, Allee effects and genetic effects (Morris & Doak 2002; Cassey *et al.* 2014). We might expect the precise mechanisms underlying the relationship between numbers of individuals and persistence to differ for native and alien populations. For example, the importance of environmental heterogeneity or Allee effects may differ for species new to a location vs. species with a long evolutionary history in that environment. The role of genetic effects may also differ for populations structured by natural vs. anthropogenic processes. To date, few studies have explored the influence of propagule pressure on alien population establishment in ways that allow us to discriminate between the actions of these different processes.

Here, we review evidence that helps us to understand how the broad positive relationship between numbers and invasion success might be driven, considering all stages in the invasion process. In the spirit of

Baker & Stebbins (1965), we highlight how numbers might interact with genetic effects where possible. However, as Ernst Mayr noted in his concluding remarks to the Proceedings, 'I am sure every ecologist here realizes that he (*sic*) really ought to know more about genetics', and that very much applies to the three of us.

NUMBERS, TRANSPORT AND RELEASE

The early stages of the invasion pathway concern which species are transported beyond the limits of their native geographic ranges and which of these species are subsequently liberated into new environments. In many cases, the first evidence that species have been transported and released outside their native ranges comes when free-living individuals are observed within a new environment. Hence, most studies of these early stages of invasion concatenate transport and release. A basic dichotomy in classification at these early invasion stages is whether individuals are moved intentionally or unintentionally by humans (Lockwood *et al.* 2013). Either way, the number of individuals in the native population matters.

Examples of accidental transport and release include individuals caught up in the ballast (soil or water) of ships, within the packing material used for dry cargo, or as hitchhikers living beside or within a purposefully traded species (Mack 2003; Hulme *et al.* 2008; Hulme 2009). Under these circumstances, individuals of alien species find themselves entrained in a transport vector essentially at random. Species more prevalent in their native environment are by chance alone more likely to be unintentionally transported (Hulme 2009) and more likely to be present in those samples in higher numbers (Wonham *et al.* 2001). Species that have adapted to human-altered habitats may be more likely to be transported by accident than species that shun anthropogenic environments (Hufbauer *et al.* 2012), but we would still expect accidental transport to concern more abundant species in these environments. The same processes apply also to intraspecific variation, such that higher frequency genotypes (and phenotypes) are more likely to be captured for transportation and release (Nei *et al.* 1975). In sum, accidental transport and introduction filter out rarity. Random sampling processes also result in larger numbers of individuals per species being introduced as the size of the sample increases (Lockwood *et al.* 2009),

which has further consequences for the probability that an alien species will establish a viable population once released.

The same is true for many species deliberately transported and released. Species may be intentionally moved for a variety of reasons, including as game animals, ornamental plants or animals, as biocontrol agents, or for the purposes of conservation (Lockwood *et al.* 2013). Identity will clearly matter in such cases—not all plants are equally desirable as ornamentals, for example (Pyšek *et al.* 2003)—but the availability of species for capture and transport typically matters too. Thus, birds transported from the UK to New Zealand tended to be species that are abundant and resident in the UK (Blackburn & Duncan 2001). Similarly, parrots that are transported outside their native ranges tend to be widespread species, and widespread and abundant parrot species are more likely to be released or escape into novel regions (Cassey *et al.* 2004a). While nonrandomness in the taxonomic composition of species has revealed that certain types were preferentially moved (e.g. wildfowl, game birds; Blackburn & Duncan 2001), the species introduced were nevertheless those that were the most readily available and easily obtained (Fig. 1). Abundant species tend also to be widespread (Gaston & Blackburn 2000) and so likely to be available for collection at a wide range of locations. For any given species, higher frequency genotypes and phenotypes are again more likely to be moved. Thus, with the acknowledgement that some deliberately transported species are rare in their native range, we should expect commonness to also be favoured in the deliberate movements of species.

These transport filters have consequences for subsequent invasion stages, as they determine which species become exposed to novel environments (Cassey *et al.* 2004a). Common, widespread species are common and widespread for a reason. While it is still not obvious if we can identify the actual underlying processes with much confidence, the breadth or typicality of species' environmental requirements or tolerances seem likely to be important determinants of establishment success and subsequent invasive spread (Gaston 1994, 2003; Gaston & Blackburn 2000). Alternatively, species that have adapted to human-altered habitats may be more likely to be both transported and able to exploit conditions they find on release (into other human-altered habitats; Hufbauer *et al.* 2012). Either way, the early stages of the invasion process may be selecting for species that are pre-adapted to cope with conditions they

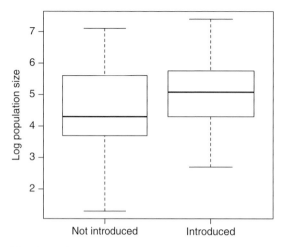

Fig. 1 Boxplot showing variation in estimates of the global population for wildfowl (Order Anseriformes) species that have or have not ever been introduced to areas beyond the limits of their native geographic range. A binomial general mixed linear model (with a random effect for genus to account for nonrandomness due to taxonomy) shows that population sizes are larger for species that have a history of introduction than for those that do not (estimate ± standard error = 0.64 ± 0.19, $N = 153$, $P < 0.001$). Data on population size were those used in Blackburn & Duncan (2001), while a list of introduced wildfowl species came from the Global Avian Invasions Atlas (GAVIA) database (E. Dyer & T. M. Blackburn, unpublished).

will encounter in the new location (Chapple *et al.* 2012). This may in part explain why establishment success is surprisingly high in at least some groups of alien species (e.g. Williamson 1996; Gaston *et al.* 2003; Jeschke 2008). Random sampling processes also result in larger numbers of individuals per species being introduced as the size of the sample increases, which has further consequences for the probability that an alien species will establish a viable population once released (Lockwood *et al.* 2009).

NUMBERS AND ESTABLISHMENT SUCCESS

Once an alien species is released into a novel environment, the individuals must found a self-sustaining population to be considered 'established' (Lockwood *et al.* 2013). The probability that this will happen is higher for alien populations that are founded by rela-

tively large numbers of individuals. If these individuals are released over more than one location, or at more than one time, the probability of establishment may also be higher. These relationships result because the perils of small population size tend to ensure that populations with few founders will eventually become extinct (Lockwood *et al.* 2005; Hayes & Barry 2008; Blackburn *et al.* 2009; Simberloff 2009). Demographic stochasticity, environmental stochasticity, Allee effects and genetic effects are all likely to play a role in increasing the chances that a small population will fail to establish. However, the actual contribution of each of these processes in the context of invasions is as yet unresolved. We are nevertheless gaining insights into these relationships from the increasing application of theoretical models of invasion dynamics to empirical data. These models show that different processes are expected to produce different relationships between propagule pressure and establishment success.

Duncan *et al.* (2014) derived the expected relationship between establishment probability and the number of individuals released for populations under the influence of demographic stochasticity, Allee effects and among-population environmental heterogeneity in establishment conditions. They assumed that founding populations initially were composed of far fewer individuals than the location's carrying capacity, and noted that a population will establish if at least one individual leaves a surviving lineage (Caswell 2001; Fox 2005). Under demographic stochasticity alone, the probability of establishment, P_{est}, for a newly introduced population of size N_0 is as follows:

$$P_{est} = 1 - (1 - P)^{N_0} \qquad (1)$$

where P is the probability that each individual leaves a surviving lineage. Demographic stochasticity affects all populations, and so Duncan *et al.* (2014) used eqn 1 as the base to which to add additional effects. They incorporated Allee effects by adding a term that models changes in the birth rate at different population sizes: a disproportionate decline in birth rate at low population sizes is expected under Allee effects. They incorporated among-population environmental heterogeneity by modelling variation in the probability of individual establishment, P, across different locations as drawn from a beta distribution.

Duncan *et al.* (2014) tested the fit of these different models to data for 55 experimental releases of the alien psyllid *Arytainilla spartiophila* to New Zealand for the purposes of biocontrol (Memmott *et al.* 2005). The data were best fit by the model of establishment success as a function of demographic stochasticity plus Allee effects, although the model of demographic stochasticity plus among-population heterogeneity also fitted the data reasonably well. Establishment success was relatively poorly predicted by demographic stochasticity alone. However, the best fitting model revealed that establishment probability per individual was actually proportionately lower at *large* population sizes, not at small population sizes as expected under a classic Allee effect. Duncan *et al.* (2014) found similar effects for global data on the outcome of bird species introductions (using data in Sol *et al.* 2012), with disproportionately lower per individual success rates when large numbers of birds were released.

These models suggest that variation in establishment success can broadly be explained by two processes. First, the decline in success at small propagule pressures (the left hand side of Fig. 2) is consistent with the effects of demographic stochasticity. Second, the disproportionate decline in success (per individual released) for larger releases suggests that success here is being driven by factors largely unrelated to the initial size of a population. This would be expected if populations are being introduced to areas that are unsuitable for their establishment, regardless of how many individuals are involved. Interestingly, Memmott *et al.*'s (2005) data showed substantial variation in the probability that each individual leaves a surviving lineage (*P*; see inset panel in Fig. 2), with many sites having a very low probability of establishment. This observation implies that even large populations of psyllids were destined to go extinct at some of the release sites. Indeed Memmott *et al.* (2005) noted that while small introduced psyllid populations tended to go extinct very quickly (consistent with demographic stochasticity), surviving populations were then prone to extinction due to site destruction, which affected populations regardless of their size.

Duncan *et al.* (2014) explored establishment probability as a function of the number of individuals released, but ignored the fact that this number can be

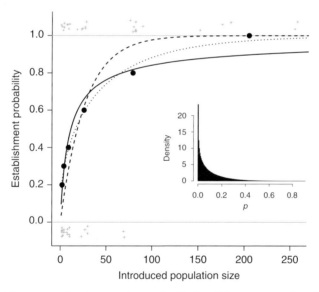

Fig. 2 Establishment probability vs. introduced population size for 55 of psyllid populations released in New Zealand. Introductions spanned a range of population sizes (10 introductions of 2, 4, 10, 30 and 90 psyllids and 5 introductions of 270 psyllids) and were considered successful if populations were still present after 5 years. Grey crosses are the raw data showing successful (*y*-axis values >1) and unsuccessful (*y*-axis values <0) establishment as a function of introduced population size, while the filled circles show the proportion of populations that established for each population size. The curved lines show the maximum-likelihood fits of different models to the data: dashed line = demographic stochasticity alone; dotted line = demographic stochasticity plus Allee effects; solid line = demographic stochasticity plus among-population heterogeneity. The inset panel shows the distribution of probabilities that each individual leaves a surviving lineage, *P*, for different populations modelling among-population heterogeneity. From Duncan *et al.* (2014), based on data in Memmott *et al.* (2005).

arrived at in a number of different ways. In particular, N total individuals may derive from one large or several smaller release events (Lockwood *et al.* 2005). Different release configurations will clearly influence the relative impacts of demographic stochasticity vs. Allee effects (and environmental suitability), but the precise outcome is likely to be influenced by how these effects are manifested. Hopper & Roush (1993) suggested that multiple, small releases may be more likely to establish than a single large one under environmental heterogeneity, because increasing the number of releases increases the probability that one of those will coincide temporally or spatially with favourable environmental conditions. This argument has subsequently been confirmed by a variety of models (e.g. Haccou & Iwasa 1996; Grevstad 1999; Haccou & Vatutin 2003). Conversely, simulations by Cassey *et al.* (2014) found that the probability of establishment was negatively correlated with the number of separate release events, and the time between them, even under conditions of extreme (interannual) environmental variability. They attributed their results to the fact that a single, large release will grow more quickly in population size, and hence is more capable of riding out harsh environmental conditions, while less likely to be reduced to a level where demographic and typical Allee effects are relevant. However, Cassey *et al.* (2014) modelled releases

distributed in time, but not in space. Multiple releases to different locations may enhance the probability that some of those released individuals encounter a favourable environment simply because these conditions are more variable across space than through time at a single location (cf. Haccou & Iwasa 1996; Haccou & Vatutin 2003; Duncan *et al.* 2014).

The number of individuals released, and how they are released, is also likely to determine the impacts of genetic stochasticity in alien populations. Smaller releases are likely to have lower genetic diversity, and higher likelihoods of population bottlenecks, genetic drift and inbreeding, all of which can cause declines in mean fitness (Fig. 3; Frankham *et al.* 2004). These founder effects will be exacerbated if the population remains small for a number of generations (Nei *et al.* 1975). All of these effects may decrease the probability that an alien species will establish a self-sustaining population.

Releasing a given number of individuals in several small releases distributed across space (or time) may exacerbate these problems by forcing the population through a series of smaller bottlenecks. Alternatively, it has been suggested that multiple releases may promote establishment by providing a 'genetic rescue effect' (*sensu* Carlson *et al.* 2014) by supplementing genetic diversity, especially in cases where supplementary

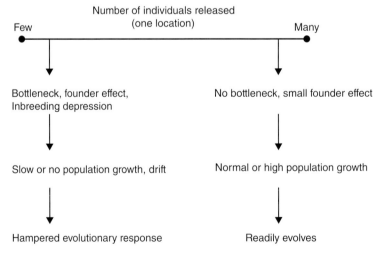

Fig. 3 Flow chart of the cause–effect sequence of events whereby low propagule pressure results in low genetic variation, slow population growth causing further genetic erosion and finally a dampened ability to evolve to new conditions in the alien range. Contrast this outcome to the sequence of events expected when propagule pressure is high. Any one of the events in this sequence can add to the likelihood that an initially small founder population will probably go extinct due to genetic issues.

individuals come from different source populations (Sakai *et al.* 2001; Brook 2004). While alien populations are expected to sample only a proportion of the genetic variation present in their native range (see e.g. Dlugosch & Parker 2008), there are prominent examples where local alien populations are genetically more diverse than local source populations (e.g. Kolbe *et al.* 2004). Individuals deriving from diverse donor locations may not only increase genetic diversity and reduce the likelihood of bottlenecks, but may also trigger novel outcrossing events that can increase the adaptive potential of an introduction (Novak & Mack 2005), or import novel genetic variation that allows evolutionary rescue. Conversely, they may also lead to outbreeding depression.

Dlugosch & Parker (2008) reviewed studies of genetic variation in introduced vs. native ranges, finding data for 80 alien species from a range of taxa. While they found some examples where genetic diversity was higher in the alien range, in most populations it was lower. The average loss of diversity was estimated at between 5.8% and 32.7%, depending on the molecular marker. However, the extent of this loss was smaller for populations that had resulted from multiple introduction events. These patterns would be expected under a genetic rescue effect, but could also be explained if multiple introductions tended to have higher propagule pressures. This is certainly the case for bird introductions to New Zealand, where number of individuals of a species released at a location is highly positively correlated with the number of releases ($r = 0.73$, $N = 92$, $P < 0.001$; from data used by Blackburn *et al.* 2013). Dlugosch & Parker (2008) do not control for this effect, but a subsequent meta-analysis of animal and plant introductions showed that genetic diversity tends to be higher for alien populations deriving from multiple introductions, controlling for the number of individuals introduced (Uller & Leimu 2011). Nevertheless, these analyses do not inform about the influence of genetic effects on establishment success, as they do not include genetic data for failed introductions (these are considerably more difficult to come by). A field experiment by Ahlroth *et al.* (2003) does suggest that genetic composition may be important: they found that the likelihood of successful colonization increased with propagule pressure for introduced water striders populations, but that colonization success was higher, for a given propagule pressure, when founders came from two vs. one source populations. Nevertheless, the maximum number of founders introduced by Ahlroth *et al.* (2003) was only 16. Even releases involving few individuals can sample much native genetic diversity (Novak & Mack 2005; see also Roman & Darling 2007), suggesting that genetic effects may be relevant only to the smallest releases.

The population model of establishment success explored by Cassey *et al.* (2014) supports this interpretation. They included genetic effects by modelling inbreeding depression and identified four factors that were influential in determining establishment in their simulations: net reproductive rate per female, the number of individuals released, the influence of extreme environmental events and the strength of inbreeding effects. The genetic effects were the smallest of the four. Inbreeding had negligible effects on establishment probability in cases where success or failure is more or less guaranteed on the basis of other modelled parameters, but could tip the balance towards failure for populations with intermediate chances of establishment. Overall, their models found that a single release was always more successful than multiple releases, all else being equal, even under conditions of extreme environmental variability. One large release grew more quickly, was less likely to be reduced to a population size where demographic, genetic and Allee effects come into play and could exploit favourable conditions when they occurred. All that spreading out the release of more individuals did, in terms of establishment success, was to delay their reproductive contribution to future generations (Cassey *et al.* 2014). However, their model did not incorporate negative density dependence, and so it is possible that multiple releases may matter more for populations that are highly constrained in size.

Alien populations are not fixed entities, but can evolve to meet challenges of novel environments to which they are not pre-adapted (Sakai *et al.* 2001). Adaptation that occurs rapidly enough that a population recovers from environmentally induced demographic effects that otherwise would have caused extinction is termed 'evolutionary rescue' (Gomulkiewicz & Holt 1995; Gonzalez *et al.* 2013; Carlson *et al.* 2014). The recent growth of interest in evolutionary rescue is largely driven by attempts to understand the likely responses of species to rapid, anthropogenic environmental change (Gonzalez *et al.* 2013), but it is clearly also relevant to biological invasions (e.g. Holt *et al.* 2005). The likelihood that evolutionary rescue occurs will also be influenced by propagule pressure, because the size of a population is generally positively related to the rate at which it can adapt, and to the maximal rate of environmental change to which it can adapt (Lanfear *et al.* 2014). Larger populations may also take more time to decline to the size at which extinction due to stochastic processes is likely, although this will depend also on the degree of maladaptation (Holt *et al.* 2005), and hence the rate of population declines (Carlson *et al.* 2014). Moreover, the likelihood of evolutionary rescue is higher for populations with greater standing genetic variation available for selection to act upon (Bell 2013; Carlson *et al.* 2014), which also should be positively related to propagule pressure.

Evolutionary rescue may be influenced by how individuals are released. Multiple releases may promote evolutionary rescue if the additional propagules import novel genetic material or increase opportunities for mutations (Carlson *et al.* 2014). Once again, however, it is not obvious that these benefits would be greater than if the population derived from a single release of the same total number of individuals. Conversely, multiple releases may hamper evolutionary rescue if the immigrants bring maladaptive genes into the population (e.g. Schiffers *et al.* 2013), which could eliminate fitness gains made from adaptation in the original release. This suggests that, in some cases, multiple releases may actually reduce the likelihood of establishment by an alien population, although we are not aware of any examples of this.

Models and analyses of historical data on establishment success have been informative about the potential and actual influence of numbers introduced, but we are now at a point where the most useful development would be more experimental tests of the processes concerned. There is a small but growing number of such studies (Grevstad 1999; Ahlroth *et al.* 2003; Drake *et al.* 2005; Memmott *et al.* 2005; Fauvergue *et al.* 2007; Bailey *et al.* 2009; Bell & Gonzalez 2009; Gertzen *et al.* 2011; Hufbauer *et al.* 2013; Szucs *et al.* 2014). Experiments that manipulated the genetic composition of individuals introduced to locations (e.g. microcosms, enclosures) in different numbers and multiples of events could be particularly rewarding in distinguishing the influence of genetic, demographic and Allee effects. Recent studies by Hufbauer *et al.* (2013) and Szucs *et al.* (2014) that manipulated the numbers and genetic composition of insects introduced to experimental arenas are an excellent first step. For example, Szucs *et al.* (2014) demonstrated that establishment success in populations of *Tribolium*

beetles depended on founder size but not on their genetic provenance (inbred to outbred), although subsequent population growth was depressed at low founder sizes for inbred lines.

NUMBERS AND ALIEN SPECIES SPREAD

The influence of numbers on the process of invasion continues beyond the establishment phase to influence how far alien species spread across the new environment. Here, numbers affect spread (invasiveness) in two broad ways.

First, populations that produce more offspring in the new environment are more likely to spread more widely across it (Caswell *et al.* 2003). For example, successful plant invaders tend to be more fecund compared to their native congeners or related taxa, and to alien congeners with different degrees of invasiveness (Pyšek & Richardson 2008). Bird species with life history traits associated with higher fecundity (and higher rates of population growth) have larger alien geographic range sizes in both New Zealand (Duncan *et al.* 1999) and Australia (Duncan *et al.* 2001). This relationship may arise if the process of spread in a new environment is functionally equivalent to a sequence of establishment events (Blackburn *et al.* 2011). Under this analogy, an alien species that has established a sustainable population at a 'beachhead' in a new environment then spreads by establishing further populations at new locations, at each of which the same challenges that faced the original introduction are overcome. Just as the number of individuals in the initial introduction is of fundamental importance to that first establishment, so too is the number of individuals that reach subsequent locations a key determinant of the likelihood of establishes there. All else being equal, increasing local abundance boosts the number of propagules available for dispersal to new, unoccupied locations. If higher fecundity increases a population's local abundance, then a relationship between fecundity and spread would be expected.

Second, populations introduced to the new environment in larger numbers are not just more likely to establish a viable population there, but are also more likely to spread more widely across it (e.g. Duncan *et al.* 1999, 2001). Thus, the influence of propagule pressure appears to extend beyond the establishment phase, also to determine alien range size. Propagule pressure is argued to affect establishment success because it helps populations to overcome the consequences of Allee effects, and of demographic, environmental and genetic stochasticity (Fig. 3, see above). Population and conservation biology tell us that the first three of these processes quickly lose their threat to a population's persistence as it grows away from small numbers. The residual influence of propagule pressure therefore seems most likely to act through the continued impacts of genetic effects.

Alien species typically pass through a small population bottleneck on the pathway to establishment, and the concomitant declines in genetic diversity and increases in the likelihood of inbreeding can affect fitness over extended time frames (Fig. 3). For example, a review of genetic variation in established alien bird species found that more severe bottlenecks reduced genetic variability in the resulting populations, relative to the native range (Merilä et al. 1996). Briskie & Mackintosh (2004) used data on the breeding success of alien bird species established in New Zealand to demonstrate that rates of hatching failure were a negative function of the number of individuals introduced. Their data suggest that increases in failure rates are

mainly expressed in populations for which fewer than 100 individuals were introduced (Fig. 4). Given that the species analysed by Briskie & Mackintosh (2004) were introduced in the middle of the nineteenth century, their data suggest that the fitness consequences of passing through a population bottleneck are still being expressed more than a century after the bottleneck occurred. An alternative outcome is that prolonged inbreeding will allow harmful genetic mutations to be expressed and that these mutations will then be 'purged' from the population via selection (Frankham et al. 2004). If this situation pertains in alien invasions, we should expect that some alien populations established with few individuals, and whose populations remained low but persistent for long periods, will eventually begin to express higher fitness. For example, Facon et al. (2011) showed that inbred individuals of the ladybird *Harmonia axyridis* from the native range showed greater inbreeding depression than individuals from alien populations, suggesting that recessive deleterious mutations had been purged from the latter.

The fitness consequences of bottlenecks for alien species may affect the dynamics of their populations. For example, the experimental work of Szucs et al.

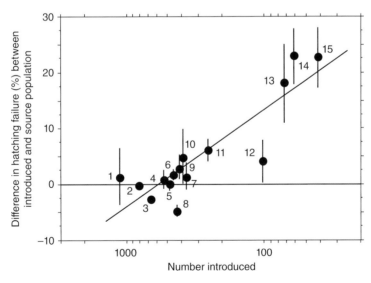

Fig. 4 The relationship between the number of birds introduced and relative hatching failure rate (the difference between the failure rate in the alien and native source populations) for alien bird species established in New Zealand. Species introduced in smaller numbers show a larger increase in failure rates in the alien relative to the native population. Species are as follows: (1) *Callipepla californica*; (2) *Turdus merula*; (3) *Sturnus vulgaris*; (4) *Fringilla coelebs*; (5) *Carduelis carduelis*; (6) *Carduelis flammea*; (7) *Passer domesticus*; (8) *Alauda arvensis*; (9) *Turdus philomelos*; (10) *Emberiza citrinella*; (11) *Prunella modularis*; (12) *Chloris chloris*; (13) *Acridotheres tristis*; (14) *Branta canadensis*; and (15) *Corvus frugilegus*. From Briskie & Mackintosh (2004).

(2014) showed that populations of *Tribolium* that establish from low propagule pressures grow more slowly if those founders are inbred, while dispersal rates in these populations increase with genetic diversity. Their results imply that population growth and spread following establishment can both be driven by genetic processes. Similarly, Signorile *et al.* (2014) studied alien grey squirrel (*Sciurus carolinensis*) populations at four locations in Europe. They found that genetic variation across these populations increased with founder population size and that there was a positive relationship between founder population size and the rate of population spread. A common feature of the population growth curves of alien populations is a lag phase, defined as a period of slow population growth followed by a marked increase in the rate of growth (Shigesada & Kawasaki 1997; Williamson *et al.* 2005; Aikio *et al.* 2010; Aagaard & Lockwood 2014). The lag phase may be a simple consequence of the form of population growth curves, but may also reflect the time taken for the population to produce the necessary adaptations to allow spread (Crooks & Soulé 1999). If so, losses of genetic variation that result from low propagule pressure and subsequent slow population growth may promote longer lag phases by reducing the genetic diversity available for selection to act upon.

The sampling effect of bottlenecks may also reduce the likelihood that individuals with appropriate adaptations to allow spread are introduced to a new location (McCauley 1991) (Fig. 4). For example, Zenni *et al.* (2014) showed that range expansions by alien *Pinus taeda* resulted from an interaction between the genetic provenance of the introduced individuals and the climate in the alien range. These invasions are led by plants with a genotype that conveyed higher fitness in the alien range, which suggests that the invasions would have at best proceeded more slowly had this genotype not been initially introduced. Such effects may also influence the extent of the alien range size.

The impacts of introduction history can have long-term effects through genetic stochasticity, but as with the effects of demographic and environmental stochasticity, these impacts are likely to decrease as an alien population grows and spreads. Selection may, depending on the precise conditions, purge exposed deleterious recessive alleles, weakening inbreeding depression over time (Frankham *et al.* 2004). Dlugosch & Parker (2008) found some evidence that the proportional change in genetic diversity in alien relative to native populations shows a U-shaped relationship to the length of time that an alien species has been established, at least for species established from multiple introduction events (but see Uller & Leimu 2011). Allelic richness decreased with time across populations up to around 80–100 years after first introduction, but then started to increase again. Dlugosch & Parker (2008) argued that drift and strong selection were likely to have caused a loss of within-population genetic diversity in the initial phase of establishment, for all initially established alien populations, given their associated slow population growth. However, as these alien populations increased in numbers and became more connected through dispersal (integrating more across multiple populations at multiple introduction sites), genetic diversity would begin to increase again (Fig. 5). This relatively higher genetic diversity can then be preserved during population expansion (when populations might be expected to go through a series of bottlenecks with each new colonization event or suffer from gene surfing by deleterious alleles; Edmonds *et al.* 2003), probably as a result of frequent long-distance dispersal events (e.g. Berthouly-Salazar *et al.* 2013).

Based on the evidence we present above, we suggest that the degree to which alien species will show a U-shaped pattern in diversity is dependent on at least two factors: (i) numbers of individuals initially released and (ii) the degree of genetic structure in the native range and how the transport process 'samples' this variation (Fig. 5). These two factors essentially determine the 'down' and 'up' of the U-shape, respectively. All available evidence suggests that the loss of genetic variation at the time of founding is dependent on propagule pressure. This manifests through founder effects and bottlenecks, but can also result from the synergistic effect of these factors dictating long periods of slow population growth leading to drift. Thus, the smaller the propagule pressure for any of the initially established populations of an alien species, the steeper and deeper the drop into the bottom of the U-shape. Once these independently established populations begin to exchange individuals, they establish gene flow, effectively homogenizing any existing differences in genetic variation across populations. If the initial populations are effectively drawn from a single panmictic native source, the rise in diversity in the alien range from gene flow will be minor; gene flow in this case will only be overcoming the effects of randomly sampling alleles from a single large population. However, if the initial populations are founded by individuals taken from

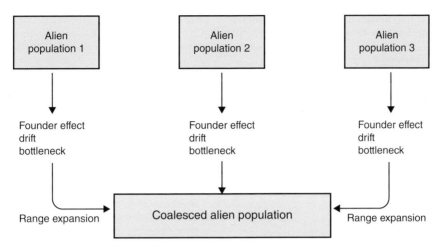

Fig. 5 A visual explanation of the U-shaped genetic diversity pattern shown by Dlugosch & Parker (2008). Multiple, independently founded alien populations lose genetic diversity (the down-slope of the U), but eventually coalesce to form one large, panmictic population, thereby increasing the total genetic diversity (the up-slope of the U). The depth of the U is dependent on the number of individuals released at each location and the degree to which the transport mechanism 'sampled' individuals representing divergent genetic backgrounds.

across the native range, and there is a high degree of geographic genetic structure in this range, the rise up the U-shape will be steeper. In this case, gene flow effectively represents admixture whereby alleles from divergent sources are intermixing in a single population.

CONCLUSIONS

Looking through the older literature on invasions, it is not that surprising that none of the contributors to Baker & Stebbins (1965), or any of the other early works on invasion biology (e.g. Elton 1958; Drake *et al.* 1989), recognized the importance of numbers of individuals on the dynamics of biological invasions. The arguments we proffer here result from an intense interest in small population dynamics within conservation biology and the ability to track the loss (and gain) of genetic diversity that accompanied transformative genetic technologies. Each of these lines of inquiry really only produced their insights from the late 1980s and early 1990s. What they bring to invasion biology is the fundamental insight that numbers matter at all stages of the invasion process. We have reviewed evidence for these effects above, but we also hopefully have highlighted gaps in our understanding of them, especially as they relate to the influence of genetic diversity. Below we summarize what we see are the

three next steps to pushing our understanding past its current position and associated debates.

First, empirical studies clearly show the profound influence that the transportation and release stage of invasions have on the genetic composition of alien species and that the number of individuals transported strongly mediates these effects. However, such studies are few in number, and their insights have not apparently penetrated a good portion of the invasion literature as yet. For example, Dlugosch & Parker (2008) were able to source information on differences in genetic diversity between 80 alien and native populations. They found some consistent patterns, but there was also much variation present. Knowledge of the transport dynamics behind each of these introduction events, including the number of individuals released across spatial and temporal scales, could explain much of this variation. As yet, there have been very few attempts to make these connections. Given the potential for knock-on effects of low genetic diversity across all invasion stages, we suggest that these connections will prove insightful relative to our understanding of invasion dynamics.

Second, the general influence of propagule pressure on establishment success is well supported across species and ecosystems (Lockwood *et al.* 2005; Colautti *et al.* 2006; Hayes & Barry 2008; Simberloff 2009). The more individuals released, the more likely an incipient alien population is to persist. The mechanisms behind this relationship are easy to speculate upon, and we detail them above, but they have only recently been actively under investigation. For example, recent modelling and empirical evidence strongly suggests that the relationship between numbers released and establishment is an increasing but asymptotic function (see e.g. Fig. 2). One under-explored possibility is that variation is influenced by genetic variability on population growth, either manifest as inbreeding depression (very low numbers) or admixture (very high numbers). The existence and location of these effects would be likely to vary across species, for example according to their life history traits. There is a clear need to explore genetic and other mechanisms behind the role of numbers released on establishment success, including the synergistic interactions of these mechanisms with one another.

Finally, the link between propagule pressure and subsequent invasive spread sits squarely within the realm of genetics. There is some evidence that low genetic variability within an alien population can limit its potential to adapt to novel conditions in the new range, cause lasting negative effects on survival and reproduction or trigger a behavioural change that is maladaptive in the new range. There is also evidence that some invasive populations have benefited from the purging of deleterious alleles as a consequence of passing through a bottleneck at introduction, while some invasive populations have higher genetic diversity than in the native range as a result of admixture and hybridization. There is considerable room for exploration in this area, and we suggest that future investigations explicitly consider the link between 'standing' genetic variation in an alien species and the number of individuals (and the source of these individuals) in producing this variation. Making these connections will provide crucial insight into our basic understanding of the role of genetic variation in species' range sizes, but more practically into our ability to predict which newly introduced alien species may go on to become invasive.

ACKNOWLEDGEMENTS

We thank Katrina Dlugosch, Richard Duncan, Ruth Hufbauer and Marianna Szucs for helpful discussions and for drawing our attention to key literature. This study was supported by the King Saud University Distinguished Scientist Fellowship Program and John Wiley & Sons Inc. (TMB), and by an ARC Discovery Grant (DP140102319) and Future Fellowship (FT0991420) to PC.

REFERENCES

Aagaard K, Lockwood JL (2014) Exotic birds show lags in population growth. *Diversity and Distributions*, **20**, 547–554.

Ahlroth P, Alatalo RV, Holopainen A, Kumpulainen T, Suhonen J (2003) Founder population size and number of source populations enhance colonization success in waterstriders. *Oecologia*, **137**, 617–620.

Aikio S, Duncan RP, Hulme PE (2010) Lag-phases in alien plant invasions: separating the facts from the artefacts. *Oikos*, **119**, 370–378.

Bailey SA, Vélez-Espino LA, Johannsson OE, Koops MA, Wiley CJ (2009) Estimating establishment probabilities of *Cladocera* introduced at low density: an evaluation of the proposed ballast water discharge standards. *Canadian Journal of Fisheries and Aquatic Sciences*, **66**, 261–276.

Baker HG, Stebbins GL (1965) *The Genetics of Colonizing Species*. Academic Press, New York, New York.

Bell G (2013) Evolutionary rescue and the limits of adaptation. *Philosophical Transactions of the Royal Society of London. Series B, Biological Sciences*, **368**, 20120080.

Bell G, Gonzalez A (2009) Evolutionary rescue can prevent extinction following environmental change. *Ecology Letters*, **12**, 942–948.

Berthouly-Salazar C, Hui C, Blackburn TM et al. (2013) Long distance dispersal maximizes evolutionary potential during rapid geographic range expansion. *Molecular Ecology*, **22**, 5793–5804.

Birch LC (1965) Evolutionary opportunity for insects and mammals in Australia. In: *The Genetics of Colonizing Species* (eds Baker HG, Stebbins GL), pp. 197–211. Academic Press, New York, New York.

Blackburn TM, Duncan RP (2001) Establishment patterns of exotic birds are constrained by non-random patterns in introduction. *Journal of Biogeography*, **28**, 927–939.

Blackburn TM, Lockwood JL, Cassey P (2009) *Avian Invasions. The Ecology and Evolution of Exotic Birds*. Oxford University Press, Oxford.

Blackburn TM, Pyšek P, Bacher S et al. (2011) A proposed unified framework for biological invasions. *Trends in Ecology & Evolution*, **26**, 333–339.

Blackburn TM, Prowse TAA, Lockwood JL, Cassey P (2013) Propagule pressure as a driver of establishment success in deliberately introduced exotic species: fact or artefact? *Biological Invasions*, **15**, 1459–1469.

Briskie JV, Mackintosh M (2004) Hatching failure increases with severity of population bottlenecks in birds. *Proceedings of the National Academy of Sciences, USA*, **101**, 558–561.

Brook BW (2004) Australian bird invasions: accidents of history? *Ornithological Science*, **3**, 33–42.

Carlson SM, Cunningham CJ, Westley PAH (2014) Evolutionary rescue in a changing world. *Trends in Ecology & Evolution*, **29**, 521–530.

Cassey P, Blackburn TM, Russell G, Jones KE, Lockwood JL (2004a) Influences on the transport and establishment of exotic bird species: an analysis of the parrots (Psittaciformes) of the world. *Global Change Biology*, **10**, 417–426.

Cassey P, Blackburn TM, Sol D, Duncan RP, Lockwood J (2004b) Introduction effort and establishment success in birds. *Proceedings of the Royal Society of London. Series B: Biological Sciences*, **271**, S405–S408.

Cassey P, Blackburn TM, Duncan RP, Gaston KJ (2005) Causes of exotic bird establishment across oceanic islands. *Proceedings of the Royal Society of London. Series B: Biological Sciences*, **272**, 2059–2063.

Cassey P, Prowse TAA, Blackburn TM (2014) A population model for predicting the successful establishment of introduced bird species. *Oecologia*, **175**, 417–428.

Caswell H (2001) *Matrix Population Models*. Sinauer Associates Inc, Sunderland, Massachusetts.

Caswell H, Lensink R, Neubert MG (2003) Demography and dispersal: life table response experiments for invasion speed. *Ecology*, **84**, 1968–1978.

Caughley G (1994) Directions in conservation biology. *Journal of Animal Ecology*, **63**, 215–244.

Chang AL, Grossman JD, Spezio TS et al. (2009) Tackling aquatic invasions: risks and opportunities for the aquarium fish industry. *Biological Invasions*, **11**, 773–785.

Chapple DG, Simmonds SM, Wong B (2012) Can behavioral and personality traits influence the success of unintentional species introductions? *Trends in Ecology & Evolution*, **27**, 57–64.

Colautti RI, Grigorovich IA, MacIsaac HJ (2006) Propagule pressure: a null model for biological invasions. *Biological Invasions*, **8**, 1023–1037.

Crooks JA, Soulé ME (1999) Lag times in population explosions of invasive species: causes and implications. In: *Invasive Species and Biodiversity Management* (eds Sandlund OT, Schei PJ, Viken A), pp. 103–125. Kluwer Academic Press, Dordrecht, The Netherlands.

Dlugosch KM, Parker IM (2008) Founding events in species invasions: genetic variation, adaptive evolution, and the role of multiple introductions. *Molecular Ecology*, **17**, 431–449.

Drake JA, di Castri F, Groves RH, Kruger FJ, Rejmánek M, Williamson M (eds) (1989) *Biological Invasions: A Global Perspective*. John Wiley & Sons, Chichester, UK.

Drake JM, Baggenstos P, Lodge DM (2005) Propagule pressure and persistence in experimental populations. *Biology Letters*, **1**, 480–483.

Duncan RP, Blackburn TM, Veltman CJ (1999) Determinants of geographical range sizes: a test using introduced New Zealand birds. *Journal of Animal Ecology*, **68**, 963–975.

Duncan RP, Bomford M, Forsyth DM, Conibear L (2001) High predictability in introduction outcomes and the geographical range size of introduced Australian birds: a role for climate. *Journal of Animal Ecology*, **70**, 621–632.

Duncan RP, Blackburn TM, Rossinelli S, Bacher S (2014) Quantifying invasion risk: the relationship between establishment probability and founding population size. *Methods in Ecology & Evolution*, **5**, 1255–1263.

Edmonds CA, Lillie AS, Cavalli-Sforza LL (2003) Mutations arising in the wave front of an expanding population. *Proceedings of the National Academy of Sciences, USA*, **101**, 975–979.

Elton C (1958) *The Ecology of Invasions by Animals and Plants*. Methuen, London.

Facon B, Hufbauer RA, Tayeh A et al. (2011) Inbreeding depression is purged in the invasive insect *Harmonia axyridis*. *Current Biology*, **5**, 424–427.

Fauvergue X, Malausa J-C, Giuge L, Courchamp F (2007) Invading parasitoids suffer no Allee effect: a manipulative field experiment. *Ecology*, **88**, 2392–2403.

Fenner F (1965) *Myxoma* virus and *Oryctolagus cuniculus*: two colonizing species. In: *The Genetics of Colonizing Species* (eds Baker HG, Stebbins GL), pp. 484–502. Academic Press, New York, New York.

Fox GA (2005) Extinction risk of heterogeneous populations. *Ecology*, **86**, 1191–1198.

Frankham R, Ballou JD, Briscoe DA, McInnes KH (2004) *A Primer of Conservation Genetics*. Cambridge University Press, Cambridge.

Fraser A (1965) Colonization and genetic drift. In: *The Genetics of Colonizing Species* (eds Baker HG, Stebbins GL), pp. 117–122. Academic Press, New York, New York.

Gaston KJ (1994) *Rarity*. Chapman & Hall, London.

Gaston KJ (2003) *The Structure and Dynamics of Geographic Ranges*. Oxford University Press, Oxford.

Gaston KJ, Blackburn TM (2000) *Pattern and Process in Macroecology*. Blackwell Science, Oxford.

Gaston KJ, Jones AG, Hanel C, Chown SL (2003) Rates of species introduction to a remote oceanic island. *Proceedings of the Royal Society of London. Series B: Biological Sciences*, **270**, 1091–1098.

Gertzen EL, Leung B, Yan ND (2011) Propagule pressure, Allee effects and the probability of establishment of an invasive species (*Bythotrephes longimanus*). *Ecosphere*, **2**, art30. doi:10.1890/ES10-000170.1.

Gomulkiewicz R, Holt R (1995) When does evolution by natural selection prevent extinction? *Evolution*, **49**, 201–207.

Gonzalez A, Ronce O, Ferriere R, Hochberg ME (2013) Evolutionary rescue: an emerging focus at the intersection between ecology and evolution. *Philosophical Transactions of the Royal Society of London. Series B, Biological Sciences*, **368**, 20120404.

Grevstad FS (1999) Factors influencing the chance of population establishment: implications for release strategies in biological control. *Ecological Applications*, **9**, 1439–1447.

Haccou P, Iwasa Y (1996) Establishment probability in fluctuating environments: a branching process model. *Theoretical Population Biology*, **50**, 254–280.

Haccou P, Vatutin V (2003) Establishment success and extinction risk in autocorrelated environments. *Theoretical Population Biology*, **64**, 303–314.

Harper JL (1965) Establishment, aggression and cohabitation in weedy species. In: *The Genetics of Colonizing Species* (eds Baker HG, Stebbins GL), pp. 243–265. Academic Press, New York, New York.

Hayes KR, Barry SC (2008) Are there any consistent predictors of invasion success? *Biological Invasions*, **10**, 483–506.

Holt RD, Barfield M, Gomulkiewicz R (2005) Theories of niche conservatism and evolution: could exotic species be potential tests? In: *Species Invasions: Insights into Ecology, Evolution and Biogeography* (eds Sax DF, Stachowicz JJ, Gaines SD), pp. 259–290. Sinauer Associates Inc., Sunderland, Massachusetts.

Hopper KR, Roush RT (1993) Mate finding, dispersal, number released, and the success of biological control introductions. *Ecological Entomology*, **18**, 321–331.

Hufbauer RA, Facon B, Ravigné V *et al.* (2012) Anthropogenically induced adaptation to invade (AIAI): contemporary adaptation to human-altered habitats within the native range can promote invasions. *Evolutionary Applications*, **5**, 89–101.

Hufbauer RA, Rutschmann A, Serrate B, Vermeil de Conchard H, Facon B (2013) Role of propagule pressure in colonization success: disentangling the relative importance of demographic, genetic and habitat effects. *Journal of Evolutionary Biology*, **26**, 1691–1699.

Hulme PE (2009) Trade, transport and trouble: managing invasive species pathways in an era of globalization. *Journal of Applied Ecology*, **46**, 10–18.

Hulme PE, Bacher S, Kenis M *et al.* (2008) Grasping at the routes of biological invasions: a framework for integrating pathways into policy. *Journal of Applied Ecology*, **45**, 403–414.

Jeschke JM (2008) Across islands and continents, mammals are more successful invaders than birds. *Diversity and Distributions*, **14**, 913–916.

Kolbe JJ, Glor RE, Schettino LR, Lara AC, Larson A, Losos JB (2004) Genetic variation increases during biological invasion by a Cuban lizard. *Nature*, **431**, 177–181.

Lanfear R, Kokko H, Eyre-Walker A (2014) Population size and the rate of evolution. *Trends in Ecology & Evolution*, **29**, 33–41.

Lockwood JL, Cassey P, Blackburn TM (2005) The role of propagule pressure in explaining species invasion. *Trends in Ecology and Evolution*, **20**, 223–228.

Lockwood JL, Cassey P, Blackburn TM (2009) The more you introduce the more you get: the role of colonization and propagule pressure in invasion ecology. *Diversity and Distributions*, **15**, 904–910.

Lockwood JL, Hoopes MF, Marchetti MP (2013) *Invasion Ecology*, 2nd edn. Blackwell Publishing, Oxford.

Mack RN (2003) Global plant dispersal, naturalization, and invasion: pathways, modes and circumstances. In: *Invasive Species: Vectors and Management Strategies* (eds Ruiz GM, Carlton JT), pp. 3–30. Island Press, Washington, District of Columbia.

Mayr E (1965) Summary. In: *The Genetics of Colonizing Species* (eds Baker HG, Stebbins GL), pp. 553–562. Academic Press, New York, New York.

McCauley DE (1991) Genetic consequences of local-population extinction and recolonization. *Trends in Ecology & Evolution*, **6**, 5–8.

Memmott J, Craze PG, Harman HM, Syrett P, Fowler SV (2005) The effect of propagule size on the invasion of an alien insect. *Journal of Animal Ecology*, **74**, 50–62.

Merilä J, Bjorklund M, Baker AJ (1996) The successful founder: genetics of introduced *Carduelis chloris* (greenfinch) populations in New Zealand. *Heredity*, **77**, 410–422.

Morris WF, Doak DF (2002) *Quantitative Conservation Biology: The Theory and Practice of Population Viability Analysis.* Sinauer Associates Inc., Sunderland, Massachusetts.

Nei M, Mauyama T, Chakraborty R (1975) The bottleneck effect and genetic variability in populations. *Evolution*, **29**, 1–10.

Novak SJ, Mack RN (2005) Genetic bottlenecks in alien plant species: influence of mating systems and introduction dynamics. In: *Species Invasions: Insights into Ecology, Evolution and Biogeography* (eds Sax DF, Stachowicz JJ, Gaines SD), pp. 201–228. Sinauer Associates Inc., Sunderland, Massachusetts.

Pyšek P, Richardson DM (2008) Traits associated with invasiveness in alien plants: where do we stand? In: *Biological Invasions* (ed. Nentwig W), pp. 97–125. Springer-Verlag, Berlin.

Pyšek P, Sadlo J, Mandak B, Jarošik V (2003) Czech alien flora and the historical pattern of its formation: what came first to Central Europe. *Oecologia*, **135**, 122–130.

Richardson DM, Pyšek P, Carlton J (2011) A compendium of essential concepts and terminology in invasion ecology. In: *Fifty Years of Invasion Ecology. The Legacy of Charles Elton* (ed. Richardson DM), pp. 409–420. John Wiley & Sons Ltd, Oxford.

Roman J, Darling JA (2007) Paradox lost: genetic diversity and the success of aquatic invasions. *Trends in Ecology & Evolution*, **22**, 454–464.

Sakai KI (1965) Contributions to the problem of species colonization from the viewpoint of competition and migration. In: *The Genetics of Colonizing Species* (eds Baker HG, Stebbins GL), pp. 215–239. Academic Press, New York, New York.

Sakai AK, Allendorf FW, Holt JS *et al.* (2001) The population biology of invasive species. *Annual Review of Ecology, Evolution and Systematics*, **32**, 305–332.

Schiffers K, Bourne EC, Lavergne S, Thuiller W, Travis JMJ (2013) Limited evolutionary rescue of locally adapted populations facing climate change. *Philosophical Transactions of the Royal Society of London. Series B, Biological Sciences*, **368**, 20120083.

Signorile AL, Wang J, Lutz PWW, Bertolino S, Carbone C, Reuman DC (2014) Do founder size, genetic diversity and structure influence rates of expansion of North American grey squirrels in Europe? *Diversity and Distributions*, **20**, 918–930.

Simberloff D (2009) The role of propagule pressure in biological invasions. *Annual Review of Ecology, Evolution and Systematics*, **40**, 81–102.

Shigesada N, Kawasaki K (1997) *Biological Invasions: Theory and Practice*. Oxford University Press, Oxford.

Sol D, Maspons J, Vall-Llosera M *et al.* (2012) Unravelling the life history of successful invaders. *Science*, **337**, 580–583.

Szucs M, Melbourne BA, Tuff T, Hufbauer RA (2014) The roles of demography and genetics in the early stages of colonization. *Proceedings of the Royal Society of London. Series B: Biological Sciences*, **281**, 20141073.

Uller T, Leimu R (2011) Founder events predict changes in genetic diversity during human-mediated range expansions. *Global Change Biology*, **17**, 3478–3485.

Williamson M, Pyšek P, Jarošík V, Prach K (2005) On the rates and patterns of spread of alien plants in the Czech Republic, Britain, and Ireland. *Ecoscience*, **12**, 424–433.

Wonham MJ, Walton WC, Ruiz GM, Frese AM, Galil BS (2001) Going to the source: role of the invasion pathway in determining potential invaders. *Marine Ecology Progress Series*, **215**, 1–12.

Zenni RD, Bailey JK, Simberloff D (2014) Rapid evolution and range expansion of an invasive plant are driven by provenance-environment interactions. *Ecology Letters*, **17**, 727–735.

Chapter 3

CHARACTERISTICS OF SUCCESSFUL ALIEN PLANTS

Mark van Kleunen*, Wayne Dawson[†], and Noëlie Maurel*

*Ecology, Department of Biology, University of Konstanz, Universitätsstrasse 10, Konstanz D-78457, Germany
[†]School of Biological and Biomedical Sciences, Durham University, South Road, Durham, DH1 3LE, UK

Abstract

Herbert Baker arguably initiated the search for species characteristics determining alien plant invasion success, with his formulation of the 'ideal weed'. Today, a profusion of studies has tested a myriad of traits for their importance in explaining success of alien plants, but the multiple, not always appropriate, approaches used have led to some confusion and criticism. We argue that a greater understanding of the characteristics explaining alien plant success requires a refined approach that respects the multistage, multiscale nature of the invasion process. We present a schema of questions we can ask regarding the success of alien species, with the answering of one question in the schema being conditional on the answer of preceding questions (thus acknowledging the nested nature of invasion stages). For each question, we identify traits and attributes of species we believe are likely to be most important in explaining species success, and we make predictions as to how we expect successful aliens to differ from natives and from unsuccessful aliens in their characteristics. We organize the findings of empirical studies according to the questions in our schema that they have addressed, to assess the extent to which they support our predictions. We believe that research on plant traits of alien species has already told us a lot about why some alien species become successful after introduction. However, if we ask the right questions at the appropriate scale and use appropriate comparators, research on traits may tell us whether they are really important or not, and if so under which conditions.

Previously published as an article in *Molecular Ecology* (2015) 24, 1954–1968, doi: 10.1111/mec.13013

INTRODUCTION

Already in the Origin of Species, Darwin (1859) wrote that '*We cannot hope to explain such facts* [distributional patterns and differences in speciation rates], *until we can say why one species and not another becomes naturalized by man's agency in a foreign land*'. So, a long-standing major objective in ecology and evolution is the identification of characteristics that contribute to the success of species outside their native range. Fifty years ago, in 1965, Herbert Baker published a list of 14 characteristics that an 'ideal weed' should possess (see notes below Table 1). This list was later slightly modified by Baker (1974) himself and by Young & Evans (1976). While Baker's definition of a weed was not restricted to alien plants and focussed on plants in areas markedly disturbed by humans, most of the weeds studied by Baker were invasive alien plants. Therefore, it can be said that the search for characteristics that can explain success of alien plant species started half a century ago with the work of Herbert Baker.

The idea of the existence of invasiveness traits has received some scepticism (e.g. Thompson & Davis 2011; Moles *et al.* 2012). Moreover, it is frequently emphasized—though rarely tested—that the contribution of traits to the success of alien plants has to be context dependent (e.g. Funk 2013; Kueffer *et al.* 2014). Notwithstanding this, there is considerable evidence that invasive alien species differ in certain traits from other, either native or alien, species, at least under certain contexts. For example, invasive aliens frequently grow faster, produce more seeds and capitalize more on extra nutrients than natives or noninvasive aliens (Pyšek & Richardson 2007; van Kleunen *et al.* 2010a; Thompson & Davis 2011; Dawson *et al.* 2012; Rejmánek *et al.* 2013). Therefore, although the importance of specific traits and trait values for the success of species has to be context dependent, the overall importance of traits cannot be denied. What we need to find out, however, is which traits are important at which invasion stage, at which spatial scale and in which environmental context.

In this essay, we provide an overview of issues that, we think, should be considered when testing for traits associated with success of alien plants. Some of those issues have already been raised before, but here we collect them in one place. Furthermore, we present a schema of questions that could guide research on which traits might be important at which stage of invasion and which species might be the right comparators

to identify these traits. Because this schema of questions considers both the larger regional scale and the local-community scale, it also allows for more explicit consideration of context dependency in our search for invasion traits. We also review, though not exhaustively, studies that tested for traits associated with success of alien plants. Finally, we provide information on where the 14 ideal-weed characteristics of Baker (1965) fit within our schema of questions.

ISSUES TO BE CONSIDERED IN THE SEARCH FOR TRAITS ASSOCIATED WITH SUCCESS OF ALIEN PLANTS

An important issue to consider in our search for traits driving success of alien species is which species to compare (Hamilton *et al.* 2005; van Kleunen *et al.* 2010b). Most studies have compared invasive (i.e. highly successful) alien species to native species (see reviews by Pyšek & Richardson 2007; van Kleunen *et al.* 2010a). However, this invasive-alien-vs-native comparison might not always provide us with the answers that we are seeking. First, one should consider that some of the native species are also successful (Rejmánek 1999; Muth & Pigliucci 2006; van Kleunen *et al.* 2010b). Second, if one is interested in the question 'what determines success of alien species?', one should also include nonsuccessful or less successful alien species as comparators (Baker 1965; Rejmánek 1999; Muth & Pigliucci 2006; van Kleunen *et al.* 2010b). Thus, although failed or less successful invasions are still understudied, they are key to gaining insights into the drivers of invasion success.

A second issue to consider is the different stages of the invasion process. Richardson *et al.* (2000) developed a widely applied framework to describe this process. In this introduction–naturalization–invasion framework, a species has to cross biogeographical, environmental and reproductive barriers to achieve the status of naturalized alien (i.e. to establish a self-sustaining population), and dispersal and further environmental barriers to achieve the status of invasive alien (i.e. to spread in the landscape). For each of these transitions, one can ask which characteristics a species requires for success (Kolar & Lodge 2001; Dietz & Edwards 2006; Theoharides & Dukes 2006). The few studies that empirically tested for traits associated with different transitions indeed found that different traits

Table 1 Examples of traits that could potentially determine the answers to the questions in our schema (Fig. 1), predictions of how these traits may differ between successful and unsuccessful alien species and between successful alien and native species, and where Baker's ideal-weed characteristics fit in. Note that the predictions of potential trait differences for a specific question are conditional on whether the species passed the preceding question(s) and that the list of traits is not exhaustive

	Question	Examples of traits	Successful alien vs. unsuccessful alien	Successful alien vs. native	Baker's ideal-weed characteristics†
Region	A0 Has the species been picked up and introduced?	*Ornamental pathway**	>	> or =	
	A1 Is the appropriate environment present?	Ease of cultivation (germination characteristics, growth rate, hardiness), attractiveness to humans (size and colour of flowers, plant size)	< or >	=	
		Environmental optimum			
		Climatic optimum (photosynthetic pathway, deciduousness, water-use efficiency, flowering phenology), nutrient use (ecological indicator values, plastic root foraging, nutrient-use efficiency)		(<, > or =)‡	
		Environmental tolerance			
		Climatic range, niche width, adaptive phenotypic plasticity, fitness homeostasis or capability to capitalize on increased resources, generalization of pollination system, autonomous seed set	>	> or =	i, ii, vi, vii, ix
	A2 Can it reach the appropriate sites?	*Dispersibility*			
		Efficiency of dispersal vector	>	> or =	
		Long-distance dispersal capacity, number of propagules	>	> or =	v, viii, ix, x
		Propagule size, terminal velocity	<	< or =	
		Seed bank longevity	>	> or =	
Local community	B0 Is its required niche space currently occupied?	Same traits as listed under question A1 and traits related to resource capture (rooting depth and architecture, canopy height and architecture, growth form/functional guild, phenology)	≠	≠	
	B1 Can it quickly occupy a vacant niche?	*Dispersibility*	See question A2 for predictions		
		Same traits listed under question A2			
		Priority-effect traits			
		Timing (time to germination, time to resprouting, time to flowering)	<	< or =	ii, iv, xiii
		Germination rate, vegetative spread, fecundity (self-compatibility, capacity for autonomous self-pollination, capacity to attract pollinators)	>	> or =	i, iii, v, vi, vii, viii, xi, xii
	B2 Can it replace native occupants?	Competitive ability (allelopathy, vegetative growth, height), defence against generalist herbivores and pathogens	>	> or =	xi, xiii, xiv

*Predictions may differ for other introduction pathways.

†Baker's ideal-weed characteristics (Baker 1965) are the following: (i) has no special environmental requirements for germination; (ii) has discontinuous germination (self-controlled) and great longevity of seed; (iii) shows rapid seedling growth; (iv) spends only a short period of time in the vegetative condition before beginning to flower; (v) maintains a continuous seed production for as long as growing conditions permit; (vi) is self-compatible, but not obligatorily self-pollinated or apomictic; (vii) when cross-pollinated, this can be achieved by a nonspecialized flower visitor or by wind; (viii) has very high seed output in favourable environmental circumstances; (ix) can produce some seed in a very wide range of environmental circumstances and has high tolerance of (and often plasticity in face of) climatic and edaphic variation; (x) has special adaptation for both long-distance and short-distance dispersal; (xi) if a perennial, has vigorous vegetative reproduction; (xii) if a perennial, has brittleness at the lower nodes or of the rhizomes or rootstocks; (xiii) if a perennial, shows an ability to regenerate from severed portions of the rootstock; and (xiv) has ability to compete by special means: rosette formation, choking growth, exocrine production (but no fouling of soil for itself) etc.

‡If the environment has recently changed

may be important at the different stages (van Kleunen *et al.* 2007; Dawson *et al.* 2009; Pyšek *et al.* 2009) or even that certain traits may have opposing effects at different transitions. For example, Moodley *et al.* (2013) found for the Proteaceae family that large seeds promoted naturalization but that small seeds promote invasion. Although the introduction–naturalization–invasion framework is very useful, it is only applicable at the larger regional scale (for which species can be classified as introduced, naturalized or invasive), which hampers the inclusion of the local-community context. In other words, it does not allow explicit consideration of the local-community scale; that is the scale at which plants interact with other plants and other trophic levels (e.g. micro-organisms, herbivores, pollinators).

The third issue to consider, and which is implicit in the above-mentioned stage-like nature of invasion, is spatial scale (Ackerly & Cornwell 2007; Gurevitch *et al.* 2011), which is usually correlated with temporal scale. While establishment in a particular site is a local process (naturalization), spread of the species in the landscape, which is usually referred to as invasion (Richardson *et al.* 2000), is a regional process involving multiple local naturalization events. So, depending on whether one considers a local plant community or the flora of a larger region (e.g. a country or continent), one might expect different characteristics to be important for success. We therefore need to pitch our studies at the appropriate spatial scale according to the questions that we ask, and the stage of invasion that we are focussing on.

A SCHEMA OF QUESTIONS

Here, we develop a set of questions in a decision-tree framework that considers the different scales at which potentially important processes operate (Fig. 1). This is based on the idea that the barriers to invasion can be viewed as filters, similar to the ones considered in studies on the assembly of native communities (Belyea & Lancaster 1999; Shea & Chesson 2002). The first three questions in our schema apply to larger (regional) scales, and the last three questions apply to the local-community scale. The answers to these questions determine whether an alien species can locally establish, and ultimately, this determines whether it can establish and spread (i.e. invasion) at the regional scale. After each question in this schema, fewer and fewer species from the introduced species pool remain with the potential to establish a local population (Fig. 2). Therefore, the comparators that are used to assess which traits are associated with each question in the schema should ideally come from the pool of species that did not fail already at a preceding question (Fig. 2).

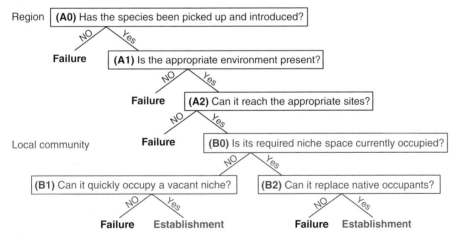

Fig. 1 Schema of questions that should guide research to find out which, if any, species traits are associated with the success of alien species, and at which stage of the invasion process. Questions A0–A2 apply to larger spatial scales, and questions B0–B2 apply to local plant communities. In Table 1, we list species traits that we think could determine the answers to these questions. In principle, there could also be a line connecting the results from question B1 to question B2. For example, if a species failed to quickly occupy a vacant niche before another species does, question B2 becomes relevant. Furthermore, if a species can first establish in a vacant niche and has a wider fundamental niche, it may replace occupants of the adjacent niche space (question B2).

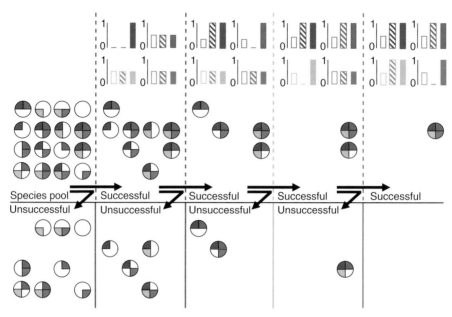

Fig. 2 Different characteristics might determine whether a species will or will not successfully pass each of the subsequent filters (questions), and consequently, the choice of the optimal comparator group for the successful species. In this example, each species (circle) can have four characteristics indicated by the four different colours, and the state of these characteristics determines whether a species can pass the filters of the same colours. For example, only species that possess the blue colour can pass the blue filter. The bar diagrams indicate the proportion of species that possess the particular colour; solid bars: species that successfully passed the respective filter, hatched bars: species that did not successfully pass the respective filter but successfully passed the preceding one(s), open bars: all the species from the initial species pool that did not pass the respective filter and also not the preceding ones. The difference in the proportion of species with each colour among successful species and unsuccessful species depends on the filter at which the colour is important and whether one also includes the unsuccessful species that failed already at one of the preceding filters. (*See insert for color representation of the figure.*)

Our schema of questions is not a risk assessment scheme (e.g. Daehler *et al.* 2004; Leung *et al.* 2012) or a classification/regression tree that predicts invasiveness based on certain characteristics or extrinsic factors (e.g. Reichard & Hamilton 1997; Widrlechner *et al.* 2004; Caley & Kuhnert 2006). Instead, our tree organizes questions that should provide guidance on which traits may be important at which stage and how we can test this. The answers to these questions will be partly determined by several factors other than species characteristics *per se*, such as environmental conditions and time since and frequency of introduction. However, even if a species has been introduced very early and very frequently, it is unlikely to establish when its trait values are not suitable for the environment in the region of introduction. In other words, the potential to pass through a certain filter should ultimately be determined by the characteristics of the species, unless all species are ecologically equivalent.

Below we make, for each question, predictions of differences we would expect in species characteristics between successful alien and unsuccessful alien species, and between successful alien and native species, at local and regional scales (Table 1). We also evaluate some of the support, particularly from the more powerful multispecies studies (van Kleunen *et al.* 2014), that these predictions have received.

Question A0: has the species been picked up and introduced?

The first basic question to ask is whether a species has crossed, with man's help, the biogeographical barriers surrounding its native range (question A0 in Fig. 1 and Table 1). In other words, has the species been picked up and introduced? This simple question is in practice frequently difficult to answer, as for many non-naturalized species it is not known with certainty whether they have been introduced or not. However, already in Europe alone, more than 20 000 taxa are commonly grown in private gardens (Cullen *et al.* 2011) and *c.* 80 000 taxa in botanical gardens (Heywood & Sharrock 2013). Therefore, given that the total number of known higher plant taxa is estimated to be *c.* 370 000 (http://www.theplantlist.org/1.1/statistics/), a considerable proportion of all plant species has been intentionally introduced by humans outside their native range.

Even if most plant species may have been introduced into regions where they are not native, it is important to consider that some have been introduced earlier and more frequently than others. The basic question (question A0 in Fig. 1 and Table 1) then transforms into 'When and how frequently has the species been introduced elsewhere?' It is likely that introduced species, and among those, species introduced earlier and more frequently, are not a random subset of all species, but have certain characteristics that make them more likely to be introduced elsewhere. In other words, an introduction bias is likely to occur. Moreover, there are various introduction pathways (Hulme *et al.* 2008), and it is likely that each pathway promotes different plant characteristics. For example, plant species that have been introduced for ornamental purposes might have been selected for characteristics that make them attractive to humans and easy to cultivate (Chrobock *et al.* 2011). One might therefore expect that such introduced aliens germinate more easily, grow faster and have more attractive flowers compared to species not introduced and to most, but not necessarily all, of the native species (Table 1). For those ornamental alien species grown in outdoor gardens, however, one would expect their cold hardiness to be similar to the cold hardiness of the natives. So, depending on the introduction pathway, the trait and the comparator group, one might expect different outcomes for comparative studies on traits associated with introduction.

Many studies have addressed how different aspects of introduction history relate to current naturalization or invasion success of species (e.g. Lonsdale 1994; Knapp & Kühn 2012). However, very few studies have addressed which traits are associated with a species' introduction history. The easiest species to study in this regard are ornamentals, as there are records of which species are used in international horticulture (e.g. https://www.rhs.org.uk/plants/) and of when and where some of them have been introduced (e.g. Goeze 1916). Two approaches have been used to shed light on the existence and magnitude of an introduction bias. First, the source-region approach (*sensu* Pyšek *et al.* 2004) tests which alien species have been introduced elsewhere and which ones not. van Kleunen *et al.* (2007) showed for a comprehensive data set of 1036 species of Iridaceae native to southern Africa that international horticultural usage was more likely for species with a larger native range, a lower maximal altitude in southern Africa and a greater height. Similarly, Pyšek *et al.* (2014) found that Central European species with a greater height and bigger propagules were more likely to be used in cultivation both in their native range and in North America. Second, the sink-region approach (*sensu* Pyšek *et al.* 2004) tests how alien species, irrespective of whether they are invasive or not, deviate from native species in the introduced region. Chrobock *et al.* (2011) compared germination characteristics between 47 introduced ornamental alien herbs and 42 native herbs in Switzerland. They found that the ornamental aliens, and particularly cultivars, germinated earlier and more successfully than the natives. This indicates that there might indeed be an introduction bias as well as effects of human-mediated selection with regard to the characteristics of introduced aliens, and this might affect the comparisons at the subsequent invasion stages.

Question A1: is the appropriate environment present?

Once a species has been introduced to a new region, it can only establish naturalized populations if there is an environment where the species is physiologically able to grow and reproduce. The second question in our schema is thus whether the appropriate environment is present in the region of introduction (question A1 in Fig. 1 and Table 1). The answer to this question will obviously depend on the region, and the environmental requirements, preferences and tolerances of the introduced alien species (e.g. the degree of climatic–niche match between the introduced and native range; Petitpierre *et al.* 2012), as well as on traits that determine whether it can reproduce (e.g. pollen vector, flowering phenology and breeding system; see question B1). In other words, the answer to this question will be highly context dependent.

Characteristics of species associated with their climatic optima—such as photosynthetic pathway, water-use efficiency and deciduousness—and with nutrient requirements—such as ecological indicator values (e.g. Ellenberg values for nitrogen), plastic root foraging and nutrient-use efficiency—may be important. The optimal trait values will obviously depend on the environmental context. However, on average, species with a wide environmental tolerance, and thus a potentially broad niche, should be more likely to encounter suitable growing conditions. A wide environmental tolerance, that is the maintenance of high values of fitness-related traits across an environmental gradient, is frequently assumed to be achieved through high levels of adaptive phenotypic plasticity in functional traits (i.e. the 'general-purpose genotype' of Baker 1965). However, it is still rarely assessed whether trait plasticity in response to the environment is really adaptive (van Kleunen & Fischer 2005) or whether homeostasis in certain traits contributes to environmental tolerance. It is therefore perhaps not surprising that recent meta-analyses on the role of plasticity in plant invasions were not very conclusive (Davidson et al. 2011; Palacio-López & Gianoli 2011).

As most native species also successfully grow and reproduce in regions invaded by aliens, successful alien species should have environmental requirements, and thus characteristics, similar to those of most native species in the region. However, if the alien species have taken advantage of environmental conditions not exploited by natives or of novel environmental conditions (e.g. due to land-use change, atmospheric nutrient deposition or climate change) to which most of the native species may not have adapted yet, the successful aliens could differ from most of the natives. This has been demonstrated in Hawai'i, where distinct leaf traits allow some invasives to occupy novel, human-created biogeochemical niches not exploited by native species (Peñuelas et al. 2010). Furthermore, this could also underlie the finding that invasive alien species in the Iberian Peninsula have higher photosynthetic nitrogen-use efficiencies than native species (Godoy et al. 2012). Therefore, we predict that in regions that have recently undergone environmental change, there are stronger differences between successful aliens and natives.

Among the introduced alien species, those without the characteristics allowing growth and reproduction in any of the environments in a region will not establish. Therefore, one would expect clear differences between successful and unsuccessful alien species. In line with this, a recent study by Dostál et al. (2013) found that among 264 Central European plants, the ones from more productive habitats are more invasive at a global scale. In a similar vein, Dawson et al. (2012) found in a study on 18 alien species in Switzerland that the invasive ones capitalize more strongly on fertilizer addition than the noninvasive ones. These results suggest that many successful aliens have taken advantage of the global increase in atmospheric nitrogen deposition (Galloway et al. 2008). However, for other traits that determine the environmental optimum of a species, the direction of the trait-value differences between successful and unsuccessful alien species will be context dependent (i.e. region specific). For example, species from either Arctic or tropical regions will most likely not find suitable environments in temperate regions. This means that the successful species do not necessarily have universally higher or lower trait values than the average unsuccessful ones, but that their values might be in between the values of unsuccessful species. To the best of our knowledge, no study has yet tested for unimodal (or inverse unimodal) relations between establishment success of aliens and their trait values.

On the other hand, for traits that determine environmental tolerance (i.e. niche width), one would expect on average higher values in successful than in unsuccessful species, as tolerance increases the likelihood that the fundamental niche of a species overlaps with the available niche space in a region. Indeed, the study by Dostál et al. (2013) on 264 Central European plants found that species primarily occurring in nutrient-poor habitats but also found in more productive habitats (i.e. species that have a wide productivity niche) are globally more invasive. This finding supports the idea that environmental tolerance increases the establishment success of introduced alien species.

Question A2: can it reach the appropriate sites?

When a suitable environment is present in a region, the introduced alien species can obviously only establish if it manages to get to some of the suitable sites. Moreover, to spread rapidly in the landscape, the species needs to reach many suitable sites. The third question in our schema is therefore whether the introduced alien species can reach the appropriate sites, given that there are such sites (question A2 in Fig. 1 and Table 1). The likelihood that a species will reach a suitable site will, in addition to extrinsic factors such as the frequency and spatial configuration of these sites, depend upon the mobility of the species. Therefore, dispersal-related traits, such as the dispersal vector and the number, weight and terminal velocity of propagules (e.g. seeds or fruits), are likely to be important.

Some unsuccessful aliens may also be very mobile, but their failure might be due to the lack of suitable environments in the region (i.e. they might have failed already at question A1; also see Fig. 2). Therefore, successful alien species should have higher values of traits promoting dispersibility than unsuccessful alien species that could potentially live in the same sites. Because species with smaller, and frequently more seeds (Moles & Westoby 2006), may disperse further, many studies have compared seed size between more and less successful aliens. Rejmánek & Richardson (1996) found that among 24 *Pinus* species, the invasive ones indeed have smaller seeds than noninvasive ones. However, Castro-Díez *et al.* (2011) found that among 85 Australian *Acacia* species, the ones that have become invasive elsewhere do not differ in seed mass from the ones that failed to become invasive. On the other hand, Lake & Leishman (2004) found that among 57 exotic species in urban bushland in Sydney, Australia, the invasive species tended to have heavier seeds than the noninvasive species. Results regarding seed size are thus variable, but given that seed size does not only affect dispersibility but also other ecological functions, the inconsistent results are not surprising. While small-seeded species may be better dispersed, once a site is reached, larger-seeded species may have an advantage through more stored resources and faster growth (Turnbull *et al.* 2008). The multiple ecological functions of seed size could thus result in species with intermediate seed sizes having the highest establishment success, but such a unimodal relationship has not yet been tested. Furthermore, the above-mentioned invasion studies may not have consistently defined unsuccessful species; some of them may have failed to reach suitable sites or failed to establish after dispersal.

Other studies have looked at traits other than seed size that are more directly related to dispersibility or at estimates of spread rate. For example, Bucharová & van Kleunen (2009) found that seed-spread rate among 192 North American tree species is positively associated with their naturalization success in Europe, although such a relationship is absent among 86 North American shrubs. Moravcová *et al.* (2010) found that among 93 neophytes in the Czech Republic, the invasive ones have a lower seed terminal velocity (i.e. a higher capacity for wind dispersal) than noninvasive ones. Murray & Phillips (2010) found that among 88 naturalized plants in southeastern Australia, invasive species invest more in seed dispersal appendages. So, overall, these results suggest that many successful alien species are better dispersers than nonsuccessful alien species.

Among native plant species, one would also expect that the ones with a high dispersibility are more successful than the ones with a low dispersibility. Consequently, if one would compare successful aliens with a mixture of successful and unsuccessful natives, the successful aliens should, on average, have higher values. However, if one would compare them to successful natives, such a difference should be absent. So, the choice of the native comparator group is also important. Daws *et al.* (2007) compared individual seed mass of 225 native and 33 invasive Asteraceae and of 74 native and 44 invasive Poaceae in California and found that the invasive species had on average higher values. These results suggest that invasives may have a lower dispersibility than related natives with potentially similar niche requirements. Possibly, seed size is not the main driver of dispersibility in these two families but has other important ecological functions. Moreover, one potential reason why studies may sometimes fail to find dispersal traits that explain spread rates of alien plants lies in the fact that they ignore long-distance dispersal events, which may be particularly important in fragmented landscapes, but are very unpredictable (Nathan 2006; Nathan *et al.* 2008).

When a species disperses, its seeds might end up in sites that are currently not suitable for establishment. Dispersal in time by means of the accumulation of a persistent soil seed bank could allow the species to wait for the right environmental conditions (Gioria *et al.* 2012). Moreover, even when the current environment is already suitable for establishment, a persistent soil seed bank can rescue populations that are small and vulnerable to demographic and environmental stochasticity. The latter is likely to be important in environments with strong, but spatio-temporally patchy, disturbance, which are frequently dominated by invasive aliens. Recently, Gioria *et al.* (2012) showed in a database study that the capacity to build up a persistent seed bank is more frequent among 32 invasive than among 39 naturalized and 92 casual species in the Czech Republic. However, the frequency of species with a persistent seed bank among the invasive aliens was similar to the frequency among 185 natives. This emphasizes that successful aliens frequently should differ from unsuccessful ones, but may be similar to natives.

Question B0: is the required niche space currently occupied?

Once the propagules of an introduced alien species have reached a potentially suitable site in the non-native region, the next question in our schema (question B0 in Fig. 1 and Table 1) is whether its niche space is already occupied or not. In both cases, population establishment might be possible, but this will depend, at least partly, on a different set of traits (questions B1 and B2 in Fig. 1 and Table 1). Because at large spatial scales both alien and native species have to pass the same environmental filters, the environmental requirements of both must match the range of environmental conditions that are locally available. Thus, the traits determining whether the required niche space in the local community is already occupied should be, in the first place, the same traits determining whether the suitable environment is present in a region (question

A1 in Fig. 1; i.e. traits that determine the environmental optimum and traits that determine the environmental tolerance of a species). For these traits, the alien species should have trait values that fall within the range of those of the natives, but its environmental niche space could still be available if the alien species does not have an equivalent native species with similar trait values in the community. As the vacant niche space might be at any position within the overall environmental space, the successful alien does not necessarily differ from the average native species (Fig. 3). In other words, the difference between successful aliens and natives in the community will be context dependent. However, a successful alien should be more different from the most similar native species in that site than an alien that reached the site but did not establish in it.

Even when the environmental requirements of the alien species overlap those of the native species in the local community, there could still be a vacant niche if the alien species has traits that are novel to the community and allow the species to access part of the environment to which the natives do not have access. Traits that might be important here are those that determine how species exploit their environment. While resource (i.e. light, water, nutrients) requirements are likely to overlap between the alien and native species in a

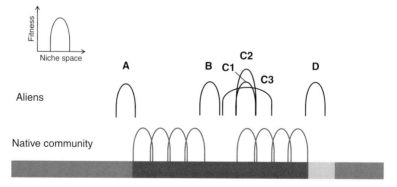

Fig. 3 Potential establishment of different alien species into a native community. The niche requirements of aliens A and D fall outside the current available environmental space (dark grey horizontal band) and have trait values that are either smaller or larger than those of the natives (and the successful aliens). However, if due to recent environmental change available environmental space is extended (light grey horizontal band), alien D will be able to establish and will have trait values that differ from those of the natives. Aliens B and C both fall in the range of trait values shown by the natives and in this case, have values close to the average value of the native community. However, alien C has the same environmental optimum as one of the native species, while alien B is more dissimilar to the most similar native species. As a consequence, alien B will have the highest chance to establish. Alien C could only invade if it had trait values resulting in greater fitness (C2) than the native species with a similar niche optimum, allowing it to outcompete the native, or if it has a wider niche (C3), allowing it to occupy part of the vacant niche space.

community, species might still differ in traits involved in resource capture. This can be illustrated with nutrient acquisition. An experimental comparison between one invasive and two native annual species of the Mojave desert showed that the invasive had a greater root surface area and exploited deeper soils (DeFalco *et al.* 2003). In a nutrient-limited experimental system, Drenovsky *et al.* (2008) found that four invasive species allocated proportionally more root length to nutrient-rich microsites in the heterogeneous nutrient distribution treatment compared to the control with a homogeneous nutrient distribution, than four native species did. These studies at the community level are typically focused on one or a few species, but further investigation across multiple invasive species and multiple invaded communities should help to understand how fine-scale niche differentiation may promote the establishment of alien species in local plant communities.

Question B1: can it quickly occupy a vacant niche?

When propagules of the alien species have reached a suitable site, and the required niche space is not occupied yet by a native species, the next question is whether the alien species can quickly occupy the vacant niche before another newcomer does (question B1 in Fig. 1 and Table 1). First, traits that increase the influx of propagules to the site should be advantageous, as these would help to overcome Allee effects and stochastic effects. These would be the traits that increase dispersibility, mentioned under question A2. When the same number of propagules has arrived at a site, then other characteristics that contribute to priority effects (Morin 1999) should be advantageous. These are traits that (i) allow a species to occupy the available space earlier than other species (e.g. early germination, early sprouting) and that (ii) allow a species to grow there in large numbers (e.g. a high germination rate, production of numerous seeds, vegetative proliferation). Associated with the latter are traits that make a species less reliant on external factors that could delay its reproduction (self-compatibility, autonomous self-pollination). The successful alien species should differ in these traits from unsuccessful aliens and native species with similar niche requirements, but not necessarily when compared to native or unsuccessful alien species in general (Table 1).

There are indeed multiple empirical studies suggesting that traits contributing to priority effects are associated with success of alien species. van Kleunen & Johnson (2007a) showed that among 60 South African Iridaceae, the ones that have become naturalized elsewhere germinated faster and more profusely. Recently, in a study on 28 grassland species, Wilsey *et al.* (2014) also showed that alien species germinated faster and more profusely than natives and that these aliens as a consequence had stronger priority effects on species sown later. Moravcová *et al.* (2010) found that among 93 neophytes in the Czech Republic, the invasive ones had a higher fecundity. Self-compatibility and autonomous seed set may contribute to fecundity and are indeed frequently associated with alien plant success (Rambuda & Johnson 2004; van Kleunen & Johnson 2007b; van Kleunen *et al.* 2008; Küster *et al.* 2008; Hao *et al.* 2011). This frequent finding is also in line with Baker's idea, published 10 years before his list of ideal-weed characteristics, that organisms capable of uniparental reproduction are more likely than obligate outcrossers to establish populations after long-distance dispersal (Baker 1955). So, overall, it seems that traits providing species with a priority-effect advantage are associated with success of alien plants.

Question B2: can it replace native occupants?

When the propagules of an alien species have reached a site that would in principle be suitable, the required niche space may already be occupied by a native species. The question then is whether the alien species can replace (or coexist with) this native species (question B2 in Fig. 1 and Table 1). Therefore, introduced alien species with trait values that increase competitive ability—such as a tall stature, fast vegetative growth or the production of allelopathic chemicals—might be more likely to establish. It should furthermore be advantageous for the alien species to have a higher resistance against generalist herbivores and pathogens than the native species that is occupying its niche, as this will also give the alien species a competitive advantage. The successful alien species should have higher values of these traits than the native species that it is replacing, but not necessarily when compared to the other native species in the community that do not use the same niche space (Fig. 2). Unsuccessful alien species with similar niche requirements should have lower values of these traits than the successful alien species. In other words, if alien species are to successfully outcompete and replace native species, they should have traits promoting a higher fitness compared to unsuccessful aliens and to natives, despite having similar niche-related traits (Fig. 3).

Although it is frequently implied that invasive alien plants outcompete native plants, relatively few studies have directly assessed the competitive ability of successful alien plants. Vilà & Weiner (2004) concluded in a review that the effect of invasive species on native species is usually stronger than vice versa. However, in a large experiment with 12 invasive species in the Czech Republic, Dostál (2011) showed that the intensity of competition was not affected by whether the competing neighbour was a native or an invasive alien species. Furthermore, although allelopathy has been suggested to drive the success of several highly invasive species, such as *Centaurea diffusa* (Callaway & Aschehoug 2000) and *Alliaria petiolata* (Prati & Bossdorf 2004; Stinson *et al.* 2006), Dostál (2011) also found no evidence for the role of allelopathy in competitive interactions of the 12 invasive species in his study. However, he found that the best predictor of competitive intensity was plant size. Therefore, taller aliens might be more successful. On the other hand, other studies revealed mixed evidence for the role of stature. Bucharová & van Kleunen (2009) found that naturalization success of North American trees in Europe was associated with maximum height, but that this was not the case for North American shrubs. Hamilton *et al.* (2005) did not find an association between plant height and invasiveness of alien species in Australia. Speek *et al.* (2011) found a correlation between plant height and regional frequency of alien plants, but not between plant height and local dominance, which is actually the most relevant spatial scale for this trait to be important for replacement of native species in invaded communities. This mixed evidence of plant height as an important trait promoting successful replacement of natives could be due to plant height increasing competitive ability only in certain communities and habitats, emphasizing again context dependency. Moreover, it could be that some of the tall unsuccessful aliens in these studies had failed at one of the earlier stages and therefore were not optimal comparators.

WHERE DO BAKER'S IDEAL-WEED CHARACTERISTICS FIT IN THE SCHEMA OF QUESTIONS?

When devising the characteristics of an ideal weed (see notes below Table 1), Baker (1965) defined weeds as having '*populations* [that] *grow entirely or predominantly in situations markedly disturbed by man*'. Because of this restriction to disturbed habitats, it is perhaps unsurprising that the majority of Baker's ideal-weed characteristics can be assigned to question B1 in our schema ('Can it quickly occupy a vacant niche?'; Fig. 1; Table 1). Characteristics of species that should be able to quickly occupy and increase in abundance in a vacant niche include the following ideal-weed characteristics (Table 1): no special germination requirements, rapid growth, an ability to produce seeds throughout the growing period and in high volumes in favourable environments, little or no dependence on (specialized) pollinators for seed production, and vegetative reproduction. Some characteristics might also act as priority-effect traits, with invaders able to germinate (controlled seed germination), flower (short time to flowering) and establish (through vegetative propagation) earlier than native species when a vacant space arises.

Before a species is able to occupy a vacant niche in a local community, the local community should first be in an appropriate environment (question A1). However, a number of Baker's ideal-weed characteristics that could enable a vacant niche to be occupied would also result in greater environmental tolerance, therefore increasing the probability of an appropriate environment being present (no special germination requirements, self-controlled germination and seed longevity, self-compatibility/lack of dependence on pollinators being present, plant tolerance to and seed production under varying climatic and edaphic environments). But even if an appropriate environment is present, propagules must be able to disperse there (question A2) and to vacant niches in a community (question B1), which should be more likely for species with adaptations for short- and long-distance dispersal, another of Baker's ideal-weed characteristics. In addition, continuous seed production throughout the growing season, high seed production in favourable environments and an ability to produce seeds in multiple environments are characteristics that will increase the probability of seeds reaching suitable environments at larger scales and vacant niches at smaller scales.

Given Baker's focus on weeds in ruderal and agricultural habitats, it is not surprising that relatively few of the ideal-weed characteristics would be implicated in replacement of native species (question B2); competition is presumably (and at least initially) of low importance in such open habitats. Baker does, however, suggest that an ideal weed would have an ability to compete by 'special means' including direct competition via allelopathic effects on natives. 'Vigorous vegetative reproduction' in an invader might also result in native species being outcompeted and replaced. However, Baker's concentration on ruderal and agricultural weeds likely means that the ideal-weed characteristics are only one possible syndrome of successful alien species.

DISCUSSION

The intensive search for traits explaining success of alien plant species and invasions has not yet led to the identification of a universal syndrome of invasion traits. This may partly be because invasions are complex processes with multiple stages, occurring at different spatial and temporal scales, and these complexities have not always been considered in studies searching for invasiveness traits. Thus, the inability to find a universal suite of traits that consistently differ between successful alien and unsuccessful alien or native species could be because there are multiple successful suites of traits, depending on context. This does not mean that traits are unimportant; however, we need to work out which sets of traits are most important under different contexts.

Our schema of questions (Fig. 1) organizes a series of questions at regional and community scales that could guide research on traits associated with success of alien plants. Based on this schema, we have made predictions regarding how we expect successful aliens, unsuccessful aliens and natives to differ in their traits, and which traits are most relevant for each question asked (Table 1). When testing a hypothesis based on this schema of questions, it is important to consider that the group of species to which each question applies is conditional upon the preceding questions (Fig. 2). This means that if one wants to test, for example, whether successful alien species are more mobile than unsuccessful alien species, one should ideally select successful and unsuccessful aliens that all have been introduced and all have suitable

environments in the region (Fig. 2). The reason is that an unsuccessful alien that did not pass one of the earlier questions, for example because its climatic requirements are not fulfilled, may nevertheless be very mobile. As a consequence, the importance of traits that allow species to pass through the later filters of the invasion process may be obscured or diluted if one would take a random sample of all unsuccessful aliens.

While there is already empirical support for some of the predictions that we based on our schema of questions (e.g. the association of success with high environmental tolerance, high dispersibility and traits related to priority effects), others are not or infrequently supported. There are multiple possible reasons why studies may have failed to find associations between traits and success of alien species. One reason could be that the trait of interest is truly not important for success. However, there could also be many other reasons: (i) the contribution of a trait is obscured by extrinsic drivers of success, such as introduction frequency. (ii) The comparator group may have been inappropriate. Most studies on traits associated with invasiveness still use native species as noninvasive controls, although in many cases the traits of successful aliens should be similar to those of the natives (Table 1). A comparator problem could also arise when the successful aliens and the comparators belong to completely different taxonomic groups. Consequently, many studies have avoided this problem by making comparisons within specific families or corrected for it by applying a phylogenetic correction. (iii) As explained in the preceding paragraph and in Fig. 2, if one compares a successful alien to any nonsuccessful alien, the latter might already have failed at an earlier step for which the trait of interest was not relevant. (iv) There may be a nonlinear (e.g. unimodal) instead of a linear relation between trait values and success. (v) The spatial scale at or the context in which the trait should be important was not considered. (vi) Interactions among traits were not considered (see Küster *et al.* 2008 for a notable exception). Certain traits may only contribute to the success of an alien give the presence of another characteristic. Our schema of questions (Fig. 1) might help to understand such interactions. For example, environmental tolerance increases the likelihood that an appropriate environment is present (question A1) and time to germination whether it can quickly occupy a vacant niche (question B1). Then among all introduced alien plants, time to germination might only contribute to invasiveness for species that also have a wide environmental tolerance. (vii) There are trade-offs between potentially important traits (e.g. competitive ability and resistance against generalist herbivores) or between multiple functions of a single trait (e.g. seed size, as discussed above under question A2) that may also act across different invasion stages. Most of these issues can be avoided, and we hope that our schema of questions might contribute to this.

The Darwinian demon, that is a species that is superior in any trait and in all possible contexts, does not exist as a consequence of the above-mentioned trade-offs (Law 1979). Therefore, it is unlikely to find a species that possesses all characteristics for being invasive in any context. Indeed, in an attempt by Williamson (1996) to test the importance of Baker's ideal-weed characteristics, he found that none of the 49 annual plant species that he had selected possessed more than seven of the 14 ideal-weed characteristics. Interestingly, however, it seems that many of the ideal-weed characteristics—such as self-fertilization, clonal reproduction, fast growth and environmental tolerance (Levin 2002)—come together in polyploid plants. Indeed, it has been reported that polyploid plants are more likely to be invasive than diploid ones (Pandit et al. 2011, 2014; te Beest et al. 2012). Nevertheless, although some polyploid plants may approach the Darwinian demon, it is still difficult to point out which characteristic of polyploids is driving their success.

Most of the studies that test for associations between traits and success of alien plants draw inferences based on large-scale geographical patterns or use a classification of aliens as non-naturalized, naturalized or invasive at those scales. While these studies provide important insights into traits that might be important for the questions that are relevant at the regional scale (questions A0–A2 in Fig. 1), they do not necessarily provide insight into the questions that are relevant at the community level (questions B0–B2 in Fig. 1). While it is possible to identify some introduced aliens that are not successful for a larger region, this is more difficult for a local community. There we know which alien species have established, but not which ones managed to get there and failed to establish. Controlled experimental introductions of novel species into native communities could provide important insights in which traits might be important for establishment in a specific community.

Several studies have performed such experimental introductions. However, most of them focussed on extrinsic factors, such as disturbance and species richness, rather than on intrinsic species traits (Robinson et al. 1995; Naeem et al. 2000; Seabloom et al. 2003; Maron et al. 2013). Moreover, most of these studies introduced only one or few species, which precludes comparisons of characteristics associated with establishment success. A notable exception is a study by Burke & Grime (1996); they experimentally introduced 54 nonlocal, though native, species into a single limestone grassland in the UK. They found that particularly species with large seeds and fast and profuse germination under different temperature and light conditions were successful during the first 2 years of the study. In a later assessment of the same experiment, Thompson et al. (2001) reported that for the change in cover of the species during the subsequent 3 years, none of the tested traits had a significant effect. This, however, was after applying the very conservative Bonferroni correction (Moran 2003); without that correction, the positive effects of competitive ability and the negative effect of palatability would have been significant. More recently, Kempel et al. (2013) introduced 48 alien and 45 native plant species into eight disturbed (tilled) and eight nondisturbed grassland sites in Switzerland, and in parallel, experiments assessed functional traits of the species. They also found that the importance of the different traits changed over time, but that at the end of the 3-year study, particularly native species and—like in the study of Thompson et al. (2001)—species with a high resistance against a generalist herbivore were successful. Moreover, perennial species were more successful, and this was particularly the case in the disturbed grassland plots, indicating context dependency. A limitation of these two multispecies introduction experiments is that that they did not consider how the introduced species differed from the native resident species. Fargione et al. (2003), however, did this, although only for a single characteristic, the functional guild. By introducing a total of 27 native and exotic species belonging to four different guilds—C3 grasses, C4 grasses, legumes and nonlegume forbs—into experimental grassland plots of different species richness, they showed that establishment of introduced species was inhibited by the presence of resident species from the same guild. So, multispecies experimental introductions might provide important insights into which characteristics are important for establishment in specific local communities.

CONCLUSION

In the 50 years after the publication of Baker's list of the ideal-weed characteristics in 1965, many studies have tested for traits associated with the success of alien plants. Although there is support for the importance of some of the characteristics listed by Baker (1965), such as fast growth, self-compatibility and high seed output, there also appear to be many exceptions or inconsistent results. In the face of the frequently emphasized context dependency, we should move away from the search for a 'one-size-fits-all' combination of traits that explains success. Instead, we might do better to focus on identifying the different syndromes of traits possessed by successful aliens when different sets of abiotic and biotic filters are imposed. Our schema of questions can help in organizing efforts to identify such syndromes of successful alien species by (i) acknowledging the importance of scale when asking questions about the success of alien species, moving from larger to smaller (community) scales as one progresses through the schema; (ii) sorting the traits that are most relevant to species success at each question; and (iii) identifying the best comparator groups. So, if we ask the right questions at the appropriate scale and use appropriate comparators, research on traits may tell us whether they are important or not.

ACKNOWLEDGEMENTS

We would like to thank the organizers of the *The Baker and Stebbins Legacy Symposium* for inviting us to write this study, and Rob Colautti, Keneth Whitney, Marcel Rejmánek and an anonymous reviewer for helpful comments on a previous version. MvK and NM thank the DFG for funding (Grant KL 1866/5-1). WD also thanks the DFG for funding (Grant DA 1502/1-1).

REFERENCES

Ackerly DD, Cornwell WK (2007) A trait-based approach to community assembly: partitioning of species trait values into within- and among-community components. *Ecology Letters*, **10**, 135–145.

Baker HG (1955) Self-compatibility and establishment after 'long-distance' dispersal. *Evolution*, **9**, 347–349.

Baker HG (1965) Characteristics and modes of origin of weeds. In: *The Genetics of Colonizing Species* (eds Baker HG, Stebbins GL), pp. 147–172. Academic Press, New York.

Baker HG (1974) The evolution of weeds. *Annual Review of Ecology and Systematics*, **5**, 1–23.

te Beest M, Le Roux JJ, Richardson DM *et al.* (2012) The more the better? The role of polyploidy in facilitating plant invasions. *Annals of Botany*, **109**, 19–45.

Belyea LR, Lancaster J (1999) Assembly rules within a contingent ecology. *Oikos*, **86**, 402–416.

Bucharová A, van Kleunen M (2009) Introduction history and species characteristics partly explain naturalization success of North American woody species in Europe. *Journal of Ecology*, **97**, 230–238.

Burke MJW, Grime JP (1996) An experimental study of plant community invasibility. *Ecology*, **77**, 776–790.

Caley P, Kuhnert PM (2006) Application and evaluation of classification trees for screening unwanted plants. *Austral Ecology*, **31**, 647–655.

Callaway RM, Aschehoug ET (2000) Invasive plants versus their new and old neighbors: a mechanism for exotic invasion. *Science*, **290**, 521–523.

Castro-Díez P, Godoy O, Saldaña A, Richardson DM (2011) Predicting invasiveness of Australian acacias on the basis of their native climatic affinities, life history traits and human use. *Diversity and Distributions*, **17**, 934–945.

Chrobock T, Kempel A, Fischer M, van Kleunen M (2011) Introduction bias: cultivated plant species germinate faster and more profusely than native species in Switzerland. *Basic and Applied Ecology*, **12**, 244–250.

Cullen J, Knees SG, Cubey HS, eds (2011) *The European Garden Flora*. Cambridge University Press, Cambridge, UK.

Daehler CC, Denslow JS, Ansari S, Kuo H-C (2004) A risk-assessment system for screening out invasive pest plants from Hawaii and other Pacific islands. *Conservation Biology*, **18**, 360–368.

Darwin C (1859) *The Origin of Species*. John Murray, London.

Davidson AM, Jennions M, Nicotra AB (2011) Do invasive species show higher phenotypic plasticity than native species and, if so, is it adaptive? A meta-analysis. *Ecology Letters*, **14**, 419–431.

Daws MI, Hall J, Flynn S, Pritchard HW (2007) Do invasive species have bigger seeds? Evidence from intra- and inter-specific comparisons. *South African Journal of Botany*, **73**, 138–143.

Dawson W, Burslem DFRP, Hulme PE (2009) Factors explaining alien plant invasion success in a tropical ecosystem differ at each stage of invasion. *Journal of Ecology*, **97**, 657–665.

Dawson W, Fischer M, van Kleunen M (2012) Common and rare plant species respond differently to fertilisation and competition, whether they are alien or native. *Ecology Letters*, **15**, 873–880.

DeFalco LA, Bryla DR, Smith-Longozo V, Nowak RS (2003) Are Mojave Desert annual species equal? Resource acquisition and allocation for the invasive grass *Bromus madritensis* subsp. *rubens* (Poaceae) and two native species. *American Journal of Botany*, **90**, 1045–1053.

Dietz H, Edwards PJ (2006) Recognition that causal processes change during plant invasion helps explain conflicts in evidence. *Ecology*, **87**, 1359–1367.

Dostál P (2011) Plant competitive interactions and invasiveness: searching for the effects of phylogenetic relatedness and origin on competition intensity. *American Naturalist*, **177**, 655–667.

Dostál P, Dawson W, van Kleunen M, Keser LH, Fischer M (2013) Central European plant species from more productive habitats are more invasive at a global scale. *Global Ecology and Biogeography*, **22**, 64–72.

Drenovsky RE, Martin CE, Falasco MR, James JJ (2008) Variation in resource acquisition and utilization traits between native and invasive perennial forbs. *American Journal of Botany*, **95**, 681–687.

Fargione J, Brown CS, Tilman D (2003) Community assembly and invasion: an experimental test of neutral versus niche processes. *Proceedings of the National Academy of Sciences*, **100**, 8916–8920.

Funk JL (2013) The physiology of invasive plants in low-resource environments. *Conservation Physiology*, **1**, doi: 10.1093/conphys/cot026.

Galloway JN, Townsend AR, Erisman JW et al. (2008) Transformation of the nitrogen cycle: recent trends, questions, and potential solutions. *Science*, **320**, 889–892.

Gioria M, Pyšek P, Moravcová L (2012) Soil seed banks in plant invasions: promoting species invasiveness and long-term impact on plant community dynamics. *Preslia*, **84**, 327–350.

Godoy O, Valladares F, Castro-Díez P (2012) The relative importance for plant invasiveness of trait means, and their plasticity and integration in a multivariate framework. *New Phytologist*, **195**, 912–922.

Goeze E (1916) Liste der seit dem 16. Jahrhundert bis auf die Gegenwart in die Gärten und Parks Europas eingeführten Bäume und Sträucher. *Mitteilungen der Deutschen Dendrologischen Gesellschaft*, **25**, 129–201.

Gurevitch J, Fox GA, Wardle GM, Inderjit Taub D (2011) Emergent insights from the synthesis of conceptual frameworks for biological invasions. *Ecology Letters*, **14**, 407–418.

Hamilton MA, Murray BR, Cadotte MW et al. (2005) Life-history correlates of plant invasiveness at regional and continental scales. *Ecology Letters*, **8**, 1066–1074.

Hao JH, Qiang S, Chrobock T, van Kleunen M, Liu QQ (2011) A test of Baker's Law: breeding systems of invasive species of Asteraceae in China. *Biological Invasions*, **13**, 571–580.

Heywood VH, Sharrock S (2013) *European Code of Conduct for Botanic Gardens on Invasive Alien Species*. Council of Europe, Strasbourg, Botanic Gardens Conservation International, Richmond.

Hulme PE, Bacher S, Kenis M et al. (2008) Grasping at the routes of biological invasions: a framework for integrating pathways into policy. *Journal of Applied Ecology*, **45**, 403–414.

Kempel A, Chrobock T, Fischer M, Rohr RP, van Kleunen M (2013) Determinants of plant establishment success in a multispecies introduction experiment with native and alien species. *Proceedings of the National Academy of Sciences*, **110**, 12727–12732.

van Kleunen M, Fischer M (2005) Constraints on the evolution of adaptive phenotypic plasticity in plants. *New Phytologist*, **166**, 49–60.

van Kleunen M, Johnson SD (2007a) South African Iridaceae with rapid and profuse seedling emergence are more likely to become naturalized in other regions. *Journal of Ecology*, **95**, 674–681.

van Kleunen M, Johnson SD (2007b) Effects of self-compatibility on the distribution range of invasive European plants in North America. *Conservation Biology*, **21**, 1537–1544.

van Kleunen M, Johnson SD, Fischer M (2007) Predicting naturalization of southern African Iridaceae in other regions. *Journal of Applied Ecology*, **44**, 594–603.

van Kleunen M, Manning JC, Pasqualetto V, Johnson SD (2008) Phylogenetically independent associations between autonomous self-fertilization and plant invasiveness. *American Naturalist*, **171**, 195–201.

van Kleunen M, Weber E, Fischer M (2010a) A meta-analysis of trait differences between invasive and non-invasive plant species. *Ecology Letters*, **13**, 235–245.

van Kleunen M, Dawson W, Schlaepfer DR, Jeschke JM, Fischer M (2010b) Are invaders different? A conceptual framework of comparative approaches for assessing determinants of invasiveness. *Ecology Letters*, **13**, 947–958.

van Kleunen M, Dawson W, Bossdorf O, Fischer M (2014) The more the merrier: multi-species experiments in ecology. *Basic and Applied Ecology*, **15**, 1–9.

Knapp S, Kühn I (2012) Origin matters: widely distributed native and non-native species benefit from different functional traits. *Ecology Letters*, **15**, 696–703.

Kolar CS, Lodge DM (2001) Progress in invasion biology: predicting invaders. *Trends in Ecology and Evolution*, **16**, 199–204.

Kueffer C, Pyšek P, Richardson DM (2014) Integrative invasion science: model systems, multi-site studies, focused meta-analysis and invasion syndromes. *New Phytologist*, **200**, 615–633.

Küster EC, Kühn I, Bruelheide H, Klotz S (2008) Trait interactions help explain plant invasion success in the German flora. *Journal of Ecology*, **96**, 860–868.

Lake JC, Leishman MR (2004) Invasion success of exotic plants in natural ecosystems: the role of disturbance, plant attributes and freedom from herbivores. *Biological Conservation*, **117**, 215–226.

Law R (1979) Optimal life histories under age-specific predation. *American Naturalist*, **114**, 399–417.

Leung B, Roura-Pascual N, Bacher S *et al.* (2012) TEASIng apart alien species risk assessments: a framework for best practices. *Ecology Letters*, **15**, 1475–1493.

Levin D (2002) *The Role of Chromosomal Change in Plant Evolution*. Oxford University Press, New York.

Lonsdale WM (1994) Inviting trouble: introduced pasture species in northern Australia. *Australian Journal of Ecology*, **19**, 345–354.

Maron JL, Waller LP, Hahn MA *et al.* (2013) Effects of soil fungi, disturbance and propagule pressure on exotic plant recruitment and establishment at home and abroad. *Journal of Ecology*, **101**, 924–932.

Moles AT, Westoby M (2006) Seed size and plant strategy across the whole life cycle. *Oikos*, **113**, 91–105.

Moles AT, Flores-Moreno H, Bonser SP *et al.* (2012) Invasions: the trail behind, the path ahead, and a test of a disturbing idea. *Journal of Ecology*, **100**, 116–127.

Moodley D, Geerts S, Richardson DM, Wilson JRU (2013) Different traits determine introduction, naturalization and invasion success in woody plants: Proteaceae as a test case. *PLoS One*, **8**, e75078.

Moran MD (2003) Arguments for rejecting the sequential Bonferroni in ecological studies. *Oikos*, **100**, 403–405.

Moravcová L, Pyšek P, Jarošik V, Havlíčková V, Zákravský P (2010) Reproductive characteristics of neophytes in the Czech Republic: traits of invasive and non-invasive species. *Preslia*, **82**, 365–390.

Morin PJ (1999) *Community Ecology*. Wiley Blackwell, Malden, Massachusetts.

Murray BR, Phillips ML (2010) Investment in seed dispersal structures is linked to invasiveness in exotic plant species of south-eastern Australia. *Biological Invasions*, **12**, 2265–2275.

Muth NJ, Pigliucci M (2006) Traits of invasives reconsidered: phenotypic comparisons of introduced invasive and introduced noninvasive plant species within two closely related clades. *American Journal of Botany*, **93**, 188–196.

Naeem S, Knops JMH, Tilman D, Howe KM, Kennedy T, Gale S (2000) Plant diversity increases resistance to invasion in the absence of covarying extrinsic factors. *Oikos*, **91**, 97–108.

Nathan R (2006) Long-distance dispersal of plants. *Science*, **313**, 786–788.

Nathan R, Schurr FM, Spiegel O, Steinitz O, Trakhtenbrot A, Tsoar A (2008) Mechanisms of long-distance seed dispersal. *Trends in Ecology and Evolution*, **23**, 638–647.

Palacio-López K, Gianoli E (2011) Invasive plants do not display greater phenotypic plasticity than their native or noninvasive counterparts: a meta-analysis. *Oikos*, **120**, 1393–1401.

Pandit MK, Pocock MJ, Kunin WE (2011) Ploidy influences rarity and invasiveness in plants. *Journal of Ecology*, **99**, 1108–1115.

Pandit MK, White SM, Pocock MJ (2014) The contrasting effects of genome size, chromosome number and ploidy

level on plant invasiveness: a global analysis. *New Phytologist*, **203**, 697–703.

Peñuelas J, Sardans J, Llusia J *et al.* (2010) Faster returns on 'leaf economics' and different biogeochemical niche in invasive compared with native plant species. *Global Change Biology*, **16**, 2171–2185.

Petitpierre B, Kueffer C, Broennimann O, Randin C, Daehler C, Guisan A (2012) Climatic niche shifts are rare among terrestrial plant invaders. *Science*, **335**, 1344–1348.

Prati D, Bossdorf O (2004) Allelopathic inhibition of germination by *Alliaria petiolata* (Brassicaceae). *American Journal of Botany*, **91**, 285–288.

Pyšek P, Richardson DM (2007) Traits associated with invasiveness in alien plants: where do we stand? In: *Biological Invasions* (ed. Nentwig W), pp. 97–125. Springer, New York.

Pyšek P, Richardson DM, Williamson M (2004) Predicting and explaining plant invasions through analysis of source area floras: some critical considerations. *Diversity and Distributions*, **10**, 179–187.

Pyšek P, Jarošik V, Pergl J *et al.* (2009) The global invasion success of Central European plants is related to distribution characteristics in their native range and species traits. *Diversity and Distributions*, **15**, 891–903.

Pyšek P, Manceur AM, Alba C *et al.* (2014) Naturalization of central European plants in North America: species traits, habitats, propagule pressure, residence time. *Ecology*, doi: org/10.1890/14-1005.1.

Rambuda TD, Johnson SD (2004) Breeding systems of invasive alien plants in South Africa: does Baker's rule apply? *Diversity and Distributions*, **10**, 409–416.

Reichard SH, Hamilton CW (1997) Predicting invasions of woody plants introduced into North America. *Conservation Biology*, **11**, 193–203.

Rejmánek M (1999) Invasive plant species and invasible ecosystems. In: *Invasive Species and Biodiversity Management* (eds Sandlund OT, Schei PJ, Viken Å), pp. 79–102. Kluwer Academic Publishers, Dordrecht, the Netherlands.

Rejmánek M, Richardson DM (1996) What attributes make some plants species more invasive? *Ecology*, **77**, 1655–1661.

Rejmánek M, Richardson DM, Pyšek P (2013) Plant invasions and invisibility of plant communities. In: *Vegetation Ecology*, 2nd edn (eds van der Maarel E, Franklin J), pp. 332–355. John Wiley & Sons, Chichester, UK.

Richardson DM, Pyšek P, Rejmánek M, Barbour MG, Panetta FD, West CJ (2000) Naturalization and invasion of alien plants: concepts and definitions. *Diversity and Distributions*, **6**, 93–107.

Robinson GR, Quinn JF, Stanton ML (1995) Invasibility of experimental habitat islands in a California winter annual grassland. *Ecology*, **76**, 786–794.

Seabloom EW, Harpole WS, Reichman OJ, Tilman D (2003) Invasion, competitive dominance, and resource use by exotic and native California grassland species. *Proceedings of the National Academy of Sciences*, **100**, 13384–13389.

Shea K, Chesson P (2002) Community ecology theory as a framework for biological invasions. *Trends in Ecology and Evolution*, **17**, 170–176.

Speek TAA, Plotz LAP, Ozinga WA, Tamis WLM, Schaminée JHJ, van der Putten WH (2011) Factors relating to regional and local success of exotic plant species in their new range. *Diversity and Distributions*, **17**, 542–551.

Stinson K, Campbell SA, Powell JR *et al.* (2006) Invasive plant suppresses the growth of native Tree seedlings by disrupting belowground mutualisms. *PloS Biology*, **4**, 727–731, e140.

Theoharides KA, Dukes JS (2006) Plant invasion across space and time: factors affecting nonindigenous species success during four stages of invasion. *New Phytologist*, **176**, 256–273.

Thompson K, Davis MA (2011) Why research on traits of invasive plants tells us very little. *Trends in Ecology and Evolution*, **26**, 155–156.

Thompson K, Hodgson JG, Grime JP, Burke MJW (2001) Plant traits and temporal scale: evidence from a 5-year invasion experiment using native species. *Journal of Ecology*, **89**, 1054–1060.

Turnbull LA, Paul-Victor C, Schmid B, Purves DW (2008) Growth rates, seed size, and physiology: do small-seeded species really grow faster? *Ecology*, **89**, 1352–1363.

Vilà M, Weiner J (2004) Are invasive plant species better competitors than native plant species? —evidence from pair-wise experiments. *Oikos*, **105**, 229–238.

Widrlechner MP, Thompson JR, Iles JK, Dixon PM (2004) Models for predicting the risk of naturalization of non-native woody plants in Iowa. *Journal of Environmental Horticulture*, **22**, 23–31.

Williamson M (1996) *Biological Invasions*. Chapman & Hall, London, UK.

Wilsey BJ, Barber K, Martin LM (2014) Exotic grassland species have stronger priority effects than natives regardless of whether they are cultivated or wild genotypes. *New Phytologist*, doi: 10.1111/nph.13028.

Young JA, Evans RA (1976) Responses of weed populations to human manipulations of the natural environment. *Weed Science*, **24**, 186–190.

Chapter 4

EVOLUTION OF THE MATING SYSTEM IN COLONIZING PLANTS

John R. Pannell

Department of Ecology and Evolution, University of Lausanne, Biophore Building, 1015 Lausanne, Switzerland

Abstract

Colonization is likely to be more successful for species with an ability to self-fertilize and thus to establish new populations as single individuals. As a result, self-compatibility should be common among colonizing species. This idea, labelled 'Baker's law', has been influential in discussions of sexual-system and mating-system evolution. However, its generality has been questioned, because models of the evolution of dispersal and the mating system predict an association between high dispersal rates and outcrossing rather than selfing, and because of many apparent counter examples to the law. The contrasting predictions made by models invoking Baker's law versus those for the evolution of the mating system and dispersal urges a reassessment of how we should view both these traits. Here, I review the literature on the evolution of mating and dispersal in colonizing species, with a focus on conceptual issues. I argue for the importance of distinguishing between the selfing or outcrossing rate and a simple ability to self-fertilize, as well as for the need for a more nuanced consideration of dispersal. Colonizing species will be characterized by different phases in their life pattern: dispersal to new habitat, implying an ecological sieve on dispersal traits; establishment and a phase of growth following colonization, implying a sieve on reproductive traits; and a phase of demographic stasis at high density, during which new trait associations can evolve through local adaptation. This dynamic means that the sorting of mating-system and dispersal traits should change over time, making simple predictions difficult.

Previously published as an article in *Molecular Ecology* (2015) 24, 2018–2037, doi: 10.1111/mec.13087

INTRODUCTION

Colonization involves the establishment of a new population in habitat unoccupied by its species, usually by a small number of colonizers. On the one hand, reduced crowding in the new habitat may allow an increase in per-individual reproductive success that could compensate against the loss of local adaptation; on the other hand, it means that sexual individuals will have fewer potential mates. In the extreme, where colonization has been effected by a single individual, mating opportunities will have been lost entirely – unless that individual is able to mate with itself via self-fertilization, or await the later arrival of compatible mates. Where dispersal events are rare, as might be the case for dispersal over large distances, an ability to self-fertilize will usually represent the only basis on which a new sexual population might become successfully established. If colonization by long-distance dispersal typically involves single or small numbers of individuals, we should expect populations that then become established to comprise self-fertile individuals.

This simple and intuitive line of reasoning was exposed by Herbert Baker (1955) to explain his observation that populations that had been established by putatively long-distance dispersal indeed showed a capacity for self-fertilization (and uniparental reproduction in general). Baker (1955) had been thinking mainly about plants, but it was the publication of similar patterns found in animals by Longhurst (1955) that prompted his paper; the pattern seemed to be general to both plants and animals. Two years later, reflecting on just how general the pattern seemed to be, G. Ledyard Stebbins (1957) suggested that the idea should be labelled 'Baker's law'.

By elevating Baker's idea to the status of 'law', Stebbins' (1957) paper was bound to attract debate and disagreement. Indeed, almost immediately the generality of Baker's law was questioned by the botanist Carlquist (1966), who drew attention to examples of obligate outcrossers on oceanic islands, notably dioecious plants that seemed to be even more common than one should expect on the basis of their frequency elsewhere. Carlquist (1966) suggested abandoning the idea altogether. The idea has not (yet) been abandoned, but discussion about the appropriateness of its label has rumbled on to the present, 50 years later (Baker 1967; Carr *et al.* 1986; Mackiewicz *et al.* 2006). The proposition that long-distance dispersal might influence the distribution of mating system and reproductive traits has inspired both theoretical analysis and empirical tests in biological contexts that go beyond Baker's initial articulation of his idea. The notions implied in Baker's idea, for example, have been discussed and analysed not only in the context of long-distance dispersal to oceanic islands (McMullen 1987; Barrett 1996; Bernardello *et al.* 2006), but also of dispersal in range expansions (Barrett & Husband 1990; Randle *et al.* 2009; Wright *et al.* 2013), species invasions (Van Kleunen & Johnson 2007; Van Kleunen *et al.* 2008; Barrett 2011; Ward *et al.* 2012; Rodger *et al.* 2013) and metapopulation dynamics (Pannell 1997a; Pannell & Barrett 1998; Dornier *et al.* 2008; Schoen & Busch 2008).

That colonization might act as a selective sieve on the trait combinations displayed by a population (or ensemble of populations) is an important idea, but it has been profitable to consider it more broadly than in the terms initially discussed by Baker (1955). Whereas Baker focused his attention specifically on the effect that the selective sieve imposed by long-distance colonization should have on the *capacity* of successful colonizers to self-fertilize, colonization might affect the geographical distribution of reproductive traits for other reasons, too. These include effects of changes in pollinator availability, the evolution of inbreeding depression and dispersal, and affects that the mating system might have on a capacity for local adaptation. Such effects take place in a sequence of phases that all follow from colonization, but which should have different implications for the traits we ought to see. Specifically, it may be useful to conceptualize these various aspects of colonization in terms of three separate phases, which encompass dispersal and colonization, initial establishment following colonization, and the longer term evolution of trait combinations after establishment. This conceptualization highlights the fact some trait combinations can be a direct result of dispersal and reproductive sieves on pre-existing trait variation, which we might thus view as exaptations, while others will be the result of new adaptations (see Box 1).

Baker (1955) focussed his discussion on the colonization of oceanic islands, but the extension of his explanation to contexts that were not explicitly envisaged by him, particularly recurrent dispersal in metapopulations and range expansions, have both yielded important new insights and raised new questions. Here, I review these advances and address questions that remain. I first consider the evolution of colonizing

Box 1 Why should colonization affect plant mating?

Baker (1955) drew attention to the evolution of a capacity to self-fertilize in organisms prone to long-distance dispersal, but we might expect selection during and after colonization to influence plant mating for a number of reasons. Baker drew attention to one of these reasons in his initial study: the likelihood that colonization would not only take populations through a severe demographic bottleneck (of a single individual), but that it would also likely disrupt the native association with pollinators, with concomitant effects on the evolution of the mating system (Cheptou & Massol 2009). Other potential causes of an association between the mating system and colonization include: the loss of genetic diversity at loci that affect levels of inbreeding depression (Pujol *et al.* 2009; Peischl *et al.* 2013), a variable known to be important for the evolution of self-fertilization (Lande & Schemske 1985; Goodwillie *et al.* 2005); the loss of self-incompatibility alleles, otherwise maintained in large populations by negative frequency-dependent selection, with a resulting decrease in mate availability even after populations have grown (Vekemans *et al.* 1998; Brennan *et al.* 2003, 2006; Busch & Schoen 2008; Young & Pickup 2010); the effect that (repeated) colonization might have on the joint evolution of the mating system and dispersal – as opposed to the evolution of the mating system within the context of a particular (fixed) syndrome of dispersal (Cheptou & Massol 2009; Massol & Cheptou 2011a); the interactive effect that bottlenecks and the mating system have on levels of genetic diversity and linkage disequilibrium for loci that might be important in local adaptation to novel environments (Stebbins 1957; Allard 1965); and the extent to which

self-fertilization tends to isolate populations from gene flow that might compromise local adaptation at range margins, thereby preventing further range expansion.

The diverse ways in which colonization might affect the evolution of the mating system and dispersal traits might be conceptualized in terms of three phases of colonization in its broad sense, as depicted in the figure below.

• Phase A concerns the fact that colonization requires dispersal. There will thus be an ecological sieve on variation among and within species that selects on dispersal and reproductive traits already represented in the pool from which potential colonists are drawn. Traits that confer success on colonization are thus best viewed as exaptations.

• Phase B concerns establishment immediately upon dispersal. Here, success will be favoured for lineages with exaptations in terms of their reproductive system (e.g. perenniality, a capacity for self-fertilization or uniparental reproduction in general, or a capacity to be pollinated by generalist pollinators). The sieves associated with phases A and B may thus fashion combinations of dispersal and reproductive exaptations, but such associations do not require the joint evolution of these traits as adaptations.

• Phase C concerns evolution of traits subsequent to establishment, in the new ecological setting (with, for example, new and generalist pollinators and seed dispersers). How populations respond to selection will depend on genetic diversity that survives the colonization bottleneck (in the short term) or on new mutations (in the longer term). An immediate consequence of a genetic bottleneck will be a decline in inbreeding depression, which can set up conditions for the evolution of selfing. See text for details.

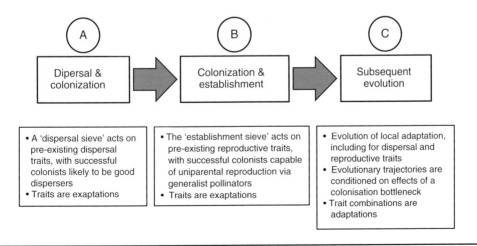

- • A 'dispersal sieve' acts on pre-existing dispersal traits, with successful colonists likely to be good dispersers
- • Traits are exaptations

- • The 'establishment sieve' acts on pre-existing reproductive traits, with successful colonists capable of uniparental reproduction via generalist pollinators
- • Traits are exaptations

- • Evolution of local adaptation, including for dispersal and reproductive traits
- • Evolutionary trajectories are conditioned on effects of a colonisation bottleneck
- • Trait combinations are adaptations

species under selection for reproductive assurance and then explore in turn the importance of the joint evolution of the mating system and inbreeding depression and that of the mating system and dispersal. I go on to emphasize the distinction between selection on the mating system in terms of a selfing rate and a simple capacity for occasional self-fertilization. I ask to what extent is it appropriate to consider (e.g. in a modelling context) the impact of a single event of colonization of an oceanic island, for example, within the context of a scenario where colonization is repeated and ongoing, such as in range expansions or metapopulations of ruderal weeds. Finally, I end by reflecting on the contributions made on these issue by authors of the landmark book 'The Genetics of Colonizing Species' (Baker & Stebbins 1965), which we are here commemorating.

COLONIZATION AND THE BENEFITS OF REPRODUCTIVE ASSURANCE

A striking feature of plant reproduction is the extent to which the mating system (in particular, the selfing or outcrossing rate) varies among species and even among populations of the same species (Barrett 2002; Harder & Barrett 2006). To a large extent, research has focused on understanding the apparent enigma of intermediate selfing rates, where self-compatible plants both outcross and self-fertilize their progeny (Goodwillie *et al.* 2005). Most of this work has considered selection within demographically stable populations, but intermediate selfing rates could also be the result of selection in habitually colonizing species, in which selfing rates might fluctuate from high values during colony establishment and lower values when mates (or pollinators) are more abundant in large established populations. The consideration of how dispersal and fluctuating pollinator availability might select for intermediate selfing rates has been considered by Cheptou & Massol (2009) and Massol & Cheptou (2011a), and metapopulation models such as those of Pannell (1997a), Pannell & Barrett (2001) and Schoen & Busch (2008), reviewed below, point to this possibility, although more work is needed on mixed mating in metapopulation models (see review by Barrett & Pannell 1999).

At the extremes, the variation in the mating system observed among species and populations likely reflects transitions between predominant outcrossing, often secured by way of molecular self-incompatibility (SI)

systems (Hiscock & McInnis 2003; Takayama & Isogai 2005), and predominant self-fertilization, which may evolve via the spread of mutations that cause a breakdown of the SI system (Barrett 1988; Uyenoyama *et al.* 2001; Busch & Schoen 2008). Indeed, the shift from outcrossing to selfing has been perhaps the most common major evolutionary transition to have taken place during the course of plant evolution (Stebbins 1950, 1974), with two main explanations advanced to explain it (Goodwillie *et al.* 2005; Busch & Delph 2012). Fisher's 'automatic transmission hypothesis' posits that mutations that increase the selfing rate should enjoy an automatic and immediate selective advantage, because they should be transmitted not only through the ovules of affected individuals, but also by the pollen grains (Fisher 1941). Such mutations might be expected to spread particularly rapidly if they affect the male component of self-rejection, that is if genes expressed in or on pollen grains fail to be recognized as self, because such pollen grains not only benefit by siring the ovules on the plant that produces them, but will be compatible with all other individuals in the population so that they quickly become expressed in the progeny of other individuals. Empirical evidence supports this logic: in those cases for which the mutation causing self-compatibility (SC) has been found in natural populations, all are in the male-acting component of the self-incompatibility locus (S-locus) (Tsuchimatsu *et al.* 2012). (Interestingly, in crops, where self-fertility will presumably have been favoured by humans choosing the more highly fertile selfing variants in populations that produce more seed, mutations in the female-acting component are more common (Tsuchimatsu *et al.* 2010)).

The second major hypothesis for the evolution of selfing is the 'reproductive assurance hypothesis' (Darwin 1876; Lloyd 1979, 1992). This hypothesis predicts that selfing will evolve from outcrossing in response to selection for an ability to self-fertilize when mates and/or pollinators are absent. Although it is easy to see the advantage of mechanisms that confer reproductive assurance when opportunities for outcrossing are limited, clear experimental support for the reproductive assurance hypothesis has been surprisingly elusive (Herlihy & Eckert 2002; Busch & Delph 2012). Evidence for it is still largely based on associations between self-fertilization and either life histories for which reproductive assurance would seem likely to be more often important (e.g. annual or ephemeral plants) or a geographical distribution that is marginal to a species' range (e.g. because both mates and pollinators might be scarcer in marginal habitats) (Fausto *et al.* 2001).

The hypothesis encompassed by Baker's law is, in an important respect, an analogue of the reproductive assurance hypothesis (Schoen & Busch 2008) and a potential example of a strong Allee effect, where population growth is negative below a threshold population size (Allee *et al.* 1949; Stephens *et al.* 1999). Here, an ability to self-fertilize will be strongly favoured over obligate outcrossing in colonizing species, because colonizers will often lack compatible mates and may have to rely on less abundant, unspecialized (and thus less effective) pollinators. In the extreme situation of colonization by a single individual (which Baker (1955) and Stebbins (1957) both emphasized), the benefits of an ability to self are obvious and almost do not need to be tested. For instance, Hesse & Pannell (2011) showed that isolated self-compatible monoecious individuals of the wind-pollinated annual plant *Mercurialis annua* set maximal seed, whereas isolated females of the same species did not (Fig. 1). How often populations are colonized by single as opposed to multiple individuals is not known and unfortunately very difficult to gauge directly, because populations often only become evident when they are already large. Indirect estimates

based on genetic diversity do not allow discrimination between a scenario of colonization by multiple individuals and a gradual accumulation of diversity over time as migrants join previously established populations. Mate limitation can also occur in the less severe situation where colonization is by more than one individual that, however, are of the same gender (in dioecious species) or carry the same S-alleles – the so-called S-Allee effect (Wagenius *et al.* 2007) (see Box 2).

Several models have considered various aspects of the benefits of reproductive assurance during colonization, typically in the context of metapopulation models in which stochastic extinction is balanced by frequent recolonization of patches by dispersal. Pannell (1997a) modelled the maintenance of males or females with hermaphrodites and showed that a unisexual (obligate outcrossing) strategy was increasingly disfavoured by selection at the metapopulation level as the rate of population turnover increased and/or the mean number of individuals that colonized available habitat patches decreased. He found that unisexuals were much more quickly lost from a metapopulation with increasing population turnover when the hermaphrodites were SC than when they were SI, in which case unisexuals could be maintained with SI hermaphrodites even to

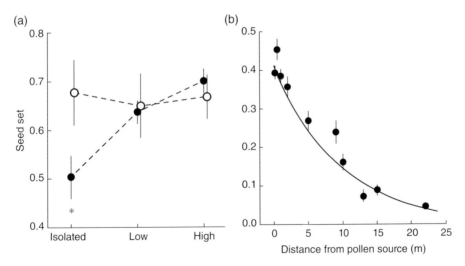

Fig. 1 (a) Mean seed set for females (closed circles) and self-compatible hermaphrodites (open circles) of *Mercurialis annua* from high-density (approx. 70 plants per m²) and low-density populations (approx. 8 plants per m²) and isolated individuals (at least 2 m from the nearest non-self-pollen donor). Only the isolated females set significantly less than what appeared to be maximal seed set. (b) The negative dependence of seed set by females on the distance to the nearest male in an experimental population of *M. annua*. Reproduced from Hesse & Pannell (2011). A similar dependence of seed set on distance to neighbours has recently been found for *Echinacea angustifolia* (Ison & Wagenius 2014).

Box 2 Mate limitation when there is more than one colonizing individual

Baker (1955) emphasized the advantages of an ability to self-fertilize in cases of colonization by a single individual. However, mate limitation can also occur when more than one individual colonizes a new habitat, if the individuals are of the same gender or are self-incompatible with one another, for example by sharing the same self-incompatibility alleles. In large populations of self-incompatible (SI) species, mate availability (i.e. the probability that a given mating partner is cross-compatible) is maintained at high levels by negative frequency-dependent selection on the S-locus (Wright 1964; Schierup 1998). Typically, SI acts such that individuals that share an S-allele will tend to be cross-incompatible (although the details will depend on whether SI is gametophytic or sporophytic). This means that individuals carrying a rare S-allele will be self-incompatible but cross-compatible with a large majority of the rest of the population, whereas those carrying common S-alleles will have greater cross-incompatibility within a population. The result is that rare alleles tend to increase in frequency and are protected from loss through genetic drift. Nevertheless, colonization, especially when it involves repeated genetic bottlenecks that would occur during a protracted range expansion, will strongly increase the effect of drift, and S-alleles can be lost (Wagenius et al. 2007). We might thus expect that populations established via long-distance colonization or range expansion to have a small number of more common S-alleles and relatively low mate availability, even after population growth.

It is not known how often SI colonizers have low S-allele diversity, but there is accumulating evidence that S-allele diversity can indeed be much lower in small, isolated populations than in larger ones. In the successful colonizer Senecio squalidus, which has spread rapidly throughout much of the United Kingdom following its introduction from Sicily, UK populations appear to harbour fewer S-alleles than their Italian counterparts, even though population sizes are much larger in the introduced range (Brennan et al. 2013). In the herbaceous perennial Rutidosis leptorrhynchoides, small populations not only have lower S-allele diversity, but also appear to suffer from the reduced mate availability by producing fewer seeds (Young et al. 2000). An experimental manipulation of both population size in the relatedness among individuals within populations of the invasive species Raphinus sativus has shown that such affects can be directly attributed to genetic causes (including S-allele interactions) and are not just a biproduct of pollination failure in small populations per se (Elam et al. 2007). The direct impact on seed production of mate availability as a result of shared S-alleles was also demonstrated for the threatened European species Biscutella neustriaca using experimental manipulations (Leducq et al. 2010). Computer simulations similarly point to a direct effect of S-allele diversity on mate availability (Young & Pickup 2010). All these studies suggest that diversity at the S-locus could have an important impact on a species' colonization success.

---

the point of metapopulation extinction (Pannell 1997a). The model made predictions linking the evolution of the sexual and mating system with the proportion of occupied habitat in the landscape, which can potentially be measured (Fig. 2a). In particular, the proportion of occupied sites should increase with the amount of dispersal across the metapopulation (which also limits advantages of reproductive assurance, because populations can be more often colonized by more than one individual) and decrease with the rate of local turnover (which also favours a selfing ability). The proportion of occupied sites can thus be used as a useful signature of the underlying metapopulation dynamics, as they affect the evolution of an ability to self. These predictions were corroborated in a test linking high versus low rates of habitat occupancy with the occurrence of dioecy versus monoecy, respectively

(Eppley & Pannell 2007b; and Fig. 2b). Here, the predictions depended strongly on the expected number of individuals that founded new colonies.

In a subsequent model, Pannell & Barrett (1998) examined in further detail the persistence of SI vs. SC individuals in a metapopulation as a function of the proportion of occupied habitat (again, assuming stochastic extinction and migration that involved a given number of individuals). A key insight from this model is that although individuals with a selfing ability will often enjoy the expected advantage of reproductive assurance during colonization, the maintenance of SI in some species subject to metapopulation processes, such as ruderal weeds, should not come as a surprise. It was reasonably expected, for example, that the highly successful colonizer of human-disturbed habitat in the UK, Senecio squalidus, might have evolved increased

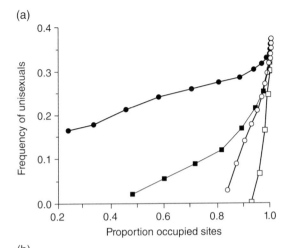

(a)

(b)

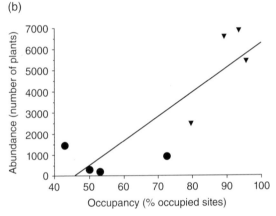

Fig. 2 (a) The frequency of unisexual individuals (males or females) maintained in a metapopulation with either SC hermaphrodites (open symbols) or SI hermaphrodites (closed symbols). Squares and circles denote low (0.05) and high population turnover rates (0.2), respectively. Each point corresponds to simulations with a given mean number of colonizing individuals, with higher values to the right on each curve. Reproduced from Pannell (1997a). (b) Mean population size plotted against the percentage of habitat occupied by *Mercurialis annua* across eight regions (metapopulations) with different sexual systems in the Iberian Peninsula: triangles represent populations that are dioecious or androdioecious (males and hermaphrodites); circles represent populations of self-fertile hermaphrodites. Reproduced from Eppley & Pannell (2007b).

levels of SC during its invasive spread (Abbott & Forbes 1993), but its strong SI system has been maintained throughout. The metapopulation models of Pannell (1997a) and Pannell & Barrett (1998) suggest, perhaps

counter-intuitively, that it is precisely in good colonizing species such as *S. squalidus* that we might expect to see SI maintained. This is because colonization will often be by more than one individual when abundance and occupancy rates are high. On the basis of a related model, Dornier *et al.* (2008) also concluded that a high number of propagules of SI species can compensate for an incapacity to found new colonies as single individuals. All these models highlight the importance of considering not only population turnover, but also the 'propagule pressure' involved in recolonization. Indeed, Salisbury (1953), cited in Baker (1965), suggested the need for a certain density of propagule production before the explosive spread of *S. squalidus* throughout Britain. Of course, the effect of propagule pressure on mate availability will necessarily be modulated by the dispersal mode (endozoochory, epizoochory, wind-dispersal, etc.), particularly in dioecious or self-incompatible species, because of its influence on the relatedness of individuals within a dispersing propagule and the extent to which they will be mutually compatible with one another.

EFFECT OF COLONIZATION ON INBREEDING DEPRESSION AND THE MATING SYSTEM

The most likely reason for the maintenance of outcrossing in hermaphrodites despite the automatic transmission advantage of selfing is that selfed progeny will often express inbreeding depression. Indeed, simple models that account for gene transmission via both seeds and pollen predict that selfing can only invade and spread in a population if the levels of inbreeding depression suffered by selfed progeny are <0.5 (Lande & Schemske 1985; Goodwillie *et al.* 2005). To understand how the mating system should evolve in colonizing species, it would thus seem important to incorporate notions of inbreeding depression. Relevant models vary substantially in how they have dealt with this issue (Goodwillie *et al.* 2005).

Empirical work has established quite firmly that most inbreeding depression expressed within populations is the result of the expression of deleterious recessive alleles in progeny rendered more homozygous by inbreeding (Charlesworth & Charlesworth 1987; Charlesworth & Willis 2009). We also know from theory (Lande & Schemske 1985; Charlesworth *et al.* 1990; Uyenoyama & Waller 1991; Porcher & Lande

2005), from the observed association between the mating system and levels of inbreeding depression (Byers & Waller 1999), and from experiments (Crnokrak & Barrett 2002) that inbreeding depression due to deleterious recessives will evolve with the mating system and cannot realistically be assumed to be fixed. Specifically, inbreeding allows the deleterious recessive alleles responsible for inbreeding depression to be purged from the population, so that inbreeding populations end up with little or no inbreeding depression, even though there may have been substantial inbreeding depression before a transition to self-fertilization began (Crnokrak & Barrett 2002).

Although a great deal has been learned about mating system evolution from single-population models that assume a fixed value of inbreeding depression, it seems difficult to extract meaningful predictions from such models about inbreeding depression and the mating system in metapopulations. Single-population models with fixed inbreeding depression are useful because they can inform us about the boundary conditions for the invasion of a mutant strategy into a population, but they are unable to predict the trajectories that populations will take as they approach a new equilibrium. This is particularly important in the context of metapopulations, because the genetic architecture of each deme or subpopulation is likely to change over the course its existence from the time of its colonization until its extinction; the genetic architecture of the whole metapopulation will be thus the accumulated outcome of multiple (ongoing) trajectories (see Box 3). The relationship between the selfing rate

---

Box 3 Models of inbreeding depression and the mating system in metapopulations

Models of the evolution of the mating system in metapopulations vary in how they have dealt with the issue of inbreeding depression. The models made by Pannell (1997a) and Pannell & Barrett (1998) did not address the question of how inbreeding depression might affect the maintenance of different mating strategies in a metapopulation. In a sense, Pannell & Barrett (1998) avoided this issue by comparing the fate of self-compatible and self-incompatible species separately, that is their model was not concerned with the evolution of the mating system, but with the conditions under which one or other system might be maintained on its own. The effects of inbreeding depression, should it occur, could thus be seen to be incorporated implicitly into the intrinsic population growth rates of selfers vs. outcrossers. Dornier et al. (2008) did incorporate inbreeding depression as a fixed parameter into their metapopulation model and showed that, under their assumptions, inbreeding depression could paradoxically favour the evolution of self-fertilization in a metapopulation by diminishing local population densities sufficiently. Later models by Cheptou and Massol (Cheptou & Massol 2009; Massol & Cheptou 2011a) also incorporated fixed inbreeding depression into their analysis of how variation in the pollination environment might affect the joint evolution of dispersal and the mating system and predicted an association between outcrossing and dispersal, although colonization was not part of their models.

To see why metapopulation models that assume fixed inbreeding depression might be unrealistic in important ways, consider inbreeding depression in a recently colonized population (the argument will apply to any scenario of population growth from a narrow bottleneck). Because individuals in a deme will all tend to be related by descent from the one or few colonizers, selfed progeny will tend to differ genetically very little from locally outcrossed individuals. This has at least two important consequences. First, a population established by one or few self-fertilizing individuals simply cannot express much inbreeding depression in the sense that is usually required in mating system models: outcrossing within such a population does not avoid the cost of inbreeding any more than self-fertilization does. And second, because selfing and local outcrossing will tend to suffer equivalently from the inheritance of locally fixed deleterious alleles (or to benefit equivalently from their local purging), self-fertilization should enjoy the benefits from the automatic transmission advantage without paying the relative costs (Uyenoyama 1986). Any analysis of the evolution of the mating system in a metapopulation as a function of inbreeding depression that does not explicitly allow for the dynamic evolution of inbreeding depression clearly needs to be interpreted critically. More work needs to be done to explore mating-system evolution in metapopulations in which inbreeding depression is maintained in a balance between mutation (or migration into sinks from demographically stable source populations) and its purging by inbreeding and purifying selection.

and the level of inbreeding depression predicted for single populations, as used in the appendix of Dornier *et al.* (2008) model, thus seems unlikely to apply to a metapopulation scenario in a straightforward manner. Rather, models need to account for the fact that inbreeding depression can vary greatly between long-established and recently colonized populations, in a way that might foster the evolution of self-fertilization in the latter (Box 3). Some progress has been made in this direction through the theoretical and empirical population genetic analysis of range expansions, which involve the repeated colonization of available habitat and which thus have much in common with metapopulation scenarios.

EVOLUTION OF INBREEDING DEPRESSION AND THE MATING SYSTEM AT RANGE EDGES

In species range expansions, the wave of colonizations and repeated genetic bottlenecks that establish populations in new habitat are expected to bring about reduced levels of genetic diversity (Petit *et al.* 2002; Eckert *et al.* 2008). This affects not only neutral loci, but also loci subject to selection (Pujol & Pannell 2008; Pujol *et al.* 2009). While the increased levels of inbreeding towards the range edge can potentially allow the purging of deleterious recessive alleles, deleterious alleles can also be taken to high frequency by drift, resulting in a so-called expansion load in these populations (Peischl *et al.* 2013). This additional genetic load carried by colonized populations reduces mean fitness, but the combined effect of purging deleterious mutations at some loci and purging them at others means that inbreeding depression within populations should be lower towards range edges than at the core of a species' distribution. To test this prediction, Pujol *et al.* (2009) measured inbreeding depression in 16 populations of *Mercurialis annua* by comparing fitness components of selfed and outcrossed progeny. They found a sharp decline in inbreeding depression in populations with distance from the putative pre-expansion core of the species range, as predicted (Fig. 3).

It has long been observed that populations on the edge of a species' range are more likely to be self-fertilizing than those towards its core (Randle *et al.* 2009). This pattern has been attributed to potential selection for reproductive assurance, either during

colonization at the leading edge of a range expansion (as expected under Baker's law) or because range-edge populations occur in marginal habitat where population densities are low and/or pollinators are scarce (Randle *et al.* 2009). Another possible reason for an enrichment of selfing towards range margins is that selfing may allow range-edge populations to evolve local adaptation more freely than outcrossing ones, because they will be less affected by the migration load caused by suboptimal alleles dispersing from populations in less marginal populations (Kirkpatrick & Barton 1997; Sexton *et al.* 2009). This perspective is similar to that taken by Allard (1965)

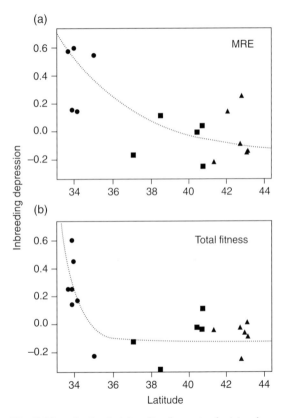

Fig. 3 The reduction in inbreeding depression for (a) male reproductive effort and (b) total fitness in populations of the wind-pollinated annual colonizer *Mercurialis annua* from populations that migrated from southern Spain and northern Morocco, the species' putative refugium (the lowest latitudes shown) into northern Spain (the higher latitudes). Symbols refer to populations sampled in Morocco (circles), north-eastern Iberia (squares) and north-western Iberia (triangles). Figure taken from Pujol *et al.* (2009).

in his reflections on the advantages that selfers would have in occupying novel habitats in a heterogeneous environment. Finally, in discussing the results of their study of inbreeding depression, Pujol *et al.* (2009) posed yet another hypothesis: if range-edge populations express lower inbreeding depression, we should expect mutations that increase the selfing rate to spread because of their automatic transmission advantage.

With four distinct hypotheses for an increased incidence of self-fertilization at species range edges, we face something of an embarrassment of riches. Usefully, however, they differ in their implications for subsequent evolution once selfing has evolved, notably for the likelihood that selfing could spread throughout the species range. If selfing evolves at range edges because their habitats are marginal (Allard 1965), we would predict the maintenance of selfing in marginal habitats and the persistence of outcrossing where populations are denser and pollinators more abundant. This appears to be the case, for example, in *Clarkia xantiana* (Moeller & Geber 2005). If selfing has evolved because only selfing (or asexual) populations are able to continue their expansion without being held back by the effects of gene flow from nonadapted populations towards the species' core, then we might expect that selfing populations come to occupy much larger areas than their outcrossing counterparts, as is the case for the selfing ruderal weed *Capsella rubella* (Foxe *et al.* 2009; Guo *et al.* 2009). In contrast, the benefits of reproductive assurance during the demographic bottlenecks inherent to Baker's law will be transitory and should quickly be lost once populations grow in size. An observation of a strategy that reflects a capacity to self-fertilize in the absence of mates but potentially high outcrossing rates in established populations would be consistent with this explanation, as seems to be the case for metapopulations of the self-compatible *M. annua*, in which outcrossing rates are positively density dependent (Eppley & Pannell 2007a; Dorken & Pannell 2008) and are typically high in established populations (see Fig. 4).

The long-term implications of the evolution of selfing in range-edge populations with depleted inbreeding depression are less clear. On the one hand, the reduction in inbreeding at range margins ought to be transitory, as shown by the simulations of Pujol *et al.* (2009), because genetic load can re-establish itself both through new mutations and migration from the

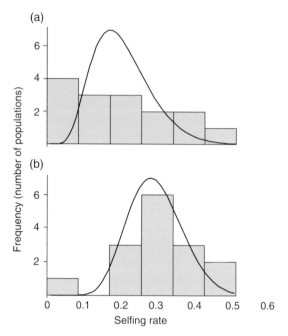

Fig. 4 The distribution of selfing rates in natural populations of the wind-pollinated annual herb *Mercurialis annua* for: (a) androdioecious populations in which males co-occurred with hermaphrodites and (b) hermaphroditic populations in which males were absent. Although *M. annua* hermaphrodites set full seed upon autonomous selfing in the absence of mates, for example during colonization (Pannell 1997b; Eppley & Pannell 2007a), these data indicate that natural established populations have a mixed mating system with outcrossing rates that range from near zero to one. From Korbecka *et al.* (G. Korbecka, J. P. David, J. L. García-Castaño and J. R. Pannell, unpublished).

species core. On the other hand, the evolution of selfing in populations with low inbreeding depression would prevent the accumulation of genetic load, thereby further maintaining conditions for the persistence of selfing (Lande & Schemske 1985). The long-term maintenance of selfing at the range margin would thus seem assured. This intuitive prediction was confirmed in simulations conducted by Encinas-Viso *et al.* (F. Encinas-Viso, J.R. Pannell and A.G. Young, unpublished), which assess evolution of the mating system in a range expansion with fully dynamic inbreeding depression. Interestingly, their simulations indicate that the long-term maintenance

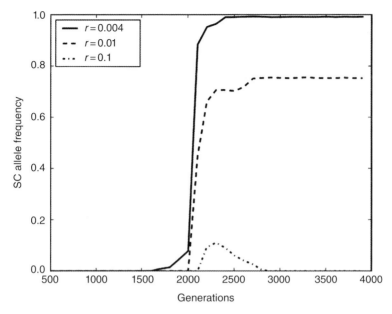

Fig. 5 The spread of self-compatibility in a range expansion as a result of selection for selfing when inbreeding depression is depleted by colonization bottlenecks. The graph shows the trajectory of three different simulations of a range expansion, under different rates of recombination between loci (r, shown). Simulations began as a metapopulation of self-incompatible hermaphrodites at equilibrium between largely recessive deleterious mutations at multiple loci, purifying selection, and drift. The loaded loci were partially linked to a gametophytic self-incompatibility locus at which multiple S-alleles were maintained by negative frequency-dependent selection. Following equilibrium in the species core distribution, the metapopulation was allowed to expand its range by stepping-stone migration, a process which gradually purged inbreeding depression, as previously shown by Pujol *et al.* (2009). Initially, recurrent mutations at the S-locus to alleles conferring self-compatibility were always prevented from spreading by selection in a background of high inbreeding depression. But with declining inbreeding depression, the SC allele spread, occupying range-edge populations only. With high recombination between the S-locus and viability loci, the evolution of self-fertilization in range-edge populations was short-lived. However, with increased linkage between the S-locus and viability loci, selfing was either maintained at the edge populations or could even spread into the core of the species' distribution. Results from Encinas-Viso *et al.* (F. Encinas-Viso, J.R. Pannell and A.G.Young, unpublished).

of outcrossing in the core is also assured as long as there is some recombination between the mating system locus and viability loci causing inbreeding depression, that is the evolved selfing strategy cannot spread back into the core (Fig. 5). This, too, is quite intuitive: an allele for increased selfing could only spread into core populations if it were linked to a non-loaded genome; recombination into the background of an outcrossing population would associate it with inbreeding depression, and selection would eliminate it. Encinas-Viso *et al.*'s simulations also confirm this prediction: with sufficiently low recombination rates, selfing spreads into the core and throughout the species range (Fig. 5). Species with a reduced number of

chromosomes, as is found in some colonizing species [e.g. genes from 20 chromosomes appear to have been brought together on only four chromosomes in the ruderal *Eupatorium microstemon* (Baker 1965)], would presumably ease the conditions for the spread of selfing throughout the species range.

JOINT EVOLUTION OF DISPERSAL AND THE MATING SYSTEM

Almost by definition, colonizing species will be those that have evolved some capacity to disperse. Any consideration of the mating system of colonizing species would thus seem to require incorporation of the evolution of dispersal, too. Baker's explanation for the high frequency of colonizing species with a capacity to self-fertilize did not account for the evolution of dispersal, and simply took it for granted that long-distance dispersal occasionally took place (or, in some species such as ruderal annuals, is habitual). This absence of any account of the joint evolution of dispersal and the mating system in models or discussions of the biology implied in Baker's law has recently attracted considerable criticism, not least because models that do so make predictions that appear to run counter to the patterns Baker was attempting to explain (Cheptou & Massol 2009; Massol & Cheptou 2011a). It is argued in these models that whereas Baker predicted an association between self-fertilization and dispersal (and thus colonization), we should in fact expect a propensity towards outcrossing and inbreeding avoidance to be associated with dispersal.

It would seem that these arguments take out of context the points that Baker (1955) was attempting to make. For example, in response to criticism by Busch (2011) that Cheptou and Massol had misconstrued the essence of Baker's argument in discussing their models, Massol & Cheptou (2011b) replied that there are two ideas encompassed in Baker's law: the idea that species with a capacity to self-fertilize would be favoured when colonizing new habitats following long-distance dispersal as single individuals; and the idea that newly established populations might often lack appropriate pollinators. It is true that both these points were made by Baker in his original 1955 study. However, Stebbins was referring to the first of these two ideas, which was indeed dominant in Baker's study, when he labelled it 'Baker's law':

> The correlation just mentioned [the association between colonization and an ability to self-fertilize] occurs so widely and has such great significance for studies of the origin and migration of genera of flowering plants and probably of other groups, including some animals (Baker 1955), that it deserves recognition as Baker's law. It was logically and reasonably explained by its author on the assumption that accidental long distance dispersal of a single propagule can lead to

establishment of a colony only in a species capable of self fertilization.

More interesting than disagreements over emphasis and definition is the question of how one might reconcile Baker's and Stebbins' explanation of an association between a selfing ability and colonization with models that predict that (i) dispersal and outcrossing should be linked and that (ii) a syndrome of self-fertilization should be associated with a stay-at-home strategy (Auld & de Casas 2013). Cheptou and Massol's models (Cheptou & Massol 2009; Massol & Cheptou 2011a), as the authors point out, join a large corpus of theory on the evolution of dispersal. This literature, which has been amply reviewed elsewhere (Clobert *et al.* 2001; Auld & de Casas 2013), emphasizes dispersal as a strategy that evolves in response to selection to avoid inbreeding and inbreeding depression (Perrin & Goudet 2001; Auld & de Casas 2013) and was not motivated by questions relevant to reproductive assurance and colonization *per se*. In other words, the models that might appear to contradict the ideas of Baker (1955, 1967) and Stebbins (1957) are modelling a different process. To an important extent, the confusion has arisen because similar terms have been used too liberally to refer to fundamentally different things, both in terms of the characterization of the mating system and what authors have meant by an ability to self-fertilize, and in terms of the extent to which metapopulation-like processes are truly able to represent Baker's initial idea. I address these issues in the following two sections.

THE MATING SYSTEM VERSUS AN ABILITY TO SELF-FERTILIZE

In reconciling models that predict an advantage to selfing in colonizing species with those that predict a syndrome of dispersal with outcrossing, it is important to distinguish between a simple ability to self-fertilize, and the habitual mating system of the species or populations in question. Models linking dispersal with inbreeding avoidance and outcrossing typically refer to the latter concept and are coined in terms of selfing or outcrossing rates, and/or inbreeding coefficients (Perrin & Goudet 2001; Cheptou & Massol 2009;

Massol & Cheptou 2011a). In contrast, Baker and Stebbins, in their original studies, were more concerned with a species' *capacity* for self-fertilization. This is explicit in Stebbins' paragraph cited at length above, in which Baker's law was first proposed and where he refers to the 'establishment of a colony only in a species *capable of self fertilization*' (my italics). Stebbins is true to the sense of Baker's original study, whose title refers to self-compatibility rather than self-fertilization. The crucial point that Baker makes in his study is expressed thus:

> With self-compatible individuals a single propagule is sufficient to start a sexually-reproducing colony, making its establishment much more likely than if the chance growth of two self-incompatible yet cross-compatible individuals sufficiently close together spatially and temporally is required. In addition, self-compatible flowering plants are usually able to form some seed in the absence of visits from specialized pollinating insects, which may be absent from the new situation.

The second sentence in this passage has been interpreted by Massol & Cheptou (2011b) as a second element to Baker's law, in which selection in large populations facing low pollinator visitation might favour evolution of high selfing rates. Rather, that sentence emphasizes the fact that self-compatible species are 'usually able to form *some* seed' (my italics) under pollinator limitation. There is no sense here of a prediction that long-distance dispersal should be associated with high selfing rates (although that is not ruled out) – only that species that could produce a few seeds by selfing would be more likely to become successful colonizers than those that are completely self-incompatible.

It may be tiresome to quibble about words, but the distinction between an ability to self-fertilize (simply by being self-compatible) and a strategy of self-fertilization is conceptually important. One of the hallmarks of plant sexuality is its plastic nature, as are minor deviations from otherwise fixed reproductive strategies (Lloyd & Bawa 1984; Ehlers & Bataillon 2007). The phenomenon of pseudo-self-compatibility is a case in point: in some species, individuals that are basically self-incompatible allow a degree of self-seed set, particularly at the end of the flowering period if ovules have not been pollinated (Levin 1996; Stephenson *et al.* 2000). Such behaviour has been interpreted as a reproductive assurance device and might aid colonization. Critically, an ability to self-fertilize in this case does not imply a selfing strategy, and the mating system of such species would properly be characterized as outcrossing.

Incomplete separation of the sexes in dioecious species provides another example of an outcrossing strategy that accommodates an ability to self-fertilize, and indeed one that has been much discussed in the context of Baker's law. It is well known that individuals of dioecious species often display 'leaky' gender expression by producing a few flowers of the opposite sex (Lloyd & Bawa 1984; Delph 2003; Delph & Wolf 2005). Indeed, Baker (1967) himself drew attention to this tendency in a lengthy passage in his response to Carlquist (1966), who had argued that the high frequency of dioecy on oceanic islands such as Hawaii ran counter to the prediction of Baker's law. (The same argument continues to be raised by detractors of Baker's law (Massol & Cheptou 2011b)). Here, Baker (1967) was making the simple point that leakiness in gender would be sufficient to allow colonization by single individuals, particularly in perennial species, as are all Hawaiian plants. In the subsequent paragraphs of his study, he then argued that the high frequency of dioecy on the islands of Hawaii may also be the outcome of repeated instances of selection to maintain outcrossing in taxa that had arrived 'unarmed with any self-incompatibility' (i.e. following the principle of Baker's law). Thus, whether dioecious lineages arrived as such (with colonization aided by leakiness in gender expression) or whether separate sexes evolved subsequent to colonization (from colonizers that lacked self-incompatibility; see Box 1), Baker (1967) did not agree that the high incidence of dioecy on islands provided a counter-example to Baker's law. Critically, he expected the colonizers to have maintained or evolved outcrossing mechanisms, not a selfing syndrome. The interesting tension that results from selection for an ability to self-fertilize in a species otherwise selected for outcrossing is well illustrated by the maintenance of males with hermaphrodites in androdioecious metapopulations (see Box 4).

Box 4 Selection of an ability to self-fertilize in the context of the maintenance of outcrossing: the case of androdioecious metapopulations

The idea that outcrossing may be maintained alongside an *ability* to self-fertilize applies not only to the once-off colonization by a lineage on an oceanic island, but also to the recurrent colonization that takes place in colonizing species that occupy ephemeral habitats, as might be characterized in terms of a metapopulation. In the wind-pollinated ruderal plant *Mercurialis annua*, for example which colonizes disturbed and ephemeral habitats, self-compatible hermaphroditism has evolved from dioecy (Pannell *et al.* 2008). Interestingly, hermaphrodites of this species often co-occur with males in 'androdioecious' populations (Pannell 1997b). It is known from sexual-system theory that androdioecious populations can only be maintained if outcrossing rates are high (because selfing by hermaphrodites would seriously compromise the siring opportunities of males) (Lloyd 1975; Charlesworth & Charlesworth 1978; Charlesworth 1984). Recent estimates of the mating system in wild populations of *M. annua* (G. Korbecka, J. P. David, J. L. García-Castaño· and J. R. Pannell, unpublished results; Fig. 4) have indeed confirmed that outcrossing rates are often high, consistent with this prediction. Yet isolated hermaphrodites of *M. annua* set abundant seed through selfing (Pannell 1997b; Eppley & Pannell 2007a; Hesse & Pannell 2011), allowing them to colonize available habitat as single individuals. Here, it would seem that an ability to self-fertilize has been selected in hermaphrodites, but that males can persist

in the metapopulation by migrating into established populations in which outcrossing rates are in fact high (Pannell 2001).

The metapopulation model for the maintenance of sexual-system variation in *M. annua* seems to apply to other androdioecious species, too (Pannell 2002). The plants *Datisca glomerata* and *Schizopepon bryoniaefolius* are both colonizers of disturbed and ephemeral habitats and comprise males that coexist with self-compatible hermaphrodites populations that are largely outcrossing (Fritsch & Rieseberg 1992; Akimoto *et al.* 1999). A particularly interesting example is provided by the Branchiopod crustacean genus *Eulimnadia*, in which self-compatible hermaphrodites have evolved from gonochoristic (i.e. dioecious) ancestors more than once (Sassaman & Weeks 1985; Weeks *et al.* 2006). Indeed, Baker was inspired to write his original 1955 study by reading about the evolution of hermaphroditism from gonochory in freshwater crustaceans. Here, the animals colonize ephemeral freshwater ponds that frequently dry out during drought. Their ability to self-fertilize almost certainly improves their colonization abilities, but they tend preferentially to outcross with males when males are present (Hollenbeck *et al.* 2002). This example is especially revealing for the current discussion, because hermaphrodites in populations that lack males are incapable of outcrossing at all and thus display complete self-fertilization.

SINGLE-EVENT VERSUS RECURRENT LONG-DISTANCE COLONIZATION

Just as a plant's mating system might encompass both its ability to self-fertilize and its selfing rate, so the term 'dispersal' can refer to different phenomena. For instance, dispersal scenarios can vary in terms of the frequency or probability with which organisms disperse from their natal patch, the mean distance over which they do so, the full distribution of dispersal distances, and the composition of the dispersing propagule or group of individuals (Ronce *et al.* 2001). This diversity of perspectives on dispersal is also relevant for discussions of models of colonization. To what extent does selection acting on dispersal traits influence long-distance colonization? Can traits for long-distance colonization be selected directly? Despite substantial effort spent in modelling the evolution of dispersal and,

to some extent, its links with the mating system, there is still a dearth of theory on selection directly on colonization, and there is surely scope for more work here. In a stimulating article, Ronce *et al.* (2001) laid out a wide range of questions that still need to be addressed, some of which are directly relevant to colonization. A particularly important question, for example, concerns the extent to which selection favouring increased dispersal at short spatial scales will affect the frequency and success of long-distance dispersal. Another question concerns whether rare long-distance dispersal, of the sort that establishes new species on oceanic islands, can be related to, and even modelled by, recurrent dispersal in a metapopulation (Ronce *et al.* 2001).

From a conceptual point of view, there are at least two factors that distinguish dispersal that leads to colonization from dispersal over short distances. First, an important facet of dispersal is that it leads to an immediate rupture from genetic exchange and any possibilities of mating with individuals that have not dispersed. Depending on the organisms involved, such a rupture could take place over distances that are, in absolute terms, quite small. And second, because dispersal kernels are inevitably leptokurtic, with long thin tails, long-distance dispersal will be rare (Nathan & Muller-Landau 2000). Thus, we might expect selection to act overwhelmingly on the basis of the fate of short-distance events.

The importance of a continuing connection to a gene pool under selection is nicely illustrated by the contrast between the expected evolution of dispersal traits on oceanic islands and metapopulations. On oceanic islands, selection on dispersal is, in a sense, entirely negative: because dispersers leave the island and are unlikely to return (even as descendants of the dispersers), oceanic island populations should lose the very dispersal traits that allowed them to colonize the island when they first arrived. This is presumably one reason for the evolution of flightlessness in many animal island inhabitants (Darwin 1859; Slikas *et al.* 2002), and the evolution of traits associated with reduced seed dispersal in island plants (Carlquist 1966; Cody & Overton 1996). In the context of a subdivided population with colonization of empty habitat patches, this notion has been labelled the 'metapopulation effect' (Olivieri *et al.* 1997; Ronce & Olivieri 1997): it explains why the initial colonists of a patch are likely to possess dispersal traits, and why, over time, patch residents gradually lose those traits. The observed decline with patch age in the proportion of dispersing achenes in Asteraceae inflorescences illustrates this phenomenon well (Olivieri *et al.* 1990). The predicted and observed evolution of dispersal traits following colonization of both oceanic islands and demes or patches in a metapopulation show that models that seek a single dispersal rate will mislead us. Significantly, just as we might find that populations established via long-distance colonization will end up as (almost) obligate outcrossers with a (residual) ability to self-fertilize, so traits that allow colonization by long-distance dispersal may soon become lost.

Because some mating system and dispersal traits will evolve more quickly after colonization than others, careful thought is required when choosing traits for tests of Baker's law. The most revealing trait variation is probably that which Baker (1955) first focussed on: self-incompatibility and its loss. First, although SI can be leaky, typically it is strong, and SI individuals are thus likely to be poor colonizers as single individuals in comparison with SC individuals. And second, the loss of SI tends to be definitive; once lost, SI systems are very unlikely to re-evolve (Igic *et al.* 2006; Goldberg & Igic 2012). This means that subsequent evolution of reproductive traits will not obscure the relationship predicted by Baker's law, irrespective of whether a syndrome of near complete selfing evolves (e.g. with the loss of inbreeding depression), or whether some other mechanism of outcrossing evolves (e.g. dioecy). The frequency of self-compatibility among colonizers relative to their noncolonizer relatives should therefore be a clearer indication of a selective sieve for a selfing ability than measures of the mating system (i.e. selfing or outcrossing rates).

Although metapopulation models seem able to reflect certain aspects of once-off colonization of islands (the selective advantage of a selfing ability is one of them; the metapopulation effect and the evolution of reduced dispersal on islands is another), they also differ in an important respect: a genetic connection is potentially maintained between the demes of a metapopulation by migration, whereas oceanic island colonization disrupts all ongoing connection with other populations. The connection among demes of a metapopulation by migrants effectively allows selection to act on individuals or strategies at the metapopulation level, whereas selection on newly established populations on oceanic islands will be able to act only on the island. This difference will not always be critical: the selective advantage of a selfing ability seems to apply to both scenarios, and the metapopulation effect, cited above, is to some extent analogous to the loss of dispersal traits on oceanic islands. However, from other points of view the different probably matters, and metapopulation models will then be poor analogues for once-off dispersal. This probably applies to the joint selection of the mating system and dispersal, because its outcome is contingent on the structure and dynamics of the metapopulation as a whole. The evolution of conditional altruism, phenotypic plasticity and sex allocation would be other cases in which it is probably important to distinguish between single-event and recurrent dispersal and colonization. In the present context, the evolution of sex allocation is of particular note, because sex allocation is known to be selected differently in outcrossing vs. inbreeding species, and because colonization gives rise to high levels of inbreeding at the metapopulation level, even where mating in each local population is random (see Box 5).

Box 5 Evolution of sex allocation under the influence of recurrent colonization in a metapopulation

The genetic bottlenecks brought about by colonization in a metapopulation cause recurrent bouts of local inbreeding, even in otherwise outcrossing species. Hamilton (1967) first showed that selection under inbreeding will favour female-biased sex allocation, because of 'local mate competition', a phenomenon that explains the low polen:ovule ratios in selfing populations of hermaphroditic plants (Cruden 1977). Theory thus predicts a positive association between the inbreeding coefficient within populations, typically F_{IS}, and the proportion of reproductive resources allocated to female function (West 2009). As we have discussed, although metapopulation dynamics are expected to favour a capacity for self-fertilization, outcrossing within populations following colonization is often likely to be high; in this case, the local inbreeding coefficient F_{IS} will on average tend to be close to zero. But inbreeding may nevertheless continue to be high across the metapopulation if most mating takes place among the descendants of the single (or few) colonists of each deme. This type of metapopulation-wide inbreeding is measured as F_{ST}.

It is well known that population turnover in a metapopulation gives rise to elevated F_{ST} (Wade & McCauley 1988; Pannell & Charlesworth 2000) – unless demes are colonized by more than one colonist drawn from more than one source deme (Slatkin 1977; Whitlock & McCauley 1990). The relationship between F_{ST} and sex allocation in a metapopulation has to my knowledge hitherto not been explored, but simulations investigating the evolution of sex allocation in hermaphroditic metapopulations have indeed found the expected relationship (J.R. Pannell and C. Roux, unpublished): population turnover selects for female-biased sex allocation across the metapopulation, the level of which is well predicted by F_{ST} (but not by F_{IS}; see Fig. 7). This is of course not something we would expect to find in the context of oceanic island colonization, because selection on the sex allocation of the colonized population will occur only locally. We should thus be cautious when using metapopulation models to explore processes that occur at spatial and/or temporal scales that do not correspond.

PERSPECTIVES FROM 'THE GENETICS OF COLONIZING SPECIES'

Before concluding, it is worth reflecting on the extent to which the ideas presented and discussed above were featured in the chapters of 'The Genetics of Colonising Species' and in the discussions that took place at Asilomar in 1964, which the current volume is commemorating. Both Baker and Stebbins, the book's editors, contributed chapters to the book. Although neither chapter focused specifically on questions concerning Baker's law, both authors made comments and observations relevant to it. Baker (1965) devoted his chapter to comparisons of the characteristics of ruderals and 'agrastals' (weeds that colonize disturbed waste places and agricultural fields, respectively), with several examples taken from the Asteraceae, a large family comprising both species with strong sporophytic self-incompatibility and highly self-fertile species. He reported that the evolution of weediness in this family (the ability to be a successful colonizer of disturbed, ephemeral habitats) frequently coincided with a shift away from self-incompatible perenniality towards a life history typified by rapid development, rapid flowering,

increased plasticity and self-compatibility. In describing the colonizing weed *Ageratum conysoides*, it is noteworthy that he typified its breeding system as 'thoroughly self-compatible, even self-fertilizing', revealing Baker's view that self-compatibility was the critical trait, not necessarily a syndrome of self-fertilization, common though that appeared to be in many weeds, too. Yet, later in his chapter (p. 165), Baker (1965) stated 'self-pollination or even apomixis is likely to be important for establishment after long-distance dispersal'. We see here, in his reference to 'self-pollination' (rather than a capacity to self-fertilize), germs of the confusion that has plagued recent discussion of Baker's law.

In his chapter, Stebbins (1965) in turn directed his attention towards an analysis of the weeds of the Californian flora, searching for generalizations that might be made about what makes certain species successful colonizers of human-disturbed habitats. On the basis of his survey, in which he found an equal frequency of self-fertility among annual weeds as in their nonweedy annual relatives, Stebbins (1965) concluded (p. 181) that 'no particular type of mating system or chromosomal condition is either necessary or generally favourable for preadapting a group of species to evolve in the direction of weediness'. Here, Stebbins distinguished between annual and perennial species, arguing that 'the most adaptive condition for an annual is self-fertilization and in a perennial obligate outcrossing is most often favoured by selection'. The importance of distinguishing between annual and perennial colonizers is a theme running through several of the other chapters of the volume. For instance, Ehrendorfer (1965) felt that it is sensible to look for the hallmarks of colonizing species particularly in annuals, notably because a perennial habit removes some of the urgency with which plants must self-fertilize to reproduce. In an interesting analysis of variation in dispersal, chromosomal and breeding system traits in Mediterranean Dipsacaceae species, he picked up the familiar theme regarding the importance of an ability to self-fertilize. Thus, Ehrendorfer (1965) noted that in selfing species, 'seed production is safeguarded by autogamy making possible reproduction even of single founder individuals'. In predominantly outbreeding species, he observed, 'self-incompatibility very rarely seems to be complete, so that at least occasionally self-fertilization is nearly always possible' (p. 348–349).

The flexibility shown by plants in their reproductive systems was discussed at some length by delegates of the Asilomar meeting, not only the ability of outcrossers to set occasional seed, as just noted, but also the tendency of habitual selfers towards occasional outcrossing. Allard (1965) noted, for example, that 'although barley is commonly considered to be one of the most highly self-pollinated among the cereal grasses, [...] some outcrossing appears almost every generation' – as is the case for many other selfing species, including the much studied *Arabidopsis thaliana* (Abbott & Gomes 1989; Platt *et al.* 2010). Although this mixed strategy seems unlikely to have evolved in response to selection within populations and is more likely a residue from a history of greater outcrossing prior to a transition to selfing, Allard (1965) argued that it allowed colonizers to enjoy the best of both worlds. On the one hand, predominant selfing conferred not only the advantage of reproductive assurance, but also benefits of high homozygosity, which allows 'the perpetuation of the presently best adapted genotypes for various specific microenvironments'; on the other hand, occasional outcrossing and recombination among the otherwise selfing populations should maintain substantial variation among subpopulations that inhabit different microenvironments (Allard

1965). The idea that selfing allows adaptation to local microsites by avoiding the contaminating effects of massive gene flow from nonadapted populations is perhaps one reason for which selfing lineages have been able to expand their species' range limits to end up with geographically marginal distributions.

Finally, although expressed in very general terms, the model presented by Lewontin (1965) provides a helpful perspective for discussing the evolution of the mating system in colonizing species and conceptually formalizes some of the points made in the preceding sections here. Lewontin (1965, p. 78) made the valuable point that instead of referring to 'colonizing species', it would be more generally interesting to consider 'colonizing episodes' for any species. Ultimately, as Lewontin (1965) observed, all species are colonizers – in the sense that, at some point, they will have colonized the site they currently occupy. But roadside weeds are colonizers in a sense that giant redwood forests are not, because the former represent species 'whose entire life pattern is one of colonization', characterized by relative time spent in a phase of density-independent (or log-phase) growth, whereas the life pattern of the latter is characterized by demographic stability (at a density-dependent carrying capacity). Lewontin's (1965) emphasis on the relative importance of colonizing episodes brings to mind Sewell Wright's cartoon (Fig. 6) of a subdivided population with frequent population turnover, which we would now recognize as a metapopulation.

In establishing his case for the importance of considering colonizing episodes, Lewontin (1965) drew attention to three aspects of selection that might characterize colonizers. First, selection shaping colonizers should take place predominantly during the exponential colonization phase, that is from the point of colonization to the point where population density begins to slow growth. This is where an ability to self-fertilize might be favoured, where inbreeding depression might be purged, and it is where selection on traits such as sex allocation will tend to differ most from that expected in populations at (perhaps outcrossing) demographic stability. Second, selection during the growth phase may actually be in a direction opposite to that on a population at demographic equilibrium. For example, while an ability to self-fertilize (and a female-biased sex allocation, Fig. 7) might be favoured during colony growth, outcrossing (and an equal sex allocation) might subsequently be favoured. And third, it is the peculiarity of differential local population growth after

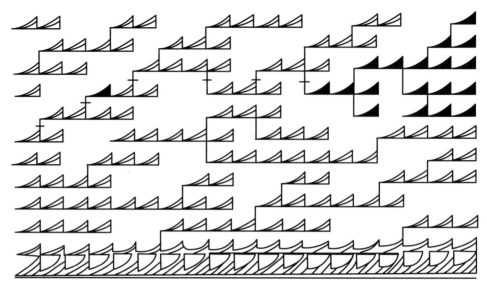

Fig. 6 Sewell Wright's (1940) cartoon depicting the dynamics of a subdivided population with frequent extinctions and recolonizations. Each row of 'saw-teeth' represents the dynamics of the subpopulation (deme) occupying a single habitat patch; the horizontal axis represents the passage of time; and the vertical axis represents population size for each deme. The cartoon emphasizes the fact that, for most of the time, lineages will find themselves in a population during density-independent (log-phase) growth. Wright's sketch was intended to illustrate the effect of population turnover on a species' effective population size. However, it also illustrates the repeated phases during which selection on the mating system and sex allocation will differ from that in populations maintained at carrying capacity. Reproduced with permission.

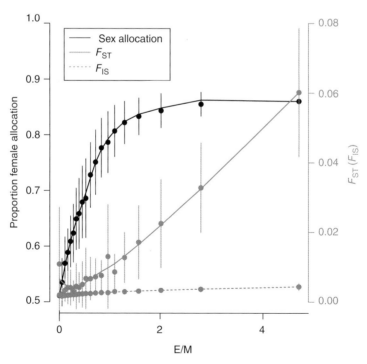

Fig. 7 The effect of the ratio of population turnover (the extinction rate, E) to propagule pressure (the migration rate, M) on the selection of female-biased sex allocation in a metapopulation (left axis) and on the corresponding values of F_{ST} and F_{IS} (right axis). Points and bars are the means and standard deviations over 3000 independent simulations of each scenario; lines are the loess regression fits for each relationship. We assumed a metapopulation of 200 demes, each with a carrying capacity of 500 individuals. Migration and colonization followed an island model. We assumed that extinction was followed immediately by recolonization by a single individual, with migration into extant patches that varied among simulations. From Pannell and Roux (J.R. Pannell and C. Roux, unpublished).

colonization that makes most likely the possibility of interdeme selection. The role of group selection in the evolution of metapopulations has been controversial, but similarities between kin selection in subdivided populations (put into terms of relatedness coefficients) and group selection render this controversy partially semantic (Grafen 1984; West *et al.* 2007).

Lewontin's (1965) analysis, although highlighting the fact that colonizing species will typically not remain in a colonizing phase for long, captures the essence of the implications of demographic fluctuations through a single colonization event, but it misses the importance of gene flow among demes of a metapopulation. We may retain the useful perspective of two phases, but incorporation of the notion of migration splits them somewhat differently. In a metapopulation of demes connected by gene flow, colonization establishes a pattern of tight genetic relationships among individuals that will persist during population growth (and potentially continue into Lewontin's phase of density-dependent stasis). It is thus not demographic stasis that ends the phase, but rather the erosion of patterns of relatedness by migration from other demes. Ultimately, the extent to which population turnover will matter in a genetic sense must depend of the rate of extinction and colonizations relative to that of migration. This intuition is supported by formal metapopulation models (Slatkin 1977; Pannell & Charlesworth 1999).

CONCLUDING REMARKS

The overwhelming feeling one gets from the literature on the evolution of reproductive and dispersal traits in colonizing species is one of cryptic complexity. Terms such as colonization, self-fertilization and dispersal roll easily off the tongue (or pen), but each of these terms encompasses a hazardously broad range of possible meanings. The simplest of them is perhaps colonization, or what we mean by a colonizing species. Lewontin (1965) made a helpful observation that what probably matters most in discussions about the evolution of traits of colonizing species is the extent to which such species actually find themselves in a state of colonization (the colonization bottleneck and the period of log-phase growth) as opposed to a state of demographic status following colonization. This perspective is heuristically useful both for repeated colonization in metapopulations or range expansions as well as for single-event colonization of oceanic islands, as it prompts

the question: how much trait evolution might have occurred to erase the mark left by colonization? The answer will depend on the trait involved. The loss of self-incompatibility is unlikely to be reversed, whereas species on islands may evolve both reduced dispersal and increased outcrossing following their arrival.

In this study, I have stressed importance of clarifying what we mean by the mating system in the context of colonizing species, but the same applies to dispersal. The complexity of the mating system is due largely to the fact that plants are often plastic in their sex expression, so that outcrossers and selfers may self or outcross, respectively, from time to time. It is also due to the fact that outcrossing rates are often density-dependent, so that the mating system during the colonization phase may differ from that in established populations. An ability to self-fertilize is likely to be a substantial advantage to many species establishing new populations following long-distance dispersal, as noted by Baker (1955), but this does not imply an association between colonization and a mating system characterized by high selfing rates generally.

Dispersal, too, is a term loaded with complexity that encompasses a distribution of distances moved as well as rates of proportions of individuals that leave their natal patch. Critically, there is no necessarily clear link between dispersal and colonization rates or distances. Explorations of the links between dispersal and the mating system have been valuable, not least in exposing possible reasons for which organisms have evolved to take on the risk of leaving the safety of their natal patch. However, the sort of dispersal that leads to long-distance colonization may be an incidental outcome of selection for dispersal over much smaller distances, so we may ask how relevant models for the evolution of dispersal rates are for colonization; this relationship deserves attention. Stebbins' (1957) view, in coining Baker's law, was that colonization is essentially 'accidental'. This would suggest that species that have evolved towards a syndrome of low dispersal may, occasionally (but significantly), be subject to the accidents of colonization that establish the patterns Baker (1955) sought to explain. Of course, many plants possess no clear dispersal mechanisms, but they may nevertheless be good colonizers.

Despite the complexities of mating and dispersal, new insights have been gained by models that consider their joint evolution, but, to date, no models address questions concerning dispersal as part of a process of colonization through genetic and demographic bottlenecks. This would seem to be an obvious direction for future theoretical work. Another obvious direction to take would be the incorporation of dynamic inbreeding depression into such models. Ultimately, though, the crucial questions are empirical. Despite substantial effort, we still have a poor understanding of the distribution of reproductive and dispersal traits in colonizing species, and further tests based on reliable data will be valuable. Nevertheless, Baker's (1955) observation of a paucity of self-incompatible species in island situations would seem to have withstood the test of time so far (Igic & Busch 2013).

ACKNOWLEDGEMENTS

I am grateful to the organizers of the Asilomar meeting for inviting me to attend, to my collaborators F. Encinas-Viso, A. Young, C. Roux, G. Korbecka and P. David for permission to include our unpublished data, to the members of the NESCent working group on Baker's law for stimulating discussions, and to the Swiss National Science Foundation and the University of Lausanne for funding.

REFERENCES

Abbott RJ, Forbes DG (1993) Outcrossing rate and self-incompatibility in the colonising species *Senecio squalidus*. *Heredity*, **71**, 155–159.

Abbott RJ, Gomes MF (1989) Population genetic structure and outcrossing rate of *Arabidopsis thaliana* (L) Heynh. *Heredity*, **62**, 411–418.

Akimoto J, Fukuhara T, Kikuzawa K (1999) Sex ratios and genetic variation in a functionally androdioecious species, *Schizopepon bryoniaefolius* (Cucurbitaceae). *American Journal of Botany*, **86**, 880–886.

Allard RW (1965) Genetic systems associated with colonizing ability in predominantly self-pollinated species. In: *The Genetics of Colonizing Species* (eds Baker HG, Stebbins GL), pp. 49–75. Academic Press, London.

Allee WC, Emersen AE, Park O, Park T, Schmidt KP (1949) *Principles of Animal Ecology*. W.B. Saunders Company, Philadelphia.

Auld JR, de Casas RR (2013) The correlated evolution of dispersal and mating-system traits. *Evolutionary Biology*, **40**, 185–193.

Baker HG (1955) Self-compatibility and establishment after "long-distance" dispersal. *Evolution*, **9**, 347–348.

Baker HG (1965) Characteristics and modes of origin of weeds. In: *The Genetics of Colonising Species* (eds Baker HG, Stebbins GL), pp. 141–172. Academic Press, London.

Baker HG (1967) Support for Baker's Law - as a rule. *Evolution*, **21**, 853–856.

Baker HG, Stebbins GL (eds.) (1965) *The Genetics of Colonizing Species*. Academic Press, London.

Barrett SCH (1988) The evolution, maintenance, and loss of self-incompatibility systems. In: *Plant Reproductive Ecology* (ed. Lovett Doust J), pp. 98–124. Oxford University Press, Oxford.

Barrett SCH (1996) The reproductive biology and genetics of island plants. *Philosophical Transactions of the Royal Society of London Series B-Biological Sciences*, **351**, 725–733.

Barrett SCH (2002) The evolution of plant sexual diversity. *Nature Reviews Genetics*, **3**, 274–284.

Barrett SCH (2011) Why reproductive systems matter for the invasion biology of plants. In: *Fifty Years of Invasion Biology: The Legacy of Charles Elton* (ed. Richardson DM), pp. 175–193. Wiley-Blackwell, Oxford.

Barrett SCH, Husband BC (1990) Variation in outcrossing rates in *Eichhornia paniculata*: the role of demographic and reproductive factors. *Plant Species Biology*, **5**, 41–55.

Barrett SCH, Pannell JR (1999) Metapopulation dynamics and mating-system evolution in plants. In: *Molecular Systematics and Plant Evolution* (eds Hollingsworth P, Bateman R, Gornall R), pp. 74–100. Chapman and Hall, London.

Bernardello G, Anderson GJ, Stuessy TF, Crawford DJ (2006) The angiosperm flora of the Archipelago Juan Fernandez (Chile): origin and dispersal. *Canadian Journal of Botany-Revue Canadienne De Botanique*, **84**, 1266–1281.

Brennan AC, Harris SA, Hiscock SJ (2003) The population genetics of sporophytic self-incompatibility in *Senecio squalidus* L. (Asteraceae): avoidance of mating constraints imposed by low S-allele number. *Philosophical Transactions of the Royal Society of London Series B-Biological Sciences*, **358**, 1047–1050.

Brennan AC, Harris SA, Hiscock SJ (2006) The population genetics of sporophytic self-incompatibility in *Senecio squalidus* L. (Asteraceae): the number, frequency, and dominance interactions of S alleles across its British range. *Evolution*, **60**, 213–224.

Brennan AC, Harris SA, Hiscock SJ (2013) The population genetics of sporophytic self-incompatibility in three hybridizing *Senecio* (Asteraceae) species with contrasting population histories. *Evolution*, **67**, 1347–1367.

Busch JW (2011) Demography, pollination and Baker's law. *Evolution*, **65**, 1511–1513.

Busch JW, Delph LF (2012) The relative importance of reproductive assurance and automatic selection as hypotheses for the evolution of self-fertilization. *Annals of Botany*, **109**, 553–562.

Busch JW, Schoen DJ (2008) The evolution of self-incompatibility when mates are limiting. *Trends in Plant Science*, **13**, 128–136.

Byers DL, Waller DM (1999) Do plant populations purge their genetic load? Effects of population size and mating history on inbreeding depression. *Annual Review of Ecology and Systematics*, **30**, 479–513.

Carlquist S (1966) The biota of long distance dispersal. *Evolution*, **20**, 30–48.

Carr GD, Powell EA, Kyhos DW (1986) Self-incompatibility in the Hawaiin Madiinae (Compositae): an exception to Baker's Rule. *Evolution*, **40**, 430–434.

Charlesworth D (1984) Androdioecy and the evolution of dioecy. *Biological Journal of the Linnean Society*, **23**, 333–348.

Charlesworth D, Charlesworth B (1978) A model for the evolution of dioecy and gynodioecy. *The American Naturalist*, **112**, 975–997.

Charlesworth D, Charlesworth B (1987) Inbreeding depression and its evolutionary consequences. *Annual Review of Ecology and Systematics*, **18**, 273–288.

Charlesworth D, Willis JH (2009) The genetics of inbreeding depression. *Nature Reviews Genetics*, **10**, 783–796.

Charlesworth D, Morgan MT, Charlesworth B (1990) Inbreeding depression, genetic load, and the evolution of outcrossing rates in a multilocus system with no linkage. *Evolution*, **44**, 1469–1489.

Cheptou PO, Massol F (2009) Pollination fluctuations drive evolutionary syndromes linking dispersal and mating system. *The American Naturalist*, **174**, 46–55.

Clobert J, Danchin E, Dhondt AA, Nichols JD (eds) (2001) *Dispersal*. Oxford University Press, Oxford.

Cody ML, Overton JM (1996) Short-term evolution of reduced dispersal in island plant populations. *Journal of Ecology*, **84**, 53–61.

Crnokrak P, Barrett SCH (2002) Purging the genetic load: a review of the experimental evidence. *Evolution*, **56**, 2347–2358.

Cruden RW (1977) Pollen-ovule ratios: a conservative index of breeding systems in flowering plants. *Evolution*, **31**, 32–46.

Darwin C (1859) *The Origin of Species*, 1st edn. Murray, London.

Darwin C (1876) *The Effects of Cross- and Self-fertilization in the Vegetable Kingdom*. John Murray, London.

Delph LF (2003) Sexual dimorphism in gender plasticity and its consequences for breeding system evolution. *Evolution and Development*, **5**, 34–39.

Delph LF, Wolf DE (2005) Evolutionary consequences of gender plasticity in genetically dimorphic breeding systems. *New Phytologist*, **166**, 119–128.

Dorken ME, Pannell JR (2008) Density-dependent regulation of the sex ratio in an annual plant. *The American Naturalist*, **171**, 824–830.

Dornier A, Munoz F, Cheptou P-O (2008) Allee effect and self-fertilization in hermaphrodites: reproductive assurance in a structured metapopulation. *Evolution*, **62**, 2558–2569.

Eckert CG, Samis KE, Lougheed SC (2008) Genetic variation across species' geographical ranges: the central-marginal hypothesis and beyond. *Molecular Ecology*, **17**, 1170–1188.

Ehlers BK, Bataillon T (2007) 'Inconstant males' and the maintenance of labile sex expression in subdioecious plants. *New Phytologist*, **174**, 194–211.

Ehrendorfer F (1965) Dispersal mechanisms, genetic systems, and colonizing abilities in some flowering plant families. In: *The Genetics of Colonising Species* (eds Baker HG, Stebbins GL), pp. 331–351. Academic Press, London.

Elam DR, Ridley CE, Goodell K, Ellstrandt NC (2007) Population size and relatedness affect fitness of a self-incompatible invasive plant. *Proceedings of the National Academy of Sciences of the United States of America*, **104**, 549–552.

Eppley SM, Pannell JR (2007a) Density-dependent self-fertilization and male versus hermaphrodite siring success in an androdioecious plant. *Evolution*, **61**, 2349–2359.

Eppley SM, Pannell JR (2007b) Sexual systems and measures of occupancy and abundance in an annual plant: testing the metapopulation model. *The American Naturalist*, **169**, 20–28.

Fausto JA, Eckhart VM, Geber MA (2001) Reproductive assurance and the evolutionary ecology of self-pollination in *Clarkia xantiana* (Onagraceae). *American Journal of Botany*, **88**, 1794–1800.

Fisher RA (1941) Average excess and average effect of a gene substitution. *Annals of Eugenics*, **11**, 53–63.

Foxe JP, Slotte T, Stahl E, Neuffer B, Hurka H, Wright SI (2009) Recent speciation associated with the evolution of selfing in *Capsella*. *Proceeding of the National Academy of Sciences of the United States of America*, **106**, 5241–5245.

Fritsch P, Rieseberg LH (1992) High outcrossing rates maintain male and hermaphrodite individuals in populations of the flowering plant *Datisca glomerata*. *Nature*, **359**, 633–636.

Goldberg EE, Igic B (2012) Tempo and mode in plant breeding system evolution. *Evolution*, **66**, 3701–3709.

Goodwillie C, Kalisz S, Eckert CG (2005) The evolutionary enigma of mixed mating systems in plants: occurrence, theoretical explanations, and empirical evidence. *Annual Review of Ecology Evolution and Systematics*, **36**, 47–79.

Grafen A (1984) Natural selection, kin selection and group selection. In: *Behavioral Ecology: An Evolutionary Approach* (eds Krebs JR & Davies NB), pp. 62–84. Blackwell Scientific, Oxford.

Guo YL, Bechsgaard JS, Slotte T *et al.* (2009) Recent speciation of *Capsella rubella* from *Capsella grandiflora*, associated with loss of self-incompatibility and an extreme bottleneck. *Proceedings of the National Academy of Sciences of the United States of America*, **106**, 5246–5251.

Hamilton WD (1967) Extraordinary sex ratios. *Science*, **156**, 477–488.

Harder LD, Barrett SCH (eds.) (2006) *Ecology and Evolution of Flowers*. Oxford University Press, Oxford.

Herlihy CR, Eckert CG (2002) Genetic cost of reproductive assurance in a self-fertilizing plant. *Nature*, **416**, 320–323.

Hesse E, Pannell JR (2011) Density-dependent pollen limitation and reproductive assurance in a wind-pollinated herb with contrasting sexual systems. *Journal of Ecology*, **99**, 1531–1539.

Hiscock SJ, McInnis SM (2003) The diversity of self-incompatibility systems in flowering plants. *Plant Biology*, **5**, 23–32.

Hollenbeck VG, Weeks SC, Gould W, Zucker N (2002) Maintenance of androdioecy in the freshwater shrimp *Eulimnadia texana*: sexual encounter rates and outcrossing success. *Behavioral Ecology*, **13**, 561–570.

Igic B, Busch JW (2013) Is self-fertilization an evolutionary dead end? *New Phytologist*, **198**, 386–397.

Igic B, Bohs L, Kohn JR (2006) Ancient polymorphism reveals unidirectional breeding system shifts. *Proceedings of the National Academy of Sciences of the United States of America*, **103**, 1359–1363.

Ison JL, Wagenius S (2014) Both flowering time and distance to conspecific plants affect reproduction in *Echinacea angustifolia*, a common prairie perennial. *Journal of Ecology*, **102**, 920–929.

Kirkpatrick M, Barton NH (1997) Evolution of a species' range. *The American Naturalist*, **150**, 1–23.

Lande R, Schemske DW (1985) The evolution of self-fertilization and inbreeding depression in plants: I. Genetic models. *Evolution*, **39**, 24–40.

Leducq JB, Gosset CC, Poiret M, Hendoux F, Vekemans X, Billiard S (2010) An experimental study of the S-Allee effect in the self-incompatible plant *Biscutella neustriaca*. *Conservation Genetics*, **11**, 497–508.

Levin DA (1996) The evolutionary significance of pseudo-self-fertility. *The American Naturalist*, **148**, 321–332.

Lewontin RC (1965) Selection for colonizing ability. In: *The Genetics of Colonizing Species* (eds Baker HG, Stebbins GL), pp. 77–91. Academic Press, London.

Lloyd DG (1975) The maintenance of gynodioecy and androdioecy in angiosperms. *Genetica*, **45**, 325–339.

Lloyd DG (1979) Some reproductive factors affecting the selection of self-fertilization in plants. *The American Naturalist*, **113**, 67–79.

Lloyd DG (1992) Self- and cross-fertilization in plants. II. The selection of self-fertilization. *International Journal of Plant Science*, **153**, 370–380.

Lloyd DG, Bawa KS (1984) Modification of the gender of seed plants in varying conditions. *Evolutionary Biology*, **17**, 255–338.

Longhurst AR (1955) Evolution in the Notostraca. *Evolution*, **9**, 84–86.

Mackiewicz M, Tatarenkov A, Taylor DS, Turner BJ, Avise JC (2006) Extensive outcrossing and androdioecy in a vertebrate species that otherwise reproduces as a self-fertilizing hermaphrodite. *Proceedings of the National Academy of Sciences of the United States of America*, **103**, 9924–9928.

Massol F, Cheptou P-O (2011a) Evolutionary syndromes linking dispersal and mating system: the effect of autocorrelation in pollination conditions. *Evolution*, **65**, 591–598.

Massol F, Cheptou P-O (2011b) When should we expect the evolutionary association of self-fertilization and dispersal? *Evolution*, **65**, 1217–1220.

McMullen CK (1987) Breeding systems of selected Galapagos-island angiosperms. *American Journal of Botany*, **74**, 1694–1705.

Moeller DA, Geber MA (2005) Ecological context of the evolution of self-pollination in *Clarkia xantiana*: population size, plant communities, and reproductive assurance. *Evolution*, **59**, 786–799.

Nathan R, Muller-Landau HC (2000) Spatial patterns of seed dispersal, their determinants and consequences for recruitment. *Trends in Ecology & Evolution*, **15**, 278–285.

Olivieri I, Couvet D, Gouyon PH (1990) The genetics of transient populations: research at the metapopulation level. *Trends in Ecology and Evolution*, **5**, 207–210.

Olivieri I, Gouyon PH, Gilpin ME (1997) Evolution of migration rate and other traits: the metapopulation effect. In: *Metapopulation Biology: Ecology, Genetics, and Evolution* (ed. Hanski I), pp. 293–323. Academic Press, San Diego.

Pannell J (1997a) The maintenance of gynodioecy and androdioecy in a metapopulation. *Evolution*, **51**, 10–20.

Pannell J (1997b) Widespread functional androdioecy in *Mercurialis annua* L. (Euphorbiaceae). *Biological Journal of the Linnean Society*, **61**, 95–116.

Pannell JR (2001) A hypothesis for the evolution of androdioecy: the joint influence of reproductive assurance and local mate competition in a metapopulation. *Evolutionary Ecology*, **14**, 195–211.

Pannell JR (2002) The evolution and maintenance of androdioecy. *Annual Review of Ecology and Systematics*, **33**, 397–425.

Pannell JR, Barrett SCH (1998) Baker's Law revisited: reproductive assurance in a metapopulation. *Evolution*, **52**, 657–668.

Pannell JR, Barrett SCH (2001) Effects of population size and metapopulation dynamics on a mating system polymorphism. *Theoretical Population Biology*, **59**, 145–155.

Pannell JR, Charlesworth B (1999) Neutral genetic diversity in a metapopulation with recurrent local extinction and recolonization. *Evolution*, **53**, 664–676.

Pannell JR, Charlesworth B (2000) Effects of metapopulation processes on measures of genetic diversity. *Philosophical Transactions of the Royal Society of London Series B-Biological Sciences*, **355**, 1851–1864.

Pannell JR, Dorken ME, Pujol B, Berjano R (2008) Gender variation and transitions between sexual systems in *Mercurialis annua* (Euphorbiaceae). *International Journal of Plant Sciences*, **169**, 129–139.

Peischl S, Dupanloup I, Kirkpatrick M, Excoffier L (2013) On the accumulation of deleterious mutations during range expansions. *Molecular Ecology*, **22**, 5972–5982.

Perrin N, Goudet J (2001) Inbreeding, kinship, and the evolution of natal dispersal. In: *Dispersal* (eds Clobert J, Danchin E, Dhondt AA, Nichols JD), pp. 123–142. Oxford University Press, Oxford.

Petit RJ, Brewer S, Bordacs S *et al.* (2002) Identification of refugia and post-glacial colonisation routes of European

Evolution of the mating system in colonizing plants 79

white oaks based on chloroplast DNA and fossil pollen evidence. *Forest Ecology and Management*, **156**, 49–74.

Platt A, Horton M, Huang YS *et al.* (2010) The scale of population structure in *Arabidopsis thaliana*. *Plos Genetics*, **6**, e1000843.

Porcher E, Lande R (2005) The evolution of self-fertilization and inbreeding depression under pollen discounting and pollen limitation. *Journal of Evolutionary Biology*, **18**, 497–508.

Pujol B, Pannell JR (2008) Reduced responses to selection after species range expansion. *Science*, **321**, 96.

Pujol B, Zhou SR, Sahchez-Vilas J, Pannell JR (2009) Reduced inbreeding depression after species range expansion. *Proceeding of the National Academy of Sciences of the United States of America*, **106**, 15379–15383.

Randle AM, Slyder JB, Kalisz S (2009) Can differences in autonomous selfing ability explain differences in range size among sister-taxa pairs of *Collinsia* (Plantaginaceae)? An extension of Baker's Law. *New Phytologist*, **183**, 618–629.

Rodger JG, van Kleunen M, Johnson SD (2013) Pollinators, mates and Allee effects: the importance of self-pollination for fecundity in an invasive lily. *Functional Ecology*, **27**, 1023–1033.

Ronce O, Olivieri I (1997) Evolution of reproductive effort in a metapopulation with local extinctions and ecological succession. *The American Naturalist*, **150**, 220–249.

Ronce O, Olivieri I, Cobert J, Danchin E (2001) Perspectives on the study of dispersal evolution. In: *Dispersal* (eds Clobert J, Danchin E, Dhondt AA, Nichols JD), pp. 123–142. Oxford University Press, Oxford.

Salisbury EJ (1953) A changing flora as shown in the study of weeds of arable lands and waste places. In: *The Changing Flora of Britain* (ed. Lousley JE), pp. 130–139. Botanical Society of the British Isles, Oxford.

Sassaman C, Weeks SC (1985) Adaptation to ephemeral ponds in the Conchostracan *Eulimnadia texana*. *American Zoologist*, **25**, A61.

Schierup MH (1998) The number of self-incompatibility alleles in a finite, subdivided population. *Genetics*, **149**, 1153–1162.

Schoen DJ, Busch JW (2008) On the evolution of self-fertilization in a metapopulation. *International Journal of Plant Sciences*, **169**, 119–127.

Sexton JP, McIntyre PJ, Angert AL, Rice KJ (2009) Evolution and ecology of species range limits. *Annual Review of Ecology, Evolution and Systematics*, **40**, 415–436.

Slatkin M (1977) Gene flow and genetic drift in a species subject to frequent local extinction. *Theoretical Population Biology*, **12**, 253–262.

Slikas B, Olson SL, Fleischer RC (2002) Rapid, independent evolution of flightlessness in four species of Pacific Island rails (Rallidae): an analysis based on mitochondrial sequence data. *Journal of Avian Biology*, **33**, 5–14.

Stebbins GL (1950) *Variation and Evolution in Plants*. Columbia University Press, New York.

Stebbins GL (1957) Self-fertilization and population variability in the higher plants. *The American Naturalist*, **91**, 337–354.

Stebbins GL (1965) Colonizing species of the native California flora. In: *The Genetics of Colonising Species* (eds Baker HG, Stebbins GL), pp. 173–191. Academic Press, London.

Stebbins GL (1974) *Flowering Plants: Evolution above the Species Level*. Harvard University Press, Cambridge, MA.

Stephens PA, Sutherland WJ, Freckleton RP (1999) What is the Allee effect? *Oikos*, **87**, 185–190.

Stephenson AG, Good SV, Vogler DW (2000) Interrelationships among inbreeding depression, plasticity in the self-incompatibility system, and the breeding system of *Campanula rapunculoides* L. (Campanulaceae). *Annals of Botany*, **85**, 211–219.

Takayama S, Isogai A (2005) Self-incompatibility in plants. *Annual Review of Plant Biology*, Annual Reviews, Palo Alto, **56**, 467–489.

Tsuchimatsu T, Suwabe K, Shimizu-Inatsugi R *et al.* (2010) Evolution of self-compatibility in *Arabidopsis* by a mutation in the male specificity gene. *Nature*, **464**, 1342–1346.

Tsuchimatsu T, Kaiser P, Yew C-L, Bachelier JB, Shimizu KK (2012) Recent loss of self-incompatibility by degradation of the male component in allotetraploid *Arabidopsis kamchatica*. *Plos Genetics*, **8**, e1002838.

Uyenoyama MK (1986) Inbreeding and the cost of meiosis: the evolution of selfing in populations practicing biparental inbreeding. *Evolution*, **40**, 388–404.

Uyenoyama MK, Waller DM (1991) Coevolution of self-fertilization and inbreeding depression. 3. Homozygous lethal mutations at multiple loci. *Theoretical Population Biology*, **40**, 173–210.

Uyenoyama MK, Zhang Y, Newbigin E (2001) On the origin of self-incompatibility haplotypes: transition through self-compatible intermediates. *Genetics*, **157**, 1805–1817.

Van Kleunen M, Johnson SD (2007) Effects of self-compatibility on the distribution range of invasive European plants in North America. *Conservation Biology*, **21**, 1537–1544.

Van Kleunen M, Manning JC, Pasqualetto V, Johnson SD (2008) Phylogenetically independent associations between autonomous self-fertilization and plant invasiveness. *The American Naturalist*, **171**, 195–201.

Vekemans X, Schierup MH, Christiansen FB (1998) Mate availability and fecundity selection in multi-allelic self-incompatibility systems in plants. *Evolution*, **52**, 19–29.

Wade MJ, McCauley DE (1988) Extinction and recolonization: their effects on the genetic differentiation of local populations. *Evolution*, **42**, 995–1005.

Wagenius S, Lonsdorf E, Neuhauser C (2007) Patch aging and the S-allee effect: breeding system effects on the demographic response of plants to habitat fragmentation. *The American Naturalist*, **169**, 383–397.

Ward M, Johnson SD, Zalucki MP (2012) Modes of reproduction in three invasive milkweeds are consistent with Baker's Rule. *Biological Invasions*, **14**, 1237–1250.

Weeks SC, Benvenuto C, Reed SK (2006) When males and hermaphrodites coexist: a review of androdioecy in animals. *Integrative and Comparative Biology*, **46**, 449–464.

West SA (2009) *Sex Allocation*. Princeton University Press, Princeton.

West SA, Griffin AS, Gardner A (2007) Social semantics: altruism, cooperation, mutualism, strong reciprocity and group selection. *Journal of Evolutionary Biology*, **20**, 415–432.

Whitlock MC, McCauley DE (1990) Some population genetic consequences of colony formation and extinction: genetic correlations within founding groups. *Evolution*, **44**, 1717–1724.

Wright S (1940) Breeding structure of populations in relation to speciation. *The American Naturalist*, **74**, 232–248.

Wright S (1964) The distribution of self-sterility alleles in populations. *Evolution*, **18**, 609–619.

Wright SI, Kalisz S, Slotte T (2013) Evolutionary consequences of self-fertilization in plants. *Proceedings of the Royal Society B-Biological Sciences*, **280**, 20130133.

Young AG, Pickup M (2010) Low S-allele numbers limit mate availability, reduce seed set and skew fitness in small populations of a self-incompatible plant. *Journal of Applied Ecology*, **47**, 541–548.

Young AG, Brown AHD, Murray BG, Thrall PH, Miller CH (2000) Genetic erosion, restricted mating and reduced viability in fragmented populations of the endangered grassland herb *Rutidosis leptorrhynchoides*. In: *Genetics, Demography and Viability of Fragmented Populations* (eds Young AG, Clarke GM), pp. 335–359. Cambridge University Press, Cambridge.

Chapter 5

THE POPULATION BIOLOGY OF FUNGAL INVASIONS

Pierre Gladieux, Alice Feurtey,* Michael E. Hood,†
Alodie Snirc,* Joanne Clavel,‡ Cyril Dutech,§
Mélanie Roy,¶ and Tatiana Giraud**

*Ecologie Systématique Evolution, Univ. Paris-Sud, CNRS, AgroParisTech, Université Paris-Saclay, 91400 Orsay, France
†Department of Biology, Amherst College, Amherst, MA 01002, USA
‡Conservation des Espèces, Restauration et Suivi des Populations – CRBPO, Muséum National d'Histoire Naturelle-CNRS-Université Pierre et Marie Curie, 55 rue Buffon, 75005, Paris, France
§BIOGECO, INRA, Université Bordeaux, 33610, Cestas, France
¶Evolution et Diversité Biologique, Université Toulouse Paul Sabatier-Ecole Nationale de Formation Agronomique-CNRS, 118 route de Narbonne, 31062, Toulouse, France

Abstract

Fungal invasions are increasingly recognized as a significant component of global changes, threatening ecosystem health and damaging food production. Invasive fungi also provide excellent models to evaluate the generality of results based on other eukaryotes. We first consider here the reasons why fungal invasions have long been overlooked: they tend to be inconspicuous, and inappropriate methods have been used for species recognition. We then review the information available on the patterns and mechanisms of fungal invasions. We examine the biological features underlying invasion success of certain fungal species. We review population structure analyses, revealing native source populations and strengths of bottlenecks. We highlight the documented ecological and evolutionary changes in invaded regions, including adaptation to temperature, increased virulence, hybridization, shifts to clonality and association with novel hosts. We discuss how the huge census size of most fungi allows adaptation even in bottlenecked, clonal invaders. We also present new analyses of the invasion of the anther-smut pathogen on white campion in North America, as a case study illustrating how an accurate knowledge of species limits and phylogeography of fungal populations can be used to decipher the origin of invasions. This case study shows that successful invasions can occur even when life history traits are particularly unfavourable to long-distance dispersal and even with a strong bottleneck. We conclude that fungal invasions are valuable models to contribute to our view of biological invasions, in particular by providing insights into the traits as well as ecological and evolutionary processes allowing successful introductions.

Previously published as an article in *Molecular Ecology* (2015) 24, 1969–1986, doi: 10.1111/mec.13028

INTRODUCTION

Biological invasions have become a major focus of ecologists and evolutionary biologists, due to their severe negative consequences and their potential usefulness as models for studying basic processes in population biology (Sakai *et al.* 2001). However, invasions by fungi have only recently come into the limelight (Desprez-Loustau *et al.* 2007), even though most biological invasions are probably caused by micro-organisms, and fungal invasions may actually outnumber invasions caused by plants and animals. Furthermore, fungi, in the broad sense of the term (moulds, yeasts, mushrooms, lichens, rusts, smuts, mildews and including the phylogenetically distant oomycetes), display remarkably diverse life histories. This group includes everything from pathogens to mutualists and many free-living saprobes. They have a major impact on many different human activities and control essential ecosystem processes, such as nutrient cycling and carbon storage (Stajich *et al.* 2009). Fungal invasions thus constitute a key component of global change that is often overlooked.

When we think about fungal invasions, the species that come to mind are generally those responsible for emerging fungal diseases that jeopardize food security or drive natural populations into decline or even extinction (Palm 2001; Strange & Scott 2005; Fisher *et al.* 2012). Jarrah dieback in Australia, caused by the oomycete *Phytophthora cinnamomi* (Hansen *et al.* 2012), and chestnut blight in North America, caused by the ascomycete *Cryphonectria parasitica* (Dutech *et al.* 2012), are textbook examples of ecological disasters caused by the establishment of pathogenic fungi outside their native range. However, it is important to recognize that nonpathogenic fungi are just as likely to have been moved across continents, even though their impact on ecosystem structure and function has rarely been quantified (Rizzo 2005). A particularly relevant example is that of mycorrhizal fungi (Schwartz *et al.* 2006), such as the infamous death-cap mushroom, *Amanita phalloides*, which was introduced into eastern North America, where it is now common (Pringle & Vellinga 2006). Besides the effects of pathogenic fungi on their hosts, little is known about the ecological impact of invasive fungi, in particular on resident microbial communities at the same trophic level. Further community-level or competition studies are therefore required, similar to those typically carried out in the invasion ecology of plants and animals

(Díez 2005; Rizzo 2005; Schwartz *et al.* 2006). Fungi play fundamental roles in terrestrial ecosystem function and composition, and there is no reason to assume that the effects of nonpathogenic invasive fungi would necessarily be negligible or positive (Desprez-Loustau *et al.* 2007).

One of the key challenges in fungal invasion ecology is understanding and predicting introduction and establishment pathways, but baseline data concerning the composition of endemic fungal communities are limited, even for well-studied environments in temperate regions, hampering efforts to monitor invasive species and to determine their origin (Pringle & Vellinga 2006). Many invasive species were actually introduced well before they were recognized as such or before their potential impact was understood (Palm 2001; Rizzo 2005). The amphibian-killing chytrid fungus, *Batrachochytrium dendrobatidis*, illustrates this point clearly. It was only after *B. dendrobatidis*, which belongs to an understudied basally branching lineage of fungi, was first described in 1998 (Berger *et al.* 1998) that efforts were made to characterize the diversity of animal-associated chytrids and their environmental reservoirs (Morgan *et al.* 2007; Fisher *et al.* 2009; Farrer *et al.* 2011; Martel *et al.* 2013; McMahon *et al.* 2013). Genetic analyses pointed to a single, recent origin of an asexual worldwide lineage of *B. dendrobatidis*, and revealed additional cryptic global lineages of the fungus (Schloegel *et al.* 2012; Farrer *et al.* 2013). However, early studies were unable to identify the origin of the disease. With the exception of a few well-known examples (Table S1, Supporting information), the sources of invasive fungi and the timing of their introduction remain a mystery. Even the invasive status of some fungi can be unclear, as problematic fungi may have been there already and remained unidentified until their negative effects have begun to become apparent and of great concern. This uncertainty, in turn, impedes our capacity to manage possible invasion pathways in a proactive manner and to implement the regulatory mechanisms required to reduce the risks of further disasters (Sakalidis *et al.* 2013). Current efforts to catalogue fungal diversity will allow increasingly precise and scientifically sound risk assessments, with the aim of preventing new introductions (Palm 2001).

Another important challenge in the field of invasion biology is determining whether adaptive evolution occurs in the invaded range or whether the successful invaders were already well suited for establishment and spread before their introduction (Facon *et al.* 2006), and which traits (*e.g.* asexuality) or situations (*e.g.* presence of naive hosts) facilitate invasions. Many invasive fungal populations are under sustained pressure to adapt to environments distinct from their endemic ranges. The response to these novel environments can lead to invasive fungi emerging on new hosts (Giraud *et al.* 2010), colonizing new varieties of their hosts (Brown 1994; Guérin *et al.* 2007), or re-emerging with greater pathogenicity (Hovmøller *et al.* 2008).

Here, we summarize and discuss recent studies on the ecology and population genetics of fungal invasions, with the dual aims of highlighting the features of these invasions and the ways in which they can help us to find out more about biological invasions in general. We begin by considering the reasons why fungal invasions have long been overlooked. We then examine the extrinsic factors and biological features underlying invasion success of certain fungal species. We present current knowledge about the origins and dispersal paths of invasive fungi, and documented ecological and evolutionary changes in regions where they have been introduced. We discuss how adaptation may occur so rapidly in bottlenecked, clonal fungal invaders. We also use a case study to illustrate the usefulness of baseline biodiversity data for the study of fungal invasions. For this, we present an original analysis of the invasion of the fungus causing anther-smut disease on white campion in the United States; the interaction between *Microbotryum lychnidis-dioicae* and its host *Silene latifolia* is one of the best-studied nonagricultural plant–microbe pathosystems (Bernasconi *et al.* 2009). Overall, we aim at showing that fungi provide excellent and important models for studying the patterns and processes of biological invasions, for elucidating the factors favouring successful introductions, and for understanding their evolutionary and ecological mechanisms.

FUNGAL BIOGEOGRAPHY: 'NOTHING IS GENERALLY EVERYWHERE'

The first obvious step in any study of fungal invasions is the characterization of the geographic limits of endemic species of fungi. A failure to do this correctly can severely limit the ability to detect invasions and, thus, to infer their causal factors. Like other organisms of less than a few millimetres in length, fungi have long been thought to have global distributions, contrasting the highly restricted geographic ranges of larger organisms

(Bisby 1943; Martiny *et al.* 2006; Taylor *et al.* 2006). This hypothesis was based on the observation that almost all fungi have small, powder-like propagules (typically spores <10 μm in diameter) and that many have structures favouring their dissemination (Pringle *et al.* 2005), suggesting that dispersal ability *per se* is unlikely to limit the geographic distribution of these organisms (Bisby 1943; Peay *et al.* 2010; Sato *et al.* 2012). The misconception that many fungi had global distributions also resulted from the use of morphological species recognition criteria that are largely inappropriate for fungi, providing an erroneous picture of fungal diversity, distributions and ecologies (Swann *et al.* 1999; Taylor *et al.* 2006). Obviously, if all fungi were ubiquitous, there would be no reason for writing this article, as no fungal species would be exotic anywhere. As startling as it may sound to researchers studying plants and animals, it is only recently that most fungi have been shown to have distributions similar to those of their distant macrobial relatives (Green & Bohannan 2006; Thorsten Lumbsch *et al.* 2008). They are, for example, rarely present across entire continents (Tedersoo *et al.* 2012), and they satisfy the expectations of island biogeography theory (MacArthur & Wilson 2001; Peay *et al.* 2007).

The use of DNA markers have made it possible to identify new species that were previously cryptic, as these methods are more discriminating in fungi than traditional approaches based on morphological phenotypes or mating success (Giraud *et al.* 2008a). Modern evolutionary concepts of species are based on the lack of gene flow, with species being lineages that evolve independently (De Queiroz 2007; Giraud *et al.* 2008a), and DNA markers allow direct assessment of allelic exchange. In particular, genealogical concordance phylogenetic species recognition (GCPSR) is rapidly becoming the gold-standard approach for delimiting species (Henk *et al.* 2011; Hibbett & Taylor 2013). GCPSR uses breaks in the concordance of phylogenies for multiple genes to determine the limits of recombining entities (Avise & Wollenberg 1997; Taylor *et al.* 2000). It is one of the most objective, reproducible, achievable and widely applicable criteria for species identification (Giraud *et al.* 2008a). This approach appears to be the most powerful and accurate in fungi, leading to the identification of larger numbers of species than previously estimated (Giraud *et al.* 2008a; Stewart *et al.* 2014).

The cryptic species identified using DNA markers usually have narrower distributions in terms of geography range or host diversity than had been suggested in the prior literature (Taylor & Fisher 2003). The use of DNA markers has thus revealed that major geographic features (*e.g.* continental breaks, mountains) can act as effective barriers to fungal dispersal and that limited dispersal, combined with competitive advantage due to the colonization of resources ahead of competitors (Kennedy 2010), is a major driver of fungal endemicity (Peay *et al.* 2010). In fungal symbionts, strong specialization on different hosts and host characteristics (*e.g.* life history traits and distribution) appears to be key elements underlying biogeographic patterns (Vellinga *et al.* 2009; Hood *et al.* 2010; Schulze-Lefert & Panstruga 2011).

WHICH BIOLOGICAL FEATURES UNDERLIE THE SUCCESS OF INVASIVE FUNGAL SPECIES?

Most fungi thus did not have cosmopolitan distributions, but an increasing number of fungi are spreading around the world, some with dramatic effects on plant or animal populations (Fisher *et al.* 2012). It is therefore of paramount importance to examine the extrinsic factors and biological features underlying invasion success of certain fungal species. The success of invasive species is dependent on their ability to overcome a series of barriers, from introduction to establishment and invasion in a novel environment (Kolar & Lodge 2001), which involve stochastic and selection events in addition to initial transport by humans (Fig. 1). Many fungal species are highly prolific, producing large numbers of spores (Peay *et al.* 2012; Prussin *et al.* 2014), thereby increasing propagule pressure, and potentially facilitating dispersal and persistence in the new environment, even if the initial degree of adaptation is low (Giraud *et al.* 2010).

Unlike many invasive plant and animal species that were initially deliberately introduced into new areas for economic or other purposes (*e.g.* recreational), fungal introductions are mostly an unintended consequence of human-mediated movement and trade (Desprez-Loustau *et al.* 2007). Such 'hitchhiking' dispersal is favoured by some typical fungal features, including their inconspicuousness and the production of numerous, small propagules. The trade of animal and plant goods is a major pathway of fungal introductions, due to the close association of many fungi with plant or animal hosts (in mutualistic, pathogenic or saprotrophic

interactions; Desprez-Loustau *et al.* 2010; Liebhold *et al.* 2012). Even though some fungi provide striking examples of long-distance aerial dispersal (Brown & Hovmøller 2002) and of global distributions (Pringle *et al.* 2005; Henk *et al.* 2011), it is generally very difficult to discount a role for human-mediated long-range movement in current distributions or migration pathways. Dispersal rates of several kilometres or even up to tens of kilometres per day over a season have nevertheless been reported for some species (Aylor 2003; Mundt *et al.* 2013). Dispersal capacity, including fecundity and spore traits in particular, is critical in determining colonization success following the introduction of a species into a new environment and has been shown to affect variation in invasion success between species in a large data set of fungal tree pathogens (Philibert *et al.* 2011). Mode of dispersal may also account for the overrepresentation of aerial pathogens among invasive plant pathogenic fungi (Desprez-Loustau *et al.* 2010).

As with other organisms, the successful establishment of fungal species outside their native range can often be accounted for by some level of pre-adaptation: the presence of attributes allowing these species to survive elsewhere, in novel environments (Mayr 1965; Janzen 1985), possibly due to traits displaying relatively high phenotypic plasticity or conferring a broad ecological niche, *for example*, in terms of temperature or possible host range (Parker & Gilbert 2004; Agosta & Klemens 2008; Hulme & Barrett 2013). Pre-adaptation has been proposed to explain the ability of fungal species to colonize extreme environments (Robert & Casadevall 2009; Gostinčar *et al.* 2010). An example of phenotypic plasticity that has been experimentally evaluated for the potential to facilitate invasion is the breadth of potential host range in pathogens. Pathogenicity tests have indeed shown that fungal pathogens of plants are often able to infect hosts closely related to their original host, although initially with low aggressiveness (Table 1; de Vienne *et al.* 2009; Gilbert & Webb 2007).

Alternatively, successful invaders may adapt to the new environment after their introduction, and rapid and drastic evolutionary changes may be facilitated by the huge census population size and mixed mating systems of fungi. Many fungal species can undergo both sexual and asexual reproduction, which could facilitate establishment from few founding propagules and accelerate adaptation in new environments. Theoretical studies have shown that species with mixed mating systems (both asexual reproduction

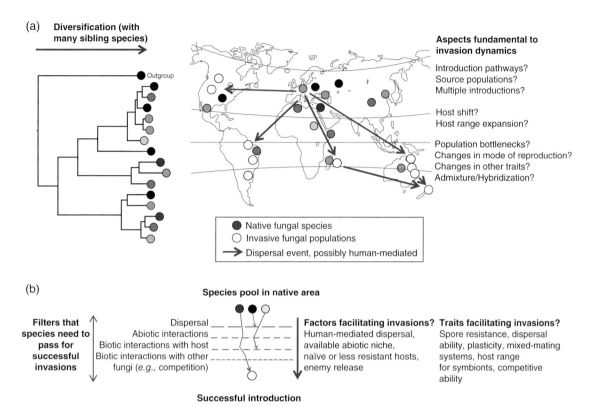

Fig. 1 An invasion scenario typical for fungi and essential questions to illuminate the process. (a) The figure shows the phylogenetic diversity and geographic distribution of a fungal group from which a species has invaded other regions and lists the questions fundamental to invasion dynamics: recognition of local species diversity in the native range, with the delimitation of sibling species, identification of the traits facilitating invasion, source populations, introduction pathways, occurrence of bottlenecks or of multiple introductions, possible evolutionary or ecological changes, such as host shift, changes in mode of reproduction or in other traits. The red circles represent introduced populations while the nonred circles represent native fungal species. (b) Successive filters that species must pass for invasions to be successful (dispersal, abiotic and biotic interactions) and possible mechanisms facilitating invasions (Human-mediated dispersal, available abiotic and biotic niche, enemy release, spore resistance, dispersal ability, plasticity, mixed mating systems, host range for symbionts or competitive ability). (*See insert for color representation of the figure.*)

Table 1 Glossary

Term	Definition
Host range expansion	Colonization of a new host species, while retaining the ability to infect the ancestral host
Host shift	Colonization of a new host species associated with a loss of the ability to infect the ancestral host
Host tracking	Codivergence of the host and the pathogen during crop domestication, with a lineage of fungal pathogens differentiating from that infecting wild crop progenitors, specialization on the domesticated species and the subsequent invasion of crops worldwide
Effectors	Secreted proteins that manipulate host innate immunity, enabling infection to occur (Dodds & Rathjen 2010)
Virulence	Ability of a pathogen to cause disease on a host (qualitative measure of pathogenicity)
Aggressiveness	Degree of damage caused by a pathogen on a host (quantitative measure of pathogenicity)

and rare sex events) have the highest potential for invasion success in terms of probability and time to establishment (Bazin *et al.* 2014). Mitotic spores and vegetative reproduction promote rapid increases in population size and the spread of the fittest genotypes, and several cycles of asexual reproduction may be completed in a single season (*e.g.* polycyclic plant pathogens). Meiotic spores promote the creation of novel genotypes through recombination, and they are mostly used as dispersal and survival structures (Taylor *et al.* 1999).

Several traits have been shown to be predictive of invasion success for forest pathogenic fungi, including traits relating to long-distance dispersal, sexual reproduction combined with asexual reproduction, spore shape and size, number of cells in spores, optimal temperature for growth and broad host range for pathogens (host range and infected organs; Hayes & Barry 2008; Philibert *et al.* 2011; Pyšek *et al.* 2010). The traits cited above are intuitively appealing as facilitators of introduction and establishment, but other, more complex factors, such as spore resistance, competitive ability, host range for symbionts and evolutionary or plasticity potential, are also probably important. Indeed, fungal invaders successfully establish themselves in new environments, to which they probably must adapt despite their low genetic diversity. This expectation to evolve despite the genetic resources is often described as the 'genetic paradox of invasions'. Below, we call into question the reality of this paradox in fungal invasions, and we review some of the solutions adopted by fungi.

THE FUNGAL SOLUTION TO THE GENETIC PARADOX OF INVASIONS

Is reduced genetic diversity commonplace in invasive fungi?

Colonization bottlenecks are generally thought to decrease adaptive potential. Nevertheless, founder events have been shown to contribute to invasion success in some cases, by increasing the genetic variance through the conversion of epistatic to additive variance (Naciri-Graven & Goudet 2003) or by reducing the genetic load (Facon *et al.* 2006). Admixture between populations from multiple introduction sources can also promote adaptation by rapidly creating novel allelic combinations (Prentis *et al.* 2008).

To assess whether bottlenecks or admixture occurred, one first needs to determine the origin of the invasive species and the genetic make-up of the source populations. For fungal symbionts of domesticated agricultural and forest plants, it is tempting to speculate that the origin of an invasive population is the geographic origin of the host. This has been shown to be the case for many crop pathogens (Table S1), through the demonstration that the centre of diversity of the fungus coincides with the centre of origin of its host plant (Gladieux *et al.* 2008; Munkacsi *et al.* 2008). However, some fungal pathogens have been found to have a higher level of diversity outside the centre of origin of their current host, and this has been taken as indirect evidence for diversification following host-shift/range expansion (Zaffarano *et al.* 2009; Ali *et al.* 2014). Advances in clustering methods and approximate Bayesian computation have made it possible to carry out rigorous comparisons of complex scenarios modelling genealogical relationships between populations and their geographic and demographic dynamics (*e.g.* Barrès *et al.* 2012; Dilmaghani *et al.* 2012; Dutech *et al.* 2012; Goss *et al.* 2014). Despite these methodological advances, it remains difficult to identify the origin of invasive symbiotic fungi because they are inconspicuous, hindering delimitation of the possible regions, substrates and hosts of origin. This uncertainty raises possibilities for endless disputes between experts (Goss *et al.* 2014) and hampers our understanding of the initial steps leading to invasion, such as the prevalence of major niche shifts.

In most cases in which genetic diversity has been compared between introduced and native fungal populations, a bottleneck signal has been detected, but such footprints have rarely been quantified in terms of effective population size ratios (Fig. 2; Table S1). Highly contrasting situations have been reported, from the introduction of a single genetic lineage (*e.g.* Raboin *et al.* 2007; Hovmøller *et al.* 2008) to greater genetic diversity in the invaded area than in the native range due to multiple introductions from diverse sources (Dilmaghani *et al.* 2012; Fig. 2; Table S1). Admixture between heterogeneous genetic pools has been detected in several studies on invasive fungi (*e.g.* Ahmed *et al.* 2012; Dilmaghani *et al.* 2012; Dutech *et al.* 2012; Jezic *et al.* 2012), but the source populations have rarely been identified (Fig. 2; Table S1). The detection of older admixture events requires the use of coalescent theory-based methods if several rounds of reproduction have erased allele frequency differences between the founding populations.

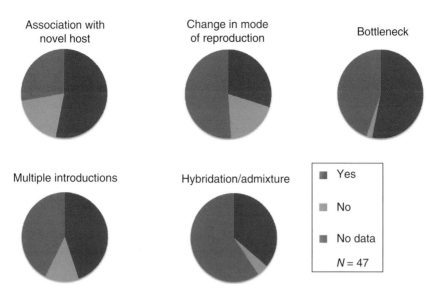

Fig. 2 Pie charts representing the frequency of documented cases of associations with novel hosts, bottlenecks, multiple introductions, hybridization/admixture and changes in mode of reproduction (shift to clonality), in a sample of $N = 47$ invasive fungi or oomycetes. These are the 47 cases, found in an exhaustive literature search, for which data were available on the population genetics of invasion or on the occurrence of genetic changes in introduced regions. Details and references are presented in Table S1. For each of the 47 cases, the following information was recorded when available: species name, taxonomic affiliation (*e.g.* Basidiomycota, Oomycota), lifestyle (*e.g.* tree mutualist, animal pathogen, crop pathogen, tree pathogen), native region, introduced area, host in native range, host in introduced range, change in mode of reproduction (shift to clonality), change in other life history traits (*e.g.* increase in aggressiveness, new virulence traits, adaptation to higher temperatures), putative introduction pathway (*e.g.* trade of host plants), evidence for the occurrence of a bottleneck and its strength, evidence for the occurrence of multiple introductions in at least one introduced area, evidence for the occurrence of hybridization between species or admixture between gene pools and the corresponding references. Admixture refers to cases where the mixing of two distinct lineages has been shown in clustering analyses or when it has been explicitly tested using simulations (*i.e.* the occurrence of multiple introduction *per se* is not considered as sufficient to infer 'admixture'). When the native region has not been identified, an emergence without invasion (*e.g.* by evolution of higher virulence or host shift) can often not be excluded.

Interspecific hybridization: increased genetic diversity promoting rapid evolutionary changes

Interspecific hybridization can also promote rapid adaptation (Prentis *et al.* 2008) and has been reported between native and invasive species in fungi (*e.g.* Bovers *et al.* 2008; Brasier 2001; Gonthier *et al.* 2007; Lin *et al.* 2007; Montarry *et al.* 2010; Stukenbrock *et al.* 2012; Fig. 2; Table S1). In some cases, hybridization has been shown to be a powerful mechanism of fungal adaptation to new hosts and environments, allowing the acquisition of a missing mating type, for example, in *Ophiostoma novo-ulmi* (Paoletti *et al.* 2006). Interspecific hybridization, as a process allowing the adaptation of organism to new environments, may be

facilitated in fungi by the large number of sibling species with incomplete reproductive isolation (Le Gac & Giraud 2008) and by the possibility of non-Mendelian forms of exchange, such as hyphal anastomosis that allows, for instance, the transfer of supplementary chromosomes (Roper *et al.* 2011; Gladieux *et al.* 2014).

Changes in life history traits during fungal invasions: mode of reproduction, virulence, host range and adaptation to abiotic factors

Investigators are generally interested in the questions discussed above, regarding levels of genetic diversity in invasive populations, because they implicitly assume a direct relationship between genetic variation and adaptive potential and a need for evolutionary change in introduced ranges. Many changes in life history traits during fungal invasions have actually been documented, with most studies focusing on ecological interactions of symbiotic fungi and particularly on plant pathogens. These changes have involved the infection of a new host population or species, adaptation to abiotic factors (*e.g.* temperature) and changes in reproductive mode (Fig. 2; Table S1). New ecological associations may result from fungi exhibiting wide ecological niches breadth, plasticity (*i.e.* phenotypic changes without evolution) or, alternatively, may have involved adaptation or trade-offs, but these possibilities are rarely tested experimentally. For instance, the infection of a new host by a fungal population may involve a pre-existing broad host range, a host shift, a host-range expansion or plasticity, but the relative importance of these processes is difficult to establish unequivocally, as the pathogenicity of fungal populations infecting ancestral and new hosts has rarely been compared directly. For simplicity, we will use the term 'host shift' below, but it should be borne in mind that it has not been clearly demonstrated whether these phenomena correspond to host shifts in the narrow sense (*i.e.* with evolution, allowing infection of a new host and a decrease in the ability to infect the original host), host range expansion (*i.e.* with evolution increasing host range), plasticity (*i.e.* expansion of observed host range without genetic change) or an initially broad host range (*i.e.* with no expansion of observed host range).

Several well-documented cases of emerging plant diseases have been attributed to such infection of a new host by introduced fungal pathogens (Desprez-Loustau *et al.* 2007; Stukenbrock & McDonald 2008; Giraud *et al.* 2010). Examples include wild tree or shrub pathogens, with the devastating shifts of *Cryphonectria parasitica* from the Asian to European and North American chestnuts, *Cronartium ribicola* from Asian to American and European pines, and *Discula destructiva* from Asian to American *Cornus* spp. (Palm & Rossman 2003; and references therein). Other examples, for crop pathogens, include the rice blast fungus *Magnaporthe oryzae*, which originated on *Setaria* millet (Couch *et al.* 2005), or the wheat pathogen *Zymoseptoria tritici*, which is derived from wild species infecting other grasses (Stukenbrock *et al.* 2007). The Irish potato famine pathogen was also recently shown to have emerged from populations infecting wild relatives of potato in Mexico (Goss *et al.* 2014). These host shifts may be due to recent contact with naïve host or

may have involved evolutionary changes to permit infection of the new host. Such shifts to a new host would be facilitated by certain life history traits of fungal pathogens favouring the colonization of new hosts, such as mating within hosts and strong selective pressure on a limited number of genes (Giraud *et al.* 2010; Gladieux *et al.* 2011). Host shifts have actually been shown to be the main mode of diversification in fungal pathogens (Giraud *et al.* 2008a, 2010; de Vienne *et al.* 2013b).

Associations with new hosts after introductions into new geographic regions have been less frequently documented for mutualists. There are nonetheless several examples in ectomycorrhizal fungi, such as *Laccaria* spp. in Costa Rica (Mueller & Strack 1992) and the emblematic *Amanita muscaria*, where introductions of these fungi appear to have occurred with their host trees species, followed by the formation of new associations with native hosts (Bagley & Orlovich 2004; Díez 2005; Rizzo 2005; Vellinga *et al.* 2009; Jairus *et al.* 2011).

Aside from host shifts or host range expansions occurring in introduced ranges due to contact with new hosts, several invasive crop pathogens have also emerged through host tracking (Table 1), a process by which host and pathogen codiverge during crop domestication and the development of agro-ecosystems (Stukenbrock & McDonald 2008). Movement of the pathogen from the wild progenitor of the crop to incipient domesticated populations can be followed by invasion at a worldwide scale associated with agricultural trade and product distributions. Examples include corn smut, caused by *Ustilago maydis* (Munkacsi *et al.* 2008), and ascochyta blight of chickpea, caused by *Didymella rabiei* (Frenkel *et al.* 2010). An interesting case is *Venturia inaequalis*, causing apple scab, which tracked its host tree during domestication in Asia and along the Silk Route to Europe, where host range expansion to the European wild apple occurred, with a certain degree of pathogen specialization on apple varieties (Gladieux *et al.* 2008, 2010; Le Van *et al.* 2012). In this case, cross-inoculation experiments indicated the occurrence of evolutionary change, in particular the increased aggressiveness and the acquisition of new virulence traits (Le Van *et al.* 2012).

Only a few studies have investigated whether changes in life history traits have occurred in introduced fungal populations, and whether such changes were due to adaptation in particular. However, some evidence has been obtained for adaptation in the form of ecological traits that function to enhance fitness in the new environments. Such changes include an increase in aggressiveness and/or larger numbers of virulence traits in introduced regions (Table 1; Fig. 2; Table S1; Enjalbert *et al.* 2005; Gilbert & Parker 2010; Le Van *et al.* 2012; Markell & Milus 2008; Milus *et al.* 2009; Rouxel *et al.* 2013). The development of greater aggressiveness or a broader virulence spectrum may be a general feature of crop pathogens, connected to the agricultural practice of very dense, homogeneous populations of host plants growing on a vast scale (Stukenbrock & McDonald 2008; McDonald 2010). Indeed, the presence in the environment of numerous individual hosts with the same genotype strengthens the selection on the pathogen for the ability to infect this genotype. Furthermore, the availability of numerous, regularly replanted host individuals should select for higher aggressiveness, because the pathogen would not benefit from a prudent host exploitation strategy when there are constantly new host individuals to infect in its environment (Frank 1996).

Another frequent change in life history traits of introduced fungi is the loss of sexual reproduction. Many studies have documented clonal population structures throughout the entire introduced range (Table S1; Fig. 2; Ali *et al.* 2010; Couch *et al.* 2005; Dobrowolski *et al.* 2003; Duan *et al.* 2010; Enjalbert *et al.* 2005; Farrer *et al.* 2011; Fisher *et al.* 2009; Goodwin *et al.* 1994; Goss *et al.* 2009; Ivors *et al.* 2006; Mboup *et al.* 2007, 2009; Saleh *et al.* 2012a,b; Singh *et al.* 2011; Smart & Fry 2001; Stajich *et al.* 2009) or over parts of that range (Gladieux *et al.* 2008; Dutech *et al.* 2010; Dilmaghani *et al.* 2012). Evidence has also been provided for an evolution towards a lower capacity for sexual reproduction in introduced regions (Couch *et al.* 2005; Enjalbert *et al.* 2005; Mboup *et al.* 2007, 2009; Ali *et al.* 2010; Duan *et al.* 2010; Saleh *et al.* 2012a,b). F_{ST} versus Q_{ST} comparisons have suggested that this pattern is indeed an adaptation, rather than merely the consequence of genetic drift (Ali *et al.* 2010).

There may initially be several proximal causes for a loss of sexual reproduction: (i) a lack of appropriate mates where asexual propagation is possible, as in the case of *Phytophthora infestans*, *P. cinnamomi*, *P. ramorum* and *O. novo-ulmi*, for which a single mating type was present for long periods in introduced ranges (Goodwin *et al.* 1994; Brasier 2001; Dobrowolski *et al.* 2003; Hardham 2005); (ii) a hybridization event, *for example*, in the human pathogen *Cryptococcus neoformans*

(Lin *et al.* 2007); and (iii) a lack of the alternate host on which sex occurs in the case of complex life cycles, *for example*, in the rusts *Puccinia striiformis* f. *sp.tritici* and *Melampsora larici-populina* (Mboup *et al.* 2007, 2009; Barrès *et al.* 2008).

There may also be diverse evolutionary causes for the selection against sexual reproduction (Bazin *et al.* 2014): (i) reduced advantages of sex due to the release of an enemy, such as viral pathogens of fungi, or the lower host resistance variation in the case of pathogens introduced into large, genetically homogeneous crops; (ii) in heterokaryotic fungi and diploid oomycetes, preservation of the initial heterozygosity, which might otherwise rapidly convert to homozygosity in small founder populations, leading to inbreeding depression; (iii) a release of constraints not related to recombination, maintaining sex in the short term in native ranges, such as the need for sexual spores for winter survival or long-distance dispersal (de Vienne *et al.* 2013a), which is particularly likely for introduced pathogens on crops, due to the availability of large, homogeneous host populations; and (iv) clonality may allow the locally adapted genotypes to be propagated without breaking up adaptive allelic combinations; again, this is advantageous on genetically homogeneous crops planted over vast areas.

Finally, abiotic factors can indeed limit the success of introductions or require adaptation, as a tolerance to high temperatures has been shown to have evolved in the introduced ranges of some fungi. In models of distribution range, temperature has been shown to be a critical factor for successful establishment of invasive fungi (Wolfe *et al.* 2010). Some populations of wheat yellow rust in the Mediterranean basin and south-eastern regions of the USA have evolved temperature-specific adaptations (Mboup *et al.* 2012), as did the *D. rabiei* population infecting domesticated chickpea (Frenkel *et al.* 2010). Furthermore, the successful establishment of exotic fungi has been shown to be sensitive to moisture conditions in attempts to deliberately introduce edible fungi such as truffles (Vellinga *et al.* 2009).

Invasive fungi, particularly pathogens of agricultural and forest plants, thus provide many examples of environmental change or colonization of a new niche acting as the primary drivers of post-colonization adaptive evolution. It is clearly important to understand the biological parameters driving selection on pathogens and disease outbreaks, given their impact on ecosystem functioning, human health and agricultural production.

Fungal pathogens evolve rapidly in agricultural and forest ecosystems

The potential for a fast pace of evolution in invasive fungal pathogens is a major impediment to maintaining ecosystem health and sustainable agricultural production. Measurements of genetic diversity and population structure are commonly used to draw inferences regarding the evolutionary potential of these pathogens, based on the assumption that adaptation can proceed quickly only from standing variation (McDonald & Linde 2002; McDonald 2014). Life history strategies (*e.g.* transmission and dispersal, virulence, mode of reproduction) are considered to be important determinants of diversity and, thus, for the potential of pathogens to colonize and spread on new host genotypes (Barrett *et al.* 2008). While standing genetic variation in a population is clearly a key influence on its ability to adapt in response to an environmental challenge, the general validity of predictive frameworks based on measured diversity patterns is questionable, given that evolution occurs rapidly in many invasive plant pathogens despite relatively modest levels of genetic diversity. For instance, some of the textbook examples of rapidly adapting fungal pathogens, such as lettuce downy mildew and wheat yellow rust, are also notoriously clonal and harbour little genetic diversity (Valade 2012; Ali *et al.* 2014).

The classical model of positive selection assumes that adaptation proceeds principally from *de novo* mutations, in a mutation-limited regime (Maynard-Smith *et al.* 1974). Under this regime, the lag time to adaptation depends on the product of the rate of mutation to the adaptive variant at the relevant locus and effective population size (N_e; Charlesworth 2009; Garud *et al.* 2013). Many plant pathogens have a small N_e as estimated from levels of standing neutral variation (*e.g.* Zaffarano *et al.* 2009; Dilmaghani *et al.* 2012; Ali *et al.* 2014). This apparent barrier to adaptation is because the 'diversity N_e' values are primarily determined by long-term population dynamics, which are dominated by extreme episodes of population crashes (bottlenecks or founder events; Lewontin 1974), hallmarks coincidentally of both disease cycles and invasions dynamics. If we consider the mutation rate per site to be between 10^{-8} and 10^{-10} (Kasuga *et al.* 2002), the adaptation of invasive fungi should, therefore, be strongly mutation limited, *that is* relatively slow and involving a single adaptive variant (a 'hard selective sweep').

At least two factors may nevertheless allow the rapid adaptation of invasive fungi, despite their relatively small 'diversity N_e'. First, recent studies have shown that predicting the pattern of adaptation (speed of adaptation and number of adaptive alleles) is more accurate if N_e is estimated over timescales relevant to adaptation (Karasov *et al.* 2010). For an invasive fungal pathogen, this may be the timescale over which a new fungicide is applied or a new resistant variety is released. Despite the regular crashes of fungal populations (*e.g.* during winters), their short-term N_e can be expected to be much higher than their long-term N_e, particularly as they generally have periods of huge census size. In such very large populations, the distinction between standing variation and *de novo* mutations becomes blurred, because almost all the possible mutations of the target gene are presented in the population (Barton 2010; Karasov *et al.* 2010).

Second, adaptation may be faster if the mutational targets are larger than a single nucleotide (Pritchard *et al.* 2010). The mutational target size is the number of base pairs within an allele for which mutations would alter a particular phenotype and increase fitness in the new environment. For fungal secreted virulence determinants (effectors, Table 1), for instance, the mutational target size may be of the order of hundreds of base pairs within the gene, rather than one: changing any one among hundreds of nucleotide, or even a large deletion in fungal effectors, may allow escaping host recognition (Rafiqi *et al.* 2012). Adaptive alleles would thus be expected to evolve repeatedly and relatively quickly.

Recent case studies in invasive fungal pathogens have revealed several examples of adaptive mutations of multiple, independent *de novo* origins sweeping through populations (so-called soft selective sweeps), providing evidence that adaptation is not necessarily mutation-limited in these organisms. For instance, the molecular characterization of virulence alleles of effector genes in *Leptosphaeria maculans* (stem canker of oilseed rape) and *Rhynchosporium secalis* (barley leaf scald) revealed a diversity of mutational events consistent with a model of adaptation via soft sweeps (Schürch *et al.* 2004; Gout *et al.* 2007; Fudal *et al.* 2009; Daverdin *et al.* 2012). Another example is provided by fungicide resistance in *Plasmopara viticola* (grapevine downy mildew), which has arisen from several independent mutations (Chen *et al.* 2007).

CASE STUDY OF ANTHER-SMUT FUNGUS ON WHITE CAMPION

The insights into fungal invasions using comprehensive analyses of species boundaries, population structure and life history traits can be illustrated using the anther-smut disease of Caryophyllaceae, caused by the basidiomycete *Microbotryum violaceum sensu lato*. The anther-smut fungi produce spores in anthers of diseased plants and are mostly transmitted by pollinators. A sex event is required before infection of a new plant, and the mating system is highly selfing (Giraud *et al.* 2008b). *Microbotryum* pathogens on the Caryophyllaceae are relatively uniform morphologically and were long thought to be a single, widely distributed species. However, GCPSR has shown that the old epithet actually consists of a large number of different, host-specialized species (Lutz *et al.* 2005; Le Gac *et al.* 2007) that diversified following host shifts (Refregier *et al.* 2008). One of the defined species, *Microbotryum lychnidis-dioicae*, and its particular host plant, *Silene latifolia*, are native to Europe. Based on a collection of samples of unprecedented density and geographic scale for a pathogen, cluster analyses revealed the existence of strong subdivisions within the geographic range of this fungal species, reflecting recolonization from southern glacial refugia (Vercken *et al.* 2010; Gladieux *et al.* 2013). The genetic structure of the pathogen populations was found to be generally congruent with the genetic structure of the host, *S. latifolia* (Taylor & Keller 2007), both exhibiting three genetic clusters distributed, respectively, across western Europe, eastern Europe and Italy. This congruence of population structures suggests that the migration pathway of the fungus had been dependent on its host over a time scale of hundreds or thousands of years. However, *M. lychnidis-dioicae* exhibited a much stronger structure than observed for its host, each of the three main fungal clusters being subdivided into further genetic clusters, with more restricted geographic distributions (Vercken *et al.* 2010). This suggested that *M. lychnidis-dioicae* has recolonized central and northern Europe from more numerous, smaller refugia than sheltered its plant host and that it has experienced fewer large-scale dispersal events.

Both the pathogen and its host were introduced relatively recently in North America, and the invasion history of *S. latifolia* has been extensively studied (Keller *et al.* 2009, 2012). The diversity of the host plant, *S. latifolia*, was found to be high in the introduced range, with the introduction of multiple lineages from several European genetic clusters that are now spread across North America (Keller *et al.* 2009, 2012). The anther-smut disease appeared much more recently on *S. latifolia* in North America, and with a more eastern restricted distribution, than observed for the host plant (Antonovics *et al.* 2003; Fontaine *et al.* 2013). The delimitation of anther-smut species confirms that the disease on the introduced *S. latifolia* was the result of a later introduction by its specialist pathogen, *M. lychnidis-dioicae*, instead of the host shift of a *Microbotryum* species endemic to North American on other *Silene* species or from another European host species (Fontaine *et al.* 2013). The very strong genetic structure of *M. lychnidis-dioicae* in Europe made it possible to identify the United Kingdom as the most probable source of the limited set of genotypes introduced into the central East Coast of North America. Given the assignment results and the limited number of genotypes in North America, a single introduction event appeared the most likely scenario (Fontaine *et al.* 2013). However, only a few samples from the United Kingdom were available for study at that time, and we have since identified an additional diseased population in the state of Massachusetts.

We therefore sampled further populations in the United Kingdom and from the newly identified diseased regions in North America, in an effort to ascertain the origin of the invasion more precisely and to check whether independent introductions may be responsible for the invasions of central East Coast and more northern populations. The samples analysed here represent an exhaustive sampling of the extant North American populations of *M. lychnidis-dioicae*, and a representative subset of the European populations previously analysed (Vercken *et al.* 2010; Fontaine *et al.* 2013). Fungal isolates were genotyped and analysed as previously described (Fontaine *et al.* 2013), with the inclusion here of more microsatellite markers (fourteen, see data on dryad). Fourteen multilocus genotypes were found in North America, representing a strong bottleneck compared with the diversity of *M. lychnidis-dioicae* across Europe, where 174 genotypes were detected in the present sample. The values of allelic richness, a standardized measure of diversity corrected for sample size differences, also pointed to a strong bottleneck: allelic richness was 3.78 in North America versus 7.46 in Europe and 5.55 in the UK alone. Genotypes were highly homozygous in all geographic regions, and no admixture was detected, indicating that the fungus has retained its highly selfing mating system in the introduced range (Giraud *et al.* 2008b). Sexual reproduction has also remained obligate for plant infection in the introduced range (MEH, personal observation).

We used STRUCTURE assignment analyses to investigate the origin of invasive fungal populations. This provided evidence for an additional, independent introduction of *M. lychnidis-dioicae* in North America (Fig. 3a; Fig. S1, Supporting information). The previously sampled

(a)

(b)

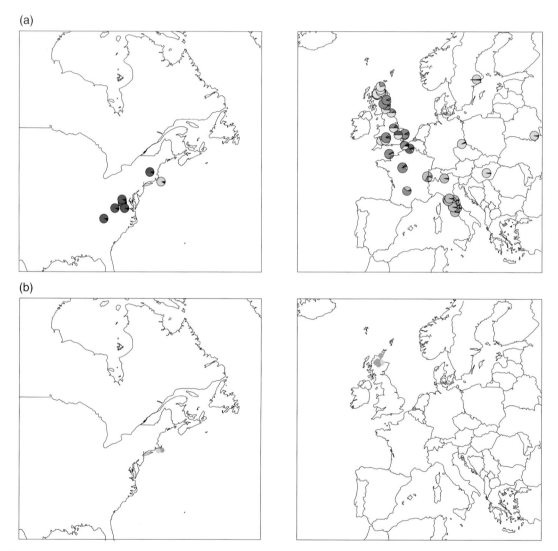

Fig. 3 Assignment of North American fungal *Microbotryum lychnidis-dioicae* samples, collected on *Silene latifolia*, to European populations. (a) Map of the samples collected in the United States and Europe (*N* = 328 individuals), with colours representing the mean membership proportions inferred by STRUCTURE for *K* = 6; higher *K* did not allow assigning more precisely American isolates to European populations, see Fig. S1. (b) Map of the samples assigned to the yellow cluster in (a) and with identical genotypes in the United States and in Scotland; a specific colour (yellow or brown) has been given for each of the two multilocus genotypes of the yellow cluster in the USA. Colours represent the inferred ancestry from *K* ancestral populations. The percentages indicated are the proportions of runs that found the main solution shown here. (*See insert for color representation of the figure.*)

populations in Virginia and one of the populations from Massachusetts still appeared, as in our previous study (Fontaine *et al.* 2013), to have originated in the Lothian (Edinburgh) region in southern Scotland (red cluster in Fig. 3a). However, the second Massachusetts population was assigned to another genetic cluster (yellow cluster in Fig. 3a), which included populations spread across the UK and could not be separated by further increasing the *K* parameter in STRUCTURE analyses (number of clusters). The examination of individual genotypes, however, allowed narrowing down the possible sources of this second introduction. Genotypes that were identical to the ones found in the North American yellow cluster (Fig. 3a) were also found in the north of Scotland, in the Moray council area and in the Highland close to cities of Inverness and Thurso (Fig. 3b). The finding of two distinct genetic clusters in North American populations of *M. lychnidis-dioicae*, found in different locations and assigned to clusters originating from different regions in Scotland, altogether strongly suggests two or more independent introductions of the anther-smut fungus.

The fungal invasion history still represents a much smaller number of introductions than for its host plant species (Keller *et al.* 2009, 2012), probably reflecting a restriction by the pathogen's life history traits. Indeed, this obligate, specialized pathogen is neither transmitted by the seeds of its host nor persistent in the environment. It can be transmitted only from a living plant to another living plant. This life history trait limits the probability of introduction with its host plants, as weed introduction often occurs through seeds. The question of which traits prevent or facilitate invasions, during the several steps of dispersal and establishment, is of particular importance for fungal invasions in general. The *Microbotryum* case nevertheless shows that even the specialized, obligate fungus unable to disperse by wind (being mostly insect-borne) or seeds, or to persist in contaminated soil, can successfully invade a new continent, and multiple times independently. This case study further illustrates the importance of delimiting cryptic species in the native range and the usefulness of comprehensive sampling for performing relevant population structure analyses and identifying the locations of invasion source populations. It also highlights the possibility of successful invasion without change in mode of reproduction and despite a strong bottleneck.

CONCLUSION AND FUTURE PROSPECTS

In conclusion, fungal invasions are valuable yet often overlooked models to contribute to our view of biological invasions. In particular, studies on large data sets including many fungal tree pathogens have been successful at identifying the traits favouring fungal invasions, *for example*, fecundity and spore traits favouring dispersal, as well as mixed mating systems, with both clonal and sexual reproduction. Furthermore, the ecological and evolutionary processes favouring successful introductions could be revealed in several case studies, such as adaptation to higher temperatures, increased aggressiveness, acquisition of novel virulences, phenotypic plasticity, novel host associations, and admixture under multiple introductions or hybridization. Sophisticated methods of population genetic inference have elucidated the dispersal pathways and native sources of several fungal invasions, pinpointing in particular host tracking for pathogens and the possibility of invasions even with low genetic variability, *that is*, following strong bottlenecks or loss of sex. The rapid adaptation of invasive fungi, in particular of crop pathogens, often despite limited standing variation, is likely facilitated by their huge population sizes and the large genomic targets of selection, where any change within an effector gene for instance may allow escaping host recognition. Altogether, these insights broaden our understanding of biological invasions in eukaryotes.

Fungal invasion biology is a dynamic, recently emergent area of research. Many fungal invasions, possibly even the majority that exist, remain to be discovered, and there is much yet to be learned. The use of high-throughput sequencing technologies and the availability of universal primers for the sequencing of a single fungal barcode, the ITS (White *et al.* 1990), are greatly advancing our understanding of fungal diversity (Peay *et al.* 2008; Talbot *et al.* 2014) and increasing the power to detect invasive fungi, which are often concealed within the soil or in plant tissues. The use of sequencing technologies and substantial sampling efforts should shed light on the processes underlying invasions by nonpathogenic fungi, about which we know very little, despite their tremendous importance in ecosystem functioning, as decomposers for example. More generally, technological advances in the characterization of fungal populations and progress in inference methods (Stukenbrock

& McDonald 2008; Neher 2013) are creating new, unprecedented opportunities to study fungal ecology and evolution, and fungi are proving to be excellent organisms for challenging the universality of expectations derived from plants and animals (Gladieux *et al.* 2014).

Progress has recently been made towards determining the origins and pathways of fungal invasions, but a more comprehensive understanding of the processes underlying fungal invasions will require integrating population and comparative genomics within an experimental frameworks as has been applied to plant models. Phenotypic differences between introduced and native populations occur under nonequilibrium demographic conditions and may be affected by chance events, evolutionary history and adaptive responses to divergent selection (Taylor & Keller 2007). Experimental comparisons of species from their native and introduced ranges, testing hypotheses of adaptation against null models of neutral phenotypic evolution, should make it possible to determine whether the causal mechanisms of phenotypic change are adaptive. It is now increasingly feasible to push such studies further with population genetics to identify potentially adaptive genomic features and then to make use of bioinformatics and reverse genetics tools to connect selected genomic features to important phenotypes. Full-genome sequencing of multiple strains is now routinely carried out in fungi. Improvements in the functional annotation of fungal genomes, and their amenability to experimental manipulation, are making it possible to implement a genuine 'reverse ecology' approach involving the identification of genes with unusual patterns of evolution and environmental parameters consistent with the function of the genes and then testing hypotheses experimentally (Ellison *et al.* 2011, 2014).

Adaptive evolution in invasive fungi may pose a major threat to plant and animal health, but it also provides us with interesting opportunities to improve our understanding of the evolution of biotic interactions (McDonald 2010). We must not waste this opportunity to learn from the natural experiments represented by fungal invasions if we are to develop sustainable regulatory mechanisms based on evolutionary models, using predictive frameworks to control the pace of pathogen evolution in natural and agro-ecosystems (Parker & Gilbert 2004; Thrall *et al.* 2007, 2011).

ACKNOWLEDGEMENTS

This work was funded by ANR grants 07-BDIV-003-Emerfundis and 12-ADAP-0009-02-GANDALF, the National Science Foundation grant DEB-1115765, a Marie Curie postdoctoral fellowship awarded to PG (FP7-PEOPLE-2010-IOF-No.273086) and the ERC starting grant GenomeFun 309403 awarded to TG. We acknowledge the REID for funding attendance at meetings and workshops. We thank Marie-Laure Desprez-Loustau for her considerable assistance with this manuscript and for discussions. We thank all the people who have helped with completing our *Microbotryum* collection from the UK: Deborah Charlesworth, Jacky Muscott, Clare Gachon and Damien Hicks, and we thank Britta Büker and Thomas Meagher for identifying locations in Massachusetts.

REFERENCES

Agosta SJ, Klemens JA (2008) Ecological fitting by phenotypically flexible genotypes: implications for species associations, community assembly and evolution. *Ecology Letters*, **11**, 1123–1134.

Ahmed S, de Labrouhe DT, Delmotte F (2012) Emerging virulence arising from hybridisation facilitated by multiple introductions of the sunflower downy mildew pathogen *Plasmopara halstedii*. *Fungal Genetics and Biology*, **49**, 847–855.

Ali S, Leconte M, Walker A-S, Enjalbert J, de Vallavieille-Pope C (2010) Reduction in the sex ability of worldwide clonal populations of *Puccinia striiformis f.sp tritici*. *Fungal Genetics and Biology*, **47**, 828–838.

Ali S, Gladieux P, Leconte M *et al.* (2014) Origin, migration routes and worldwide population genetic structure of the wheat yellow rust *Puccinia striiformis f. sp tritici*. *PLoS Pathogens*, **10**, e1003903.

Antonovics J, Hood ME, Thrall PH, Abrams JY, Duthie GM (2003) Herbarium studies on the distribution of anther-smut fungus (*Microbotryum violaceum*) and *Silene* species (Caryophyllaceae) in the eastern United States. *American Journal of Botany*, **90**, 1522–1531.

Avise JC, Wollenberg K (1997) Phylogenetics and the origin of species. *Proceedings of the National Academy of Sciences, USA*, **94**, 7748–7755.

Aylor DE (2003) Spread of plant disease on a continental scale: role of aerial dispersal of pathogens. *Ecology*, **84**, 1989–1997.

Bagley SJ, Orlovich DA (2004) Genet size and distribution of *Amanita muscaria* in a suburban park, Dunedin, New Zealand. *New Zealand Journal of Botany*, **42**, 939–947.

Barrès B, Halkett F, Dutech C et al. (2008) Genetic structure of the poplar rust fungus Melampsora larici-populina: evidence for isolation by distance in Europe and recent founder effects overseas. Infection, Genetics and Evolution, 8, 577–587.

Barrès B, Carlier J, Seguin M et al. (2012) Understanding the recent colonization history of a plant pathogenic fungus using population genetic tools and Approximate Bayesian Computation. Heredity, 109, 269–279.

Barrett LG, Thrall PH, Burdon JJ, Linde CC (2008) Life history determines genetic structure and evolutionary potential of host-parasite interactions. Trends in Ecology & Evolution, 23, 678–685.

Barton N (2010) Understanding adaptation in large populations. PLoS Genetics, 6, e1000987.

Bazin E, Mathe-Hubert H, Facon B, Carlier J, Ravigne V (2014) The effect of mating system on invasiveness: some genetic load may be advantageous when invading new environments. Biological Invasions, 16, 875–886.

Berger L, Speare R, Daszak P et al. (1998) Chytridiomycosis causes amphibian mortality associated with population declines in the rain forests of Australia and Central America. Proceedings of the National Academy of Sciences, USA, 95, 9031–9036.

Bernasconi G, Antonovics J, Biere A et al. (2009) Silene as a model system in ecology and evolution. Heredity, 103, 5–14.

Bisby GR (1943) Geographical distribution of fungi. The Botanical Review, 9, 466–482.

Bovers M, Hagen F, Kuramae EE, Boekhout T (2008) Six monophyletic lineages identified within Cryptococcus neoformans and Cryptococcus gattii by multi-locus sequence typing. Fungal Genetics and Biology, 45, 400–421.

Brasier CM (2001) Rapid evolution of introduced plant pathogens via interspecific hybridization. BioScience, 51, 123–133.

Brown JKM (1994) Chance and selection in the evolution of barley mildew. Trends in Microbiology, 2, 470–475.

Brown JKM, Hovmøller MS (2002) Aerial dispersal of pathogens on the global and continental scales and its impact on plant disease. Science, 297, 537–541.

Charlesworth B (2009) Effective population size and patterns of molecular evolution and variation. Nature Reviews Genetics, 10, 195–205.

Chen W-J, Delmotte F, Cervera SR et al. (2007) At least two origins of fungicide resistance in grapevine downy mildew populations. Applied and Environmental Microbiology, 73, 5162–5172.

Couch BC, Fudal I, Lebrun MH et al. (2005) Origins of host-specific populations of the blast pathogen Magnaporthe oryzae in crop domestication with subsequent expansion of pandemic clones on rice and weeds of rice. Genetics, 170, 613–630.

Daverdin G, Rouxel T, Gout L et al. (2012) Genome structure and reproductive behaviour influence the evolutionary potential of a fungal phytopathogen. PLoS Pathogens, 8, e1003020.

De Queiroz K (2007) Toward an integrated system of clade names. Systematic Biology, 56, 956–974.

Desprez-Loustau M-L, Robin C, Buee M et al. (2007) The fungal dimension of biological invasions. Trends in Ecology & Evolution, 22, 472–480.

Desprez-Loustau M-L, Courtecuisse R, Robin C et al. (2010) Species diversity and drivers of spread of alien fungi (sensu lato) in Europe with a particular focus on France. Biological Invasions, 12, 157–172.

Díez J (2005) Invasion biology of Australian ectomycorrhizal fungi introduced with eucalypt plantations into the Iberian Peninsula. Biological Invasions, 7, 3–15.

Dilmaghani A, Gladieux P, Gout L et al. (2012) Migration patterns and changes in population biology associated with the worldwide spread of the oilseed rape pathogen Leptosphaeria maculans. Molecular Ecology, 21, 2519–2533.

Dobrowolski MP, Tommerup IC, Shearer BL, O'Brien PA (2003) Three clonal lineages of Phytophthora cinnamomi in Australia revealed by microsatellites. Phytopathology, 93, 695–704.

Dodds PN, Rathjen JP (2010) Plant immunity: towards an integrated view of plant–pathogen interactions. Nature Reviews Genetics, 11, 539–548.

Duan X, Tellier A, Wan A et al. (2010) Puccinia striiformis f.sp tritici presents high diversity and recombination in the over-summering zone of Gansu, China. Mycologia, 102, 44–53.

Dutech C, Fabreguettes O, Capdevielle X, Robin C (2010) Multiple introductions of divergent genetic lineages in an invasive fungal pathogen, Cryphonectria parasitica, in France. Heredity, 105, 220–228.

Dutech C, Barres B, Bridier J et al. (2012) The chestnut blight fungus world tour: successive introduction events from diverse origins in an invasive plant fungal pathogen. Molecular Ecology, 21, 3931–3946.

Ellison CE, Hall C, Kowbel D et al. (2011) Population genomics and local adaptation in wild isolates of a model microbial eukaryote. Proceedings of the National Academy of Sciences, USA, 108, 2831–2836.

Ellison CE, Kowbel D, Glass NL, Taylor JW, Brem RB (2014) Discovering functions of unannotated genes from a transcriptome survey of wild fungal isolates. MBio 5, e01046–13.

Enjalbert J, Duan X, Leconte M, Hovmøller MS, De Vallavieille-Pope C (2005) Genetic evidence of local adaptation of wheat yellow rust (Puccinia striiformis f. sp tritici) within France. Molecular Ecology, 14, 2065–2073.

Facon B, Genton BJ, Shykoff J et al. (2006) A general eco-evolutionary framework for understanding bioinvasions. Trends in Ecology & Evolution, 21, 130–135.

Farrer RA, Weinert LA, Bielby J et al. (2011) Multiple emergences of genetically diverse amphibian-infecting chytrids include a globalized hypervirulent recombinant lineage. *Proceedings of the National Academy of Sciences, USA*, **108**, 18732–18736.

Farrer RA, Henk DA, Garner TWJ *et al.* (2013) Chromosomal copy number variation, selection and uneven rates of recombination reveal cryptic genome diversity linked to pathogenicity. *PLoS Genetics*, **9**, e1003703.

Fisher MC, Garner TWJ, Walker SF (2009) Global emergence of *Batrachochytrium dendrobatidis* and amphibian chytridiomycosis in space, time, and host. *Annual Review of Microbiology*, **63**, 291–310.

Fisher MC, Henk DA, Briggs CJ *et al.* (2012) Emerging fungal threats to animal, plant and ecosystem health. *Nature*, **484**, 186–194.

Fontaine MC, Gladieux P, Hood ME, Giraud T (2013) History of the invasion of the anther smut pathogen on *Silene latifolia* in North America. *New Phytologist*, **198**, 946–956.

Frank SA (1996) Models of parasite virulence. *Quarterly Review of Biology*, **71**, 37–78.

Frenkel O, Peever TL, Chilvers MI *et al.* (2010) Ecological genetic divergence of the fungal pathogen *Didymella rabiei* on sympatric wild and domesticated *Cicer* spp. (Chickpea). *Applied and Environmental Microbiology*, **76**, 30–39.

Fudal I, Ross S, Brun H *et al.* (2009) Repeat-Induced Point Mutation (RIP) as an alternative mechanism of evolution toward virulence in *Leptosphaeria maculans*. *Molecular Plant-Microbe Interactions*, **22**, 932–941.

Garud NR, Messer PW, Buzbas EO, Petrov DA (2013) Soft selective sweeps are the primary mode of recent adaptation in *Drosophila melanogaster*. *arXiv preprint arXiv:1303.0906*.

Gilbert GS, Parker IM (2010) Rapid evolution in a plant-pathogen interaction and the consequences for introduced host species. *Evolutionary Applications*, **3**, 144–156.

Gilbert GS, Webb CO (2007) Phylogenetic signal in plant pathogen-host range. *Proceedings of the National Academy of Sciences, USA*, **104**, 4979–4983.

Giraud T, Refregier G, Le Gac M, de Vienne DM, Hood ME (2008a) Speciation in fungi. *Fungal Genetics and Biology*, **45**, 791–802.

Giraud T, Yockteng R, Lopez-Villavicencio M, Refregier G, Hood ME (2008b) Mating system of the anther smut fungus *Microbotryum violaceum*: selfing under heterothallism. *Eukaryotic Cell*, **7**, 765–775.

Giraud T, Gladieux P, Gavrilets S (2010) Linking the emergence of fungal plant diseases with ecological speciation. *Trends in Ecology & Evolution*, **25**, 387–395.

Gladieux P, Zhang XG, Afoufa-Bastien D *et al.* (2008) On the origin and spread of the scab disease of apple: out of Central Asia. *PLoS ONE*, **3**, e1455.

Gladieux P, Zhang XG, Roldan-Ruiz I *et al.* (2010) Evolution of the population structure of *Venturia inaequalis*, the apple scab fungus, associated with the domestication of its host. *Molecular Ecology*, **19**, 658–674.

Gladieux P, Guerin F, Giraud T *et al.* (2011) Emergence of novel fungal pathogens by ecological speciation: importance of the reduced viability of immigrants. *Molecular Ecology*, **20**, 4521–4532.

Gladieux P, Devier B, Aguileta G, Cruaud C, Giraud T (2013) Purifying selection after episodes of recurrent adaptive diversification in fungal pathogens. *Infection Genetics and Evolution*, **17**, 123–131.

Gladieux P, Ropars J, Badouin H *et al.* (2014) Fungal evolutionary genomics provides insight into the mechanisms of adaptive divergence in eukaryotes. *Molecular Ecology*, **23**, 753–773.

Gonthier P, Nicolotti G, Linzer R, Guglielmo F, Garbelotto M (2007) Invasion of European pine stands by a North American forest pathogen and its hybridization with a native interfertile taxon. *Molecular Ecology*, **16**, 1389–1400.

Goodwin SB, Cohen BA, Fry WE (1994) Panglobal distribution of a single clonal lineage of the Irish potato famine fungus. *Proceedings of the National Academy of Sciences, USA*, **91**, 11591–11595.

Goss EM, Carbone I, Grunwald NJ (2009) Ancient isolation and independent evolution of the three clonal lineages of the exotic sudden oak death pathogen *Phytophthora ramorum*. *Molecular Ecology*, **18**, 1161–1174.

Goss EM, Tabima JF, Cooke DEL *et al.* (2014) The Irish potato famine pathogen *Phytophthora infestans* originated in central Mexico rather than the Andes. *Proceedings of the National Academy of Sciences, USA*, **201401884**.

Gostinčar C, Grube M, De Hoog S, Zalar P, Gunde-Cimerman N (2010) Extremotolerance in fungi: evolution on the edge. *FEMS Microbiology Ecology*, **71**, 2–11.

Gout L, Kuhn ML, Vincenot L *et al.* (2007) Genome structure impacts molecular evolution at the AvrLm1 avirulence locus of the plant pathogen *Leptosphaeria maculans*. *Environmental Microbiology*, **9**, 2978–2992.

Green J, Bohannan BJM (2006) Spatial scaling of microbial biodiversity. *Trends in Ecology & Evolution*, **21**, 501–507.

Guérin F, Gladieux P, Le Cam B (2007) Origin and colonization history of newly virulent strains of the phytopathogenic fungus *Venturia inaequalis*. *Fungal Genetics and Biology*, **44**, 284–292.

Hansen EM, Reeser PW, Sutton W (2012) Phytophthora beyond agriculture. *Annual Review of Phytopathology*, **50**, 359–378.

Hardham AR (2005) Phytophthora cinnamomi. *Molecular Plant Pathology*, **6**, 589–604.

Hayes KR, Barry SC (2008) Are there any consistent predictors of invasion success? *Biological Invasions*, **10**, 483–506.

Henk DA, Eagle CE, Brown K *et al.* (2011) Speciation despite globally overlapping distributions in Penicillium chrysogenum: the population genetics of Alexander Fleming's lucky fungus. *Molecular Ecology*, **20**, 4288–4301.

Hibbett DS, Taylor JW (2013) Fungal systematics: is a new age of enlightenment at hand? *Nature Reviews Microbiology*, **11**, 129–133.

Hood ME, Mena-Ali JI, Gibson AK *et al.* (2010) Distribution of the anther-smut pathogen *Microbotryum* on species of the Caryophyllaceae. *New Phytologist*, **187**, 217–229.

Hovmøller MS, Yahyaoui AH, Milus EA, Justesen AF (2008) Rapid global spread of two aggressive strains of a wheat rust fungus. *Molecular Ecology*, **17**, 3818–3826.

Hulme PE, Barrett SCH (2013) Integrating trait-and niche-based approaches to assess contemporary evolution in alien plant species. *Journal of Ecology*, **101**, 68–77.

Ianzen DH (1985) On ecological fitting. *Oikos*, **45**, 308–310.

Ivors K, Garbelotto M, Vries IDE *et al.* (2006) Microsatellite markers identify three lineages of Phytophthora ramorum in US nurseries, yet single lineages in US forest and European nursery populations. *Molecular Ecology*, **15**, 1493–1505.

Jairus T, Mpumba R, Chinoya S, Tedersoo L (2011) Invasion potential and host shifts of Australian and African ectomycorrhizal fungi in mixed eucalypt plantations. *New Phytologist*, **192**, 179–187.

Jezic M, Krstin L, Rigling D, Curkovic-Perica M (2012) High diversity in populations of the introduced plant pathogen, *Cryphonectria parasitica*, due to encounters between genetically divergent genotypes. *Molecular Ecology*, **21**, 87–99.

Karasov T, Messer PW, Petrov DA (2010) Evidence that adaptation in Drosophila is not limited by mutation at single sites. *PLoS Genetics*, **6**, e1000924.

Kasuga T, White TJ, Taylor JW (2002) Estimation of nucleotide substitution rates in eurotiomycete fungi. *Molecular Biology and Evolution*, **19**, 2318–2324.

Keller SR, Sowell DR, Neiman M, Wolfe LM, Taylor DR (2009) Adaptation and colonization history affect the evolution of clines in two introduced species. *New Phytologist*, **183**, 678–690.

Keller SR, Gilbert KJ, Fields PD, Taylor DR (2012) Bayesian inference of a complex invasion history revealed by nuclear and chloroplast genetic diversity in the colonizing plant, *Silene latifolia*. *Molecular Ecology*, **21**, 4721–4734.

Kennedy P (2010) Ectomycorrhizal fungi and interspecific competition: species interactions, community structure, coexistence mechanisms, and future research directions. *New Phytologist*, **187**, 895–910.

Kolar CS, Lodge DM (2001) Progress in invasion biology: predicting invaders. *Trends in Ecology & Evolution*, **16**, 199–204.

Le Gac M, Giraud T (2008) Existence of a pattern of reproductive character displacement in *Homobasidiomycota* but not in *Ascomycota*. *Journal of Evolutionary Biology*, **21**, 761–772.

Le Gac M, Hood ME, Fournier E, Giraud T (2007) Phylogenetic evidence of host-specific cryptic species in the anther smut fungus. *Evolution*, **61**, 15–26.

Le Van A, Gladieux P, Lemaire C *et al.* (2012) Evolution of pathogenicity traits in the apple scab fungal pathogen in response to the domestication of its host. *Evolutionary Applications*, **5**, 694–704.

Lewontin RC (1974) *The Genetic Basis of Evolutionary Change.* Columbia University Press, New York.

Liebhold AM, Brockerhoff EG, Garrett LJ, Parke JL, Britton KO (2012) Live plant imports: the major pathway for forest insect and pathogen invasions of the US. *Frontiers in Ecology and the Environment*, **10**, 135–143.

Lin X, Litvintseva AP, Nielsen K *et al.* (2007) αADα hybrids of *Cryptococcus neoformans*: evidence of same-sex mating in nature and hybrid fitness. *PLoS Genetics*, **3**, e186.

Lutz M, Göker M, Piatek M *et al.* (2005) Anther smuts of Caryophyllaceae: molecular characters indicate host-dependent species delimitation. *Mycological Progress*, **4**, 225–238.

MacArthur RH, Wilson EO (2001) The theory of island biogeography. *Princeton Landmarks in Biology*, **224**.

Markell SG, Milus EA (2008) Emergence of a novel population of *Puccinia striiformis* f. sp. *tritici* in eastern United States. *Phytopathology*, **98**, 632–639.

Martel A, Spitzen-van der Sluijs A, Blooi M *et al.* (2013) Batrachochytrium salamandrivorans sp. nov. causes lethal chytridiomycosis in amphibians. *Proceedings of the National Academy of Sciences, USA*, **110**, 15325–15329.

Martiny JBH, Bohannan BJM, Brown JH *et al.* (2006) Microbial biogeography: putting microorganisms on the map. *Nature Reviews Microbiology*, **4**, 102–112.

Maynard-Smith J, Smith NH, O'Rourke M, Spratt BG (1974) The hitch-hiking effect of a favourable gene. *Genetical Research*, **23**, 23–35.

Mayr E (1965) The nature of colonization of birds. In: *The Genetics of Colonizing Species* (eds Baker HG, Stebbins GL), pp. 29–44. Academic Press, New York, London.

Mboup M, Leconte M, De Vallavieille Pope C, Enjalbert J (2007) Evidence of genetic recombination in Chinese wheat yellow rust populations. *Phytopathology*, **97**, S72–S73.

Mboup M, Leconte M, Gautier A *et al.* (2009) Evidence of genetic recombination in wheat yellow rust populations of a Chinese oversummering area. *Fungal Genetics and Biology*, **46**, 299–307.

Mboup M, Bahri B, Leconte M *et al.* (2012) Genetic structure and local adaptation of European wheat yellow rust populations: the role of temperature-specific adaptation. *Evolutionary Applications*, **5**, 341–352.

McDonald B (2010) How can we achieve durable disease resistance in agricultural ecosystems? *New Phytologist*, **185**, 3–5.

McDonald BA (2014) Using dynamic diversity to achieve durable disease resistance in agricultural ecosystems. *Tropical Plant Pathology*, **39**, 191–196.

McDonald BA, Linde C (2002) Pathogen population genetics, evolutionary potential, and durable resistance. *Annual Review of Phytopathology*, **40**, 349–379.

McMahon TA, Brannelly LA, Chatfield MWH *et al.* (2013) Chytrid fungus *Batrachochytrium dendrobatidis* has nonamphibian hosts and releases chemicals that cause pathology in the absence of infection. *Proceedings of the National Academy of Sciences, USA*, **110**, 210–215.

Milus EA, Kristensen K, Hovmøller MS (2009) Evidence for increased aggressiveness in a recent widespread strain of

Puccinia striiformis f. sp. tritici causing stripe rust of wheat. *Phytopathology*, **99**, 89–94.

Montarry J, Andrivon D, Glais I et al. (2010) Microsatellite markers reveal two admixed genetic groups and an ongoing displacement within the French population of the invasive plant pathogen *Phytophthora infestans*. *Molecular Ecology*, **19**, 1965–1977.

Morgan JAT, Vredenburg VT, Rachowicz LJ et al. (2007) Population genetics of the frog-killing fungus *Batrachochytrium dendrobatidis*. *Proceedings of the National Academy of Sciences, USA*, **104**, 13845–13850.

Mueller GM, Strack BA (1992) Evidence for a mycorrhizal host shift during migration of *Laccaria trichodermophora* and other agarics into neotropical oak forests. *Mycotaxon*, **45**, 249–256.

Mundt CC, Wallace LD, Allen TW et al. (2013) Initial epidemic area is strongly associated with the yearly extent of soybean rust spread in North America. *Biological Invasions*, **15**, 1431–1438.

Munkacsi AB, Stoxen S, May G (2008) *Ustilago maydis* populations tracked maize through domestication and cultivation in the Americas. *Proceedings of the Royal Society B-Biological Sciences*, **275**, 1037–1046.

Naciri-Graven Y, Goudet J (2003) The additive genetic variance after bottlenecks is affected by the number of loci involved in epistatic interactions. *Evolution*, **57**, 706–716.

Neher RA (2013) Genetic draft, selective interference, and population genetics of rapid adaptation. *arXiv preprint arXiv:1302.1148*.

Palm ME (2001) Systematics and the impact of invasive fungi on agriculture in the United States. *BioScience*, **51**, 141–147.

Palm ME, Rossman AY (2003) Invasion pathways of terrestrial plant-inhabiting fungi. In: *Bioinvasions: Pathways, Vectors, and Management Strategies* (eds Ruiz GM, Carlton JT), pp. 31–43. Island Press, New York, New York.

Paoletti M, Buck KW, Brasier CM (2006) Selective acquisition of novel mating type and vegetative incompatibility genes via interspecies gene transfer in the globally invading eukaryote *Ophiostoma novo-ulmi*. *Molecular Ecology*, **15**, 249–262.

Parker IM, Gilbert GS (2004) The evolutionary ecology of novel plant-pathogen interactions. *Annual Review of Ecology, Evolution and Systematics*, **35**, 675–700.

Peay KG, Bruns TD, Kennedy PG, Bergemann SE, Garbelotto M (2007) A strong species–area relationship for eukaryotic soil microbes: island size matters for ectomycorrhizal fungi. *Ecology Letters*, **10**, 470–480.

Peay KG, Kennedy PG, Bruns TD (2008) Fungal community ecology: a hybrid beast with a molecular master. *BioScience*, **58**, 799–810.

Peay KG, Bidartondo MI, Arnold AE (2010) Not every fungus is everywhere: scaling to the biogeography of fungal-plant interactions across roots, shoots and ecosystems. *New Phytologist*, **185**, 878–882.

Peay KG, Schubert MG, Nguyen NH, Bruns TD (2012) Measuring ectomycorrhizal fungal dispersal: macroecological patterns driven by microscopic propagules. *Molecular Ecology*, **21**, 4122–4136.

Philibert A, Desprez-Loustau M-L, Fabre B et al. (2011) Predicting invasion success of forest pathogenic fungi from species traits. *Journal of Applied Ecology*, **48**, 1381–1390.

Prentis PJ, Wilson JRU, Dormontt EE, Richardson DM, Lowe AJ (2008) Adaptive evolution in invasive species. *Trends in Plant Science*, **13**, 288–294.

Pringle A, Vellinga EC (2006) Last chance to know? Using literature to explore the biogeography and invasion biology of the death cap mushroom *Amanita phalloides* (Vaill. ex Fr.: Fr.) Link. *Biological Invasions*, **8**, 1131–1144.

Pringle A, Baker D, Platt J et al. (2005) Cryptic speciation in the cosmopolitan and clonal human pathogenic fungus *Aspergillus fumigatus*. *Evolution*, **59**, 1886–1899.

Pritchard JK, Pickrell JK, Coop G (2010) The genetics of human adaptation: hard sweeps, soft sweeps, and polygenic adaptation. *Current Biology*, **20**, R208–R215.

Prussin AJ, Szanyi NA, Welling PI, Ross SD, Schmale III DG (2014) Estimating the production and release of ascospores from a field-scale source of *Fusarium graminearum* inoculum. *Plant Disease*, **98**, 497–503.

Pyšek P, Jarošík V, Hulme PE et al. (2010) Disentangling the role of environmental and human pressures on biological invasions across Europe. *Proceedings of the National Academy of Sciences, USA*, **107**, 12157–12162.

Raboin L-M, Selvi A, Oliveira KM et al. (2007) Evidence for the dispersal of a unique lineage from Asia to America and Africa in the sugarcane fungal pathogen *Ustilago scitaminea*. *Fungal Genetics and Biology*, **44**, 64–76.

Rafiqi M, Ellis JG, Ludowici VA, Hardham AR, Dodds PN (2012) Challenges and progress towards understanding the role of effectors in plant–fungal interactions. *Current Opinion in Plant Biology*, **15**, 477–482.

Refregier G, Le Gac M, Jabbour F et al. (2008) Cophylogeny of the anther smut fungi and their caryophyllaceous hosts: prevalence of host shifts and importance of delimiting parasite species for inferring cospeciation. *BMC Evolutionary Biology*, **8**, 100.

Rizzo DM (2005) Exotic species and fungi: interactions with fungal, plant and animal communities. *The Fungal Community: Its Organization and Role in the Ecosystem*, pp. 857–880. CRC Press, Boca Raton, Florida.

Robert VA, Casadevall A (2009) Vertebrate endothermy restricts most fungi as potential pathogens. *Journal of Infectious Diseases*, **200**, 1623–1626.

Roper M, Ellison C, Taylor JW, Glass NL (2011) Nuclear and genome dynamics in multinucleate ascomycete fungi. *Current Biology*, **21**, R786–R793.

Rouxel M, Mestre P, Comont G et al. (2013) Phylogenetic and experimental evidence for host-specialized cryptic species in a biotrophic oomycete. *New Phytologist*, **197**, 251–263.

Sakai AK, Allendorf FW, Holt JS *et al.* (2001) The population biology of invasive species. *Annual Review of Ecology & Systematics*, **32**, 305–332.

Sakalidis ML, Slippers B, Wingfield BD, Hardy GESJ, Burgess TI (2013) The challenge of understanding the origin, pathways and extent of fungal invasions: global populations of the *Neofusicoccum parvum–N. ribis* species complex. *Diversity and Distributions*, **19**, 873–883.

Saleh D, Milazzo J, Adreit H, Tharreau D, Fournier E (2012a) Asexual reproduction induces a rapid and permanent loss of sexual reproduction capacity in the rice fungal pathogen *Magnaporthe oryzae*: results of in vitro experimental evolution assays. *BMC Evolutionary Biology*, **12**, 42.

Saleh D, Xu P, Shen Y *et al.* (2012b) Sex at the origin: an Asian population of the rice blast fungus Magnaporthe oryzae reproduces sexually. *Molecular Ecology*, **21**, 1330–1344.

Sato H, Tsujino R, Kurita K, Yokoyama K, Agata K (2012) Modelling the global distribution of fungal species: new insights into microbial cosmopolitanism. *Molecular Ecology*, **21**, 5599–5612.

Schloegel LM, Toledo LF, Longcore JE *et al.* (2012) Novel, panzootic and hybrid genotypes of amphibian chytridiomycosis associated with the bullfrog trade. *Molecular Ecology*, **21**, 5162–5177.

Schulze-Lefert P, Panstruga R (2011) A molecular evolutionary concept connecting nonhost resistance, pathogen host range, and pathogen speciation. *Trends in Plant Science*, **16**, 117–125.

Schürch S, Linde CC, Knogge W, Jackson LF, McDonald BA (2004) Molecular population genetic analysis differentiates two virulence mechanisms of the fungal avirulence gene NIP1. *Molecular Plant-Microbe Interactions*, **17**, 1114–1125.

Schwartz MW, Hoeksema JD, Gehring CA *et al.* (2006) The promise and the potential consequences of the global transport of mycorrhizal fungal inoculum. *Ecology Letters*, **9**, 501–515.

Singh RP, Hodson DP, Huerta-Espino J *et al.* (2011) The emergence of Ug99 races of the stem rust fungus is a threat to world wheat production. *Annual Review of Phytopathology*, **49**, 465–481.

Smart CD, Fry WE (2001) Invasions by the late blight pathogen: renewed sex and enhanced fitness. *Biological Invasions*, **3**, 235–243.

Stajich JE, Berbee ML, Blackwell M *et al.* (2009) The fungi. *Current Biology*, **19**, R840–R845.

Stewart JE, Timmer LW, Lawrence CB, Pryor BM, Peever TL (2014) Discord between morphological and phylogenetic species boundaries: incomplete lineage sorting and recombination results in fuzzy species boundaries in an asexual fungal pathogen. *BMC Evolutionary Biology*, **14**, 38.

Strange RN, Scott PR (2005) Plant disease: a threat to global food security. *Annual Review of Phytopathology*, **43**, 83–116.

Stukenbrock EH, McDonald BA (2008) The origins of plant pathogens in agro-ecosystems. *Annual Review of Phytopathology*, **46**, 75–100.

Stukenbrock EH, Banke S, Javan-Nikkhah M, McDonald BA (2007) Origin and domestication of the fungal wheat pathogen I via sympatric speciation. *Molecular Biology and Evolution*, **24**, 398–411.

Stukenbrock EH, Christiansen FB, Hansen TT, Dutheil JY, Schierup MH (2012) Fusion of two divergent fungal individuals led to the recent emergence of a unique widespread pathogen species. *Proceedings of the National Academy of Sciences, USA*, **109**, 10954–10959.

Swann EC, Frieders EM, McLaughlin DJ (1999) *Microbotryum, Kriegeria* and the changing paradigm in basidiomycete classification. *Mycologia*, **91**, 51–66.

Talbot JM, Bruns TD, Taylor JW *et al.* (2014) Endemism and functional convergence across the North American soil mycobiome. *Proceedings of the National Academy of Sciences, USA*, **111**, 6341–6346.

Taylor JW, Fisher MC (2003) Fungal multilocus sequence typing—it's not just for bacteria. *Current Opinion in Microbiology*, **6**, 351–356.

Taylor DR, Keller SR (2007) Historical range expansion determines the phylogenetic diversity introduced during contemporary species invasion. *Evolution*, **61**, 334–345.

Taylor JW, Jacobson D, Fisher M (1999) The evolution of asexual fungi: reproduction, speciation and classification. *Annual Review of Phytopathology*, **37**, 197–246.

Taylor JW, Jacobson DJ, Kroken S *et al.* (2000) Phylogenetic species recognition and species concepts in fungi. *Fungal Genetics and Biology*, **31**, 21–32.

Taylor JW, Turner E, Townsend JP, Dettman JR, Jacobson D (2006) Eukaryotic microbes, species recognition and the geographic limits of species: examples from the kingdom Fungi. *Philosophical Transactions of the Royal Society B-Biological Sciences*, **361**, 1947–1963.

Tedersoo L, Bahram M, Toots M *et al.* (2012) Towards global patterns in the diversity and community structure of ectomycorrhizal fungi. *Molecular Ecology*, **21**, 4160–4170.

Thorsten Lumbsch H, Buchanan PK, May TW, Mueller GM (2008) Phylogeography and biogeography of fungi. *Mycological Research*, **112**, 423–424.

Thrall PH, Hochberg ME, Burdon JJ, Bever JD (2007) Coevolution of symbiotic mutualists and parasites in a community context. *Trends in Ecology & Evolution*, **22**, 120–126.

Thrall PH, Oakeshott JG, Fitt G *et al.* (2011) Evolution in agriculture: the application of evolutionary approaches to the management of biotic interactions in agro-ecosystems. *Evolutionary Applications*, **4**, 200–215.

Valade R (2012) Potentiel évolutif et adaptation des populations de l'agent du mildiou de la laitue, Bremia lactucae, face aux pressions de sélection de la plante hôte, *Lactuca sativa*.

Vellinga EC, Wolfe BE, Pringle A (2009) Global patterns of ectomycorrhizal introductions. *New Phytologist*, **181**, 960–973.

Vercken E, Fontaine MC, Gladieux P *et al.* (2010) Glacial refugia in pathogens: European genetic structure of anther smut pathogens on *Silene latifolia* and *Silene dioica*. *PLoS Pathogens*, **6**, e1001229.

de Vienne DM, Hood ME, Giraud T (2009) Phylogenetic determinants of potential host shifts in fungal pathogens. *Journal of Evolutionary Biology*, **22**, 2532–2541.

de Vienne DM, Giraud T, Gouyon P-H (2013a) Lineage selection and the maintenance of sex. *PLoS ONE*, **8**, e66906.

de Vienne DM, Refregier G, Lopez-Villavicencio M *et al.* (2013b) Cospeciation vs host-shift speciation: methods for testing, evidence from natural associations and relation to coevolution. *New Phytologist*, **198**, 347–385.

White TJ, Bruns T, Lee S, Taylor JW (1990) Amplification and direct sequencing of fungal ribosomal RNA genes for phylogenetics. *PCR Protocols: A Guide to Methods and Applications*, **18**, 315–322.

Wolfe BE, Richard F, Cross HB, Pringle A (2010) Distribution and abundance of the introduced ectomycorrhizal fungus Amanita phalloides in North America. *New Phytologist*, **185**, 803–816.

Zaffarano PL, McDonald BA, Linde CC (2009) Phylogeographical analyses reveal global migration patterns of the barley scald pathogen *Rhynchosporium secalis*. *Molecular Ecology*, **18**, 279–293.

DATA ACCESSIBILITY

Information on sampling locations as well as the microsatellite genotypic data has been deposited at Dryad: doi:10.5061/dryad.k5786.

SUPPORTING INFORMATION

Additional supporting information can be found at doi:10.1111/mec.13028

Fig. S1 Population structure of fungus *Microbotryum lychnidis-dioicae* collected on *Silene latifolia* ($N = 328$ individuals), inferred with STRUCTURE, for $K = 2$ to $K = 8$. Each vertical line represents an individual. Individuals are grouped on the horizontal axis by country.

Table S1 Cases of invasive fungi or oomycetes for which data are available on the population genetics of invasion or on the occurrence of genetic changes in introduced regions, such as on the evolution of mode of reproduction, the occurrence of bottlenecks or of hybridization.

Chapter 6

CONTEMPORARY EVOLUTION DURING INVASION: EVIDENCE FOR DIFFERENTIATION, NATURAL SELECTION, AND LOCAL ADAPTATION

Robert I. Colautti* and Jennifer A. Lau[†]

*Department of Biology, Queen's University, Kingston, ON K7L 3N6, Canada
[†]Kellogg Biological Station and Department of Plant Biology, Michigan State University, 3700 E. Gull Lake Dr., Hickory Corners, MI 49060, USA

Abstract

Biological invasions are 'natural' experiments that can improve our understanding of contemporary evolution. We evaluate evidence for population differentiation, natural selection and adaptive evolution of invading plants and animals at two nested spatial scales: (i) among introduced populations (ii) between native and introduced genotypes. Evolution during invasion is frequently inferred, but rarely confirmed as adaptive. In common garden studies, quantitative trait differentiation is only marginally lower (~3.5%) among introduced relative to native populations, despite genetic bottlenecks and shorter timescales (i.e. millennia vs. decades). However, differentiation between genotypes from the native vs. introduced range is less clear and confounded by nonrandom geographic sampling; simulations suggest this causes a high false-positive discovery rate (>50%) in geographically structured populations. Selection differentials ($|s|$) are stronger in introduced than in native species, although selection gradients ($|\beta|$) are not, consistent with introduced species experiencing weaker genetic constraints. This could facilitate rapid adaptation, but evidence is limited. For example, rapid phenotypic evolution often manifests as geographical clines, but simulations demonstrate that nonadaptive trait clines can evolve frequently during colonization (~two-thirds of simulations). Additionally, Q_{ST}-F_{ST} studies may often misrepresent the strength and form of natural selection acting during invasion. Instead, classic approaches in evolutionary ecology (e.g. selection analysis, reciprocal transplant, artificial selection) are necessary to determine the frequency of adaptive evolution during invasion and its influence on establishment, spread and impact of invasive species. These studies are rare but crucial for managing biological invasions in the context of global change.

Previously published as an article in *Molecular Ecology* (2015) 24, 1999–2017, doi: 10.1111/mec.13162

Invasion Genetics: The Baker and Stebbins Legacy, First Edition. Edited by Spencer C. H. Barrett, Robert I. Colautti, Katrina M. Dlugosch, and Loren H. Rieseberg.

INTRODUCTION

Biological invasions provide opportunities to study evolution over contemporary timescales in novel environments increasingly affected by global climate change, pollution and habitat fragmentation (Sax *et al.* 2007; Moran & Alexander 2014). Contemporary evolution in natural populations can occur both as an adaptive response to natural selection and through stochastic changes resulting from introduction history, founder effects and genetic drift. Evolution is considered to be adaptive when it increases survival or reproduction, which in turn affects population growth rates (λ). Identifying the causes and fitness consequences of adaptive evolution in natural populations is therefore informative for understanding long-term viability of populations experiencing novel environments.

Species colonizing new continents or spreading over large geographical areas are certain to experience environmental conditions that are novel in some way. Even if the climatic environment is similar to the native range, an introduced species still faces novel edaphic and biotic environments that may include different microbial, herbivore or predator communities, distinct competitive regimes, a lack of mutualists, or novel prey species. Adaptive evolution of traits that increase survival and reproduction in these novel environments will facilitate establishment and proliferation, increasing the number of potential colonists to new areas (García-Ramos & Rodríguez 2002). Thus, the rate and extent of adaptive evolution may be crucial factors affecting the establishment and spread of invading species. However, the majority of studies examining biological invasions have focused on testing ecological theory with little or no consideration of evolution.

The lack of evolutionary theory in invasion biology may owe to the strong influence of Elton's (1958) volume 'The Ecology of Invasions by Animals and Plants', which briefly mentions polyploidy and pesticide resistance but otherwise does not consider evolutionary explanations for the establishment and spread of invasive species. The long absence of evolutionary theory from invasion biology may be surprising given that Darwin (1859) included several examples of 'naturalizations' to support his theory of evolution by natural selection in 'The Origin of Species' (Ludsin & Wolfe 2001). Evolution was also a major focus of Baker and Stebbins (1965) 'The Genetics of Colonizing Species', which was published just a few years after Elton's book (Barrett 2015). However, contributors to this volume overlooked the possibility of rapid adaptive evolution during colonization, perhaps because the dominant view at the time mirrored Darwin's (1859) original concept of a 'slow and gradual accumulation of numerous slight, yet profitable, variations'. This view is clear in the introduction to Stebbins' (1965) contribution, which despite explicitly examining genetic and adaptive characteristics of weeds associated with human activity concludes: 'Consequently, weeds of one sort or another have probably existed since the middle of the Pleistocene epoch, and the major groups of weedy species may well have virtually completed their course of evolution by the time the first great civilizations reached their climax 2500–3000 years ago in the Old World and 1000–1500 years ago in the New World centers'.

In contrast to the long-held view of adaptations evolving over millenia, recent studies demonstrate that evolution can occur fast enough to influence ecological dynamics (reviewed in Hendry & Kinnison 1999; Reznick & Ghalambor 2001; Stockwell *et al.* 2003; Schoener 2011). Biologists have only recently begun to examine adaptive evolution during biological invasions, and the extent to which adaptive evolution influences ecological dynamics of natural populations in general remains poorly understood. Here we use a combination of meta-analyses, simulations and case studies to review evidence for rapid adaptive evolution in invasive species; we focus on its three key components: differentiation, natural selection and adaptation. In each case, we compare genotypes from the native vs. introduced range as well as among populations within the introduced range. Our approach illustrates both the value of using biological invasions to understand rapid evolutionary responses to novel environments and the potential influence of evolution on invasion dynamics. We conclude with suggestions for future research examining how evolution influences the establishment, spread and impacts of invading species, and potential differences in the response of invasive vs. native species to global change.

DIFFERENTIATION

Divergence is simply the process by which two groups of organisms become genetically differentiated over time (Gulick 1888; Futuyma 2013), but this concept can be complicated to apply to invasive species. In classic models, divergence begins with a common ancestral population, but contemporary populations of invasive species may include different admixtures of genes or genotypes. Reconstructing the history of colonization and admixture is not a simple endeavour and can require extensive sampling, depending on the extent of geographical structuring in the native range (Keller & Taylor 2008; Cristescu 2015). Given the complications of accurately reconstructing multiple introductions and long-distance dispersal events during spread, it may be more practical in many cases to simply examine how introduced populations are differentiated (i.e. genetically different). Here, we consider population differentiation measured for noncoding molecular markers and quantitative traits across two hierarchical spatial scales: (i) between the native and introduced range–usually separate continents–and (ii) among populations within the introduced range.

Between ranges

Neutral molecular markers

The conventional view is that introduced populations will inevitably possess lower levels of genetic diversity than populations from the native range. But evidence from neutral genetic markers indicates that diversity is only moderately reduced during invasion (Roman & Darling 2007; Dlugosch & Parker 2008), with an average within-population reduction of about 10–20% in both average heterozygosity and allelic richness ($N \sim 80$ spp.). This should be considered a mild reduction from an evolutionary perspective given that neutral genetic variation is reduced by bottlenecks much more rapidly than is genetic variation for quantitative traits (Lande 1988). Moreover, admixture following multiple introductions from divergent source regions represents a fundamental restructuring of genetic variation from differentiation among populations into standing genetic variation within populations. A rich body of research examining inter- and intraspecific hybridization suggests at least three scenarios in which genetic restructuring could affect the evolution of invasive species (see Lee 2002; Seehausen 2004; Roman & Darling 2007; Rius & Darling 2014). These are briefly outlined below (see also Bock et al. 2015).

First, combining multiple differentiated lineages should increase standing genetic variation. An increase in genetic variation for ecologically relevant traits will generally increase the rate of adaptive evolution in response to natural selection (Fisher 1930; Falconer &

Mackay 1996; Lynch & Walsh 1998). Second, hybridization between divergent lineages often produces transgressive segregation (Rieseberg et al. 1999; Ellstrand & Schierenbeck 2000)—the creation of novel or more extreme phenotypes than those present in the parental generation (Rick & Smith 1953; deVicente & Tanksley 1993). This could create recombinant genotypes or linkage groups that increase fitness across a variety of environments in the introduced range. These gene combinations for increased invasiveness would be transient unless maintained through clonal reproduction, selfing, or suppression of recombination through genome rearrangement (e.g. chromosome inversion). Third, hybridization between divergent lineages will increase heterozygosity at loci that are alternatively fixed in the parental populations, which could increase mean fitness through overdominance (i.e. heterozygote advantage) or by masking deleterious mutations that are fixed at alternate loci in native populations. The fitness advantage of overdominance should be transient over longer timescales as allele frequencies drift towards fixation at each locus, but could be important in early stages of invasion (Drake 2006). In contrast, deleterious alleles can be purged by natural selection or fixed by genetic drift during the early stages of invasion when populations are still small, resulting in increases or decreases in mean population fitness, respectively.

Studies using molecular genetic markers frequently identify evidence for multiple introductions followed by admixture in the introduced range (Dlugosch & Parker 2008; Dlugosch et al. 2015). The three scenarios outlined above explain how multiple introductions and admixture could facilitate rapid evolution during biological invasion. However, determining whether admixture contributes significantly to the rate or extent of invasive spread first requires ecological experiments that quantify genetic variation of phenotypic traits that are important for survival and reproduction.

Common garden experiments

Common garden and reciprocal transplant experiments are classic approaches to study the genetic basis of ecologically relevant phenotypic variation within and among natural populations (Langlet 1971; Linhart & Grant 1996; Leimu & Fischer 2008; Hereford 2009) dating back to Clausen et al. (1940). By growing different genotypes, genetic families or populations in a relatively uniform environment, it is possible to quantify genetic differences for any measurable phenotypic trait. Common garden studies of contemporary evolution in invasive species have largely focused on whether introduced populations are larger, better competitors than native populations, with faster growth rates, more efficient metabolisms and weaker defences against specialist natural enemies (reviewed in Bossdorf *et al.* 2005; Colautti *et al.* 2009). The majority of studies have been motivated by Blossey & Nötzold's (1995) influential 'evolution of increased competitive ability' (EICA) hypothesis, which proposes an evolutionary shift from costly defences to growth and reproduction following escape from specialist herbivores.

Common garden comparisons between native and introduced genotypes of individual species reveal mixed results for most measured traits. For example, introduced populations of some species are larger (Blossey & Nötzold 1995; Joshi & Vrieling 2005; Yang *et al.* 2014), while others are smaller (e.g. Bossdorf *et al.* 2004; Vilà *et al.* 2005), or not different (e.g. McKenney *et al.* 2007), relative to native genotypes. However, introduced populations do appear to be significantly larger than their native conspecifics when averaged across a number of species (Blumenthal & Hufbauer 2007) and controlling for geographical clines (Colautti *et al.* 2009). This suggests a general trend towards evolution of larger size in introduced populations.

Quantitative trait differentiation observed between native and introduced genotypes may be adaptive, but stochastic processes such as founder effects and genetic drift in the introduced range also can cause differentiation of introduced populations from native ancestors. This is an important hypothesis to consider in any common garden comparison involving individual species (Keller & Taylor 2008). Stochastic processes alone are probably not responsible for the average increase in size observed in the introduced range because nonadaptive processes have an equal chance to increase or decrease size for any individual species. This difference should be close to zero when averaged across many species. Differentiation between native and introduced genotypes for size and other quantitative traits closely associated with fitness could alternatively be an effect of hybrid vigour resulting from admixture during invasion, as discussed above. A few studies have identified associations between admixture and size (Kolbe *et al.* 2007) or fecundity (Keller & Taylor 2010; Turgeon *et al.* 2011; Keller *et al.* 2014; but see Wolfe *et al.* 2014), but further studies are needed to determine whether this phenomenon is common among invading species.

A number of experimental methods exist for exploring the adaptive significance of quantitative trait differences observed in common garden studies, but these have rarely been applied to studies of invasive species. Well-established methods from ecological genetics (Conner & Hartl 2004) are available to directly measure natural selection (Lande & Arnold 1983), quantify genetic constraints (Lynch & Walsh 1998; Blows & McGuigan 2015) and to test whether evolution has been adaptive (see 'Adaptation' section, below). Combining these methods is necessary to rule out stochastic processes and identify the specific environmental factors maintaining adaptive differentiation.

Methods from ecological genetics could significantly improve tests of the multitude of ecological and evolutionary hypotheses formulated to explain how species become invasive (reviewed in Catford *et al.* 2009). For example, a conclusive demonstration of EICA should only begin with (i) a common garden study to confirm that introduced genotypes are larger but less defended. Identifying the mechanisms underlying such evolutionary change additionally requires (ii) quantitative genetics experiments to confirm a genetic trade-off (i.e. a negative genetic correlation) between size and defence and (iii) a selection analysis to demonstrate relaxed selection on defences and selection for increased size in the introduced range. Finally, (iv) a reciprocal transplant experiment is necessary to confirm that genetic differences in size and defence increase fitness in their range of origin. Such an approach could be complemented by additional experiments in the native range to bolster evidence that escape from enemies is the selective agent responsible for the evolution of increased size in the introduced range. For example, artificial selection experiments in the presence and absence of herbivores could be used for testing genetic constraints and adaptive responses to selection following release from natural enemies (Uesugi & Kessler 2013). Similarly, conducting selection analyses in experiments manipulating the presence of natural enemies in field experiments would also quantify shifts in natural selection definitively associated with changes in the herbivore community (Wade & Kalisz 1990). Similar experiments could be used to test other popular adaptive hypotheses in invasion biology including the evolution of novel weapons (Callaway & Ridenour 2004) and the evolution of reduced competitive ability (Bossdorf *et al.* 2004). In addition to improving tests for adaptive differentiation resulting from ecological differences between the native and introduced range, similar ecological genetics experiments could improve understanding of the evolutionary significance of quantitative trait differentiation among populations within the introduced range, as described in the next section.

Within the introduced range

Many common garden studies identify population differentiation for quantitative traits among introduced populations of plants (reviewed in Colautti *et al.* 2009) and animals (e.g. Huey 2000; Phillips *et al.* 2006; Gomi 2007). Frequently, population differentiation in quantitative traits form geographically structured clines along latitudinal or altitudinal gradients. For example, introduced populations of *Drosophila subobscura* (Huey 2000) and *Lythrum salicaria* (Montague *et al.* 2008) have each evolved latitudinal clines in size that parallel those observed in the native range. In both cases, introduced populations evolved within a few decades. Is rapid differentiation among introduced populations unique to a few case studies like these, or is it a common characteristic of introduced species?

We examined genetic differentiation in ecologically important quantitative traits among native and introduced populations in 35 common garden studies of 22 plant species using data from a previous meta-analysis (Colautti *et al.* 2009), updated in 2011 (Ross & Auge 2008; van Kleunen & Fischer 2008; Henery *et al.* 2010; Schlaepfer *et al.* 2010; see Appendix S1, Supporting information). We focus here on plants because few studies of animals include common garden comparisons of native and introduced populations (but see Huey 2000; Seiter & Kingsolver 2013). Multiple traits were measured on each species, which we combined into a single primary principal component (PC) 'trait' for each of four trait categories: physiological and growth rates, plant size, herbivore defences, and survival and reproduction (see Colautti *et al.* 2009 for further detail). We quantified the extent of differentiation as the standard deviation among population means, calculated separately among native and introduced populations for each trait category in each species. We also considered the extent of geographical sampling by calculating the standard deviation among the latitudes of population origin. Our intention is not to provide a detailed meta-analysis that considers all of the factors affecting rapid evolution. Instead, we test generally how differentiation among introduced populations compares to populations in the native range.

Overall, we found that population differentiation was higher in the native range, on average, than in the

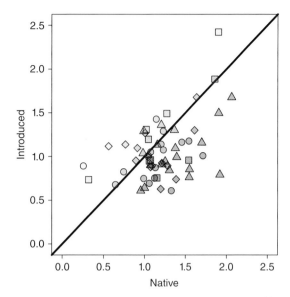

Fig. 1 Average genetic differentiation for ecologically relevant traits measured among native (*x*-axis) and introduced (*y*-axis) populations in 35 common garden studies of 22 plant species. Average population differentiation is quantified as standard deviation of principal component (PC) scores calculated for traits grouped into one of four categories: (diamonds) physiology and growth rate, (triangles) plant size, (squares) defence traits and (circles) survival and reproduction. Shading indicates deviations from equal differentiation among populations in each range (grey line), ranging from dark grey (introduced > native) to light grey (native > introduced).

introduced range (Fig. 1). Reduced quantitative trait differentiation is perhaps not surprising given that introduced populations are younger and frequently experience genetic bottlenecks (Dlugosch & Parker 2008). However, the difference is not large (−3.5%), ranging from a −21% (size) to +12% (defence), and in one-third of cases (20 of 60), trait differentiation was higher in the introduced range. This result may be somewhat counterintuitive given a difference in timescales between ranges that is typically two or three orders of magnitude: tens of thousands of years following deglaciation (or longer) in the native range vs. decades of spread and establishment in the introduced range. Although traits are correlated with latitude in several species (see Colautti *et al.* 2009), the variation in latitude of population origin (i.e. SD of latitude) had no effect on the extent of population differentiation ($P = 0.65$). Our analysis does not distinguish the effects of natural selection, founder effects and genetic drift. Rather, our results demonstrate the magnitude and speed with which these processes can cause rapid evolution and population differentiation in the introduced range.

The level of quantitative trait differentiation observed among introduced populations in Fig. 1 has two important consequences. First, it poses significant challenges for identifying and interpreting trait differentiation observed between native and introduced ranges. This is because quantitative trait variation is usually spatially clustered (e.g. isolation by distance) or geographically structured (e.g. latitudinal clines) and sampling locations are not randomly chosen (Box 1).

Box 1 Nonrandom sampling in geographically structured populations

Nonrandom sampling of populations can complicate studies of trait differentiation between native and introduced ranges when traits are geographically structured. To test for effects of nonrandom sampling, we simulated a quantitative trait (e.g. flowering time) that tracks environmental characteristics correlated with latitude (e.g. photoperiod) but otherwise does not differ between the native and introduced range. We used latitudes from sampling locations reported in each of 33 published studies of 21 species (excluding studies with fewer than five populations) and calculated the range effect size. We calculated the average difference between native and introduced populations using a t-test of population means. This is analogous to treating population as a random factor in a mixed model but there is no error associated with the calculation of population means. We also simulated a null distribution by resampling the same number of populations in each study but with latitudes chosen randomly from all of the study sites in the meta-analysis. Comparison of these null expectations with results from sampling latitudes used in individual studies can test for effects of nonrandom spatial sampling (e.g. clustering).

Simulations using sampling sites from individual studies frequently resulted in false-positive differences between ranges, both at significant (79% $P<0.05$) and highly significant (48% $P<0.001$) levels. Only 7 of 31 tests of the effect size correctly failed to reject the null hypothesis (Fig. 1). The average difference between the introduced and native range (i.e. effect size) of our hypothetical trait was small (−0.07) and nonsignificant ($P=0.362$) when all sample sites were pooled in our analysis ($N=654$). When fewer locations were randomly chosen across all studies, they showed the expected number of significant (~5% $P<0.05$) and highly significant results (~0.1% $P<0.001$), regardless of the number of populations sampled per range (shaded regions in Fig. 1). Even the largest studies in our meta-analysis (Maron et al. 2004; Henery et al. 2010) resulted in false-positive significant differences due to spatially structured sampling. In most cases (20 of 26), false-positive significant results from individual studies fell within null expectations (i.e. red and orange dots within shaded areas of

Fig. 1) — an expected result of nonrandom residual error terms in spatially clustered samples. These results suggest that meaningful estimates of phenotypic differentiation between native and introduced ranges may require dozens to hundreds of widely distributed populations to account for spatial nonindependence of samples and the potential effects of environmental gradients.

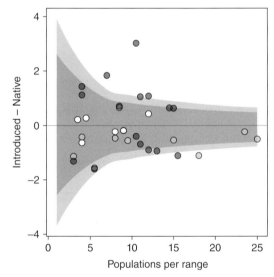

Box 1, Fig. 1 Simulated effect sizes quantifying the average difference between native and introduced genotypes (y-axis) for a quantitative trait (e.g. flowering time) in common garden comparisons of N populations from each range (x-axis). Shaded regions show the 95% (dark grey) and 99% (light grey) expectation of effect size (y-axis) for populations chosen at random from each range. Circles show the effect sizes obtained for population locations used in 33 published studies of 21 species, with shades indicating statistical nonsignificance (white), significance at $P<0.05$ (grey) and significance at $P<0.001$ (black). The effect size averaged across all population locations is −0.07 ($P=0.362$; horizontal line).

A comprehensive study of differentiation between native and introduced genotypes may therefore require large sample sizes analysed using spatially explicit statistics. Second, if population differentiation is a result of contemporary evolution in response to environmental variation within the introduced range, this could directly affect establishment and spread. Field measurements of natural selection are important for understanding the extent to which population differentiation is adaptive.

NATURAL SELECTION

Introduced species generally colonize environments that are novel in some way. Indeed, numerous hypotheses link environmental differences between the native and introduced range to the success of invasive species (reviewed in Catford *et al.* 2009). While pre-adaptation may be a key to establishment when environmental conditions in the introduced range are similar to conditions in the native range, introduced species are likely to be evolutionarily naïve to some biotic and abiotic aspects of their new environments. Even if the introduced range is within the environmental niche of the species (e.g. Alexander 2013; Petitpierre *et al.* 2013), introduced populations will almost certainly inhabit novel conditions relative to their sites of origin in the native range. Here, we review empirical measurements of natural selection in wild populations to investigate changes to the adaptive landscape occurring (i) between the native and introduced range and (ii) among introduced populations.

Between ranges

When populations are well-adapted to their local growing conditions, and environmental conditions are relatively stable, then an adaptive quantitative trait will experience strong stabilizing and weak directional selection. If introduced species colonize environments to which they are evolutionary naïve, and native populations are relatively well-adapted, then directional selection on introduced species should be stronger than for native species, at least early in the invasion process. To test whether introduced taxa experience stronger directional selection than native taxa, we classified the introduction status (native or introduced) of taxa included in a published data set of selection coefficients (Kingsolver & Diamond

2011), combined with more recent literature searches targeted at studies measuring selection on introduced species (see Appendix S2, supporting information). Selection differentials (s) measure the combined effects of direct selection acting on a focal trait and indirect selection on correlated traits. Selection gradients (β) measure direct selection on a trait of interest, after controlling for selection on correlated traits (Lande & Arnold 1983). We identified only nine studies (five plant and four animal species) that measured selection on naturally occurring introduced populations in the wild (Wittzell 1991; O'Neil 1997; Hendry *et al.* 2003; Carlson *et al.* 2004; Yeh & Price 2004; Schueller 2007; Price *et al.* 2008; Murren *et al.* 2009; O'Donnell & Pigliucci 2010). Several other studies investigated patterns of selection on introduced species, but used synthetic populations comprised of multiple source populations, experimental crosses between distant source populations, or were planted into areas or into environmental conditions where the introduced species did not naturally occur (see Table S1, Supporting information). These nine studies contributed a total of 185 linear selection differentials (s) and 152 selection gradients (β). In contrast, our data set included 149 studies that measured selection on naturally occurring native populations in the wild, yielding a total of 2581 selection differentials and 2419 selection gradients. These studies include measurements of selection in vastly different environments, on a wide variety of organisms, and on a diverse array of traits. More studies of natural selection in introduced populations are needed to test for differences in selection among particular taxa and traits, and to better understand the role evolutionary processes play in invasion success.

We used a multistage (i.e. hierarchical) bootstrap test to compare means and confidence intervals of selection differentials (s) and selection gradients (β) in native and introduced species. In each case, we were interested in the strength of selection (i.e. absolute value), rather than its direction. In each of 10000 bootstrap permutations, we randomly sampled traits from each species, and selection measurements (s or β) for each trait within each species, with replacement. We did not resample species because of the small number of studies of introduced (9 species in 9 studies) relative to native species (127 species in 104 studies). This is analogous to treating species as a fixed effect in statistical mixed models.

On average, selection differentials (s) were almost 40% stronger in introduced species (mean: 0.314;

95% CI: 0.266–0.360) than native species (mean: 0.225; 95% CI: 0.209–0.242) (Fig. 2a). In contrast, the magnitude of selection gradients (β) was similar for introduced (mean: 0.225; 95% CI: 0.175–0.278) and native taxa (mean: 0.238; 95% CI: 0.221–0.255) (Fig. 2b). Thus, while direct selection on traits does not differ significantly, the predicted evolutionary response to selection is faster in introduced taxa. This contrasting result could be an experimental artefact: β is difficult to measure accurately and biased towards zero with large standard errors (Rausher 1992; Stinchcombe *et al.* 2002; Kingsolver *et al.* 2012). However, the 95% bootstrapped confidence intervals (CIs) of introduced species were similar for |s| and |β| (± 0.05), suggesting similar measurement errors. Instead, the different results of |s| and |β| could reflect weaker genetic constraints in invading species, perhaps as a result of simplified species communities and/or changes in the genetic covariance structure of invading populations. For example, trade-offs in defences against 11 different herbivore species constrained the rate of adaptive evolution by over 60% in a native *Solanum carolinense* population (Wise & Rausher 2013). This species is native to the United States but spreading in Europe where a change in the genetic variance–covariance structure of defence traits could facilitate a more rapid adaptive evolutionary response, resulting in larger selection differentials (s) given the same selection gradients (β). Additionally, escape from a subset of herbivores during invasion would relax antagonistic selection, resulting in larger selection differentials given the same genetic covariance.

Our analysis supports stronger selection differentials in introduced relative to native species. However, this result is contingent on the relatively small ($N=9$) introduced species included in our data set. More studies measuring selection on introduced taxa are needed to determine the generality of this finding and more generally to characterize natural selection associated with range expansion.

Within the introduced range

Introduced species undergoing rapid range expansion over large geographical areas likely experience very different selection pressures across their range. This spatial variability in selection may result from variation in environmental conditions across the newly colonized range. For example, the invasive

(a)

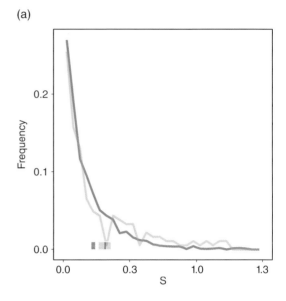

(b)

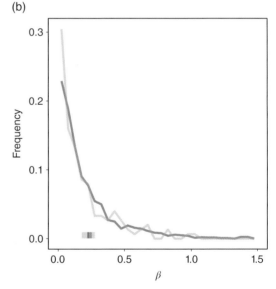

Fig. 2 Frequency distributions comparing the magnitude of (a) selection differentials (|s|) or (b) selection gradients (|β|), measured in naturally occurring populations of native (dark grey) and introduced (light grey) species. Frequency distributions are binned at 0.05 intervals. Bars show bootstrapped means (vertical lines) and 95% confidence intervals (rectangles) estimated from 10 000 bootstrap iterations. The unit of replication is an estimate of s ($N=2581$ native, 185 introduced) or β ($N=2419$ native, 152 introduced) for an individual trait, sampled hierarchically with replacement from each of 127 native and 9 introduced species.

plant *Lythrum salicaria* has colonized a broad latitudinal gradient across North America, and selection on phenological traits varies across this latitudinal gradient as a function of growing season length (Colautti *et al.* 2010; Colautti & Barrett 2013). Alternatively, spatial variability in selection may result from the invasion process itself. During an active range expansion, theory predicts the evolution of increased dispersal and reproduction concurrent with reduced inter- and intraspecific competitive ability at the invasion front as a result of both spatial sorting effects and natural selection (reviewed in Phillips *et al.* 2010a,b). This prediction is supported by work on invasive Australian cane toads (*Rhinella marina*), showing that *R. marina* individuals with longer legs move faster and are the first to arrive at invasion fronts (Phillips *et al.* 2006), leading to the evolution of increased dispersal ability at range edges (Phillips *et al.* 2010b). Few studies have explicitly quantified selection across multiple sites within an introduced range, but evolution in response to spatial variability in selection may be important to both invasion rate and impacts on native ecosystems.

ADAPTATION

Ecological differences between native and introduced ranges could result in local adaptation between the native and introduced range. For example, an evolutionary response to a shift from specialist to generalist enemies could result in local adaptation at intercontinental scales, as predicted by Blossey & Nötzold's (1995) 'evolution of increased competitive ability' hypothesis, and the related 'shifting defence hypothesis' (Müller-Schärer *et al.* 2004). Within the introduced range, rapid evolution of local adaptation can dramatically enhance survival and reproduction (Colautti & Barrett 2013), which is predicted to facilitate invasion (García-Ramos & Rodríguez 2002). Understanding the extent of local adaptation is therefore crucial to predicting the spread of invasive species.

Here, we review empirical evidence for adaptation between the native and introduced range, and among introduced populations. In addition to the quantitative trait clines in common garden experiments described previously, we consider evidence from reciprocal transplant experiments and studies comparing quantitative trait divergence with inferences from neutral markers (i.e. Q_{ST} vs. F_{ST}).

Between ranges

Adaptive evolution in response to environmental differences between the native and introduced range is extremely difficult to assess because it first requires detecting a significant increase in fitness of the local, relative to foreign, genotypes in each range, using enough populations at enough sites to account for any effects of adaptation to local conditions within each range. The use of molecular markers to match introduced populations with their putative source regions is generally not a robust alternative for quantitative traits that evolve rapidly (Box 3). Instead, many common gardens with genotypes sampled across the native and introduced range are required to understand the spatial scale of adaptation, but this is difficult for widespread invasive species. Most common garden studies do not offer a very robust test of adaptive differentiation between ranges as they typically compare the performance of introduced and native genotypes at a single site (e.g. Table 1 in Colautti *et al.* 2009), or less commonly in at least one site in each range (e.g. Maron *et al.* 2004, 2007; Genton *et al.* 2005; Williams *et al.* 2008; Keller *et al.* 2009, 2014). These studies generally do not find significantly higher fitness in the local vs foreign populations at intercontinental scales (but see Williams *et al.* 2008). As a consequence of experimental limitations, evidence for adaptive differentiation between the native and introduced range should be interpreted cautiously.

Transcontinental reciprocal transplant studies rarely measure key fitness components (i.e. survival and reproduction), substituting growth measurements as proxies of fitness. Although growth and life history traits often have strong effects on fitness, their relationships with fitness can change dramatically across environments. For example, southern genotypes of *Lythrum salicaria* are consistently larger than northern genotypes within the introduced range in eastern North America, regardless of whether they are grown in northern or southern common gardens (Colautti & Barrett 2013). In this case, local adaptation measured as shoot growth would fail to detect local adaptation, with the southern genotypes always being larger than northern ones, even though most southern genotypes do not produce any offspring in the short growing seasons of northern locales. Instead, measurements of survival and reproduction support local adaptation because a trade-off between vegetative growth and flowering time determines fitness in each environment.

Another reason to interpret trans-continental reciprocal transplant studies cautiously is that even as many as four common garden locations per continent is logistically challenging while insufficient to capture the level of environmental variation present in the native and introduced ranges of many species. For example, consider the results in Box 1 in which a phenotypic trait (e.g. flowering time) correlates with an environmental variable (e.g. photoperiod), but in this case, we assume that clines are locally adaptive, as shown in *Lythrum salicaria* (Colautti *et al.* 2010; Colautti & Barrett 2013). In this case, a trans-continental reciprocal transplant could result in apparent local adaptation, local maladaptation, superior introduced genotypes or superior native genotypes, depending on the representation of northern and southern genotypes from each range. The scope of this problem depends on the extent of local adaptation within the native and introduced ranges.

Within the introduced range

Studies of adaptive evolution in invasive species have focused primarily on comparing native and introduced ranges, but environments are often highly variable within ranges (e.g. across a continent). Differences in habitat and climate among sites within the introduced range are likely to select for different phenotypes, but evolution of local adaptation may be limited by gene flow from maladapted populations or by trade-offs among fitness-related traits. To assess the extent of local adaptation among introduced populations, we review three lines of evidence: (i) geographical clines correlated with environmental characteristics, (ii) reciprocal transplant studies showing higher fitness in local relative to foreign genotypes and (iii) divergence in quantitative traits in excess of neutral divergence inferred using neutral markers (i.e. Q_{ST}-F_{ST} comparisons).

Geographical clines

Geographical clines in quantitative traits have been identified frequently in studies of introduced animals (e.g. Huey 2000; Phillips *et al.* 2006; Gomi 2007) and plants (reviewed in Colautti *et al.* 2009; see also Alexander *et al.* 2009). Natural selection varying along

environmental gradients can create and maintain phenotype–environment correlations indicative of local adaptation (Nagylaki 1975; Slatkin 1978; Mallet & Barton 1989). However, founder effects and random genetic drift can cause the evolution of nonadaptive (or maladaptive) clines in at least two ways. First, multiple introductions of divergent lineages to different geographical locations can have a strong influence on the phylogeographic structure of introduced populations and associated phenotypes, especially when recombination is limited (Keller *et al.* 2009). Second, serial founder effects during spread can cause rapid increases in allele frequency—a process sometimes called 'allele surfing' (reviewed in Excoffier *et al.* 2009; Peischl & Excoffier 2015). In a process analogous to allele surfing and genetic drift, nonadaptive geographical clines in multilocus quantitative traits can evolve rapidly in invading species (Box 2). The possibility that geographical clines can evolve from stochastic mechanisms alone suggests an urgent need for further investigation into the adaptive (or nonadaptive) relevance of geographical clines observed in common garden studies of invading species. Reciprocal transplant experiments are perhaps the most direct test of whether clines or other patterns of population differentiation are locally adaptive.

Reciprocal transplant studies

Reciprocal transplant experiments are a conceptually simple and direct test of local adaptation, but have only rarely been used to examine adaptive evolution of invasive species (Rice & Mack 1991; Maron *et al.* 2004), until recently (Kinnison *et al.* 2008; Leger *et al.* 2009; Erfmeier & Bruelheide 2010; Ridley & Ellstand 2010; Ebeling *et al.* 2011; Colautti & Barrett 2013; Rice *et al.* 2013; Pahl *et al.* 2013; Westley *et al.* 2013; Burger & Ellstrand 2014; Li *et al.* 2015). Definitive evidence for local adaptation comes in the form of crossing reaction norms for fitness (i.e. survival and reproduction) with the local populations (or genotypes) having significantly higher fitness when compared to others at each of several common garden sites; ideally this is assessed over multiple growing seasons or generations. Applying these strict criteria, we know of only a single study confirming local adaptation at all study sites (Colautti & Barrett 2013). Several additional studies of plants (Rice & Mack 1991; Colautti & Barrett 2013; Li *et al.*

Box 2 Nonadaptive clines in a quantitative trait

Clines in quantitative traits observed in a colonizing species could evolve nonadaptively through serial founder effects. To demonstrate, we use a simple individual-based, spatially explicit simulation model of a selectively neutral trait affected by 5–500 independent loci in Hardy–Weinberg equilibrium with additive effects on phenotype. Our model simulates the rapid spread of an invasive species across a geographical gradient, beginning with a founding population with equal allele frequencies at each locus ($p = q = 0.5$). Invasive spread proceeds along a linear geographical gradient in a series of stepwise colonization events. In each colonization step, a new population is founded by sampling 5–500 individuals composed of multilocus genotypes, each randomly sampled from the nearest established population. After colonizing 20 populations, we calculate the linear regression of population phenotype with position along the geographical gradient to test for significant clines.

Our simulations reveal significant geographical clines in a quantitative trait arising frequently through serial founder effects alone (Fig. 1). These results are similar to novel mutations increasing in frequency to fixation at individual loci over hundreds to thousands of generations (i.e. allele surfing; reviewed in Excoffier *et al.* 2009; Peischl & Excoffier 2015). However, in contrast to allele surfing models, clines arise over much shorter timescales (decades)—long before alleles reach fixation. Although the average phenotype does not change significantly across 1000 simulations, about two-third of individual simulations generate significant clines ($P < 0.05$), with highly significant clines ($P < 0.001$) occurring in about half, independent of the number of founders and the number of loci. These results demonstrate how geographical clines in quantitative traits can evolve within years to decades through nonadaptive processes. In contrast to rapid evolution of local adaptation, nonadaptive clines in our simulated quantitative traits do not affect population vital rates or the rate of spread of invasive species.

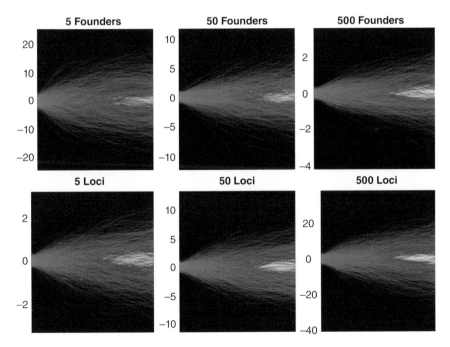

Box 2, Fig. 1 Rapid evolution of nonadaptive geographical clines in simulated invasions. The change in a population's mean phenotype for a quantitative trait (*y*-axis) in 20 populations distributed along a spatial gradient (*x*-axis) is shown for each of 1000 replicates in each panel. Highly significant clines ($P < 0.001$), based on linear regression, are shown in yellow (~50% of simulations in each panel), and simulations with non-significant clines shown are in blue. (*See insert for color representation of the figure.*)

2015) and two of fish (Kinnison *et al.* 2008; Westley *et al.* 2013) partially support the local adaptation hypothesis in the form of significantly higher fitness of the local genotypes at a subset of transplant sites, often with nonsignificant trends consistent with local adaptation at other sites. Despite a long history in ecology and evolution, field experiments using the reciprocal transplant approach to rigorously confirm adaptive hypotheses are extremely rare in invasion biology. Frequent claims that rapid adaptive evolution is common during invasions should be interpreted cautiously. Instead, more reciprocal transplant field experiments are needed to assess the degree and generality of local adaptation in invading species, particularly in animals.

One criticism of the field reciprocal transplant approach is that it does not account for stochastic effects of invasion history—for example, using neutral markers to identify the geographic source(s) of introduction. In response to this critique, we consider the possibility of local adaptation evolving through nonadaptive processes alone—and find that it is unlikely. Specifically, we consider introduction of pre-adapted genotypes from locations in the native range with local environmental conditions that parallel those in the introduced range (see Box 3, Fig. 1c). This scenario alone is unlikely, given that most invading species spread from few sites of initial introduction. Additionally, at least four unlikely conditions must be met to establish local adaptation through parallel introductions alone.

1 No locally adaptive mutations arise de novo and increase in frequency during invasion (e.g. Box 4).

2 The environment and the subset of introduced genotypes are so closely matched that adaptive evolution does not fine-tune introduced population towards their optimum in the new range.

3 Founder populations are so large that genetic drift never moves introduced populations away from their locally adaptive phenotypes.

4 The environment is stable enough to preclude contemporary adaptation towards a fluctuating optimum.

These conditions are so unlikely that adaptive evolution must have occurred to some extent wherever reciprocal transplant studies confirm local adaptation. When introduced populations are locally adapted, it is therefore rather trivial to ask whether adaptive evolution was involved. Instead, a more relevant question is, what are the relative influences of stochastic (i.e. drift, founder effects, migration) versus deterministic (i.e. natural selection) forces in shaping population differentiation and local adaptation among introduced populations? Modern tools available in molecular genetics provide a complementary approach to disentangle these processes. However, molecular markers are not viable substitutes for comprehensive field experiments.

Q_{ST}-F_{ST} comparisons

Population differentiation can be compared between quantitative traits and neutral genetic loci to infer past selection, or lack thereof (hereafter Q_{ST}-F_{ST} studies). Reciprocal transplant experiments directly test whether populations are locally adapted to current selective pressures that differ across sites, which can be measured directly in studies of natural selection. In contrast, Q_{ST}-F_{ST} comparisons test for adaptive evolution occurring at some time in the past, not whether populations are locally adapted to current conditions per se. These differences are often subtle but important as the two approaches may not agree on the form of natural selection responsible for observed patterns of population differentiation (Box. 3). For example, a recent Q_{ST}-F_{ST} study did not support divergent natural selection ($Q_{ST} > F_{ST}$) for growth and phenology of introduced *Lythrum salicaria* (Chun *et al.* 2009), even though these traits differentiate along latitudinal clines in eastern North America (Montague *et al.* 2008), with parallel clines observed among native populations in Europe (Olsson & Ågren 2002). Moreover, reciprocal transplants combined with measurements of natural selection within the introduced range confirm that these latitudinal clines are locally adaptive (Colautti & Barrett 2013). The contradictory conclusions about the role of adaptation inferred from reciprocal transplant and Q_{ST}-F_{ST} studies could simply be a legacy effect of multiple introductions and spread similar to the scenarios outlined in Box 3. Without more field experiments measuring natural selection and testing local adaptation in other invading species, it will be difficult to determine the general accuracy of Q_{ST}-F_{ST} studies. Nevertheless, the *Lythrum salicaria* example highlights some of the potential complications inherent in applying Q_{ST}-F_{ST} approaches to investigate adaptive evolution during biological invasion.

Box 3 Q_{ST}-F_{ST} comparisons: complications for introduced populations

Several factors can complicate the application of Q_{ST}-F_{ST} comparisons to test for adaptive evolution in introduced populations, even in the ideal case in which the source populations of introduced genotypes are known with absolute certainty. As the following examples demonstrate, reciprocal transplant and Q_{ST}-F_{ST} studies may not often lead to similar conclusions, and in general, the inferences from field experiments are more relevant to understanding natural selection currently acting on introduced populations and the effect of local adaptation on invasion dynamics.

First, consider the case where local adaptation evolves rapidly following either a single introduction or multiple introductions forming a single admixed population,

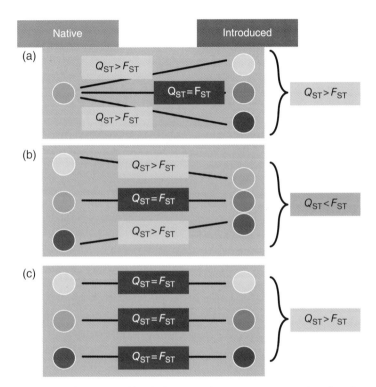

Box 3, Fig. 1 Three hypothetical scenarios of local adaptation along an environmental gradient (vertical axis) in the introduced range and inferences about natural selection using quantitative traits and neutral markers (Q_{ST} and F_{ST}, respectively). Local adaptation may evolve rapidly and be maintained by divergent natural selection following (a) spread from a single source or founder population, or (b) introduction from a wider to a more narrow range of environments, or (c) following parallel introductions to similar environments. Inferences about natural selection are shown for each scenario—no selection ($Q_{ST} = F_{ST}$), directional/divergent selection ($Q_{ST} > F_{ST}$) or stabilizing selection ($Q_{ST} < F_{ST}$); inferences depend on the genetic source(s) of introduction and whether populations are compared within the introduced range (right side) or between introduced and native source populations (middle). Third, parallel introduction of pre-adapted genotypes into similar environments re-establishes locally adaptive latitudinal clines with a weaker role for adaptive evolution (Fig. 1c). Natural selection maintains divergence of introduced populations, but a Q_{ST}-F_{ST} comparison between invaded and source populations fails to reject a neutral model of evolution ($Q_{ST} = F_{ST}$). Divergent selection ($Q_{ST} > F_{ST}$) is identified only if comparisons are made among introduced populations without accounting for their source region. Traits are under divergent selection to maintain phenotypic differentiation, so matching introduced populations to their genetic sources would erroneously reject the local adaptation hypothesis.

Box 3 (Continued)

followed by selection for locally adapted genotypes (Fig. 1a). A reciprocal transplant experiment demonstrates that populations are locally adapted, and a Q_{ST}-F_{ST} analysis identifies divergent selection acting among populations ($Q_{ST} > F_{ST}$). In this case, reciprocal transplants and Q_{ST}-F_{ST} studies reach the same conclusion: local adaptation has evolved in response to divergent natural selection. However, in practice it can be difficult to correctly reject the null hypothesis (i.e. neutral divergence) in Q_{ST}-F_{ST} studies, particularly when F_{ST} is small, when few populations are compared (<10), or when selection is weak (Whitlock & Guillaume 2009; Gilbert & Whitlock 2014).

Second, highly divergent native populations are introduced into a narrower range of environments (Fig. 1b). An adaptive evolutionary response to selection in this case causes populations to converge towards more similar phenotypes, resulting in a significant signal of stabilizing selection ($Q_{ST} < F_{ST}$) across introduced populations. When introduced populations are compared to native sources, either divergent selection ($Q_{ST} > F_{ST}$) or neutral evolution ($Q_{ST} = F_{ST}$) is identified, depending on the specific populations compared. Analysis of introduced populations suggests stabilizing selection ($Q_{ST} < F_{ST}$). However, all populations are locally adapted because divergent selection among introduced populations maintains differentiation, just not at the same magnitude as the original source populations. In this case, the Q_{ST}-F_{ST} inference of stabilizing selection is a historical signal of adaptive evolution whereas a reciprocal transplant experiment correctly identifies divergent natural selection maintaining introduced populations at their local optima. As these traits are under divergent, not stabilizing selection, reciprocal transplant experiments better reflect the evolutionary forces maintaining population differentiation.

FUTURE DIRECTIONS

A major theme of the first Invasion Genetics Symposium (Baker & Stebbins 1965) was the importance of evolution for understanding biological invasions and the utility of biological invasions for studying fundamental evolutionary processes. But whereas evolution was perceived to occur over centuries to millennia in the original symposium (Baker 1965; Stebbins 1965; Wilson 1965), our review demonstrates that evolution also occurs over years to decades during the establishment and spread of invasive species. Despite growing evidence for the influence of contemporary evolution in ecological dynamics (reviewed in Hendry & Kinnison 1999; Reznick & Ghalambor 2001; Stockwell et al. 2003; Schoener 2011), we have found few empirical studies on invasive species that definitively test which evolutionary changes are adaptive, and even fewer linking adaptive evolution to the spread and dominance of invasive species or their impacts on invaded communities and ecosystems. To address this knowledge gap, more research is needed to measure natural selection, test local adaptation and investigate the ecological consequences of adaptive evolutionary change occurring in invading populations. These studies will be important for understanding eco-evolutionary dynamics of biological invasions and predicting range expansion in the context of climate change and other aspects of global change.

Until recently, there appears to have been a strong bias against quantitative genetic studies of non-native populations. Of the 158 studies we found that measure selection in natural populations, few (~5%) have involved introduced species, and fewer have done so using genetic family structure to quantify additive genetic variation and genetic correlations. This detail of experimental design is necessary to rigorously predict future evolutionary responses (Rausher 1992; Stinchcombe et al. 2002; Morrissey et al. 2010). Unbiased measures of selection can definitively identify traits under selection in invaded populations and determine how selection may differ between invaded and native ranges.

A few but increasing number of studies have tested for local adaptation in invasive species using reciprocal transplant experiments, but survival and reproduction are rarely measured, and to our knowledge, few have linked local adaptation to invasion success (but see Colautti & Barrett 2013). Quantifying effects of phenotypic evolution on survival and reproduction in field studies within the introduced range is crucial to evaluate the extent to which adaptive evolution during invasion affects the spread and dominance of invasive species. Large reciprocal transplants are needed to rigorously test for adaptive evolutionary responses to ecological differences between the native and introduced range, such as shifts in interactions with natural enemies and competitors or differences in abiotic environments. However, interpreting results of transcontinental reciprocal transplant studies requires careful consideration of environmental variability and local adaptation among populations within the native and introduced ranges.

In addition to influencing the dynamics of invading populations, local adaptation evolving in an invading species could also influence its impact, not only by increasing its dominance within the community, but also potentially by altering nutrient dynamics and species interactions. These effects of contemporary evolution during invasion have rarely been considered (Eppinga & Molofsky 2013), but could be an important area of future research.

Eco-evolutionary processes affecting the impact of invasive species

The field of eco-evolutionary dynamics has flourished over the past decade, including a large and growing number of studies demonstrating evolutionary changes over ecological timescales (Hendry & Kinnison 1999; Reznick & Ghalambor 2001; Schoener 2011), and new approaches to quantify the relative importance of evolution vs. ecology to ecological processes (Hairston *et al.* 2005; Kinnison & Hairston 2007; Ellner *et al.* 2011). Eco-evolutionary interactions involve continuous feedbacks between evolutionary responses to environmental changes that influence population, community or ecosystem processes, which in turn alter the environment and natural selection acting on natural populations.

Contemporary evolution of local adaptation may increase the impact of an invading species simply by increasing its abundance and dominance within a community, or additionally by changing the frequency of genotypes with particular community and ecosystem impacts. Adaptive evolution increases rates of survival and reproduction, which could increase the abundance of a problematic invasive species. For example, local adaptation of Chinook salmon invading New Zealand more than doubled survival and fecundity compared to nonlocal genotypes (Kinnison *et al.* 2008). Additionally, genetic variants within a species can perform very different ecological functions (Sthultz *et al.* 2009; Crutsinger *et al.* 2014), and evolutionary changes in the frequency of these variants can alter community

and ecosystem functions. For example, Eppinga *et al.* (2011) expanded a classic resource competition model to include effects of litter dynamics and evolution on the growth of the wetland invasive *Phalaris arundinacea*. Their model illustrates how litter feedbacks and evolutionary changes to *Phalaris* resource needs (C:N ratios) could act synergistically to accelerate *Phalaris* invasion in both currently invaded and uninvaded locales. More research is needed to determine whether and how evolutionary changes in invading species affect their impacts on ecosystem function.

In addition to altering ecosystem processes, evolutionary changes in invaders may influence the evolution of native species in response to the impacts of invaders (Strauss *et al.* 2006). This creates a potential for rapid co-evolution between invaders and invaded communities (Leger & Espeland 2010). For example, research by Lankau *et al.* (2009) suggests that the invasive plant garlic mustard (*Alliaria petiolata*) is evolving reduced allelopathy in heavily invaded sites as plants shift from interspecific to intraspecific competition while populations of a native competitor (*Pilea pumila*) evolve tolerance to allelochemicals from *A. petiolata*. More generally, measurements of natural selection and reciprocal transplant experiments are needed to identify and understand eco-evolutionary dynamics between invading species and recipient communities. Understanding the causes and consequences of contemporary evolution in invading species and invaded communities will be crucial to predicting species distributions in the context of global change.

Global change, evolution and invasive species

Biological invasions may benefit from a variety of global change scenarios ranging from increased nitrogen deposition and increasing atmospheric CO_2 concentrations to habitat fragmentation, overexploitation of natural resources and increased disturbance (Dukes & Mooney 1999). The influence of contemporary evolution on species responses to global change remains relatively unknown (Visser 2008; Merilä & Hendry 2014). However, contemporary evolution may provide invasive species an evolutionary advantage over competing natives for at least four reasons.

First, invasive species may be better pre-adapted to environmental changes caused by humans because they have evolved for centuries to millenia in human-altered environments (see discussion in Stebbins 1965;

Hufbauer *et al.* 2012). Association with humans is correlated with establishment and spread of introduced species, even after accounting for higher propagule pressure expected with increasing human activity (Jeschke & Strayer 2006). An ongoing co-evolutionary history with human disturbance may serve introduced species well under future environments with greater intensity of human use, fragmentation and disturbance.

Second, admixture resulting from gene flow among historically isolated populations should be more common in introduced than native species. Multiple introductions from divergent source regions are likely to increase standing genetic variation in ecologically important traits and to produce novel phenotypes, both of which could facilitate adaptation in novel environments (Rius & Darling 2014). Some species show an association between admixture and size (Kolbe *et al.* 2007) or fecundity (Keller & Taylor 2010; Turgeon *et al.* 2011; Keller *et al.* 2014). Other studies have documented increases in molecular diversity over time (Krehenwinkel & Tautz 2013). However, the effect of admixture on the establishment and spread of invasive species is poorly understood.

Third, many invasive species reach high densities in the invaded range; theory predicts that large population size increases the probability of evolutionary rescue in changing environments by increasing both the time to extinction and the probability of adaptive mutations arising (Gomulkiewicz & Holt 1995; Box 4). Thus, the high densities achieved by many invasive species could provide both demographic and evolutionary mechanisms that reduce extinction risks in changing environments.

Finally, invaders may escape many evolutionary constraints when they leave behind their specialized natural enemies and mutualists and engage in a simpler suite of interactions in the introduced range (Strauss 2014). Rather than evolving increases in size and competitive ability per se, release from specialist herbivores could relax selection on multiple traits, allowing populations more degrees of freedom to adapt more rapidly than native species. Measurements of natural selection and quantitative genetics studies measuring constraints are needed to test this hypothesis.

In sum, conclusive evidence for adaptive evolutionary change in invaders is rare. More studies are needed because rapid adaptation during biological invasions can increase invader demographic success and alter the magnitude of impacts on native species. Distinguishing adaptive evolution from stochastic evolutionary processes occurring at cross-continental scales is challenging, but studies of population differentiation, natural selection and local adaptation can help determine the role evolution plays in the spread and ecological impact of invasive species.

ACKNOWLEDGEMENTS

We thank N. Scheepens, S. Yakimowski and the organizers and participants of the 2014 Invasion Genetics: The Baker and Stebbins Legacy Symposium for discussions

---

Box 4 Evolution from de novo mutation

Polygenic traits may be affected by dozens to hundreds of genes and therefore represent multiple targets for new adaptive mutations. When multiplied by large population sizes of established invasive species, a significant number of adaptive mutations can be expected, even in species experiencing extreme bottlenecks (see also Dlugosch *et al.* 2015). To demonstrate the high potential for de novo adaptive mutations to contribute to the spread of invasive species, consider the invasion of North America by zebra mussels (*Dreissena polymorpha*). Given an average mutation rate of 10^{-11} to 10^{-8} per bp per generation (Drake *et al.* 1998) and an average gene length of 10–50 kbp (Xu *et al.* 2006), the rate of new mutation in each gene will be somewhere between $\sim 10^{-7}$ and $\sim 10^{-3}$ per individual, per generation. About 5–10% of these mutations are expected to be adaptive (Eyre-Walker & Keightley 2007) in each of perhaps $\sim 30\,000$ genes (oyster genome, Zhang *et al.* 2012). *Dreissena polymorpha* populations typically exceed an average density of 3000 individuals per square metre (Karatayev *et al.* 2014), so every km^2 of infestation could conceivably contain $\sim 10^6$ to $\sim 10^{10}$ novel adaptive mutations. Most of these new mutations will be lost to genetic drift, but this example demonstrates the potential for novel mutations arising during invasion to contribute significantly to adaptive evolution, especially in species that experience strong bottlenecks during initial introduction.

and comments on many of the ideas presented here. RIC was supported by an NSERC Banting Postdoctoral Fellowship, and JAL was supported by the US National Science Foundation award DEB 0918963. This is Kellogg Biological Station publication number 1840.

REFERENCES

Alexander JM (2013) Evolution under changing climates: climatic niche stasis despite rapid evolution in a non-native plant. *Proceedings of the Royal Society Series B: Biological Sciences*, **280**, 20131446.

Alexander JM, Edwards PJ, Poll M, Parks CG, Hansjörg D (2009) Establishment of parallel altitudinal clines in traits of native and introduced forbs. *Ecology*, **90**, 612–622.

Baker HG (1965) Characteristics and modes of origin of weeds. In: *The Genetics of Colonizing Species* (eds Baker HG, Stebbins GL), pp. 147–172. Academic Press, New York.

Barrett SCH (2015) Foundations of invasion genetics: the Baker and Stebbins legacy. *Molecular Ecology*.

Blossey B, Nötzold R (1995) Evolution of increased competitive ability in plants: a hypothesis. *Journal of Ecology*, **83**, 887–889.

Blows MW, McGuigan K (2015) The distribution of genetic variance across phenotypic space and the response to selection. *Molecular Ecology*, **24**, 2056–2072.

Blumenthal DM, Hufbauer RA (2007) Increased plant size in exotic populations: A common-garden test with 14 invasive species. *Ecology*, **88**, 2758–2765.

Bock DG, Caseys C, Cousens RD et al. (2015) What we still don't know about invasion genetics. *Molecular Ecology*, **24**, 2277–2298.

Bossdorf O, Prati D, Auge H, Schmid B (2004) Reduced competitive ability in an invasive plant. *Ecology Letters*, **7**, 346–353.

Bossdorf O, Auge H, Lafuma L et al. (2005) Phenotypic and genetic differentiation between native and introduced plant populations. *Oecologia*, **144**, 1–11.

Burger JC, Ellstrand JC (2014) Rapid evolutionary divergence of an invasive weed from its crop ancestor and evidence for local diversification. *Journal of Systematics and Evolution*, **6**, 750–764.

Callaway RM, Ridenour WM (2004) Novel weapons: invasive success and the evolution of increased competitive ability. *Frontiers in Ecology and the Environment*, **2**, 436–443.

Carlson SM, Hendry AP, Letcher BH (2004) Natural selection acting on body size, growth rate and compensatory growth: an empirical test in a wild trout population. *Evolutionary Ecology Research*, **6**, 955–973.

Catford JA, Jansson R, Nilsson C (2009) Reducing redundancy in invasion ecology by integrating hypotheses into a single theoretical framework. *Diversity and Distributions*, **15**, 22–40.

Chun YJ, Nason JD, Moloney KA (2009) Comparison of quantitative and molecular genetic variation of native vs. invasive populations of purple loosestrife (*Lythrum salicaria* L., Lythraceae). *Molecular Ecology*, **18**, 3020–3035.

Clausen J, Keck DD, Hiesey W (1940) *Experimental Studies on the Nature of Species. I. Effects of Varied Environments on Western North American Plants*. Carnegie Institute, Washington, District of Columbia.

Colautti RI, Barrett SCH (2013) Rapid adaptation to climate facilitates range expansion of an invasive plant. *Science*, **342**, 364–366.

Colautti RI, Maron JL, Barrett SCH (2009) Common garden comparisons of native and introduced plant populations: latitudinal clines can obscure evolutionary inferences. *Evolutionary Applications*, **2**, 187–199.

Colautti RI, Eckert CG, Barrett SCH (2010) Evolutionary constraints on adaptive evolution during range expansion in an invasive plant. *Proceedings of the Royal Society of London B*, **277**, 1799–1806.

Conner JK, Hartl DL (2004) *A Primer of Ecological Genetics*. Sinauer Associates, Sunderland, Massachusetts.

Cristescu ME (2015) Genetic reconstructions of invasion history. *Molecular Ecology*, **24**, 2212–2225.

Crutsinger GM, Rodriguez-Cabal MA, Roddy AB et al. (2014) Genetic variation within a dominant shrub structures green and brown community assemblages. *Ecology*, **95**, 387–398.

Darwin C (1859) *On the Origin of Species By Means of Natural Selection, or the Preservation of Favoured Races in the Struggle for Life*. John Murray, London.

Dlugosch KM, Parker IM (2008) Founding events in species invasions: genetic variation, adaptive evolution, and the role of multiple introductions. *Molecular ecology*, **17**, 431–449.

Dlugosch KM, Anderson SR, Braasch J, Cang FA, Gillette HD (2015) The devil is in the details: genetic variation in introduced populations and its contribution to invasion. *Molecular Ecology*, **24**, 2095–2111.

Drake JM (2006) Heterosis, the catapult effect and establishment success of a colonizing bird. *Biology Letters*, **2**, 304–307.

Drake JW, Charlesworth B, Charlesworth D, Crow JF (1998) Rates of spontaneous mutation. *Genetics*, **148**, 1667–1686.

Dukes JS, Mooney HA (1999) Does global change increase the success of biological invaders? *Trends in Ecology & Evolution*, **14**, 135–139.

Ebeling SK, Stocklin J, Hensen I, Auge H (2011) Multiple common garden experiments suggest lack of local adaptation in an invasive ornamental plant. *Journal of Plant Ecology*, **4**, 209–220.

Ellner SP, Geber MA, Hairston NG (2011) Does rapid evolution matter? Measuring the rate of contemporary evolution and its impacts on ecological dynamics. *Ecology letters*, **14**, 603–614.

Ellstrand NC, Schierenbeck KA (2000) Hybridization as a stimulus for the evolution of invasiveness in plants? *Proceedings of the National Academy of Sciences, USA*, **97**, 7043–7050.

Elton CS (1958) *The Ecology of Invasions by Animals and Plants*. Methuen, London.

Eppinga MB, Molofsky J (2013) Eco-evolutionary litter feedback as a driver of exotic plant invasion. *Perspectives in Plant Ecology, Evolution and Systematics*, **15**, 20–31.

Eppinga MB, Kaproth MA, Collins AR, Molofsky J (2011) Litter feedbacks, evolutionary change and exotic plant invasion. *Journal of Ecology*, **99**, 503–514.

Erfmeier A, Bruelheide H (2010) Invasibility or invasiveness? Effects of habitat, genotype, and their interaction on invasive *Rhododendron ponticum* populations. *Biological Invasions*, **12**, 657–676.

Excoffier L, Foll M, Petit RJ (2009) Genetic consequences of range expansions. *Annual Review of Ecology, Evolution, and Systematics*, **40**, 481–501.

Eyre-Walker A, Keightley PD (2007) The distribution of fitness effects of new mutations. *Nature Reviews Genetics*, **8**, 610–618.

Falconer DS, Mackay TF (1996) *Introduction to Quantitative Genetics*, 4th edn. Longmans Green, Harlow.

Fisher RA (1930) *The Genetical Theory of Natural Selection*. Clarendon, Oxford.

Futuyma DJ (2013) *Evolution*, 3rd edn. Sinauer, Sunderland, Massachusetts.

García-Ramos G, Rodríguez D (2002) Evolutionary speed of species invasions. *Evolution*, **56**, 661–668.

Genton BJ, Kotanen PM, Cheptou P-O, Adolphe C, Shykoff JA (2005) Enemy release but no evolutionary loss of defence in a plant invasion: an inter-continental reciprocal transplant experiment. *Oecologia*, **146**, 404–414.

Gilbert KJ, Whitlock MC (2014) Q_{ST}–F_{ST} comparisons with unbalanced half-sib designs. *Molecular Ecology*, **15**, 262–267.

Gomi T (2007) Seasonal adaptations of the fall webworm *Hyphantria cunea* (Drury) (Lepidoptera: Arctiidae) following its invasion of Japan. *Ecological Research*, **22**, 855–861.

Gomulkiewicz R, Holt RD (1995) When does evolution by natural selection prevent extinction? *Evolution*, **49**, 201–207.

Gulick JT (1888) Divergent evolution through cumulative segregation. *Journal of the Linnean Society of London, Zoology*, **20**, 189–274.

Hairston NG, Ellner SP, Geber MA, Yoshida T, Fox JA (2005) Rapid evolution and the convergence of ecological and evolutionary time. *Ecology Letters*, **8**, 1114–1127.

Hendry AP, Kinnison MT (1999) The pace of modern life: measuring rates of contemporary microevolution. *Evolution*, **53**, 1637–1653.

Hendry AP, Letcher BH, Gries G (2003) Estimating natural selection acting on stream-dwelling Atlantic salmon: implications for the restoration of extirpated populations. *Conservation Biology*, **17**, 795–805.

Henery ML, Bowman G, Mráz P *et al.* (2010) Evidence for a combination of pre-adapted traits and rapid adaptive change in the invasive plant *Centaurea stoebe*. *Journal of Ecology*, **98**, 800–813.

Hereford J (2009) A quantitative survey of local adaptation and fitness trade-offs. *The American Naturalist*, **173**, 579–588.

Huey RB (2000) Rapid evolution of a geographic cline in size in an introduced fly. *Science*, **287**, 308–309.

Hufbauer RA, Facon B, Ravigné V *et al.* (2012) Anthropogenically induced adaptation to invade (AIAI): contemporary adaptation to human-altered habitats within the native range can promote invasions. *Evolutionary Applications*, **5**, 89–101.

Jeschke JM, Strayer DL (2006) Determinants of vertebrate invasion success in Europe and North America. *Global Change Biology*, **12**, 1608–1619.

Joshi J, Vrieling K (2005) The enemy release and EICA hypothesis revisited: incorporating the fundamental difference between specialist and generalist herbivores. *Ecology Letters*, **8**, 704–714.

Karatayev AY, Burlakova LE, Pennuto C *et al.* (2014) Twenty five years of changes in *Dreissena* spp. populations in Lake Erie. *Journal of Great Lakes Research*, **40**, 550–559.

Keller SR, Taylor DR (2008) History, chance and adaptation during biological invasion: separating stochastic phenotypic evolution from response to selection. *Ecology Letters*, **11**, 852–866.

Keller SR, Taylor DR (2010) Genomic admixture increases fitness during a biological invasion. *Journal of Evolutionary Biology*, **23**, 1720–1731.

Keller SR, Sowell DR, Neiman M, Wolfe LM, Taylor DR (2009) Adaptation and colonization history affect the evolution of clines in two introduced species. *New Phytologist*, **183**, 678–690.

Keller SR, Fields PD, Berardi AE, Taylor DR (2014) Recent admixture generates heterozygosity-fitness correlations during the range expansion of an invading species. *Journal of Evolutionary Biology*, **27**, 616–627.

Kingsolver JG, Diamond SE (2011) Phenotypic selection in natural populations: what limits directional selection? *The American Naturalist*, **177**, 346–357.

Kingsolver JG, Diamond SE, Siepielski AM, Carlson SM (2012) Synthetic analyses of phenotypic selection in natural populations: lessons, limitations and future directions. *Evolutionary Ecology*, **26**, 1101–1118.

Kinnison MT, Hairston NG (2007) Eco-evolutionary conservation biology: contemporary evolution and the dynamics of persistence. *Functional Ecology*, **21**, 444–454.

Kinnison MT, Unwin MJ, Quinn TP (2008) Eco-evolutionary vs. habitat contributions to invasion in salmon: experimental evaluation in the wild. *Molecular Ecology*, **17**, 405–414.

van Kleunen M, Fischer M (2008) Adaptive rather than nonadaptive evolution of *Mimulus guttatus* in its invasive range. *Basic and Applied Ecology*, **9**, 213–223.

Kolbe JJ, Larson A, Losos JB (2007) Differential admixture shapes morphological variation among invasive populations of the lizard *Anolis sagrei*. *Molecular Ecology*, **16**, 1579–1591.

Krehenwinkel H, Tautz D (2013) Northern range expansion of European populations of the wasp spider *Argiope bruennichi* is associated with global warming-correlated genetic admixture and population-specific temperature adaptations. *Molecular Ecology*, **22**, 2232–2248.

Lande R (1988) Genetics and demography in biological conservation. *Science*, **241**, 1455–1460.

Lande R, Arnold SJ (1983) The measurement of selection on correlated characters. *Evolution*, **37**, 1210–1226.

Langlet O (1971) Two hundred years genecology. *Taxon*, **20**, 653–752.

Lankau RA, Spyreas G, Nuzzo V, Davis AS (2009) Evolutionary limits ameliorate the negative impact of an invasive plant. *Proceedings of the National Academy of Science of the United States*, **106**, 15362–15367.

Lee CE (2002) Evolutionary genetics of invasive species. *Trends in Ecology & Evolution*, **17**, 386–391.

Leger EA, Espeland EK (2010) Coevolution between native and invasive plant competitors: implications for invasive species management. *Evolutionary Applications*, **3**, 169–178.

Leger EA, Espeland EK, Merrill KR, Meyer SE (2009) Genetic variation and local adaptation at a cheatgrass (*Bromus tectorum*) invasion edge in western Nevada. *Molecular Ecology*, **18**, 4366–4379.

Leimu R, Fischer M (2008) A meta-analysis of local adaptation in plants. *PLoS ONE*, **3**, e4010.

Li X-M, She D-Y, Zhang D-Y, Liao W-J (2015) Life history trait differentiation and local adaptation in invasive populations of *Ambrosia artemisiifolia* in China. *Oecologia*, **177**, 669–677.

Linhart YB, Grant MC (1996) Evolutionary significance of local genetic differentiation in plants. *Annual Review of Ecology and Systematics*, **27**, 237–277.

Ludsin SA, Wolfe AD (2001) Biological invasion theory: Darwin's contributions from The Origin of Species. *BioScience*, **51**, 780–789.

Lynch M, Walsh B (1998) *Genetics and Analysis of Quantitative Traits*. Sinauer, Sunderland, Massachusetts.

Mallet J, Barton N (1989) Inference from clines stabilized by frequency-dependent selection. *Genetics*, **122**, 967–976.

Maron JL, Vilà M, Bommarco R, Elmendorf S, Beardsley P (2004) Rapid evolution of an invasive plant. *Ecological Monographs*, **74**, 261–280.

Maron JL, Elmendorf SC, Vilà M (2007) Contrasting plant physiological adaptation to climate in the native and introduced range of *Hypericum perforatum*. *Evolution*, **61**, 1912–1924.

McKenney JL, Cripps MG, Price WJ, Hinz HL, Schwarzländer M (2007) No difference in competitive ability between invasive North American and native European *Lepidium draba* populations. *Plant Ecology*, **193**, 293–303.

Merilä J, Hendry AP (2014) Climate change, adaptation, and phenotypic plasticity: the problem and the evidence. *Evolutionary Applications*, **7**, 1–14.

Montague JL, Barrett SCH, Eckert CG (2008) Re-establishment of clinal variation in flowering time among introduced populations of purple loosestrife (*Lythrum salicaria*, Lythraceae). *Journal of Evolutionary Biology*, **21**, 234–245.

Moran EV, Alexander JM (2014) Evolutionary responses to global change: lessons from invasive species. *Ecology Letters*, **17**, 637–649.

Morrissey MB, Kruuk LEB, Wilson AJ (2010) The danger of applying the breeder's equation in observational studies of natural populations. *Journal of Evolutionary Biology*, **23**, 2277–2288.

Müller-Schärer H, Schaffner U, Steinger T (2004) Evolution in invasive plants: implications for biological control. *Trends in Ecology & Evolution*, **19**, 417–422.

Murren CJ, Chang CC, Dudash MR (2009) Patterns of selection of two North American native and nonnative populations of monkeyflower (Phrymaceae). *New Phytologist*, **183**, 691–701.

Nagylaki T (1975) Conditions for the existence of clines. *Genetics*, **80**, 595–615.

O'Donnell KL, Pigliucci M (2010) Selection dynamics in native and introduced *Persicaria* species. *International Journal of Plant Sciences*, **171**, 519–528.

Olsson K, Ågren J (2002) Latitudinal population differentiation in phenology, life history and flower morphology in the perennial herb *Lythrum salicaria*. *Journal of Evolutionary Biology*, **15**, 983–996.

O'Neil P (1997) Natural selection on genetically correlated phenological characters in *Lythrum salicaria* L (Lythraceae). *Evolution*, **51**, 267–274.

Pahl AT, Kollmann J, Mayer A, Haider S (2013) No evidence for local adaptation in an invasive alien plant: field and greenhouse experiments tracing a colonization sequence. *Annals of Botany*, **112**, 1921–1930.

Peischl S, Excoffier L (2015) Expansion load: recessive mutations and the role of standing genetic variation. *Molecular Ecology*, **24**, 2084–2094.

Petitpierre B, Kueffer C, Broennimann O, Randin C, Daehler C, Guisan A (2013) Climatic niche shifts are rare among terrestrial plant invaders. *Science*, **335**, 1344–1348.

Phillips BL, Brown GP, Webb JK, Shine R (2006) Invasion and the evolution of speed in toads. *Nature*, **439**, 803.

Phillips BL, Brown GP, Shine R (2010a) Life-history evolution in range-shifting populations. *Ecology*, **91**, 1617–1627.

Phillips BL, Brown GP, Shine R (2010b) Evolutionarily accelerated invasions: the rate of dispersal evolves upwards during the range advance of cane toads. *Journal of Evolutionary Biology*, **23**, 2595–2601.

Price TD, Yeah PJ, Harr B (2008) Phenotypic plasticity and the evolution of a socially selected trait following colonization of a novel environment. *The American Naturalist*, **172**, S49–S62.

Rausher MD (1992) The measurement of selection on quantitative traits: biases due to environmental covariances between traits and fitness. *Evolution*, **46**, 616–626.

Reznick DN, Ghalambor CK (2001) The population ecology of contemporary adaptations: what empirical studies reveal about the conditions that promote adaptive evolution. *Genetica*, **112–113**, 183–198.

Rice KJ, Mack RN (1991) Ecological genetics of *Bromus tectorum*. *Oecologia*, **88**, 91–101.

Rice KJ, Gerlach JD, Dyer AR, McKay JK (2013) Evolutionary ecology along invasion fronts of the annual grass *Aegilops triuncialis*. *Biological Invasions*, **15**, 2531–2545.

Rick CM, Smith PG (1953) Novel variation in tomato species hybrids. *The American Naturalist*, **87**, 359–373.

Ridley CE, Ellstand NE (2010) Rapid evolution of morphology and adaptive life history in the invasive California wild radish (*Raphanus sativus*) and the implications for management. *Evolutionary Applications*, **3**, 64–76.

Rieseberg LH, Archer MA, Wayne RK (1999) Transgressive segregation, adaptation and speciation. *Heredity*, **83**, 363–372.

Rius M, Darling JA (2014) How important is intraspecific genetic admixture to the success of colonising populations? *Trends in Ecology & Evolution*, **29**, 233–242.

Roman J, Darling JA (2007) Paradox lost: genetic diversity and the success of aquatic invasions. *Trends in Ecology & Evolution*, **22**, 454–464.

Ross CA, Auge H (2008) Invasive *Mahonia* plants outgrow their native relatives. *Plant Ecology*, **199**, 21–31.

Sax DF, Stachowicz JJ, Brown JH *et al.* (2007) Ecological and evolutionary insights from species invasions. *Trends in Ecology and Evolution*, **22**, 465–471.

Schlaepfer DR, Edwards PJ, Billeter R (2010) Why only tetraploid *Solidago gigantea* (Asteraceae) became invasive: a common garden comparison of ploidy levels. *Oecologia*, **163**, 661–673.

Schoener TW (2011) The newest synthesis: understanding the interplay of evolutionary and ecological dynamics. *Science*, **331**, 426–429.

Schueller SK (2007) Island-mainland difference in *Nicotiana glauca* (Solanaceae) corolla length: a product of pollinator mediated selection? *Evolutionary Ecology*, **21**, 81–98.

Seehausen O (2004) Hybridization and adaptive radiation. *Trends in Ecology & Evolution*, **19**, 198–207.

Seiter S, Kingsolver J (2013) Environmental determinants of population divergence in life-history traits for an invasive species: climate, seasonality and natural enemies. *Journal of Evolutionary Biology*, **26**, 1634–1645.

Slatkin M (1978) Spatial patterns in the distributions of polygenic characters. *Journal of Theoretical Biology*, **70**, 213–228.

Stebbins GL (1965) Colonizing species of the native California flora. In: *The Genetics of Colonizing Species* (eds Baker HG, Stebbins GL), pp. 173–191. Academic Press, New York.

Sthultz CM, Whitham TG, Kennedy K, Deckert R, Gehring CA (2009) Genetically based susceptibility to herbivory influences the ectomycorrhizal fungal communities of a foundation tree species. *The New Phytologist*, **184**, 657–667.

Stinchcombe JR, Rutter MT, Burdick DS *et al.* (2002) Testing for environmentally induced bias in phenotypic estimates of natural selection: theory and practice. *The American Naturalist*, **160**, 511–523.

Stockwell CA, Hendry AP, Kinnison MT (2003) Contemporary evolution meets conservation biology. *Trends in Ecology & Evolution*, **18**, 94–101.

Strauss SY (2014) Ecological and evolutionary responses in complex communities: implications for invasions and eco-evolutionary feedbacks. *Oikos*, **123**, 257–266.

Strauss SY, Lau JA, Carroll SP (2006) Evolutionary responses of natives to introduced species: what do introductions tell us about natural communities? *Ecology Letters*, **9**, 357–374.

Turgeon J, Tayeh A, Facon B *et al.* (2011) Experimental evidence for the phenotypic impact of admixture between wild and biocontrol Asian ladybird (*Harmonia axyridis*) involved in the European invasion. *Journal of Evolutionary Biology*, **24**, 1044–1052.

Uesugi A, Kessler A (2013) Herbivore exclusion drives the evolution of plant competitiveness via increased allelopathy. *The New Phytologist*, **198**, 916–924.

deVicente MC, Tanksley SD (1993) QTL analysis of transgressive segregation in an interspecific tomato cross. *Genetics*, **134**, 585–596.

Vilà M, Maron JL, Marco L (2005) Evidence for the enemy release hypothesis in *Hypericum perforatum*. *Oecologia*, **142**, 474–479.

Visser ME (2008) Keeping up with a warming world; assessing the rate of adaptation to climate change. *Proceedings of the Royal Society B: Biological Sciences*, **275**, 649–659.

Wade MJ, Kalisz S (1990) The causes of natural selection. *Evolution*, **44**, 1947–1955.

Westley PAH, Ward EJ, Fleming IA (2013) Fine-scale local adaptation in an invasive freshwater fish has evolved in contemporary time. *Proceedings of the Royal Society B: Biological Sciences*, **280**, 20122327.

Whitlock MC, Guillaume F (2009) Testing for spatially divergent selection: comparing Q_{ST} to F_{ST}. *Genetics*, **183**, 1055–1063.

Williams JL, Auge H, Maron JL (2008) Different gardens, different results: native and introduced populations exhibit contrasting phenotypes across common gardens. *Oecologia*, **157**, 239–248.

Wilson EO (1965) The challenge from related species. In: *The Genetics of Colonizing Species* (eds Baker HG, Stebbins GL), pp. 8–27. Academic Press, New York.

Wise MJ, Rausher MD (2013) Evolution of resistance to a multiple-herbivore community: genetic correlations, diffuse coevolution, and constraints on the plant's response to selection. *Evolution*, **67**, 1767–1779.

Wittzell H (1991) Directional selection on morphology in the pheasant, *Phasianus colchicus*. *Oikos*, **61**, 394–400.

Wolfe LM, Blair AC, Penna BM (2014) Does intraspecific hybridization contribute to the evolution of invasiveness? An experimental test. *Biological Invasions*, **9**, 515–521.

Xu L, Chen H, Xiaohua H, Zhang R, Zhang Z, Luo ZW (2006) Average gene length is highly conserved in prokaryotes and eukaryotes and diverges only between the two kingdoms. *Molecular Biology and Evolution*, **23**, 1107–1108.

Yang X, Huang W, Tian B, Ding J (2014) Differences in growth and herbivory damage of native and invasive kudzu (*Peuraria montana* var. lobata) populations grown in the native range. *Plant Ecology*, **215**, 339–346.

Yeh PJ, Price TD (2004) Adaptive phenotypic plasticity and the successful colonization of a novel environment. *American Naturalist*, **164**, 531–542.

Zhang G, Fang X, Guo X *et al.* (2012) The oyster genome reveals stress adaptation and complexity of shell formation. *Nature*, **490**, 49–54.

DATA ACCESSIBILITY

Available at Dryad doi:10.5061/dryad.gt678

MetaPCA.txt—Text file describing common garden data.

MetaPCA_Data.csv—Data file containing standard deviations of population means and latitudes.

MetaPCA_Analysis.R—R script for the analysis of common garden data, with output for Fig. 1.

SpatialStructure.txt—Brief description of data for meta-analysis of spatial structure in common garden sampling locations.

SpatialStructure_Data.csv—Data file containing latitudes and longitudes of populations used in common garden studies.

SpatialStructure_Analysis.R—R script for the analysis of spatial structure, with output for Box 1, Fig. 1.

Selection.txt—Text file describing selection data.

Selection_Data.csv—Data file containing measurements of selection gradients and differentials.

Selection_Analysis.R—R script with analysis of selection data and producing Fig. 2.

ClinesModel.R—Stepwise colonization model demonstrating rapid evolution of nonadaptive trait clines, without output in Box 2, Fig. 1.

SUPPORTING INFORMATION

Additional supporting information can be found in the online version of the *Molecular Ecology* article.

Appendix S1 Common garden studies used in the meta-analysis of population differentiation includes studies listed in Appendix A of Colautti *et al.* (2009) with these additional studies identified through literature searches in 2011.

Appendix S2 Studies included in the analyses of selection coefficients includes studies published in the Kingsolver & Diamond (2011) dataset (http://datadryad.org/handle/10255/dryad.7996) with additional studies listed here.

Table S1 Studies included in the Kingsolver & Diamond (2011) dataset that were excluded from the analyses presented in Colautti & Lau (2015) and reason for exclusion.

Chapter 7

EXOTICS EXHIBIT MORE EVOLUTIONARY HISTORY THAN NATIVES: A COMPARISON OF THE ECOLOGY AND EVOLUTION OF EXOTIC AND NATIVE ANOLE LIZARDS

Matthew R. Helmus,*† Jocelyn E. Behm,*† Wendy A.M. Jesse,* Jason J. Kolbe,‡ Jacintha Ellers,* and Jonathan B. Losos§

* Department of Ecological Sciences, Section of Animal Ecology, Vrije Universiteit, 1081 HV Amsterdam, the Netherlands
† Center for Biodiversity, Department of Biology, Temple University, Philadelphia, PA 19122, USA
‡ Department of Biological Sciences, University of Rhode Island, Kingston, RI 02881, USA
§ Department of Organismic and Evolutionary Biology and Museum of Comparative Zoology, Harvard University, Cambridge, MA 02138, USA

Abstract
Long-distance colonization was once rare causing species within regions to be closely related. Now, in the Anthropocene, biogeographic structure is being eroded by species introductions. Here, we contrast the ecology and evolution of native versus exotic Caribbean *Anolis* lizards and show that the once strong biogeographic structure in the clade has been altered by the introduction of 22 *Anolis* species. Anole introductions are more frequent and span greater distances than natural anole colonizations. As a result, exotic anole populations in the Anthropocene often contain more genetic diversity than native populations, and anole phylogenetic diversity on islands is rapidly increasing.

Invasion Genetics: The Baker and Stebbins Legacy, First Edition. Edited by Spencer C. H. Barrett, Robert I. Colautti, Katrina M. Dlugosch, and Loren H. Rieseberg.
© 2017 John Wiley & Sons, Ltd. Published 2017 by John Wiley & Sons, Ltd.

INTRODUCTION

In the past, the rate at which species colonized new areas, such as distant islands, was relatively slow, and long-distance dispersal events were rare. As a result, species within biogeographic regions shared much of their evolutionary history; for any given species group, closely related species tended to be found in the same region. Within species, populations were further subdivided due to limited dispersal, assortative mating, and vicariance events resulting in phylogeographic patterning of populations. Today, this strong spatial patterning of evolutionary history is eroding. In an epoch in which earth systems are dominated by humans—the Anthropocene—species are frequently and often unintentionally transported far from their native regions. These long-distance dispersal events due to humans (i.e., species invasions) are occurring at rates much higher than in the past, causing the evolutionary history that underlies the geography of many organisms to become irreversibly mixed. Here, we describe this mixing of evolutionary history for a well-studied radiation of species from the lizard genus *Anolis* (or anoles). These lizards have naturally colonized Caribbean islands for millennia, and in the Anthropocene continue to be one of the most prolific groups of colonizing species.

Endemic to the Americas, over the past tens of millions of years the *Anolis* lizard radiation has produced approximately 165 Caribbean and 210 mainland species. Anoles are adapted to a multitude of tropical and subtropical habitats and exhibit a wide diversity of ecological morphologies and geographic range sizes. They are a model clade for understanding the ecological and evolutionary processes that give rise to biogeographic and phylogeographic patterns, especially across islands (Losos 2009). Before European settlement of the Americas, anoles infrequently colonized new land masses. Those colonizations that did occur were usually only from proximate areas (Williams 1969). There have been approximately 40 colonizations of Caribbean island banks over the past 60 million years as estimated by phylogenetics (Helmus & Ives 2012; Losos 2009; Mahler *et al.* 2010; Nicholson *et al.* 2005). Yet since European settlement of the Caribbean, there have been approximately 40 exotic colonizations, with the rate increasing over time. Before World War II (WWII), there were 0.07 (±0.01 SE) new exotic anole colonizations per year in the Caribbean. As of 2015, the rate is 0.96 per year (±0.12), a 14-fold increase since the end of WWII (Helmus *et al.* 2014). While making a robust comparison between past and Anthropocene colonization rates is difficult as the sources of the underlying data are different—phylogenetics versus field surveys—the message is clear: anoles are in an unprecedented age of range expansion.

Anoles have been introduced by humans across the globe and are now found far from the Americas (Fig. 1). The introduction of exotic anoles has been intentional at times, yet the majority of introductions to Caribbean islands have likely been unintentional via the shipment of live plants for landscaping and agriculture (Helmus

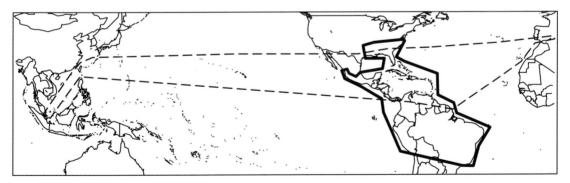

Fig. 1 The global range limits of native and exotic *Anolis* lizards. The thick solid line delineates the distribution of native *Anolis* and the dashed line roughly delineates exotic anole distribution limits as of 2015. Nations/territories with an established exotic anole are Anguilla, Aruba, Bahamas, Barbados, Belize, Bermuda, Canary Islands, Cayman Islands, Costa Rica, Curaçao, Dominica, Dominican Republic, French Guyana, Grenada, Guam, Guyana, Jamaica, Japan (Okinawa, Ogasawara Islands), Mexico, Northern Mariana Islands, Puerto Rico, St. Lucia, St. Maarten, St. Vincent and the Grenadines, St. Thomas, Singapore, Spain, Suriname, Taiwan, Trinidad and Tobago, the United States (mainland, Hawaii), and Venezuela.

et al. 2014; Kraus 2009; Losos *et al.* 1993; Perry *et al.* 2006; Powell *et al.* 2011). Before humans, anoles that established on Caribbean islands spread within intact ecosystems that typically contained few resident anole species. In contrast, exotic anoles in the Anthropocene establish into heavily disturbed natural and newly emergent anthropogenic habitats (e.g., resort gardens), encountering not only native species but also well-established exotics (Powell *et al.* 2011). The evolutionary and ecological processes that determine anole species diversity have thus greatly changed in the Anthropocene.

Here, we present a review focused on Caribbean anole species, as most exotic anoles are native to and have established in the Caribbean, not in the mainland. We have structured the review to contrast what is known regarding the natural dispersal and colonization processes of native anoles to the human-mediated translocation of exotic anoles in the Anthropocene. Our review follows the introduction process starting with transportation to a new location and ending with the stage where species adjust to local conditions. We highlight patterns of genetic variation and trait adaptation that have contributed to and are resultant of the success of native anole species and exotic populations. Because predicting the spread of exotic species is a major challenge in the Anthropocene, we follow with a description of the macroecological patterns in exotic distributions and how the evolutionary history resulting from exotic introductions differs from the natural biogeographic patterns exhibited by natives. Finally, we conclude the review with insight into the future evolution of anoles in the Anthropocene.

TRAVEL TO NEW CARIBBEAN ISLANDS

Any terrestrial species, regardless of whether they are dispersing naturally or assisted by humans, must first successfully travel over water in order to colonize a new island. For the vast majority of their evolutionary history, in the absence of human assistance, the main overwater dispersal vector for anoles was vegetative flotsam (Hedges 2001). Successful transit was dependent on prevailing currents and winds, whereby individuals had to survive desiccation, solar radiation, and starvation while clinging to vegetation that had been blown out to sea. In the Anthropocene, the majority of exotic anoles also arrive to new locations via plants that are transported by humans for agriculture and the live plant trade (e.g., Perry *et al.* 2006). One of the

earliest documented exotic colonizations in the Caribbean via live plants occurred in the early 1800s (recounted in Lazell 1972). In 1789, the British HMS *Bounty* captained by William Bligh of the Royal Navy was returning from Tahiti with a shipment of breadfruit saplings to be planted on St. Vincent island in the Caribbean. The ship never completed the journey as its crew mutinied, placing Bligh and loyal crew members on a small launch in middle of the South Pacific (i.e., The Mutiny on the *Bounty*). Bligh improbably survived and a couple years later succeeded in the voyage, establishing breadfruit on St. Vincent. A decade later, breadfruit cultivars from Bligh's St. Vincent plantation were transported to Trinidad. This live plant shipment from St. Vincent to Trinidad contained not only the fruit trees but also the St. Vincent bush anole (*Anolis trinitatis*) that is native only to St. Vincent. *Anolis trinitatis* established and became a naturalized part of the Trinidadian herpetofauna so much so that it was first described as a species endemic to Trinidad, hence the misnomer *trinitatis*.

Anolis trinitatis is one of 22 anole species that have established exotic populations in the Anthropocene (Table 1). The majority of exotic anole introductions in the Caribbean have been inadvertent via the live plant trade, although many, especially in Florida, were due to the pet trade (Krysko *et al.* 2011). Of the 38 exotic colonizations on Caribbean islands, 22 have known introduction vectors: 18 (82%) were transported by cargo shipments, primarily live plants, and 4 were intentionally released for various reasons (Table 2; see extended data table 1 in Helmus *et al.* 2014). There have also been several experimental introductions of anoles, primarily *Anolis sagrei*, to small satellite islands such as in the Bahamas, where it is native, and islands in Florida, where it is exotic, to test biodiversity theory (e.g., Kolbe *et al.* 2012a; Stuart *et al.* 2014). However, these experimental introductions relocate anoles very short distances to islands within their native or well-established exotic ranges compared to most unintentional introductions that result in the establishment of exotic anoles far from their native ranges.

Because surviving long time periods clinging to flotsam is difficult and Caribbean island banks are separated by deep water and long distances, successful among-bank natural dispersal events are relatively rare, and occurred in the past most often among proximal land masses. As a result, stepping-stone patterns of colonization and allopatric speciation are evident in the anole phylogeny (Thorpe *et al.* 2004). For

Table 1 List of *Anolis* lizard species that have had exotic populations as of 2015

Exotic *Anolis*	Ecological morphology	Number of colonizations	Native to	Genetics reference
sagrei	Trunk-ground	>10	GA	Kolbe et al. (2004)
cristatellus	Trunk-ground	≥5	GA	Eales et al. (2008)
cybotes	Trunk-ground	≤3	GA	Kolbe et al. (2007)
carolinensis	Trunk-crown	≥5	GA*	Sugawara et al. (2015)
chlorocyanus	Trunk-crown	≤3	GA	Kolbe et al. (2007)
grahami	Trunk-crown	≤3	GA	Chakravarti (1977)
maynardi	Trunk-crown	≤3	GA*	
porcatus	Trunk-crown	≤3	GA	Kolbe et al. (2007)
smaragdinus	Trunk-crown	≤3	GA*	
distichus	Trunk	≤3	GA	Kolbe et al. (2007)
equestris	Crown-giant	≥5	GA	Kolbe et al. (2007)
garmani	Crown-giant	≤3	GA	Kolbe et al. (2007)
aeneus	Small LA	≤3	LA	
pogus	Small LA	≤3	LA	
trinitatis	Small LA	≤3	LA	
wattsi	Small LA	≤3	LA	
extremus	Intermediate LA	≤3	LA	
marmoratus	Intermediate LA	≤3	LA	
bimaculatus	Large LA	≤3	LA	
leachii	Large LA	≤3	LA	
richardii	Large LA	≤3	LA	
lineatus	Unknown	≤3	SA	

The number *of colonizations* gives an estimate of the number of times each species has established a reproducing exotic population. Note that exotic *Anolis pogus* and *Anolis bimaculatus* populations are considered extirpated. *Trunk, trunk-ground, trunk-crown,* and *crown-giant* are ecomorphs that evolved across the Greater Antilles. *Small, intermediate,* and *large LA* are body sizes that evolved across the Lesser Antilles. See Losos (2009) for a review of these morphologies.
* Indicates species recently derived from a natural colonization from Cuba (i.e., the *carolinensis* clade). *Genetics reference* gives an example reference, if one exists, of an invasion genetics study on the species.

example, an ancestor from the Puerto Rican bank likely colonized the islands of the northern Lesser Antilles in a roughly stepping stone fashion from north to south, where each colonization led to an allopatric speciation event and formed part of the large-bodied *bimaculatus* clade (Stenson *et al.* 2004). Phylogenetic analyses also indicate that the anoles native to the southern Lesser Antilles were likely derived from a single colonization of an ancestor population from northern South America, which then allopatrically speciated northward up the island chain in a roughly stepping-stone fashion to Martinique (Creer *et al.* 2001; Thorpe *et al.* 2004). In comparison, exotic anole colonizations are not limited by geographic isolation. Rather, economic connectivity dictates translocation such that islands with ports that receive large amounts of shipping traffic have the most exotic anoles (Helmus *et al.* 2014).

The mainland (i.e., Mexico, Central America and South America; Fig. 1) has a higher number of native anole species compared to the Caribbean Islands (~210 vs. 165 species), yet all anole species with exotic populations are native to Caribbean islands except one, *Anolis carolinensis*. This disparity in invasion success between Caribbean and mainland anole species is likely due to two main factors. First, Caribbean anoles are more likely than mainland anoles to have native ranges that encompass shipping ports and thus are more likely to be inadvertently transported out of their native range (Latella *et al.* 2011). As an exception that proves the rule, *A. carolinensis* has a native range that encompasses several major shipping ports in the Southeastern United States (e.g., Miami and New Orleans), and it has spread to the Lesser Antilles and as far as Hawaii and Okinawa.

Table 2 Characteristics typifying native and exotic *Anolis* lizard species

	Natives	Exotics (as of 2015)
Number of species	375 total (possibly >400)	22 species have had exotic populations
	165 Caribbean species	36 extant (2 extirpated) populations across Caribbean islands
	209 in Mexico, Central America, South America	55 extant populations worldwide (rough estimate)
	1 in the United States	10 in the United States (most of any country)
Transport vector	Vegetative flotsam	Unintentional transport—frequently via the live plant trade
		Intentional transport—pet trade, biocontrol, experiments
Traits needed to survive journey	Tolerance to desiccation, starvation, and thermal extremes	Tolerance to desiccation, starvation, and thermal extremes; may differ depending on vector
Spatial extent of colonization	Proximal to source range	Can be far from native source range (e.g., Spain and Taiwan)
Temporal rate of establishment	40 in 40–60 million years (rough estimate)	1 per year in the Caribbean
Genetic diversity of populations	High genetic variation among populations despite small geographic ranges	Reduced variation through founder effects in some populations
	Low variation within populations	High variation within populations due to multiple introductions and subsequent admixture
Adaptation to local conditions	Repeated adaptation to abiotic/biotic environmental variation resulted in widespread convergent evolution of similar ecomorphs	Rapid adaptation in multiple traits: scale density, limb size, toepad size, body size, head shape, thermal tolerance, perching behavior, foraging behavior, territorial displays
Phylogenetic diversity of island assemblages	Generally low due to infrequent natural colonizations from proximate regions and subsequent speciation resulting in closely related species	Generally high due to frequent long-distance colonizations of distantly related species

The second major factor that makes Caribbean anole species more likely than mainland species to be exotic is that many Caribbean anoles have naturally evolved traits conducive to overwater colonization (Latella *et al.* 2011; Poe *et al.* 2011). For example, *A. sagrei*, the most prolific exotic anole (Table 1), has traits that confer desiccation, heat, and starvation tolerance essential for surviving the beach and dry forest scrub habitats where it is naturally found. These same traits are also essential to surviving long-distance ship travel. *Anolis sagrei* is also a natural overwater disperser. Individuals can survive for days to weeks without food, can float without rafts in seawater for long periods, and will actively swim away from inhospitable islets (Schoener & Schoener 1983a, 1984). Across species, Latella *et al.* (2011) performed a comparative analysis that included both Caribbean and mainland anoles and found that

species with exotic populations tended to be larger bodied, have larger scales, and be more sexually dimorphic in body size. Larger bodied anoles can survive longer without food and are competitively superior to smaller anoles (e.g., Pacala & Roughgarden 1985); larger scales may reduce inter-scale water loss and protect against solar radiation (Losos 2009; Wegener *et al.* 2014), while sexual dimorphism allows partitioning of resources between the sexes facilitating establishment in new environments (Schoener 1969a). Furthermore, those anole species that were good natural colonizers in the ancient past—as measured by native species distributions and an evolutionary reconstruction of anole biogeography—are the same species, or are closely related to species, that currently have exotic populations (Poe *et al.* 2011). Evolutionary history can therefore predict which anole species might become exotic; however, the overall explanatory power of phylogeny in predicting exotic success is statistically weak as there has been much convergent evolution in anole traits that erases phylogenetic signal (Latella *et al.* 2011; Mahler *et al.* 2013).

Native and exotic transport to new islands is thus very similar (Table 2). The environmental stress of transport is similar, and as a result, the traits that make a good natural colonizer are the same as those that make a good exotic colonizer. The main dispersal vectors of both natives and exotics are also the same: plants. The differences are that in the Anthropocene, the vegetation (i.e., plant cargo) in which anoles are transported arrives to new islands at a much higher rate and from longer distances than plant flotsam. These differences in distance and rate have caused anoles that were once geographically relegated to the Americas for millions of years, to have dispersed across oceans to Asia and Europe within the past century (Fig. 1).

POPULATION GENETIC STRUCTURE

Native anole species often exhibit significant phylogeographic genetic structure whereby genetic variation among populations often exceeds within population variation. For example, *Anolis cooki* has a small geographic range restricted to the southwestern coast of Puerto Rico with the most distant populations being only 50 km apart, yet it exhibits strong phylogeographic structure caused in part by a river limiting gene flow between western and eastern populations. Rodríguez-Robles *et al.* (2008) found 27 unique mitochondrial haplotypes across *A. cooki*'s range with no populations sharing any haplotypes. Similarly, other species in Puerto Rico—*Anolis poncensis* with a small range like *A. cooki*, and *Anolis cristatellus* and *Anolis krugi* with wider distributions—exhibit significant population genetic differentiation and phylogeographic structure (Jezkova *et al.* 2009; Kolbe *et al.* 2007; Rodríguez-Robles *et al.* 2010). Anole species from the Lesser Antilles also exhibit high among-population genetic diversity (Thorpe *et al.* 2015). The endemic anole *Anolis oculatus* is widespread on Dominica and has high among-population genetic variation, such that most haplotypes are restricted only to single localities, and strong phylogeographic structure likely due to past vicariance events caused by volcanism (Malhotra & Thorpe 2000). The native anole species of Martinique (*Anolis roquet*) also exhibits high genetic variation

across the island with genetically subdivided populations that correspond to precursor species that diverged allopatrically on four separate ancient islands that later coalesced into the present-day Martinique (Thorpe *et al.* 2010). Therefore, anole species exhibit strong phylogeographic structure in their native ranges, and genetic variation among populations is naturally higher than within populations.

This strong phylogeographic structure in native anoles has made it possible to trace the source populations of exotic anoles. For example, *A. sagrei* was introduced to the Florida Keys in the latter half of the nineteenth century, and by the 1970s it had naturalized and spread throughout Florida. Genetic analyses indicated that exotic *A. sagrei* populations in Florida are the result of at least eight introductions from different source populations across its native range in Cuba (Kolbe *et al.* 2004). The independent introductions subsequently admixed in Florida, greatly increasing genetic variation within exotic populations (Fig. 2). This mode of invasion—multiple colonizations from

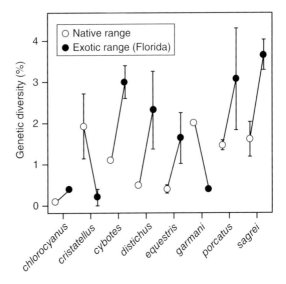

Fig. 2 The genetic diversity within exotic *Anolis* lizard populations in Florida is generally higher than the genetic diversity within populations in their native ranges. Genetic diversity is the mean pair-wise mtDNA sequence divergence of sampled individuals within populations. Points are means across populations, and standard errors are given for those means where multiple populations were sampled. Data are aggregated from Kolbe *et al.* (2004, 2007), which provide full descriptions of the data and statistical analyses.

geographically and genetically distinct source populations followed by admixture in the exotic range—has also occurred in four other exotic anole species in Florida: *A. cristatellus*, *Anolis cybotes*, *Anolis distichus*, and *Anolis equestris* (Fig. 2; Kolbe *et al.* 2007; Kolbe unpublished data). All of these species were derived from at least two introductions with *A. sagrei* derived from at least eight and *A. distichus* derived from at least four independent introductions. Similarly, the exotics *A. cristatellus* in the Dominican Republic and *A. sagrei* in Jamaica were also likely derived from at least two introductions from different parts of their native ranges (Kolbe *et al.* 2007).

The observations of high genetic variation in exotic populations and the ability of exotic populations to establish and spread are in contrast to the classically held view in invasion genetics that exotic populations should exhibit founder effects. Genetic founder effects occur when founding propagules possess a subset of the genetic variation exhibited in the ancestral population and are an expected consequence of oversea dispersal of only a few surviving individuals. Founder effects could result in reduced adaptive potential and inbreeding depression that limit population growth and ultimately the establishment success of a newly arrived species (Allendorf & Lundquist 2003). Experimental introductions of anoles to small islands have artificially induced genetic founder effects without adversely affecting population success. Schoener and Schoener (1983b) and Losos and Spiller (1999) introduced anoles to small islands in the Bahamas that were previously uninhabited by anoles. Despite the small numbers of founders (5–10 founding individuals per island), most of the populations persisted and almost all increased in population size for years and many continue to this day. In another experimental introduction, Kolbe *et al.* (2012a) founded seven small islands in the Bahamas with male–female pairs of *A. sagrei* that were randomly selected from a large island. The experiment produced strong genetic founder effects, yet all *A. sagrei* populations persisted and adapted to the new island environments (Kolbe *et al.* 2012a).

The species that naturally colonize islands may be adapted to resist the negative consequences of genetic founder effects, predisposing them to become successful exotic species (Baker 1955). For example, *A. carolinensis*, a good natural colonizer (Glor *et al.* 2005), was introduced to the Japanese island Chichi-jima in the 1960s and then spread to the nearby islands of Ani-jima and Haha-jima. Even though the Ani-jima and Haha-jima populations have genetic founder effects (i.e., reduced genetic variation), the species has established and spread across both islands (Sugawara *et al.* 2015). Similarly, *A. cristatellus*, introduced to Dominica exhibits a genetic founder effect, yet in the decade that it has colonized Dominica, it has displaced the native anole *A. oculatus* from much of southwest Dominica where it is found (Eales *et al.* 2008). *Anolis cristatellus* in Florida also exhibits a possible founder effect (Fig. 2), yet since its establishment 40 years ago, it has spread and is abundant in some areas (Krysko *et al.* 2011). *Anolis garmani* was also introduced to Florida and exhibits a founder effect (Fig. 2). While it is localized and has not yet spread widely, it has been well-established since the 1970s (Krysko *et al.* 2011). Finally, *Anolis grahami* introduced to Bermuda in 1905 exhibits a genetic founder effect (Chakravarti 1977), but it has naturalized and spread island wide. Although these preceding examples are of all successful establishments despite founder effects, it is not known if founder effects are the cause for failed anole introductions as these are rarely documented.

Successful exotic invasions in the context of single introductions with genetic founder effects may lend insight into the natural colonization process that occurred most certainly with low numbers of founding individuals. At present, the influence of founder effects on the establishment of native anole populations has not been well explored. In addition, invasion genetics in anoles needs further investigation—less than half of all exotic anoles species have had their population genetics sampled (Table 1)—but the most remarkable contrast emerging between native and exotic anoles is in their population genetic diversity (Table 2). Natives generally exhibit low within-population genetic variation, whereas exotic anoles can have very high within-population variation (Fig. 2).

RESPONDING TO ENVIRONMENTAL VARIATION

Across the Caribbean, native anole species exhibit morphological and behavioral adaptations to environmental selective pressures. Most notably, there is considerable variation in precipitation and vegetation structure within and across Caribbean islands. In the Lesser Antilles, some islands have distinct xeric coastal woodlands and wetter montane rainforests in the interior. Conspecific populations collected from xeric and rainforest sites within the same Lesser Antilles island can show divergent adaptation to this climate variation in several quantitative traits such as body shape, scalation, and hue, which are correlated with survival in their respective environmental conditions (Ogden & Thorpe 2002; Thorpe *et al.* 2015). For example, the Dominican anole *A. oculatus* exhibits morphological variation that matches elevational habitat variation despite significant gene flow between populations along the gradient (Stenson *et al.* 2000, 2002). Similarly, *Anolis marmoratus* of Guadeloupe exhibits high variation in male dewlap color and skin pattern that corresponds to a gradient of mesic and xeric habitats within the context of high levels of interpopulation gene flow (Muñoz *et al.* 2013). Because gene flow in both *A. oculatus* and *A. marmoratus* has been quantified using neutral genetic markers, the correlation between trait variation and environmental variation may indicate divergence in the functional portions of the genome related to adaptive traits. This divergence of functional genes may be maintained by assortative mating among individuals with the same ecological phenotype. In Martinique, for example, assortative mating in *A. roquet* occurs between individuals with the same ecological phenotype along the elevation gradient, rather than between individuals with shared phylogeographic ancestry as indicated by genetic markers (Thorpe *et al.* 2010).

Experimental work suggests that adaptation to environmental conditions can be rapid, which may facilitate the colonization success of introduced populations. Losos *et al.* (1997) found that *A. sagrei* populations experimentally introduced to small islands of the Bahamas had, over the course of a little more than a decade, adapted morphologically to match the vegetation structure of the islands. Later, in a similar experimental introduction, Kolbe *et al.* (2012a) found the same result: *A. sagrei* hindlimb length decreased in response to narrower vegetation on the introduced islands; however, this morphological change occurred in just 4 years. Although *A. sagrei* exhibits phenotypic plasticity in limb length in response to perch diameter (Losos *et al.* 2000), the magnitude of divergence in limb diameter due to plasticity induced in laboratory experiments is less than the divergence observed in these experimentally introduced populations. This suggests that plasticity alone is insufficient to explain the changes in limb length over time, and that adaptation likely occurred (Kolbe *et al.* 2012a).

Rapid adaptation in response to the environment has also been observed in unintentionally introduced exotic anole populations. For example, in 10 years on Dominica, exotic populations of *A. cristatellus* at high elevation sites have evolved a lower scale density possibly in response to the wet, high-elevation environments compared to the dry, low-elevation environments where it was introduced (Eales *et al.* 2010; Malhotra *et al.* 2011). However, this scalation-elevation pattern seen for *A. cristatellus* on Dominica is opposite to the average pattern observed among anole species: species naturally found in drier climates have larger and lower densities of scales than species found in wetter environments (Wegener *et al.* 2014). The difference between the causes of scale variation among and within species needs further exploration (Losos 2009).

Exotic anoles have also adapted their thermal niches. Kolbe *et al.* (2014) found that *A. sagrei* populations from southern Georgia can tolerate significantly lower temperatures than Florida populations, suggesting that *A. sagrei* is evolving to withstand colder temperatures in its expanding northern exotic range. Similarly, *A. cristatellus* was introduced to Miami in the mid-1970s and since that time has become more thermally plastic by acquiring the ability to acclimate to lower temperatures than in its native range in Puerto Rico (Kolbe *et al.* 2012b; Leal & Gunderson 2012). This adaptability of exotics stems from the evolutionary lability of anole thermal tolerances. Across the native anoles of the Greater Antilles, thermal biology traits are much more evolutionarily labile than morphological traits (Hertz *et al.* 2013). A phylogenetic study of the native *cybotes* clade on Hispaniola also showed high variability in thermal niches among species: species evolved to tolerate high temperature behaviorally by augmenting the amount of time they spend basking, but evolved to tolerate lower nighttime temperatures through physiological means (Muñoz *et al.* 2014). As a result, cold tolerance has evolved across *cybotes* clade species at a faster rate than heat tolerance.

It is clear that native anoles exhibit substantial trait adaptation in response to environmental conditions such as temperature and vegetation structure. However, the mechanisms, genetic and otherwise, behind these adaptations and the speed with which these adaptations occurred are poorly understood. Given that both experimentally and unintentionally introduced populations from multiple anole species exhibit rapid adaptation to novel environmental conditions (Table 2), studying the spread of exotic anoles should provide new insight into the adaptation process of colonizing species.

SPECIES INTERACTIONS

Species interactions, namely competition, have been an important source of selective pressure on anole functional traits as well as a major force in structuring species distribution patterns. Anoles compete strongly over prey and partition thermal and structural microhabitats (e.g., Schoener 1968). As ecological trait similarity between species increases, the strength of competition increases (see Losos 2009 for a review of the data on anole interspecific competition). For example, on St. Martin, the two native species (*Anolis pogus* and *Anolis gingivinus*) are more similar in size and ecological traits than the two native species on St. Eustatius (*Anolis schwartzi* and *Anolis bimaculatus*). In a set of experiments, Pacala and Roughgarden (1982, 1985), measured the pair-wise strength of competition between the more similar anoles on St. Martin and replicated the experiment for the two more divergent species on St. Eustatius. Because *A. pogus* and *A. gingivinus* on St. Martin have more similar body size and foraging behavior, they exhibited stronger competition: growth rates and fecundity were reduced when *A. gingivinus* was experimentally raised with *A. pogus*, while no negative effects were seen when *A. bimaculatus* was raised with *A. schwartzi*. Present-day coexistence of the ecologically similar *A. pogus* with *A. gingivinus* might be facilitated by the competitively dominant *A. gingivinus* having a higher susceptibility to lizard malaria (Schall 1992).

Strong competition between species with similar traits is likely responsible for consistent body-size distribution patterns across the Lesser Antilles. Like St. Eustatius, eight other major Lesser Antilles islands have one large and one small native species (Schoener 1970). This divergence in body size between the two species is likely due to competitive exclusion and character displacement (Giannasi *et al.* 2000; Losos 2009; Miles & Dunham 1996). The hypothesis is that only species with traits dissimilar to the traits of the resident species could successfully colonize the islands by avoiding competitive exclusion. Meaning, if a large anole species colonized first, only a small anole species could successfully colonize afterward. Following establishment, character displacement most likely occurred in a few instances to further reduce interspecific competition.

Comparatively, the majority of islands in the Lesser Antilles have only one intermediate-sized anole species. Intermediate-size is predicted to be the optimal size for anoles based on the size distributions of their prey (Roughgarden & Fuentes 1977; Schoener 1969b). If intermediate size is optimal, intermediate-sized anoles likely successfully outcompete smaller and larger colonists resulting in a single, intermediate-sized species on these islands.

Because competition is likely a major mechanism limiting the colonization success of anoles that disperse naturally, competition with resident species should also prevent the establishment of exotic anole species (i.e., biotic resistance). Indeed, there is evidence that it is less likely for exotic anoles to establish on those islands that have ecologically similar native species (Losos *et al.* 1993). Theoretically, exotic anoles that successfully establish should have different traits than resident species to avoid competition, or if the exotic and resident species have similar traits, the exotic must be a superior competitor. Empirically, there is support for both scenarios.

In several examples, exotics and/or natives have altered behavioral and morphological traits to reduce interactions. For example, on small intracoastal waterway islands in Florida where *A. sagrei* recently invaded, the native *A. carolinensis* now forages higher in the canopy away from the lower foraging *A. sagrei*. As a result of this behavioral shift, *A. carolinensis* has evolved larger toepads that improve clinging ability (Stuart *et al.* 2014). Similarly, the native *Anolis conspersus* on Grand Cayman has shifted to perch higher when sympatric with the introduced *A. sagrei* (Losos *et al.* 1993). In south Florida, exotic populations of *A. cristatellus* that are sympatric with exotic populations of *A. sagrei* have shorter, broader, and deeper heads (Losin 2012). As head shape strongly influences prey consumption, these differences in *A. cristatellus* head shape are consistent with interspecific competition from *A. sagrei*, driving ecological character displacement. Rapid change in behavioral and morphological traits may facilitate niche partitioning in order to reduce competition, allowing exotics and natives to coexist sympatrically, similar to the partitioning seen among sympatric native anoles (e.g., Schoener 1968).

While there are certainly examples of exotic anoles coexisting sympatrically with natives and other exotics, competitive exclusion also occurs. The exotic *A. cristatellus* on Dominica is competitively displacing the native *A. oculatus* from natural habitats (Eales *et al.*

2010; Malhotra *et al.* 2011). On Trinidad, *Anolis aeneus*, an exotic from Grenada, may competitively displace *A. trinitatis*, an exotic from St. Vincent, from degraded forest habitats (Hailey *et al.* 2009). An experimental introduction of *A. pogus* to Anguillita was not successful presumably due to competition from the native *A. gingivinus* (Roughgarden *et al.* 1984), and competition from *A. gingivinus* may have been responsible for the failed introduction of *A. bimaculatus* on St. Martin (Powell *et al.* 2011)—although evidence for competitive exclusion causing these failed introductions is arguably weak.

While competition is likely a strong force influencing both native and exotic anole establishment success, we have less of an understanding of how predation has influenced native and exotic anoles on ecological and evolutionary timescales. Native communities can exert biotic resistance in the form of consumer–resource interactions (e.g., predation and parasitism) on invading species, although this mechanism has received little research attention with respect to anoles. The exotic anoles in Trinidad provide a hint that predation-based biotic resistance may be operating. The two exotics, *A. trinitatis* and *A. aeneus*, are not found in intact forest habitat, and it has been suggested that this is due to high predation rates in natural Trinidadian forests (Hailey *et al.* 2009; Holt 1977). The few reports of predation on exotic anoles include the mountain wolf snake in Taiwan eating *A. sagrei* (Norval *et al.* 2007), the endangered Ogasawara buzzard eating *A. carolinensis* in the Bonin Islands near Japan (Kato & Suzuki 2005), and exotic *A. equestris* from Cuba eating exotic *A. distichus* from Hispaniola in Florida (Stroud 2013). *Anolis* antipredator defenses are largely behavioral (Leal & Rodrígues-Robles 1995); therefore, the main natural predators of anoles such as snakes, birds, and lizards may be preadapted to predate exotic anoles where exotics and natives co-occur. However, in regions like Asia where anoles are not endemic, predators must learn how to consume novel anole prey.

If native species are able to resist exotic anole establishment, then this biotic resistance should work best in environments that are intact where native competitors and predators are well-established and fill native niche space. On many islands, exotic anoles are absent from intact native habitats and are only found in anthropogenically altered habitats. Superficially, this pattern matches the outcome of biotic resistance; however, other factors such as altered microclimates and limited dispersal may be at play. For example, on Aruba

the Cuban green anole, *Anolis porcatus*, a species that requires humid microclimates, was likely introduced in a shipment of coconut palm trees planted in a resort garden (Odum & Van Buurt 2009). Resort gardens often contain exotic plant species and irrigation, which makes them highly divergent from native habitats. While *A. porcatus* is well established in the artificial habitat of the irrigated garden, it is absent from the arid natural environment of Aruba where the native *Anolis lineatus* thrives. On St. Martin, the exotic *A. cristatellus* has been on the island for years, but relegated to only a resort garden (Yokoyama 2012; Jesse unpublished data). In this case, the microclimates in the resort garden are more similar to those of the native habitat, yet the garden is isolated by inhospitable habitat (e.g., parking lots and buildings) likely limiting the dispersal of *A. cristatellus* to the native habitat.

Exotic colonization success through both character displacement and competitive exclusion of natives occurs, yet the strength and influence of competition has not been explicitly measured for most all sympatric exotic and native anole species. Further work is necessary to elucidate the role of predation in anole evolution and whether predation limited the establishment success of naturally colonizing anoles in the past and exotic anoles in the present. It is also clear that more data, and finer tests, are needed to quantify the role of biotic resistance in halting exotic establishments, although correlative data do suggest that failed exotic colonizations occur most frequently on islands with ecologically similar native anoles (Losos *et al.* 1993). However, the naturalization of exotic anoles into novel anthropogenic habitat must be accounted for when presenting evidence for biotic resistance.

ANOLE ECOMORPHS

One of the most remarkable outcomes of adaptation to environmental variation and species interactions is the convergent evolution of six distinct habitat specialist "ecomorphs" that have evolved repeatedly and independently across the Greater Antilles islands (Losos 2009; Mahler *et al.* 2013; Williams 1983). Classification of anole species into particular ecomorph categories is based on morphological traits, behavior, and habitat use. For example, species from the "trunk-crown" ecomorph have elongated bodies with relatively short fore- and hind-limbs, and larger toepads, which allow them to maneuver efficiently on narrow diameter branches in the canopy; while species from the "trunk-ground" ecomorph have longer hind-limbs that allow them to run faster on thicker diameter branches. The six ecomorphs partition resources in a manner that reduces interspecific competition among sympatric species. When two species from the same ecomorph category are sympatric, they partition resources more finely by either exploiting different climatic microhabitats, or differing substantially in body size (see the review in Losos 2009).

Ecomorphs have not evolved on the Lesser Antilles islands; however, many species are morphologically comparable to the trunk-crown ecomorph (Losos & de Queiroz 1997). In some cases, intraspecific trait variation is higher in Lesser Antilles species than Greater Antilles species. For example, *A. oculatus* from Dominica exhibits substantial interpopulation differentiation whereby different populations can be classified as either trunk-ground or trunk-crown ecomorphs (Knox *et al.* 2001). Comparatively, no Greater Antilles species can be classified into more than one ecomorph category. Further, in the Bahamas, intraspecific trait variation in limb length for *A. sagrei* follows the same phenotypic pattern as ecomorphs but to a lesser degree: *A. sagrei* populations that use broader diameter perches have longer limbs than those using smaller diameter perches (Calsbeek *et al.* 2007).

There is no clear pattern with respect to geographic origin or ecomorph category that predicts exotic success. Of the 22 exotic anole species, 12 are evolutionarily derived from the Greater Antilles, 9 are from the Lesser Antilles, and 1 is derived from a mainland South American clade (Table 1). Six exotic species are trunk-crown ecomorphs, three are trunk-ground ecomorphs, two are crown-giant ecomorphs, and one is a trunk ecomorph. There are currently no twig or grass-bush species that have exotic populations. While, as previously discussed, there is certainly evidence that particular traits make it more likely for some species to become good exotic colonizers over others (Latella *et al.* 2011; Poe *et al.* 2011), there is not a strong pattern of some ecomorphs establishing more than others (Table 1).

The absence of a consistent pattern in the ecomorphology of exotic anoles may be the result of a different phenomenon: islands once depauperate in particular ecomorphs are now acquiring them. Cayman Brac originally lacked the trunk-crown ecomorph, but now the exotic trunk-crown *Anolis maynardi* is widely established on the island. In the reverse, Grand Cayman naturally has the trunk-crown ecomorph (*A. conspersus*), but lacked a trunk-ground species until *A. sagrei* established. Note that neither of these island's single native anoles, *Anolis luteosignifer* on Cayman Brac and *A. conspersus* on Grand Cayman, exhibits ecological release such that they have high trait variation and are found in multiple habitats. Instead, the ecomorphology of each native has diverged little from their ancestor clades: the trunk-ground *sagrei* clade of Cuba in the case of *A. luteosignifer* and the trunk-crown *A. grahami* clade of Jamaica in the case of *A. conspersus*. This evolutionary conservatism seems to allow for the invasion of other ecomorphs. As a nuanced example, Jamaica already had a native trunk-ground ecomorph, *Anolis lineatopus*, when trunk-ground *A. sagrei* invaded, yet the two seem to coexist on the island perhaps due to habitat partitioning where *A. lineatopus* prefers shaded forests and *A. sagrei* dominates open areas where other native Jamaican anoles are not found (Lister 1976; Williams 1969).

Will this trend of adding new ecomorphs to islands that previously lacked them continue? Jamaica, for example, still does not have a trunk ecomorph, but the exotic *A. distichus* from Hispaniola is a trunk ecomorph that is established in Florida. With the large amount of trade between the United States and Jamaica, will *A. distichus* from the United States eventually establish a Jamaican population? Or perhaps *A. distichus* from its native range in the Bahamas or Hispaniola will eventually be transported to Jamaica.

BIOGEOGRAPHY OF ANOLES IN THE ANTHROPOCENE

While it is clear that exotic anoles can successfully survive human-assisted translocation and adapt to novel conditions, a major challenge is to understand how biogeographic patterns have been altered in the Anthropocene. The equilibrium theory of island biogeography proposes that as islands become more saturated with species, competitive interactions increase, which causes extinction rates to increase (MacArthur & Wilson 1967). As a result, it is less likely for a colonizing species to establish and spread on an island with many resident species given its area (i.e., a saturated island). The spread of exotic anoles in the Caribbean matches this expectation: the least saturated island banks (i.e., banks with the fewest native species given their area) have had the most exotic anoles colonize (Helmus *et al.* 2014). The least saturated banks are those that are geographically isolated and that have not had cladogenic (*in situ*) speciation (Losos & Schluter 2000).

This filling of the unsaturated island banks with exotic anoles has strengthened the relationship between anole species richness and area across Caribbean island banks. In the past, the relationship was two part—species richness rose modestly up to a threshold area where cladogenic speciation occurred, and then rose rapidly with increasing area. This two-part relationship is thought to be due to how area regulates speciation: small islands are not large enough for allopatric speciation, which drives the accumulation of anole species richness on the large islands (Losos & Schluter 2000). However, in the Anthropocene, the establishment of exotics on the least saturated banks has caused the Caribbean anole species–area relationship to become linear. In the past, there was also a strongly negative species–isolation relationship because the most geographically isolated banks were the least saturated, but today the negative relationship has been almost eliminated. Now it is the economic isolation of island banks (e.g., number of cargo ships docking), not geographic isolation, that determines anole colonization rates in the Anthropocene (Helmus *et al.* 2014).

In the past, Caribbean banks contained closely related species (i.e., low phylogenetic diversity) because anole species naturally accumulated through ancestral colonizations from proximate banks followed by cladogenic speciation on the larger banks (Fig. 3a; Helmus & Ives 2012; Helmus *et al.* 2014; Losos & Schluter 2000; Mahler *et al.* 2010; Thorpe *et al.* 2004). These geographic constraints also caused native anoles from the same bank to be more closely related to each other than expected if species had randomly colonized banks regardless of geographic location (Fig. 3b). In the Anthropocene, anole phylogenetic diversity is accumulating rapidly because exotic anoles are able to invade geographically distant banks that contain distantly related anoles. As a result, the total amount of evolutionary history encompassed on average by species on a Caribbean bank (i.e., both natives and exotics combined) has risen by approximately 70 million years, a 25% increase in anole evolutionary history since European settlement of the Caribbean (Fig. 3a). Anole assemblages are also more phylogenetically random, because the exotics that colonize banks come from across the Caribbean and are also distributed

widely across the *Anolis* phylogenetic tree, as opposed to all being from the same clade (Fig. 3b; Helmus *et al.* 2014).

What are the ecological consequences of increased phylogenetic diversity and randomizing anole species composition? When phylogenetically distantly related species have different traits than residents, they may be more likely to become invasive when introduced (Strauss *et al.* 2006). Therefore, the ecological consequences of *Anolis* introductions are likely a function of the traits of the resident and exotic species involved (Losos *et al.* 1993; Poe 2014). However, it is not a given that distantly related species have different traits; there is considerable trait convergence across the *Anolis* tree (Mahler *et al.* 2013). If an exotic does have different traits than the resident species, it may be at an advantage. However, there are likely far more exotic introductions than colonizations, and the species that successfully colonize likely have the traits necessary to survive in a particular island environment, which are also likely the same traits as the native species. Clearly, more work is necessary to explore the mechanisms and consequences of the pattern of increased phylogenetic diversity and randomized anole composition.

CONCLUSION

We contrasted the ecology and evolution of exotic anoles with that of native anoles (Table 2). One of the strongest patterns we found is that exotics often contain more evolutionary history—in terms of within-population genetic diversity and among-species phylogenetic diversity—than native anoles (Figs. 2, 3). Genetic diversity is higher in exotics because many exotic populations are derived from admixture of multiple colonizations from genetically distinct native range sources. Phylogenetic diversity of anole assemblages is increasing on Caribbean banks because exotic anoles are transported from geographically distant areas. This means that any given exotic anole that colonizes a Caribbean island is likely distantly related to the native anoles of the island. In the Anthropocene, the evolutionary history encompassed by anole populations and assemblages is increasing due to exotic anole colonizations.

Other patterns are also emerging (Table 2). Those natives that in the past were good at naturally colonizing and establishing in new areas are those best able to take advantage of human-aided dispersal in the

(a)

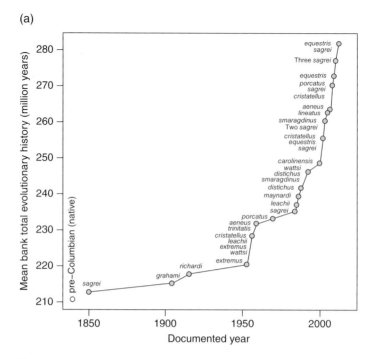

(b)

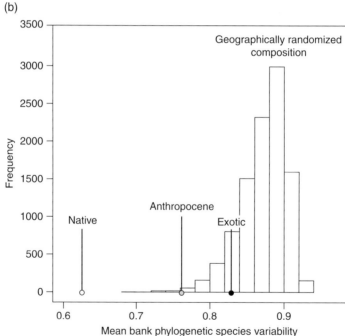

Fig. 3 Caribbean anole phylogenetic diversity is increasing in the Anthropocene. (a) The total evolutionary history of anoles on Caribbean banks (sum of phylogenetic branch lengths of species on a bank) has increased on average by approximately 25% since pre-Columbian times (also see Fig. 3 in Helmus *et al.* 2014). Points are the mean phylogenetic diversity across all Caribbean banks for a given year. Labels indicate the species that were introduced each year. Documented year is plotted because there is debate as to the actual year of some introductions. (b) Native phylogenetic species variability (average relatedness of anole species on a bank, Helmus & Ives 2012) is significantly less than expected if species randomly colonized banks regardless of geographic location ($P \ll 0.001$), while exotic anole phylogenetic composition is not ($P > 0.05$). In the Anthropocene, overall island bank composition (natives and exotics combined) is becoming less geographically structured, but still significant ($P < 0.01$). The null model histogram is of 10 000 mean values from permutations of native species among banks regardless of geographic location, but keeping the observed native bank richness constant.

Anthropocene. Exotic anoles can rapidly adapt and acclimate to new environments and species interactions, and these responses are similar to native species that are also good natural colonizers. Furthermore, native and exotics can coexist due to niche differences, phenotypic plasticity (e.g., behavioral responses), and character displacement. Islands that in the past did not contain particular ecomorphs of anoles are now gaining them. However, there is some evidence of competitive displacement of natives by naturalized exotics, and of biotic resistance of exotics by native anoles and native predators.

While native and exotic anoles can quickly adjust to new environmental conditions, the underlying architecture of genes responsible for traits related to ecological performance is unknown in many cases (Schneider 2008). Although the *Anolis* group is a well-known model system for adaptive radiation, it lags behind other adaptive radiation model systems, in part due to the comparative difficulty of rearing *Anolis* in captivity. However, the recent sequencing of the *A. carolinensis* genome (Alföldi *et al.* 2011) promises new discoveries of the genetic basis of ecologically relevant traits in the near future.

The massive introduction and naturalization of anole species in the Caribbean and across the planet is still in its exponential phase of increase and major questions remain unanswered. How will the increased genetic variation associated with Anthropocene colonizations shape future anole ecology and evolution? In the Lesser Antilles, will the islands that gained more exotic species be able to support them and maintain a higher number of species than before? Or will these islands eventually revert back to being one or two species islands? In the Greater Antilles, questions remain about ecomorph patterns. For example, Jamaica originally had only four of the six ecomorphs. Can it support all six or do some characteristics of the island limit it to only four? While we have focused on the Caribbean, what will be the eventual outcomes of those exotic anole species residing outside of the Caribbean in habitats without native anoles? Do these exotics exhibit more ecological release than exotic anoles in the Caribbean? Finally, the outcomes of these macroevolutionary questions depend largely on the magnitude of anthropogenic activities. If the translocation of species continues, island faunas may never reach an equilibrium point and could remain artificially elevated well into the future. Conversely, as humans alter the quality of habitats on islands, more invasions may be facili-

tated while simultaneously eliminating native species. One of the only certainties is that future island faunas will be the ones most tolerant to human activities.

ACKNOWLEDGMENTS

We thank R. Glor and L. Mahler for their critical feedback. M.R.H.J.E.B and W.A.M.J. were supported by the Netherlands Organisation for Scientific Research (858.14.040).

REFERENCES

Alföldi J, Di Palma F, Grabherr M *et al.* (2011) The genome of the green anole lizard and a comparative analysis with birds and mammals. *Nature*, **477**, 587–591.

Allendorf FW, Lundquist LL (2003) Introduction: population biology, evolution, and control of invasive species. *Conservation Biology*, **17**, 24–30.

Baker HG (1955) Self-compatibility and establishment after "long-distance" dispersal. *Evolution*, **9**, 347–349.

Calsbeek R, Smith TB, Bardeleben C (2007) Intraspecific variation in *Anolis sagrei* mirrors the adaptive radiation of Greater Antillean anoles. *Biological Journal of the Linnean Society*, **90**, 189–199.

Chakravarti A (1977) Genetic differentiation in the colonising lizard *Anolis grahami*. *Heredity*, **38**, 121–123.

Creer DA, de Queiroz K, Jackman TR, Losos JB, Larson A (2001) Systematics of the *Anolis roquet* series of the southern Lesser Antilles. *Journal of Herpetology*, **35**, 428–441.

Eales J, Thorpe RS, Malhotra A (2008) Weak founder effect signal in a recent introduction of Caribbean *Anolis*. *Molecular Ecology*, **17**, 1416–1426.

Eales J, Thorpe RS, Malhotra A (2010) Colonization history and genetic diversity: adaptive potential in early stage invasions. *Molecular Ecology*, **19**, 2858–2869.

Giannasi N, Thorpe RS, Malhotra A (2000) A phylogenetic analysis of body size evolution in the *Anolis roquet* group (Sauria: Iguanidae): character displacement or size assortment? *Molecular Ecology*, **9**, 193–202.

Glor RE, Losos JB, Larson A (2005) Out of Cuba: overwater dispersal and speciation among lizards in the *Anolis carolinensis* subgroup. *Molecular Ecology*, **14**, 2419–2432.

Hailey A, Quesnel VC, Boos HEA (2009) The persistence of *Anolis trinitatis* as a naturalized lizard in Trinidad against hybridization pressure with *Anolis aeneus*. *Applied Herpetology*, **6**, 275–294.

Hedges SB (2001) Biogeography of the West Indies: an overview. In: *Biogeography of the West Indies: Patterns and Perspectives* (eds. Woods CA, Sergile FE), pp. 15–33. CRC Press, Boca Raton.

Helmus MR, Ives AR (2012) Phylogenetic diversity-area curves. *Ecology*, **93**, S31–S43.

Helmus MR, Mahler DL, Losos JB (2014) Island biogeography of the Anthropocene. *Nature*, **513**, 543–546.

Hertz PE, Arima Y, Harrison A *et al.* (2013) Asynchronous evolution of physiology and morphology in *Anolis* lizards. *Evolution*, **67**, 2101–2113.

Holt RD (1977) Report on recent field research. *Anolis Newsletter III*, **30–35**.

Jezkova T, Leal M, Rodrígues-Robles JA (2009) Living together but remaining apart: comparative phylogeography of *Anolis poncensis* and *A. cooki*, two lizards endemic to the aridlands of Puerto Rico. *Biological Journal of the Linnean Society*, **96**, 617–634.

Kato Y, Suzuki T (2005) Introduced animals in the diets of the Ogasawara buzzard, an endemic insular raptor in the Pacific Ocean. *Journal of Raptor Research*, **39**, 173–179.

Knox AK, Losos JB, Schneider CJ (2001) Adaptive radiation versus intraspecific differentiation: morphological variation in Caribbean *Anolis* lizards. *Journal of Evolutionary Biology*, **14**, 904–909.

Kolbe JJ, Ehrenberger JC, Moniz HA, Angilletta MJ (2014) Physiological variation among invasive populations of the brown anole (*Anolis sagrei*). *Physiological and Biochemical Zoology*, **87**, 92–104.

Kolbe JJ, Glor RE, Rodriguez-Schettino L *et al.* (2004) Genetic variation increases during biological invasion by a Cuban lizard. *Nature*, **431**, 177–181.

Kolbe JJ, Glor RE, Rodriguez-Schettino L *et al.* (2007) Multiple sources, admixture, and genetic variation in introduced *Anolis* lizard populations. *Conservation Biology*, **21**, 1612–1625.

Kolbe JJ, Leal M, Schoener TW, Spiller DA, Losos JB (2012a) Founder effects persist despite adaptive differentiation: a field experiment with lizards. *Science*, **335**, 1086–1089.

Kolbe JJ, VanMiddlesworth PS, Losin N, Dappen N, Losos JB (2012b) Climatic niche shift predicts thermal trait response in one but not both introductions of the Puerto Rican lizard *Anolis cristatellus* to Miami, Florida, USA. *Ecology and Evolution*, **2**, 1503–1516.

Kraus F (2009) *Alien Reptiles and Amphibians: A Scientific Compendium and Analysis*. Springer, Dordrecht/London.

Krysko KL, Burgess JP, Rochford MR *et al.* (2011) Verified non-indigenous amphibians and reptiles in Florida from 1863 through 2010: outlining the invasion process and identifying invasion pathways and stages. *Zootaxa*, **3028**, 1–64.

Latella I, Poe S, Tomasz Giermakowski J (2011) Traits associated with naturalization in *Anolis* lizards: comparison of morphological, distributional, anthropogenic, and phylogenetic models. *Biological Invasions*, **13**, 845–856.

Lazell JD (1972) The anoles (Sauria: Iguanidae) of the Lesser Antilles. *Bulletin of the Museum of Comparative Zoology (Harvard)*, **143**, 1–115.

Leal M, Gunderson AR (2012) Rapid change in the thermal tolerance of a tropical lizard. *The American Naturalist*, **180**, 815–822.

Leal M, Rodrígues-Robles JA (1995) Antipredator responses of *Anolis cristatellus* (Sauria: Polychrotidae). *Copeia*, **1995**, 155–161.

Lister BC (1976) The nature of niche expansion in West Indian *Anolis* lizards I: ecological consequences of reduced competition. *Evolution*, **30**, 659–676.

Losin N (2012) The evolution and ecology of interspecific territoriality: studies of *Anolis* lizards and North American wood-warblers (PhD Dissertation), University of California, Los Angeles.

Losos JB (2009) *Lizards in an Evolutionary Tree*. University of California Press, Berkeley.

Losos JB, Creer DA, Glossip D *et al.* (2000) Evolutionary implications of phenotypic plasticity in the hindlimb of the lizard *Anolis sagrei*. *Evolution*, **54**, 301–305.

Losos JB, de Queiroz KD (1997) Evolutionary consequences of ecological release in Caribbean *Anolis* lizards. *Biological Journal of the Linnean Society*, **61**, 459–483.

Losos JB, Marks JC, Schoener TW (1993) Habitat use and ecological interactions of an introduced and a native species of *Anolis* lizard on Grand Cayman, with a review of the outcomes of anole introductions. *Oecologia*, **95**, 525–532.

Losos JB, Schluter D (2000) Analysis of an evolutionary species-area relationship. *Nature*, **408**, 847–850.

Losos JB, Spiller DA (1999) Differential colonization success and asymmetrical interactions between two lizard species. *Ecology*, **80**, 252–258.

Losos JB, Warheit KI, Schoener TW (1997) Adaptive differentiation following experimental island colonization in *Anolis* lizards. *Nature*, **387**, 70–73.

MacArthur RH, Wilson EO (1967) *The Theory of Island Biogeography*. Princeton University Press, Princeton.

Mahler DL, Ingram T, Revell LJ, Losos JB (2013) Exceptional convergence on the macroevolutionary landscape in island lizard radiations. *Science*, **341**, 292–295.

Mahler DL, Revell LJ, Glor RE, Losos JB (2010) Ecological opportunity and the rate of morphological evolution in the diversification of Greater Antillean anoles. *Evolution*, **64**, 2731–2745.

Malhotra A, Thorpe RS (2000) The dynamics of natural selection and vicariance in the Dominican anole: patterns of within-island molecular and morphological divergence. *Evolution*, **54**, 245–258.

Malhotra A, Thorpe RS, Hypolite E, James A (2011) A report on the status of the herpetofauna of the Commonwealth of Dominica, West Indies. *Conservation of Caribbean Island Herpetofaunas*, **2**, 149–166.

Miles DB, Dunham AE (1996) The paradox of the phylogeny: character displacement of analyses of body size in island *Anolis*. *Evolution*, **50**, 594–603.

Muñoz MM, Crawford NG, McGreevy TJ *et al.* (2013) Divergence in coloration and ecological speciation in the *Anolis marmoratus* species complex. *Molecular Ecology*, **22**, 2668–2682.

Muñoz MM, Stimola MA, Algar AC *et al.* (2014) Evolutionary stasis and lability in thermal physiology in a group of tropical lizards. *Proceedings of the Royal Society B: Biological Sciences*, **281**, 20132433.

Nicholson KE, Glor RE, Kolbe JJ *et al.* (2005) Mainland colonization by island lizards. *Journal of Biogeography*, **32**, 929–938.

Norval G, Huang S-C, Mao J-J (2007) Mountain wolf snake (*Lycodon r. ruhstrati*) predation on an exotic lizard, *Anolis sagrei*, in Chiayi County, Taiwan. *Herpetological Bulletin*, **101**, 13–17.

Odum RA, Van Buurt G (2009) *Anolis porcatus* (Cuban green anole). *Herpetological Review*, **40**, 450.

Ogden R, Thorpe RS (2002) Molecular evidence for ecological speciation in tropical habitats. *Proceedings of the National Academy of Sciences of the United States of America*, **99**, 13612–13615.

Pacala SW, Roughgarden J (1982) Resource partitioning and interspecific competition in two two-species insular *Anolis* lizard communities. *Science*, **217**, 444–446.

Pacala SW, Roughgarden J (1985) Population experiments with the *Anolis* lizards of St. Maarten and St. Eustatius. *Ecology*, **66**, 129–141.

Perry G, Powell R, Watson H (2006) Keeping invasive species off Guana Island, British Virgin Islands. *IGUANA*, **13**, 273.

Poe S (2014) Comparison of natural and nonnative two-species communities of *Anolis* lizards. *The American Naturalist*, **184**, 132–140.

Poe S, Giermakowski JT, Latella I *et al.* (2011) Ancient colonization predicts recent naturalization in *Anolis* lizards. *Evolution*, **65**, 1195–1202.

Powell R, Henderson RW, Farmer MC *et al.* (2011) Introduced amphibians and reptiles in the greater Caribbean: patterns and conservation implications. In: *Conservation of Caribbean Island Herpetofaunas*, Volume **1** (eds. Hailey A, Wilson BS, Horrocks JA), pp. 63–143. Brill, Leiden.

Rodríguez-Robles JA, Jezkova T, Leal M (2008) Genetic structuring in the threatened "Lagartijo del Bosque Seco" (*Anolis cooki*) from Puerto Rico. *Molecular Phylogenetics and Evolution*, **46**, 503–514.

Rodríguez-Robles JA, Jezkova T, Leal M (2010) Climatic stability and genetic divergence in the tropical insular lizard *Anolis krugi*, the Puerto Rican "Lagartijo Jardinero de la Montaña". *Molecular Ecology*, **19**, 1860–1876.

Roughgarden JD, Fuentes ER (1977) The environmental determinants of size in solitary populations of West Indian *Anolis* lizards. *Oikos*, **29**, 44–51.

Roughgarden JD, Pacala S, Rummel J (1984) Strong present-day competition between the *Anolis* lizard populations of St. Maarten (Neth. Antilles). In: *Evolutionary Ecology* (ed. Shorrocks B), pp. 203–220. Blackwell Scientific, Oxford.

Schall JJ (1992) Parasite-mediated competition in *Anolis* lizards. *Oecologia*, **92**, 58–64.

Schneider CJ (2008) Exploiting genomic resources in studies of speciation and adaptive radiation of lizards in the genus *Anolis*. *Integrative and Comparative Biology*, **48**, 520–526.

Schoener A, Schoener TW (1984) Experiments on dispersal: short-term floatation of insular anoles, with a review of similar abilities in other terrestrial animals. *Oecologia*, **63**, 289–294.

Schoener TW (1968) The *Anolis* lizards of Bimini: resource partitioning in a complex fauna. *Ecology*, **49**, 704–726.

Schoener TW (1969a) Size patterns in West Indian *Anolis* lizards: I. Size and species diversity. *Systematic Zoology*, **18**, 386–391.

Schoener TW (1969b) Models of optimal size for solitary predators. *American Naturalist*, **103**, 277–313.

Schoener TW (1970) Size patterns in West Indian *Anolis* lizards. II. Correlations with the size of particular sympatric species—displacement and convergence. *The American Naturalist*, **104**, 155–174.

Schoener TW, Schoener A (1983a) On the voluntary departure of lizards from very small islands. In: *Advances in Herpetology and Evolutionary Biology* (eds. Rhodin AG, Miyata K), pp. 491–498. Museum of Comparative Zoology Harvard University, Cambridge, MA.

Schoener TW, Schoener A (1983b) The time to extinction of a colonizing propagule of lizards increases with island area. *Nature*, **302**, 332–334.

Stenson AG, Malhotra A, Thorpe RS (2000) Highly polymorphic microsatellite loci from the Dominican anole (*Anolis oculatus*) and their amplification in other *bimaculatus* series anoles. *Molecular Ecology*, **9**, 1680–1681.

Stenson AG, Malhotra A, Thorpe RS (2002) Population differentiation and nuclear gene flow in the Dominican anole (*Anolis oculatus*). *Molecular Ecology*, **11**, 1679–1688.

Stenson AG, Thorpe RS, Malhotra A (2004) Evolutionary differentiation of *bimaculatus* group anoles based on analyses of mtDNA and microsatellite data. *Molecular Phylogenetics and Evolution*, **32**, 1–10.

Strauss SY, Webb CO, Salamin N (2006) Exotic taxa less related to native species are more invasive. *PNAS*, **103**, 5841–5845.

Stroud JT (2013) *Anolis equestris* (Cuban knight anole) and *Anolis distichus* (Hispaniolian bark anole) exotic intraguild predation. *Herpetological Review*, **44**, 661.

Stuart YE, Campbell TS, Hohenlohe PA *et al.* (2014) Rapid evolution of a native species following invasion by a congener. *Science*, **346**, 463–466.

Sugawara H, Takahashi H, Hayashi F (2015) Microsatellite analysis of the population genetic structure of *Anolis carolinensis* introduced to the Ogasawara Islands. *Zoological Science*, **32**, 47–52.

Thorpe RS, Barlow A, Malhotra A, Surget-Groba Y (2015) Widespread parallel population adaptation to climate variation across a radiation: implications for adaptation to climate change. *Molecular Ecology*, **24**, 1019–1030.

Thorpe RS, Malhotra A, Stenson AG, Reardon JT (2004) Adaptation and speciation in Lesser Antillean anoles. In: *Adaptive Speciation* (eds. Dieckmann U, Doebeli M, Metz JAJ, Tautz D), pp. 322–344. Cambridge University Press, Cambridge

Thorpe RS, Surget-Groba Y, Johansson H (2010) Genetic tests for ecological and allopatric speciation in anoles on an island archipelago. *PLoS Genetics*, **6**, e1000929.

Wegener JE, Gartner GEA, Losos JB (2014) Lizard scales in an adaptive radiation: variation in scale number follows climatic and structural habitat diversity in *Anolis* lizards. *Biological Journal of the Linnean Society*, **113**, 570–579.

Williams EE (1969) The ecology of colonization as seen in the zoogeography of anoline lizards on small islands. *The Quarterly Review of Biology*, **44**, 345–389.

Williams EE (1983) Ecomorphs, faunas, island size, and diverse end points in island radiations of *Anolis*. In: *Lizard Ecology: Studies of a Model Organism* (eds. Huey RB, Pianka ER, Schoener TW), pp. 326–370. Harvard University Press, Cambridge, MA.

Yokoyama M (2012) Reptiles and amphibians introduced on St. Martin. *IRCF Reptiles and Amphibians*, **19**, 271–279.

Chapter 8

CAUSES AND CONSEQUENCES OF FAILED ADAPTATION TO BIOLOGICAL INVASIONS: THE ROLE OF ECOLOGICAL CONSTRAINTS

Jennifer A. Lau* and Casey P. terHorst†

*Kellogg Biological Station and Department of Plant Biology, Michigan State University, 3700 E. Gull Lake Dr., Hickory Corners, MI 49060, USA
† Department of Biology, California State University, Northridge, 18111 Nordhoff Street, Northridge, CA 91330-8303, USA

Abstract
Biological invasions are a major challenge to native communities and have the potential to exert strong selection on native populations. As a result, native taxa may adapt to the presence of invaders through increased competitive ability, increased antipredator defences or altered morphologies that may limit encounters with toxic prey. Yet, in some cases, species may fail to adapt to biological invasions. Many challenges to adaptation arise because biological invasions occur in complex species-rich communities in spatially and temporally variable environments. Here, we review these 'ecological' constraints on adaptation, focusing on the complications that arise from the need to simultaneously adapt to multiple biotic agents and from temporal and spatial variation in both selection and demography. Throughout, we illustrate cases where these constraints might be especially important in native populations faced with biological invasions. Our goal was to highlight additional complexities empiricists should consider when studying adaptation to biological invasions and to begin to identify conditions when adaptation may fail to be an effective response to invasion.

Previously published as an article in *Molecular Ecology* (2015) 24, 1987–1998, doi: 10.1111/mec.13084

Invasion Genetics: The Baker and Stebbins Legacy, First Edition. Edited by Spencer C. H. Barrett, Robert I. Colautti, Katrina M. Dlugosch, and Loren H. Rieseberg.

INTRODUCTION

Biological invasions inevitably result in novel species interactions. When native community members are challenged by biological invasions, they may face novel antagonists such as predators or competitors, or they may benefit from new prey, new and underutilized host plants, or even new mutualists. Given that many types of species interactions are particularly strong selective agents, these new species interactions may lead to evolutionary change in native species, potentially resulting in novel adaptations (reviewed in Strauss *et al.* 2006a; Oduor 2013). For example, toxic invasive cane toads have caused the evolution of increased toxin tolerance and reduced preference for cane toads in black snakes (Phillips & Shine 2006) and reduced gape size in gape-limited predators – a morphological change that prevents these naïve native predators from consuming toxic cane toad prey (Phillips & Shine 2004). Similarly, native rangeland plants have evolved altered root morphologies that confer increased competitive abilities against invasive cheatgrass (Leger 2008; Rowe & Leger 2011). In both of these cases and in many of the increasing number of studies demonstrating evolutionary responses to invasions, natives are evolving in response to particularly noxious invaders that have become dominant species in the ecosystem (see table 1 in Oduor 2013). Yet, most introduced species do not reach such high abundances and may not be the predominant selective agent. Moreover, even though evidence suggests that evolutionary responses to invasion do occur, many other studies fail to detect adaptive evolutionary change or even detect maladaptive evolutionary responses; four of 14 studies included in Oduor's 2013 review of plant evolutionary responses to invasion show that native genotypes from *uninvaded* naïve populations were larger and presumably more competitive than native genotypes from invaded sites. In some cases, small size may be adaptive and most studies do not conduct reciprocal transplant studies in the field to link observed evolutionary changes to fitness, but for at least one native species, maladaptation to invader presence has been reported based on fecundity measures in a field reciprocal transplant experiment (Lau 2006; Box 1).

Both evolutionary and ecological factors can limit adaptation. Evolutionary factors, such as a lack of genetic variation, genetic constraints and conflicting patterns of selection at different life stages/fitness components, can limit adaptation of both native and inva-

sive species. These evolutionary factors are undoubtedly important for determining evolutionary trajectories, but here we focus on ecological limits to adaptation. While less studied than evolutionary constraints, ecological constraints may be especially important in species-rich, spatially and temporally complex natural communities. Multiple direct and indirect species interactions that impose selection pressure, temporal and spatial variability in selection pressures and variation in demographic parameters are ecological factors that can constrain adaptation (reviewed in Antonovics 1976; Kingsolver & Diamond 2011). Here, we review the evidence and mechanisms by which these ecological factors may constrain native species adaptation to biological invaders that harm native populations, emphasizing that these mechanisms are not independent and may frequently interact to constrain adaptation (Box 2). While ecological constraints are relevant for all taxa, we argue that some constraints may be especially important to natives facing biological invasions because of the patchy nature of many invasions, the dramatic demographic effects of some invasions, and also because ecological constraints may be stronger for natives than competing invasives. The ideas presented here build on previous reviews of native evolutionary responses to biological invasions (Strauss *et al.* 2006a; Oduor 2013) but explicitly focus on ecological factors limiting evolutionary responses. Although an increasing number of studies have documented evolutionary effects of biological invasions on native community members, the potential constraints to adaptation have rarely been considered.

ECOLOGICAL CONSTRAINTS TO ADAPTATION

Multiple species interactions

When a single trait affects multiple interacting species ('ecological pleiotropy', *sensu* Strauss & Irwin 2004), adaptation may be limited when the trait is under opposing selection from multiple selective agents. Because selection from one selective agent favours higher trait values, while selection from a second selective agent favours lower trait values, adaptation to either selective agent is constrained. A classic example is the chemical defences plants employ against herbivores. Induction of plant chemical defences in response to a generalist herbivore may

Box 1 Failed adaptation in a native plant challenged by an invasive plant competitor and exotic insect herbivores

Medicago polymorpha invaded California grasslands in the late 1800s and reduced the survival and reproductive success of the co-occurring native plant *Acmispon wrangelianus* both directly through resource competition and indirectly by increasing densities of shared herbivores (Lau & Strauss 2005; Fig. 1). *Medicago* also alters natural selection on *Acmispon* (Lau 2008), yet *Acmispon* has failed to adapt to *Medicago* invasion (Lau 2006). In fact, *Acmispon* genotypes from invaded sites outperformed genotypes from uninvaded sites when planted into uninvaded locales or experimental *Medicago* removal plots, and *Acmispon* genotypes from uninvaded sites tended to have higher fitness than genotypes from invaded sites when planted into heavily invaded locations. We summarize the potential ecological constraints limiting adaptation in this system below.

Multiple species interactions

Medicago is a strong direct competitor but also imposes large indirect effects by increasing *Hypera* herbivory. Because *Acmispon* competitive ability and defence traits are positively correlated, multiple species interactions are unlikely to constrain adaptation because in environments where competition is high

(*Medicago* is abundant), *Hypera* herbivory is also likely to be intense. It is not yet known whether flowering time, which has diverged between invaded and uninvaded populations (Lau 2006), or other traits that mediate multiple interactions could potentially be under opposing selection from different biotic agents.

Indirect effects of the invader

Medicago and insect herbivores interacted to influence the strength of selection on *Acmispon* traits (Lau 2008), resulting in nonadditive selection on resistance to herbivores (terHorst *et al.* 2015). Yet, in this system, nonadditive selection is more likely to slow adaptation, rather than result in maladaptation. Even though interactive effects of *Medicago* and *Hypera* herbivory were detected, increased herbivory increased or decreased the magnitude, but did not change the direction, of selection that *Medicago* exerted on these traits.

Spatial variation in selection

Selection favoured increased tolerance and competitive ability in invaded populations, but decreased tolerance and competitive ability in uninvaded populations (Lau 2008). Invaded and uninvaded sites are closely

(a)

(b)

Box 1, Fig. 1 *Acmispon wrangelianus* growing in uninvaded sites (a) are much larger, much more fecund (produce over 300% more seeds), and receive much less herbivore damage compared to *A. wrangelianus* individuals in sites invaded by *Medicago polymorpha* (b). (*See insert for color representation of the figure.*)

interspersed across the landscape (<<1 km); therefore, the direction of selection differs across small spatial scales, and even small amounts of gene flow could limit adaptation.

Temporal variation in selection

Across three census years, heavily invaded populations ranged from 0.29 to 0.51 proportion of leaflets damaged by *Hypera* (Lau & Strauss 2005). The strength of selection on competitive ability, tolerance to herbivory and resistance to *Hypera* herbivory increases with increasing *Hypera* herbivory. Even slight reductions in herbivory would lead to no selection or even negative selection on these traits. Low levels of herbivory observed in more recent years even in formerly heavily invaded populations are low enough to alter the direction of selection on these traits (C. P. terHorst & J. A. Lau, pers. obs).

Temporal and spatial variation in population demographics

In 2003, when strong selection for increased tolerance to herbivory was detected, only 1 of 90 study plants produced seed in one of the invaded sites (range across four study sites = 0.01 to 0.31 proportion surviving to reproduction). Given that *Acmispon* seeds may persist for decades in the seed bank and that mortality

and seed production is temporally variable, selection in invaded populations may be overwhelmed by the long-established seed bank from before *Medicago* became abundant or by high rates of seed additions to the seed bank in years when *Medicago* and/or *Hypera* is rare. Spatial variation may have similar effects, fecundity in uninvaded populations is 302% higher than fecundity in invaded populations. If even a small proportion of seeds from uninvaded sites colonize invaded sites, they may demographically overwhelm invaded *Acmispon* populations and swamp out selection in invaded sites.

Additional constraints

Acmispon occupies both serpentine and nonserpentine grasslands, but *Medicago* is largely restricted to nonserpentine habitats. Gene flow from uninvaded serpentine populations into invaded nonserpentine populations may increase maladaptation due to the added edaphic differences driving diversification between many invaded and uninvaded populations. Such a scenario requires both trade-offs between traits favoured in the presence vs absence of *Medicago* and potentially even stronger trade-offs between traits favoured in serpentine (e.g. stress tolerance traits) vs nonserpentine soils (e.g. competitive ability).

actually increase plant damage by specialist herbivores (e.g. Agrawal & Sherriffs 2001). As a result, the presence of both types of herbivores may limit adaptation to either herbivore. Similarly, increased nectar production in flowers not only attracts more pollinators, but also makes plants more susceptible to nectar robbers and other herbivores (Irwin *et al.* 2004), and advanced flowering phenologies may increase pollination and reduce attack from some seed predators but also increase attack from other seed predators (Pilson 2000). In the animal realm, increased tadpole tail depth increases burst speed, which is an effective defence against diving beetles, but makes tadpoles easier prey for bluegill sunfish (Benard 2006). These types of ecological trade-offs may be just as likely for natives challenged by invasive species. For example, the exotic predatory cladoceran *Bythotrephes* induced shifts in the vertical distributions of five of seven native prey groups tested (Bourdeau *et al.* 2011). These shifted distributions will likely alter interactions with resources, competitors and other preda-

tors, perhaps explaining why two of the prey groups have not adopted this strategy of defence.

Ecological pleiotropy could, in theory, increase the rate of adaptation to a particular invasive species, if multiple species impose selection on a trait in the same direction. However, given that many invaders fill unique ecological functions in the community, are distant relatives of native community members (and, therefore, are more likely to be ecological distinct) (Darwin 1859; Strauss *et al.* 2006b) and/or possess novel traits closely associated with their invasion success (Callaway & Ridenour 2004), selection imposed by invaders may be very different and possibly opposing to that imposed by native competitors. In addition, even for native species, negative correlations are common in the few cases where complex multispecies communities have been thoroughly investigated. For example, Wise & Rauscher (2013) find that 18 of 31 significant covariances between resistance to different herbivores were negative. These negative covariances reduced the predicted evolutionary response to the herbivore community by 60%.

Box 2 Ecological constraints on native adaptation to biological invasions and resulting ecological effects of rapid adaptation (or failed adaptation)

Several ecological factors (top row of boxes) have the potential to influence the likelihood and extent of native species adaptation vs maladaptation to biological invasions. Multispecies interactions and nonadditive selection have the potential to either facilitate or limit adaptation depending on the ecological scenario. Temporal variation in selection, especially when combined with temporal variation in demographic rates, has the potential to slow adaptation. Spatial variation coupled with modest gene flow has the potential to slow adaptation, or if coupled with spatial variation in demographic rates and asymmetric gene flow to cause

maladaptation. In turn, the extent of native adaptation to invasion will influence competitive outcomes between native and invasive taxa and native population growth and persistence. Altered native–invader competitive interactions that favour natives and increased native population growth rates in turn increase the likelihood that natives will respond less negatively (ecologically) and have greater capacity to adapt to future global change. Weight of arrows indicated hypothesized importance of that path or process. Solid arrows indicate positive effects; dashed arrows indicate negative effects on natives.

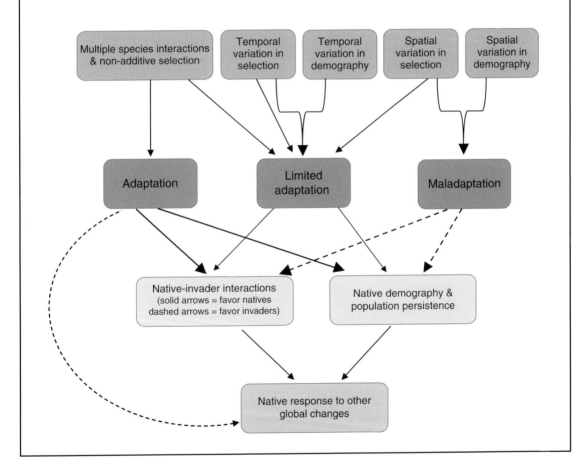

Indirect species interactions

Ecological indirect effects occur when the interaction between two species is altered by another species (Strauss 1991; Wootton 1994). Nonadditive selection results when the presence of one species alters the strength or direction of selection imposed by another species, and is driven largely by indirect ecological effects (reviewed in terHorst et al. 2015). Depending on the direction of the effect, nonadditive selection can slow or eliminate adaptation to any one biotic selective agent. For example, protozoa evolved decreased cell size in response to predators and competitors independently, but this evolutionary response did not occur when both predators and competitors were present because of strong ecological indirect effects (terHorst 2010). Given the preponderance of strong ecological indirect effects in complex communities (Menge 1995), their potential to influence fitness and evolutionary trajectories (Walsh 2013) and the potential for invasive species to indirectly affect many community members (White et al. 2006), nonadditive selection on native community members may be common in invaded communities. For example, in the case study described in Box 1, both an exotic plant competitor and insect herbivores (most of which were exotic) independently selected for increased antiherbivore resistance in a native plant; however, selection was no stronger when both the exotic plant and exotic herbivores were present, suggesting that ecological indirect effects weakened the combined selection imposed by these exotics (Lau 2008; terHorst et al. 2015). In this case, adaptation will not be prevented but will be slowed compared to the rate of expected evolutionary response under additive models.

While nonadditive selection can limit adaptation to any one selective agent, selection should still favour those genotypes that are most fit in the presence of all species, which would lead to adaptation to the invaded community more generally, even if adaptation specifically to the invasive species in isolation is limited. The evolutionary trajectory of any particular trait should depend on the magnitude and direction of the effect of that trait on both direct and indirect species interactions. Traits that minimize negative direct effects, while also maximizing positive indirect effects, will experience positive selection (Miller & Travis 1996), but traits that simultaneously minimize both negative direct effects and

positive indirect effects will experience weaker, or even opposing, selection, and instead, alternate traits may be favoured (e.g. 'effect' and 'response' traits in Miller & Travis 1996).

Temporal and spatial variation in selective agents

Temporal or spatial variation in the abundance of community members may lead to substantial variation in the strength, and potentially the direction, of selection (e.g. Schemske & Horvitz 1989; Reimchen & Nosil 2002; Vanhoenacker et al. 2013; reviewed in Siepielski et al. 2009, 2013). Spatial variation in the strength of selection appears to be common, and changes in the direction of selection across different populations also occur, but typically where selection is weakest, potentially limiting effects on longer-term evolution (Siepielski et al. 2013). Yet, in some systems, spatially variable selection combined with gene flow certainly can limit adaptation (e.g. Greya moths and their host plants (Thompson 1997)), particularly for species involved in tight co-evolutionary interactions (Thompson et al. 2002). Likewise, temporal variation in selection can be strong in some systems leading to little evolutionary change over longer timescales (e.g. Darwin's Finches, Grant & Grant 2002); however, there is little evidence that temporal variation in the direction of selection constrains adaptation in most natural populations (Kingsolver et al. 2012; Morrissey & Hadfield 2012). It is still unclear, however, when temporal variation in selection is important and whether some types of selective agents (e.g. biotic rather than abiotic agents) are more likely to lead to variation in selection across time and space.

Lag times or a storage effect may exacerbate temporal variation in selection. For example, a strong selective agent present in one year may impose strong selection on a plant species, resulting in the production of seeds that are adapted to that selective agent. However, if seeds then remain dormant in a seed bank, there is a greater chance for a change to occur in the biotic community to which the seeds are adapted, possibly resulting in maladaptive 'gene flow from the past' (Rice & Emery 2003). A similar phenomenon could occur in a metapopulation context when there is variation in selective agents among local populations. While gene flow among populations can be advantageous for adaptation because it helps spread beneficial alleles and increases standing genetic variation and population sizes, potentially leading to evolutionary rescue (e.g. Bell & Gonzalez 2011; Gonzalez et al. 2013), it can also be detrimental and lead to maladaptation by spreading locally adapted alleles to subpopulations where they are maladapted. These costs of gene flow may outweigh the benefits under certain environmental conditions, like those that characterize many invaded communities, including changing environments, where heterogeneity in selection among local populations is high (Bourne et al. 2014). Given the patchy nature of many invasions, gene flow combined with spatial variation in selection due to the presence of the invader may commonly limit native adaptation (Box 2). Although it is difficult to quantify such effects and to our knowledge, no studies have experimentally tested for effects of gene flow in natives adapting to invaders, the fact that many examples of native adaptation to biological invasions are cases where the invader has become dominant across nearly entire ecosystems (e.g. cheatgrass, garlic mustard; see Oduor 2013 table 1) may reflect cases where this constraint is escaped because nearby uninvaded areas are rare or nonexistent. Similarly, anecdotal evidence suggests that gene flow from nearby uninvaded locales has limited adaptation to invasive species in some systems (see Strauss et al. 2006a). Constraints on adaptation due to temporal and spatial variation in the strength or direction of selection are likely to be particularly important when demographic rates also vary across space or time (reviewed in Lenormand 2002; see below and Box 1).

Temporal and spatial variation in population demographics

Demographic rates commonly vary across years, or among local populations, as a result of variation in abiotic and biotic environmental conditions that affect mean reproductive output of the population. Such demographic variability may constrain adaptive evolution when years or populations with strong selection also contribute little to the population growth rate. For example, in years where an invasive antagonist is exceptionally abundant, selection on interacting native species may be very strong. Yet, if mean fitness is extremely low, reproduction in that year of intense

selection may contribute little to the population growth rate, particularly in populations with a storage effect. The strength of selection imposed by a species may be strongly correlated with the demographic effects of that species on population growth rate. Although this need not be true, a recent model and additional empirical studies suggest that, at least for antagonistic interactions (competition, herbivory, disease, etc.), the strength of the demographic effect and the strength of selection are positively correlated (Benkman 2013; Vanhoenacker et al. 2013). As a result, the years with strongest selection may also be the years contributing least to population growth. Similarly, in a metapopulation framework, the local populations that experience the strongest selection may contribute the least to the regional population growth rate. As a result, adaptation to these 'marginal habitats' may be overwhelmed by gene flow from more optimal habitats (reviewed in Lenormand 2002; Kawecki 2008), leading to maladaptation and potentially even increased extinction risks (e.g. Ronce & Kirkpatrick 2001) rather than facilitating population persistence, as in classic metapopulation theory (Pulliam 1988).

Although rarely investigated empirically, theory suggests that the interaction of selection and demography may lead to several possible evolutionary outcomes. When absolute fitness varies widely from year to year or site to site, one theoretical outcome is that temporal or spatial variation can lead to adaptation to 'optimal' environmental conditions (i.e. conditions when absolute fitness is highest) (Templeton & Levin 1979; Venable & Brown 1988; Levin 1990), particularly if generations are overlapping (Turelli et al. 2001) and gene flow is nontrivial (in the case of spatially variable selection). In the context of biological invasions, this would imply that the large evolutionary responses likely to occur in years or sites when the invader is most pernicious and likely exerting the strongest selection may be swamped out by higher rates of reproduction that occur when the invader is less abundant. In the most extreme case, where traits that are adaptive in invaded communities are negatively correlated with traits that are adaptive in uninvaded communities, a year or local population where invaders are rare and impose less selection will result in high reproductive output of traits that are maladapted to invasion. In such cases, natives will be unlikely to adapt to the invasion, instead adapting to uninvaded conditions.

A related idea is that the strength of selection and the expression of genetic variation may be coupled.

Evolutionary responses to selection (and, therefore, adaptation) depend on both the strength of selection and expressed genetic variation. In some cases, different genetic individuals may express little variation in phenotype under stressful conditions. For example, in Soay sheep, harsh environmental conditions lead to strong selection on birthweight, yet the expression of genetic variation in birthweight is very low in harsh years (Wilson *et al.* 2006). These opposing effects may result in little evolutionary response to selection, even when selection is strong. However, in cases where genetic variation is expressed most in response to the selective agent, the coupling of selection strength and the expression of genetic variation could accelerate adaptation to an invasive species. It is difficult to predict when and how selection strength and genetic variation may be coupled. On one hand, wild populations exhibit greater genetic variation on average in favourable environmental conditions (Charmantier & Garant 2005) – conditions where selection may be weakest. On the other hand, laboratory studies often detect greater expression of genetic variation under stressful conditions (reviewed in Charmantier & Garant 2005). One explanation for these contrasting results is the relationship between novelty and stress; expression of genetic variation in novel environments is often heightened because of the expression of new genes not previously under selection (Holloway *et al.* 1990; see discussion in Charmantier & Garant 2005). In laboratory studies, stress treatments may be equivalent to novel environmental conditions, whereas in field studies, stress treatments are likely within the realm of conditions experienced previously. One could make similar arguments for adaptation to biological invasions. On one hand, biological invasions may create novel environmental conditions that could increase the expression of genetic variation, particularly when the invasive species fills a novel ecological role. Alternatively, even when biological invasions substantially decrease environmental quality, they may not lead to exceptionally novel environmental conditions because native species may have experience with similar competitors, predators and mutualists. As a result, it is difficult to predict how biological invasions will influence the expression of genetic variation in native community members and whether the coupling of selection with the expression of genetic variation is most likely to facilitate or constrain adaptation to biological invasions.

CONSTRAINTS MOST PERTINENT TO NATIVE ADAPTATION TO INVASION

While each of the processes described above has the potential to constrain evolution, which constraints are most likely to limit native adaptation to invasive species requires more investigation. For example, in the case study presented in Box 1, although some evidence for nonadditive selection and ecological pleiotropy was detected in this system, these mechanisms are unlikely to fully explain the lack of adaptation to invasion, at least with respect to the suite of interactors and specific traits studied. For other invaders, particularly invaders that seem to fill unique ecological roles, multispecies interactions and nonadditive selection may be a more potent constraint. In contrast, both temporal and spatial variation in selection due to variation in the abundance of herbivore community members may constrain adaptation in the system described in Box 1, especially when combined with the demographic effects of invasion. Given the patchy distributions of most invasive species, spatial variation in selection combined with demography may be a common mechanism limiting native evolutionary responses to biological invasions. These constraints might also be overcome through evolutionary approaches to management.

Recent studies have advocated management practices that facilitate the spread of natives adapted to invasion, for example by collecting and bulking up seed for future restorations from heavily invaded populations (e.g. Leger 2008). Such a practice may help to overcome constraints imposed by spatial variation in selection and demography by essentially increasing the colonization success of natives with an evolutionary history of invasion relative to natives from uninvaded populations. Such focused colonization choices may help to negate some of the demographic advantages of uninvaded populations that could overwhelm selection in invaded environments and limit adaptation; however, the effectiveness of such strategies remains unknown.

IMPLICATIONS OF NATIVE ADAPTATION TO INVADERS

Several studies have now documented evolutionary consequences of biological invasions for native community members (Strauss *et al.* 2006a; Oduor 2013), but in other cases, natives may fail to adapt (e.g. Box 1). Identifying when adaptation is most likely to occur and the ecological consequences of evolutionary change are key next steps to understanding the long-term importance of evolution for native responses to biological invasions. The potential for adaptation, the rate of adaptation and the consequences of adaptation may be especially important for the persistence of natives in heavily invaded habitats, the long-term outcomes of native–invader interactions and native species responses to future environmental change (Box 2).

Demographic effects of invasion and evolutionary rescue

Evolutionary rescue results when adaptation occurs fast enough to reverse the negative demographic effects of environmental change and prevent populations from declining to extinction (Gonzalez *et al.* 2013). Recent theoretical and empirical work has illustrated that evolutionary rescue may help species persist in the face of rapid environmental change. Theory suggests that the likelihood of evolutionary rescue will depend on both the amount of standing genetic variation and trait heritabilities and also the magnitude of demographic effect of the environmental change (e.g. Lynch & Lande 1993; Bürger & Lynch 1995; Gomulkiewicz & Holt 1995). These models predict a classic 'U-shaped' curve in abundance over time, whereby population sizes decline before beginning to rebound. The increase in abundance occurs as rare genetic variants that are less susceptible to environmental change become more abundant in the population, or following the origin of novel mutations that decrease the negative fitness consequences of environmental change (e.g. Gomulkiewicz & Holt 1995). Importantly, evolutionary rescue is only predicted when adaptation prevents the population from declining below a critical threshold population size where demographic stochasticity is likely lead to extinction. Although models differ in the type of environmental change (abrupt vs gradual vs erratic), genetic mechanisms (few loci of large effects vs many loci of small effect) and the extent and role of gene flow, models consistently show the potential for evolutionary rescue in certain regions of parameter space (e.g. Lynch & Lande 1993; Bürger & Lynch 1995; Gomulkiewicz & Holt 1995). How commonly evolutionary rescue occurs in natural populations is unknown, in part because few evolutionary studies are conducted in the small populations most susceptible to extinction.

Empirical studies of evolutionary rescue have mostly been restricted to laboratory model systems and have illustrated how local dispersal and gradual environmental change both facilitate evolutionary rescue (e.g. Bell & Gonzalez 2011). Notably, these studies also recapture the U-shaped curve predicted by theory. Similarly, in a glasshouse study, Bodbyl Roels & Kelly (2011) showed how pollinator decline led to the evolution of increased self-fertilization which helped protect plants from the negative consequences of pollinator limitation, although reductions in genetic diversity accompanying this evolutionary change may limit adaptation to further environmental changes (see also Box 2).

Invasive species are one of the prime examples of when evolutionary rescue is likely to be important because of the combination of small population sizes and the extreme novel environmental conditions faced by these taxa during early establishment (see Gomulkiewicz & Holt 1995). Yet, evolutionary rescue may also be important for natives facing invaders. Invading species can cause or increase the rate of population declines in native taxa (e.g. Thomson 2005, 2006), and adaptive evolutionary responses should minimize these declines. There are many factors that may make evolutionary rescue likely for natives facing biological invasions. First, invasions are not instantaneous and may take many generations to reach the high densities necessary for dramatic declines in native populations. Such gradual environmental change should make evolutionary rescue more likely. If the invader does not cause immediate drastic declines in population size, more individuals will experience the selection pressure imposed by the invader and exposure to weaker stressors (e.g. small concentrations of novel allelochemicals or low densities of invasive competitors) may allow for adaptation even when more potent stressors will not (e.g. high allelochemical concentrations or intense resource competition) (see Bell 2013). Second, for many species, the selection pressures imposed by some classes of invaders, such as competition and herbivory, are not entirely novel because similar native competitors and herbivores also may exist, so standing genetic variation for traits facilitating persistence in the face of the invaders may be present. Standing genetic variation may be key for successful adaptation to novel environments and avoiding extinction (e.g. Gartside & McNeilly 1974; Bradshaw 1991). Traits increasing competitive ability against invasive plants, for example, are likely similar to traits increasing competitive ability against all competitors more generally. The notable exceptions include novel weapons, such as allelochemicals that are completely new to the native species, and novel predators with new foraging tactics. In these cases, extinction may be more likely because of a lack of standing genetic variation in traits mediating responses to these novel interactions. Indeed, most documented cases of global extinctions due to invasive species are of completely naive island bird populations impacted by novel predators (Fritts & Rodda 1998), yet smaller scale regional extinctions may be more common (Powell *et al.* 2013). Still, many factors will constrain the likelihood of evolutionary rescue in these systems, including the more general constraints to adaptation discussed above.

Native–invader interactions

Invasive species are a global change of a fundamentally different nature than many other types of global change, such as global warming, nitrogen deposition or pollutants. This is largely because invasive species are not static selective agents; they can also evolve, and these evolutionary changes in invasives have the potential to either increase or decrease the strength of selection they impose on native community members. For example, Lankau *et al.* (2009) documented that invasive garlic mustard has evolved reduced allelochemical concentrations over time because as garlic mustard becomes more abundant, it encounters conspecifics more often than heterospecifics, and increased allelochemical production is likely only favoured when interspecific competition is high (Lankau 2012). These allelochemicals are strong selective agents on native competitors, and several native species have evolved increased tolerance of garlic mustard allelochemicals (Lankau 2012; Lankau & Nodurft 2013). However, the evolution of reduced allelochemical production over time in garlic mustard is likely reducing both the negative consequences of invasion and the strength of selection imposed by garlic mustard on native species, while the evolution of increased tolerance to garlic mustard allelochemicals in natives is simultaneously reducing the fitness effects of garlic mustard invasion (Lankau 2012). In other cases, co-evolutionary responses may increase negative effects on the native. For example, native squirreltail populations have evolved increased competitive effects on invasive cheatgrass (Rowe & Leger 2011). If native squirreltail populations manage to increase in abundance as a result of this evolutionary shift, squirreltail may eventually exert selection on invasive cheatgrass, possibly selecting for increased cheatgrass competitive ability.

For other types of species interactions, these co-evolutionary interactions may be even more likely. In predator–prey systems, native prey may evolve increased defences against introduced predators, while introduced predators may evolve increased offenses against native prey. Similarly, introduced pathogens may evolve increased virulence against native hosts, while native hosts may evolve increased resistance against introduced pathogens. In these cases, co-evolution may lead to maladaptation in one or both partners, but in many circumstances, we might expect the invader to have the evolutionary advantage because of increased standing genetic variation resulting from admixture during the invasion process, large population sizes increasing the probability that beneficial mutations may arise and escape from natural enemies that may reduce constraints on evolution arising from selection from multiple interacting species (see community complexity section above) (see also Colautti & Lau, 2015).

Multiple global changes

One of the biggest obstacles for natives challenged by introduced species is that biological invasions are only one of many novel threats native species face, and the process of adapting or failing to adapt to invasive species can influence the likelihood of native adaption to future environmental change. Almost by definition, invasive species are those that are very successful at surviving novel environments. Natives are now also challenged by novel environments, but may possess few of the tricks that invaders have to succeed, including broad ecological tolerances, novel weapons, increased size or competitive ability, and escape from evolutionary constraints (Baker 1965; Williamson & Fitter 1996; Callaway & Ridenour 2004; Richards *et al.* 2006; Strauss 2014). The process of native adaptation to invaders (whether successful or unsuccessful) can influence evolutionary responses to future environmental change, both through the demographic and genetic consequences of adaptation and because natives face these new global changes while continuing to interact, and often compete, with invaders.

Facing additional environmental changes while simultaneously facing the threats of invasive species may limit adaptation to invasion much in the same manner that multispecies interactions can constrain adaptation. Adapting to multiple simultaneous stressors can be difficult, and the challenges to such adaptation have been noted for decades (e.g. Bradshaw 1991). First, strong selective events may erode genetic variation in fitness-related traits (Bradshaw 1991) as well as decrease population sizes, which can further restrict the likelihood of adaptation to future environmental change. Second, joint adaptation to multiple stressors may simply be more difficult than adaptation to a single stressor because of genetic correlations (likely resulting from antagonistic pleiotropy) and/or, relatedly, a lack of multivariate genetic variation (reviewed in Walsh & Blows 2009). Ultimately, whether evolutionary responses to invaders facilitate or constrain evolutionary responses to other types of anthropogenic environmental change likely depends on how genetic and physiological mechanisms of response to the various stressors are correlated and also on the demographic effects of invasion. In cases where demographic effects are strong and evolutionary rescue is incomplete, invasive species likely hinder adaptation to additional global changes.

A second complicating factor is that natives often respond negatively to these other global changes while competing with invaders that may respond positively to such changes (Dukes & Mooney 1999) and may have several evolutionary advantages in these novel environments (Colautti & Lau, 2015). In short, biological invasions and the failure of natives to adapt to invasions may limit native adaptation to future environmental conditions through demographic effects, genetic effects and evolutionary disadvantages compared to their invasive competitors.

CONCLUSIONS

Although an increasing number of studies have now documented evolutionary responses of natives to biological invasions (reviewed in Strauss *et al.* 2006a; Oduor 2013), we are only beginning to understand when and how adaptation occurs. Only a few studies have compared patterns of natural selection in invaded and uninvaded environments to begin to understand the traits underlying native adaptation to invasions or have tested how observed evolutionary changes influence native population responses to invasion. Notably, most studies documenting strong evolutionary changes have focused on extreme cases of invasion, in which the invader forms virtual monocultures (see table 1 in Oduor 2013). Yet, many invasives may coexist with a wide variety of native and non-native community members. Whether natives faced with the diffuse selective pressures resulting from these more complex communities also adapt to the presence of invasive species is unclear. Notably, biological invasions may be one of the most challenging types of global change to adapt to because of the constraints discussed in detail above and also because of the potential co-evolutionary dynamics that could result. Only by focusing on when and how native adaptation is likely to proceed will we begin to understand whether adaptation and the resulting potential for evolutionary rescue will help native populations persist in the face of biological invasions.

ACKNOWLEDGEMENTS

We thank the organizers and attendees of the 2014 Invasion Genetics: The Baker and Stebbins Legacy Symposium and S. Strauss for encouraging us to think more rigorously about the ideas presented in this manuscript. We also thank three anonymous reviewers for suggestions that greatly improved this manuscript. Funding was provided by the National Science Foundation under grants awarded to J.A.L. (DEB 0918963) and C.P.t (DMS 132490). This is Kellogg Biological Station publication number 1841.

REFERENCES

Agrawal AA, Sherriffs MF (2001) Induced plant resistance and susceptibility to late-season herbivores of wild radish. *Annals of the Entomological Society of America*, **94**, 71–75.

Antonovics J (1976) The nature of limits to natural selection. *Annals of the Missouri Botanical Garden*, **63**, 224–247.

Baker HG (1965) Characteristics and modes of origin of weeds. In: *The Genetics of Colonizing Species* (eds Baker HG, Stebbins GL), pp. 147–168. Academic Press, New York.

Bell G (2013) Evolutionary rescue and the limits of adaptation. *Philosophical Transactions of the Royal Society of London. Series B, Biological Sciences*, **368**, 20120080.

Bell G, Gonzalez A (2011) Adaptation and evolutionary rescue in metapopulations experiencing environmental deterioration. *Science*, **332**, 1327–1330.

Benard MF (2006) Survival trade-offs between two predator-induced phenotypes in Pacific treefrogs (*Pseudacris regilla*). *Ecology*, **87**, 340–346.

Benkman CW (2013) Biotic interaction strength and the intensity of selection. *Ecology Letters*, **16**, 1054–1060.

Bodbyl Roels SA, Kelly JK (2011) Rapid evolution caused by pollinator loss in *Mimulus guttatus*. *Evolution*, **65**, 2541–2552.

Bourdeau PE, Pangle KL, Peacor SD (2011) The invasive predator *Bythotrephes* induces changes in the vertical distribution of native copepods in Lake Michigan. *Biological Invasions*, **13**, 2533–2545.

Bourne EC, Bocedi G, Travis JMJ, Pakeman RJ, Brooker RW, Schiffers K (2014) Between migration load and evolutionary rescue: dispersal, adaptation and the response of spatially structured populations to environmental change. *Proceedings of the Royal Society B*, **281**, 20132795.

Bradshaw AD (1991) The Croonian Lecture, 1991: genostasis and the limits to evolution. *Philosophical Transactions of the Royal Society of London. Series B, Biological Sciences*, **333**, 289–305.

Bürger R, Lynch M (1995) Evolution and extinction in a changing environment: a quantitative genetic analysis. *Evolution*, **49**, 151–163.

Callaway RM, Ridenour WM (2004) Novel weapons: invasive success and the evolution of increased competitive ability. *Frontiers in Ecology & the Environment*, **2**, 436–443.

Charmantier A, Garant D (2005) Environmental quality and evolutionary potential: lessons from wild populations. *Proceedings of the Royal Society B*, **272**, 1415–1425.

Colautti RI, Lau JA (2015) Contemporary evolution during invasion: evidence for differentiation, natural selection, and local adaptation. *Molecular Ecology*, **24**, 1999–2017.

Darwin C (1859) *On the Origin of Species by Means of Natural Selection*. John Murray, London.

Dukes JS, Mooney HA (1999) Does global change increase the success of biological invaders? *Trends in Ecology & Evolution*, **14**, 135–139.

Fritts TH, Rodda GH (1998) The role of introduced species in the degradation of island ecosystems: a case history of Guam. *Annual Review of Ecology & Systematics*, **29**, 113–140.

Gartside DW, McNeilly T (1974) The potential for evolution of heavy metal tolerance in plants II. Copper tolerance in normal populations of different plant species. *Heredity*, **32**, 335–348.

Gomulkiewicz R, Holt RD (1995) When does evolution by natural selection prevent extinction. *Evolution*, **49**, 201–207.

Gonzalez A, Ronce O, Ferriere R, Hochberg ME (2013) Evolutionary rescue: an emerging focus at the intersection between ecology and evolution. *Philosophical Transactions of the Royal Society of London. Series B, Biological Sciences*, **368**, 20120404.

Grant PR, Grant BR (2002) Unpredictable evolution in a 30-year study of Darwin's finches. *Science*, **296**, 707–711.

Holloway GJ, Povey SR, Sibly RM (1990) The effect of new environment on adapted genetic architecture. *Heredity*, **64**, 323–330.

terHorst CP (2010) Evolution in response to direct and indirect effects in pitcher plant inquiline communities. *American Naturalist*, **176**, 675–685.

Irwin RE, Adler LS, Brody AK (2004) The dual role of floral traits: pollinator attraction and plant defense. *Ecology*, **85**, 1503–1511.

Kawecki TJ (2008) Adaptation to marginal habitats. *Annual Review of Ecology, Evolution & Systematics*, **39**, 321–342.

Kingsolver JG, Diamond SE (2011) Phenotypic selection in natural populations: what limits directional selection? *American Naturalist*, **177**, 346–357.

Kingsolver JG, Diamond SE, Siepielski AM, Carlson SM (2012) Synthetic analyses of phenotypic selection in natural populations: lessons, limitations, and future directions. *Evolutionary Ecology*, **26**, 1101–1118.

Lankau RA (2012) Coevolution between invasive and native plants driven by chemical competition and soil biota. *Proceedings of the National Academy of Sciences USA*, **109**, 11240–11245.

Lankau RA, Nodurft RN (2013) An exotic invader drives the evolution of plant traits that determine mycorrhizal fungal diversity in a native competitor. *Molecular Ecology*, **22**, 5472–5485.

Lankau RA, Nuzzo V, Spyreas G, Davis AS (2009) Evolutionary limits ameliorate the negative impact of an invasive plant. *Proceedings of the National Academy of Sciences USA*, **106**, 15362–15367.

Lau JA (2006) Evolutionary responses of native plants to novel community members. *Evolution*, **60**, 56–63.

Lau JA (2008) Beyond the ecological: biological invasions alter natural selection on a native plant species. *Ecology*, **89**, 1023–1031.

Lau JA, Strauss SY (2005) Insect herbivores drive important indirect effects of exotic plants on native communities. *Ecology*, **86**, 2990–2997.

Leger EA (2008) The adaptive value of remnant native plants in invaded communities: an example from the Great Basin. *Ecological Applications*, **18**, 1226–1235.

Lenormand T (2002) Gene flow and the limits to natural selection. *Trends in Ecology & Evolution*, **17**, 183–189.

Levin DA (1990) The seed bank as a source of genetic novelty in plants. *American Naturalist*, **135**, 563–572.

Lynch M, Lande R (1993) Evolution and extinction in response to environmental change. In: *Biotic Interactions and Global Change* (eds Kareiva PM, Kingsolver JG, Huey RB), pp. 234–250. Sunderland, MA, Sinauer.

Menge BA (1995) Indirect effects in marine rocky intertidal interactions webs: patterns and importance. *Ecological Monographs*, **65**, 21–74.

Miller TE, Travis J (1996) The evolutionary role of indirect effects in communities. *Ecology*, **77**, 1329–1335.

Morrissey MB, Hadfield JD (2012) Directional selection in temporally replicated studies is remarkably consistent. *Evolution*, **66**, 435–442.

Oduor AMO (2013) Evolutionary responses of native plant species to invasive plants: a review. *New Phytologist*, **200**, 986–992.

Phillips BL, Shine R (2006) An invasive species induces rapid adaptive change in a native predator: cane toads and black snakes in Australia. *Proceedings of the Royal Society B*, **273**, 1545–1550.

Phillips BL, Shine R (2004) Adapting to an invasive species: toxic cane toads induce morphological change in Australian snakes. *Proceedings of the National Academy of Sciences USA*, **49**, 17150–17155.

Pilson D (2000) Herbivory and natural selection on flowering phenology in wild sunflower, *Helianthus annuus*. *Oecologia*, **122**, 72–82.

Powell KI, Chase JM, Knight TM (2013) Invasive plants have scale-dependent effects on diversity by altering species-area relationships. *Science*, **339**, 316–318.

Pulliam HR (1988) Sources, sinks, and population regulation. *American Naturalist*, **132**, 652–661.

Reimchen TE, Nosil P (2002) Temporal variation in divergent selection on spine number in threespine stickleback. *Evolution*, **56**, 2472–2483.

Rice KJ, Emery NC (2003) Managing microevolution: restoration in the face of global change. *Frontiers in Ecology and the Environment*, **1**, 469–478.

Richards CL, Bossdorf O, Muth NZ, Gurevitch J, Pigliucci M (2006) Jack of all trades, master of some? On the role of phenotypic plasticity in plant invasions. *Ecology Letters*, **9**, 981–993.

Ronce O, Kirkpatrick M (2001) When sources become sinks: migrational meltdown in heterogeneous habitats. *Evolution*, **55**, 1520–1531.

Rowe CLJ, Leger EA (2011) Competitive seedlings and inherited traits: a test of rapid evolution of *Elymus multisetus* (big squirreltail) in response to cheatgrass invasion. *Evolutionary Applications*, **4**, 485–498.

Schemske DW, Horvitz CC (1989) Temporal variation in selection on a floral character of a Neotropical herb. *Evolution*, **43**, 461–464.

Siepielski AM, DiBattista JD, Carlson SM (2009) It's about time: the temporal dynamics of phenotypic selection in the wild. *Ecology Letters*, **12**, 1261–1276.

Siepielski AM, Gotanda KM, Morrissey MB, Diamond SE, DiBattista JD, Carlson SM (2013) The spatial patterns of directional phenotypic selection. *Ecology Letters*, **16**, 1382–1392.

Strauss SY (1991) Indirect effects in community ecology: their definition, study and importance. *Trends in Ecology & Evolution*, **6**, 206–210.

Strauss SY (2014) Ecological and evolutionary responses in complex communities: implications for invasions and eco-evolutionary feedbacks. *Oikos*, **123**, 257–266.

Strauss SY, Irwin RE (2004) Ecological and evolutionary consequences of multispecies plant-animal interactions. *Annual Review of Ecology Evolution and Systematics*, **35**, 435–466.

Strauss SY, Lau JA, Carroll SP (2006a) Evolutionary responses of natives to introduced species: what do introductions tell us about natural communities? *Ecology Letters*, **9**, 357–374.

Strauss SY, Webb CO, Salamin N (2006b) Exotic taxa less related to native species are more invasive. *Proceedings of the National Academy of Sciences USA*, **103**, 5841–5845.

Templeton AR, Levin DA (1979) Evolutionary consequences of seed pools. *American Naturalist*, **114**, 232–249.

terHorst CP (2010) Evolution in response to direct and indirect ecological effects in pitcher plant inquiline communities. *American Naturalist*, 176, 675–685.

terHorst CP, Lau JA, Cooper IA, Keller KR, La Rosa RJ, Royer AM, Schultheis EH, Suwa T, Conner JK (2015) Quantifying

nonadditive selection caused by indirect ecological effects. *Ecology*, 96, 2360–2369.

Thompson JN (1997) Evaluating the dynamics of coevolution among geographically structured populations. *Ecology*, **78**, 1619–1623.

Thompson JN, Nuismer SL, Gomulkiewicz R (2002) Coevolution and maladaptation. *Integrative and Comparative Biology*, **42**, 381–387.

Thomson DM (2005) Matrix models as a tool for understanding invasive plant and native plant interactions. *Conservation Biology*, **19**, 917–928.

Thomson DM (2006) Detecting the effects of introduced species: a case study of competition between *Apis* and *Bombus*. *Oikos*, **114**, 407–418.

Turelli M, Schemske DW, Bierzychudek P (2001) Stable two-allele polymorphisms maintained by fluctuating fitnesses and seed banks: protecting the blues in *Linanthus parryae*. *Evolution*, **55**, 1283–1298.

Vanhoenacker D, Ågren J, Ehrlén J (2013) Non-linear relationship between intensity of plant-animal interactions and selection strength. *Ecology Letters*, **16**, 198–205.

Venable DL, Brown JS (1988) The selective interactions of dispersal, dormancy, and seed size as adaptations for reducing risk in variable environments. *American Naturalist*, **131**, 360–384.

Walsh MR (2013) The evolutionary consequences of indirect effects. *Trends in Ecology & Evolution*, **28**, 23–29.

Walsh B, Blows MW (2009) Abundant genetic variation + strong selection = multivariate genetic constraints: a geometric view of adaptation. *Annual Review of Ecology, Evolution, & Systematics*, **40**, 41–59.

White EM, Wilson JC, Clarke AR (2006) Biotic indirect effects: a neglected concept in invasion biology. *Diversity and Distributions*, **12**, 443–455.

Williamson MH, Fitter A (1996) The characters of successful invaders. *Biological Conservation*, **78**, 163–170.

Wilson AJ, Pemberton JM, Pilkington JG *et al.* (2006) Environmental coupling of selection and heritability limits evolution. *PLOS Biology*, **4**, 1270–1275.

Wise MJ, Rausher MD (2013) Evolution of resistance to a multiple-herbivore community: Genetic correlations, diffuse coevolution, and constraints on the plant's response to selection. *Evolution*, **67**, 1767–1779.

Wootton JT (1994) The nature and consequences of indirect effects in ecological communities. *Annual Review of Ecology and Systematics*, **25**, 443–466.

DATA ACCESSIBILITY

Data included in the Case Study described in Box 1 were originally published in Lau & Strauss 2005; Lau 2006, 2008 and terHorst *et al.* 2015. The original data sets motivating this work are available in Dryad: doi:10.5061/dryad.f0nf4.

DISCUSSION

TIM BLACKBURN

RUTH HUFBAUER (COLORADO STATE UNIVERSITY) – The mechanisms underlying the positive effect of propagule pressure likely change over the course of an invasion; but I think there is a role for genetics even in the first generation, not just in long-term spread. Do you have any evidence from birds for that kind of thing?

TIM BLACKBURN – No, but I am keen for people to bring to my attention any evidence on this. Certainly for birds we do not have evidence, and I don't know the literature for other taxa that well. But I entirely agree. I am not saying that genetic effects are not important, I am just saying we do not have evidence for them, at least for birds.

KEN WHITNEY (UNIVERSITY OF NEW MEXICO) – I would like to point out that genotypic diversity probably increases with propagule pressure, and thus we may have to consider sampling effects and complementarity effects, as in the species diversity–ecosystem functioning literature. There is increasing evidence that genotypes use different resources, and when you put them together you get something different than you might expect from a genetically uniform population. And that can determine whether founder populations go extinct or establish. Do you think these kinds of effects might be pertinent – should we include them in our models? Or is there a sense that perhaps they are relatively unimportant?

RUSSELL LANDE – Sampling will be important if characters are involved in density-dependent and frequency-dependent interactions. This is usually the case with resource utilization, and since your question involves niche dimensions that may well be important. If selection is also density dependent, this gets into the classic trade-off between r- and K-selection in fluctuating environments. So all of these things are certainly involved.

TIM BLACKBURN – If you have larger samples from a population, then maybe there is the opportunity that you will be sampling certain genetic variants that will allow the population to have some pre-adaptation, but that will depend on population genetic structure, as well as the size of the sample.

RUSSELL LANDE – In the verbal exchanges in *The Genetics of Colonizing Species* 50 years ago, Richard Lewontin corrected Ernst Mayr's view on the founder effect, since your question partly involves this. He pointed out that in terms of heterozygosity, or additive genetic variance in quantitative traits, a single fertilized female contains three-quarters of the genetic

Invasion Genetics: The Baker and Stebbins Legacy, First Edition. Edited by Spencer C. H. Barrett, Robert I. Colautti, Katrina M. Dlugosch, and Loren H. Rieseberg.
© 2017 John Wiley & Sons, Ltd. Published 2017 by John Wiley & Sons, Ltd.

variance of the large population it was sampled from. These effects are more important over the course of multiple generations when a population remains small, or goes through a subsequent bottleneck after founding, because it is not well adapted. So I don't know to what extent you are thinking about things in a particular timeframe or not, but that is an important aspect of sampling and propagule size and genetic variation. Multiple propagules can result in genetic rescue in terms of the genetic variance *per se*, as well as adaptive potential and population size.

COREY BRADSHAW (UNIVERSITY OF ADELAIDE) – In models where you are using propagule pressure to explain invasion success, do you think that having measures of N_e (effective population size) would give more explanatory power? Effective population size is usually much lower than census size and is also highly variable. I know you probably cannot do this with birds, but are there other data sets that simply scale downwards? I am just wondering if that would improve the estimates?

TIM BLACKBURN – Does N_e vary systematically with population size? Is it a proportional decrease across the range of population sizes?

COREY BRADSHAW – N_e averages about 0.1–0.2 of census population size but is highly variable across taxa and populations, so I am just wondering if you are missing something by assuming your propagules are completely outcrossed individuals.

TIM BLACKBURN – I imagine that knowing N_e would potentially help explain some of the unexplained variance, assuming that there is no systematic bias in the difference between the introduced population size and N_e across the range of numbers of individuals introduced. I doubt that would necessarily alter, for example, the fit of different comparative models, but it might help explain some of the variance.

RUSSELL LANDE – I have a comment about the possibility of using N_e. Most methods for estimating N_e are based on neutral variation, and that changes due to the input of mutation at a rate that is orders of magnitude slower than the population size can change. So any such estimates that were made are bound to have some variation in that ratio unless population sizes remain constant, which they never do. It is also based on the assumption that all variation is neutral and not linked to anything that is

being selected. Even given all the other assumptions, this calls into question how accurate the estimates of N_e are, or what they really mean with respect to the adaptive potential of populations. So I take these things as only rough estimates at best. It probably would not pay to make a huge research effort on this if you are trying to nail things down to greater accuracy than an order of magnitude.

MARK VAN KLEUNEN

UNIDENTIFIED – I have a question regarding the idea of invasive and non-invasive as a binary effect. What we call invasive are really species with various degrees of invasiveness. Some are successful and some are failures and everything in between. This variation is context dependent. We have a lot of knowledge about intra-specific variation, but how can we incorporate this into your model when it only considers yes or no in terms of whether a species is invasive or not?

MARK VAN KLEUNEN – That is a good point and of course there is a continuum between invasive and non-invasive. We can include that information if we have good distribution data on the number of grid cells occupied, for example as we did in our study in Germany. But I think what you are mentioning is that we have questions in the binary key that are not always strictly failure or success. I think there you can define it as a yes/no or as a continuum. For the analysis, there is no problem in having either binary or continuous data.

LOREN RIESEBERG – Many countries have put together lists of traits that they try to use to predict whether to quarantine a particular plant, animal or micro-organism to keep it out of the country. These predictive models have not worked very well. The best predictor has been whether a particular organism has been a successful invader elsewhere, rather than any of the biological traits. Do you think that the tree-based or step-wise-based model that you were showing will get us to the point where we will be able to use traits to predict invasion success?

MARK VAN KLEUNEN – Most of the studies are trying to explain afterwards what could have caused invasiveness. Nevertheless, many of these predictive models, for example the wheat assessment scheme they use in Australia, seem to work quite well in predicting invasions in other regions. Several other people have tried to use this scheme to see whether it predicts, for example invasion success in Tanzania or other regions, and it seems to work quite well. But of course it is not perfect. I think that my scheme will be an improvement, but it will never be perfect because the environment is changing. One of the big challenges at the moment is that we have climate change going on, and also eutrophication, so all the rules might change.

ROB COLAUTTI – You showed very nicely in your experiments the importance of species traits, but after controlling for them it looked as though the native species did better on average than introduced species. Did I get that right? And earlier you mentioned Mark Davis' criticism of using species traits, and of course his other big criticism is whether invasion biology should even exist. Do your results support Mark Davis' positions?

MARK VAN KLEUNEN – Yes, in certain experiments we see that aliens are doing the same as natives. I think when Ken Thompson and Mark Davis wrote their paper (*Trends in Ecology & Evolution* 2011, **26**: 155–156), what they actually showed is that invasion biology had already taught us quite a lot about the importance of traits; they actually start with listing studies showing that fast growth, high fecundity and so on are correlated with invasiveness. But they then draw the conclusion, as the title says 'research on traits of invasive plants tell us very little', but that does not match what they actually say in their paper. I agree in principle that we could expect the same thing for natives – that successful natives should have similar traits – but I do not think they had the data to show this, so I think their conclusion was simply premature. Now there is more and more evidence that natives and aliens are similar, but of course there are exceptions.

JOHN PANNELL

RUSSELL LANDE – I wanted to mention a third reason why selfing often evolves at the edge of a species' range in addition to Baker's law, reproductive assurance and lower inbreeding depression, all of which you emphasized. In some cases the edge of a species' range is not expanding, which you modelled, but rather has stopped because of a strong environmental gradient so gene flow stops it, as in the Mark Kirkpatrick and Nicholas Barton paper (*American*

Naturalist 1997, **150**: 1–23). In that case, predominant selfing, asexuality and polyploidy, which are all often observed at the edge of a species' range, act as barriers to gene flow and actually promote speciation. This process can be viewed as a mechanism of range expansion by speciation.

JOHN PANNELL – That is a very good point. Thanks!

DOUGLAS GILL – Among the more famous examples of a selfer arriving somewhere is the Bee orchid (*Ophrys apifera*). Darwin described this species as an outcrosser on the European continent and selfing in the British Isles. Should we be reinterpreting the story that I grew up with, that is that following colonization of Great Britain by outcrossing forms, selfing then evolved after colonization owing to inadequate pollination? Would you perhaps suggest reinterpreting this scenario as a selfer that moved from the continent and successfully colonized Great Britain, as just one example of Baker's law?

JOHN PANNELL – Baker's law, the topic of a recent NESCent working group (see J.R. Pannell *et al.*, *New Phytologist* 2015, **208**: 656–667), can be interpreted, on the one hand, as an ecological sieve in which propagules from the point of origin are self-compatible, or have an ability to self-fertilize, and that these are more likely to become successful colonizers. On the other hand, there is also certainly a case for the selection of increased selfing ability following colonization during the establishment phase. I would not suggest abandoning either of these two models, as both can occur in the evolution of selfing.

RUTH HUFBAUER – I am fascinated by the idea that purging of deleterious alleles may help drive a successful invasion. I had not seen your data and was wondering if there are other data on this topic?

JOHN PANNELL – Perhaps I should clarify that what we have measured are the patterns of inbreeding depression across the range expansion, and I think that's what you are referring to.

RUTH HUFBAUER – Right, or the evolution of decreased inbreeding depression, which suggests purging.

JOHN PANNELL – What we are seeing is not necessarily the purging of genetic load during the range expansion, something we are likely to hear about tomorrow from Stephan Peischl. But it is important to stress that you might observe reduced inbreeding depression towards the edge of a species' range even as genetic load increases with the fixing of deleterious mutations through drift. The crucial thing is that inbreeding depression, and what matters for the selection of the mating system is within-population inbreeding depression, is reduced simply by loss of diversity at viability loci, whether through the fixation or purging of deleterious mutations. It is thus possible to observe both a decline in mean fitness towards the edge of a species' range and a decline in inbreeding depression.

RUTH HUFBAUER – I know of one example of where the flip side is true – there is evidence of purging that has led to increased fitness as opposed to a reduction in inbreeding depression due to fixation; but do you know of examples in either direction?

JOHN PANNELL – I cannot think of any empirical examples. But the increased effective inbreeding through range expansion can presumably allow for purging of genetic load as well. The recent work of Stephan Peischl and Laurent Excoffier, which we will hear about later, deals with the effect of range expansion on both deleterious and advantageous alleles. My impression is that range expansion is likely to increase the rate of fixation of deleterious alleles, and also reduce the efficacy of selection for advantageous ones.

PIERRE GLADIEUX

RIMA LUCARDI (SOUTHERN RESEARCH STATION, USDA, FOREST SERVICE) – With respect to invasions in fungi, what is a species? Is there any consensus on how to define a species within an evolutionary context?

PIERRE GLADIEUX – Although there is no consensus on species concepts in fungi, there *is* consensus on species recognition. The standard is now to use phylogenies to diagnose species because this approach is more accurate and repeatable, and you can archive data. Also, you can use phylogenetic species recognition to identify species of fungi that are asexual.

RIMA LUCARDI – There is always controversy regarding lumpers and splitters. Are there fresh thoughts regarding the phylogenetic species concept in this regard?

PIERRE GLADIEUX – For species recognition, the consensus is where one obtains concordance of multiple phylogenies. So if you have a branch that is shared among multiple phylogenies from multiple genes, but the branch holds a group within which there is no congruence among phylogenies, meaning there is recombination, then this group is considered a species.

SPENCER BARRETT – I was really struck by your statement about the dispersal routes for fungi and that a lot of migration is not necessarily by long-distance spore dispersal. I had certainly assumed that this was how many fungi were getting around. What does this say about the work done by P.H. Gregory and colleagues at Rothamsted in the United Kingdom on aerobiology? They developed models on the movement of coffee leaf rust (*Hemileia vastatrix*) from Africa to South America in air currents. Was the modelling all for nothing?

PIERRE GLADIEUX – No, there are a few examples of fungi able to travel long distances, and coffee leaf rust is one example. Also, some cereal rusts exhibit periodic migration, just like birds. Because they are not able to survive in winter, the populations have to be reestablished, in China for instance, and also in the United Kingdom which is invaded from the continent. Although there have been many publications on this topic, perhaps they are not representative.

DOUGLAS GILL (UNIVERSITY OF MARYLAND) – Your work on host race transfers in fungi reminded me of the *Rhagoletis* story on apples, and the work done to try to determine the genetic features of moving from hawthorn to apples. And yet in insects in general, host race formations are notoriously difficult; they just don't happen. Somehow I expected that they might be easier in fungi than in herbivorous insects, for example. Could you comment on parallels between host race transfers or host race formation in fungi compared to host race evolution in herbivorous insects?

PIERRE GLADIEUX – The main difference is that fungi lack active means of dispersal so they cannot move to suitable mates or choose their hosts. So that is very different from insects. Also, most insects can disperse at different stages of their life cycle so there can be migration between selection on the host and mating, so migration to the ancestral host, for instance, and then recombination on this ancestral host. This should break the association between adaptation and reproductive isolation. Whereas in fungi, at least in Ascomycetes, specifically the ones that tend to reproduce within the host, there is no migration of gametes between selection on the host and mating, so the association cannot be broken by migration. That is why ecological speciation is even more conducive in fungi compared with insects. In fungi there is no requirement for the evolution of mate choice or host preference. Actually, it is not possible to evolve host preference in fungi since they cannot choose.

DOUGLAS GILL – Do you mean that your expectation is that host race formation in fungi is more difficult than in insects?

PIERRE GLADIEUX – No, I think it's easier because genes involved in adaptation have a direct effect on mating patterns because of the lack of migration of the gametes between selection on the host and mating.

ROBERT COLAUTTI

BEN PHILLIPS (UNIVERSITY OF MELBOURNE) – With regard to the history of purple loosestrife (*Lythrum salicaria*) invasion in eastern North America; was the species introduced to the centre point of the range?

ROBERT COLAUTTI – Yes more or less, herbarium records and what limited genetic data that are available suggest multiple introductions from Europe, but to a relatively restricted range – the northeast Atlantic seaboard of the United States – followed by subsequent spread over the past century or so south and northwest into eastern Canada.

DOUGLAS GILL – I am very interested in purple loosestrife; it has been here in North America for 250 years; yet like many introduced species, it is now labelled invasive. Do you know of a single example anywhere of a long-term resident species in North America that has been impacted in a negative way by purple loosestrife? I will be very interested if you have because I have researched this extensively and could find no example of any native plant that has been negatively impacted. I found two papers where experimental removal of purple loosestrife led to a decline of the native species. So we know that in some cases it is actually enhancing some local species. From your studies, can you comment on any scientific justification for extensive extermination programmes, the 'weed warrior behaviour' of native plant societies – I am a member of many of them. That behaviour strikes me as absurd and based on some bizarre hatred of the plant, only because it was recently introduced.

ROB COLAUTTI – The short answer is no. I have not examined the impact of purple loosestrife on other species, so I cannot comment on what its specific impacts are, or when and why management might be justified. Your question gets into a difficult issue

regarding the definition of invasive species. I prefer to think about their biology and separate that from the value systems that are used to decide whether something needs to be managed/eradicated or not. That is a much harder question for me, though sometimes it seems obvious – cane toads come to mind. However, when you see these large populations of purple loosestrife, it is clear that the species is changing the ecology of North American wetland communities. So when I use the term *invasive*, I mean simply that it is non-native, spreads rapidly and becomes ecologically dominant. By virtue of becoming widespread and ecologically dominant, it must be having an impact. It may not be an impact that we should use to justify eradication. But that whole morass is an area that I have tried to stay out of, and my work has focussed more on the biology of rapid adaptation and range expansion. I think this work is important for managing invaders but agnostic in terms of when and where those management actions are justified.

JENNIFER LAU

DOUGLAS GILL – You did not mention predators in your system, can you comment on them?

JENNIFER LAU – I do have data on predators and will have to add that information to the path analysis. *Medicago* does affect the number of predators in these plots, but I do not know if that is through a bottom-up effect in which *Medicago* increases *Hypera* densities and they then increase predator densities. The *Hypera* is definitely escaping their control. The densities are amazing. I think there are potential bottom-up effects on predators, but the magnitude of top-down effects of predators on *Hypera* is not known.

REFERENCES

Kirkpatrick M, Barton N (1997) Evolution of a species' range. *American Naturalist*, **150**, 1–23.

Pannell JR, Auld JR, Brandvain Y *et al.* (2015) The scope of Baker's law. *New Phytologist*, **208**, 656–667.

Thompson K, Davis M (2011) Why research on traits of invasive plants tells us very little. *Trends in Ecology & Evolution*, **26**, 155–156.

Part 2

Evolutionary Genetics

INTRODUCTION

Robert I. Colautti* and Carol Eunmi Lee†

* Department of Biology, Queen's University, Kingston, ON K7L 3N6, Canada
† Center of Rapid Evolution (CORE), Department of Zoology, University of Wisconsin, 430 Lincoln Drive, Birge Hall, Madison, WI 53706, USA

The developmental biologist Conrad H. Waddington was among the first to apply an evolutionary genetic perspective to the study of invasion biology (Waddington 1965). While his greatest intellectual achievements were largely overlooked during the Evolutionary Synthesis of 1930s and 1940s, Waddington's conceptual innovations are much more influential today—he coined the term "epigenetics" (Waddington 1942), and his research has influenced the modern disciplines of systems biology and evolutionary developmental biology (evo-devo). He was interested in how exposure to extraordinary environmental change could expose hidden genetic variation, via de-canalization (or plasticity), which could then be subjected to natural selection (Waddington 1953). This was a process he termed "genetic assimilation." Waddington came up with the idea for the 1965 Asilomar symposium on *The Genetics of Colonizing Species* because he recognized the potential for studies of invasive species to inform basic research, as he notes in his introduction (Waddington 1965):

> ... these transplantations of species have, in effect, been a series of experiments in evolution. As such, they are potentially much more informative than most laboratory experimental work, since they have faced the introduced species, not with some simple defined change in selective conditions, but with a whole new ecological system in which the species has to find a place for itself.

In the spirit of Waddington's insight, this section explores invading species as a basis for developing theoretical models and empirical methods that explore the complex relations among genetic diversity, quantitative traits, developmental plasticity, and adaptation to novel environmental challenges.

Standing genetic variation and phenotypic plasticity are sometimes presented as mutually exclusive explanations for the success of invasive species. Chapter 9 dispels this notion as Lande reviews two key theoretical results that are often overlooked in evolutionary genetic studies of invasive species. First, introduction bottlenecks have a relatively weak effect on standing genetic variation during range expansion—for example, a single diploid colonist from a randomly mating population retains approximately 3/4 of the total additive genetic variation present in the population from which it

Invasion Genetics: The Baker and Stebbins Legacy, First Edition. Edited by Spencer C. H. Barrett, Robert I. Colautti, Katrina M. Dlugosch, and Loren H. Rieseberg.

originates. Second, phenotypic plasticity can promote "evolutionary rescue" during invasion by buffering a colonizing population against extinction in a new environment with a new phenotypic optimum. Phenotypic plasticity is thereby favored early during colonization, but this is followed by selection for genetic assimilation and the evolution of canalization as a colonizing population evolves toward its new optimum. Lande further explains why quantitative genetic theory does not predict a general tendency for increased phenotypic plasticity (or broad physiological tolerance) in colonizing populations. Mixed empirical support for the prediction of higher plasticity in introduced versus native genotypes may be explained by variation among invasive species in a number of parameters, including (i) the cost of plasticity, (ii) time since colonization, (iii) the difference in the phenotypic optima of ancestral and colonized populations, and (iv) the environmental variance of ancestral and colonized environments. To better understand the role of phenotypic plasticity in invasion success, Lande describes how empirical approaches could be improved by characterizing variance in natural selection, as well as genetic variation and phenotypic plasticity in ancestral and colonized environments. Empirical studies of natural field populations are a key area of future research on plasticity and invasion.

In Chapter 10, Kirkpatrick and Barrett investigate the evolutionary dynamics of chromosomal inversions and their implications for range expansion in large populations distributed along an environmental gradient. Recent genome-wide studies of plants and animals reveal inversions associated with phenotypes thought to be important for colonizing new habitats (e.g., Lowry and Willis 2010; Jones *et al.* 2012). Inspired by examples like these, the authors consider two scenarios in which novel inversions could influence range expansion. First, inversions spanning two loci could "capture" alleles that are locally adaptive in peripheral populations, acting as a two-mutation supergene that is protected against the disruptive influence of recombination. The authors find that this class of inversion has only a modest effect on population size. This is because in models with weak selection and high gene flow from central to marginal populations, locally adapted alleles can be lost by drift and the swamping effect of gene flow from peripheral populations. However, in models with strong selection and low gene flow, natural selection is efficient at reducing the allele frequencies of locally maladapted alleles, resulting in relatively few maladaptive genotypes in peripheral

populations even before the inversion arises. In contrast, the second case considered by Kirkpatrick and Barrett involves a novel inversion that contains alleles not present in the parental population. Such a scenario could arise via introgression of a "cassette" of locally adapted alleles from a closely related species. In this case, the gene cassette is predicted to spread rapidly, increasing population size at the range margin. From these models, it appears that genetic loci of large effect can have a significant influence on range expansion—a point revisited in Chapter 14.

In addition to structural variation in genome organization, selection on standing genetic variation for quantitative traits could have a strong influence on the rate of establishment and spread of invasive species. In Chapter 11, Blows and McGuigan explore statistical methods for characterizing pleiotropy and genetic constraints. Their methods focus on **G**, which is a symmetric $n \times n$ matrix with diagonal values representing additive genetic variance for a set of phenotypic traits, and off-diagonal components representing additive genetic covariance for each pair of traits. Biologically, the **G**-matrix is a mathematical representation of the additive genetic variation within a population. Pleiotropy causes genetic correlations among phenotypic traits, which theoretically limit the range of possible trait combinations that can evolve in response to natural selection, at least over short timescales. The authors explain how pleiotropy and genetic constraints can be explored through the "spectral distribution" of **G**—the distribution of eigenvalues from a principal components analysis. Relative to a **G**-matrix with weak trait correlations, strong trait correlations will cause a characteristic shift toward higher eigenvalues (i.e., high genetic variation) for only the first few principal component eigenvectors of **G** and toward lower eigenvalues for the other principal components. The authors find this signature in an analysis of gene expression data, supporting a pleiotropic model of genetic effects. The application of these methods to the study of invasive species should provide novel insights into genetic constraints on adaptation during invasion, particularly if estimates of **G** are combined with measurements of natural selection for the same traits in invading populations (see Chapter 6).

Statistical analyses involving the **G**-matrix can help test evolutionary hypotheses in invasive species, but many tests assume that **G** is multivariate normal. Many ecologically relevant phenotypic traits (e.g., phenology, growth, and development) are indeed normal or nearly normal following a log scale or other transformation, but discrete phenotypes and data from genetic markers violate the multivariate normal assumption. In Chapter 12, Day argues that the Shannon information (SI) entropy index can be used to characterize multivariate constraints for non-normal data. In his chapter, Day explains how this index could be used to characterize constraints across multiple loci or multiple traits by calculating a joint distribution. The resulting value represents a measure of evolvability for non-quantitative characters. The version of SI presented by Day is a slightly different formulation of the same index that is widely used to characterize biological diversity in community ecology studies. It has also been applied to quantify allelic diversity in molecular ecology, but primarily as a descriptive statistic. In contrast, Day's exploration of SI hints at a future research stream investigating the SI as a biologically informative measure of genetic constraint, for example, at range limits in invasive species. Such a research program would require not only a characterization of available genetic variation at functional loci but also an understanding of how selection acts on those loci.

While pleiotropy and standing genetic variation likely effect the rate and extent of adaptive evolution during invasion, Peischl and Excoffier (Chapter 13) describe another form of constraint on range expansion. Previous studies by Excoffier and coauthors have shown how populations at the colonization front are more prone to stochastic increases in the frequency of detrimental alleles, reducing fitness toward the invasion front (reviewed in Excoffier *et al.* 2009). This previous research focused on codominant alleles, but Chapter 13 examines the dynamics of recessive detrimental mutations. Using individual-based simulations, the authors show that a similar "expansion load" limits fitness at the colonizing front as detrimental mutations become fixed by genetic drift. In contrast to models of codominant mutations, fitness declines much more rapidly from core to peripheral populations even if the number of detrimental alleles per individual is the same throughout the range. The authors use their model to predict characteristics of the site frequency spectrum—a frequency histogram showing the proportions of alleles with different frequencies of occurrence in the population. Specifically, a reduction in rare and intermediate-frequency alleles is expected in peripheral relative to core populations, and indeed this has been found in human populations. Models like this could be expanded to better understand the effects on range dynamics of more complicated

invasion scenarios involving multiple introductions and long-distance dispersal.

Chapter 12, the final chapter in this section by Dlugosch and coauthors, provides a more general review of the role of genetic variation in the establishment and spread of invasive species. They hypothesize that alleles with large effects on phenotype are likely to play an important role in adaptation during invasion because alleles with small effects are less influenced by natural selection relative to the stochastic influence of genetic drift and founder effects. However, the authors also stress the importance of understanding the environmental context of the genotype–phenotype map, returning to Waddington's (1965) breakdown of canalization in stressful environments. Although de-canalization (plasticity) may be maladaptive in the ancestral environment, it increases heritable genetic variation available for natural selection to act upon in the new range. This "cryptic" genetic variation could include adaptive phenotypes that increase in frequency over time, followed by selection for re-canalization as proposed by Lande in Chapter 9. The authors also examine the degree of admixture in invading populations and find, somewhat surprisingly, that admixture among introduced populations is less common for species with divergent source populations. This could indicate a negative fitness consequence of admixture between highly divergent populations, perhaps due to outbreeding depression.

The chapters in this section collectively build on Waddington's insight that biological invasions present outstanding opportunities for testing, refining, and innovating theoretical and empirical methods in evolutionary genetics. These chapters demonstrate how biological invasions provide insights into a number of fundamental questions in evolutionary genetics, including (i) how the fitness advantage of phenotypic plasticity can vary over space and time (Chapter 9); (ii) how selection can act on genome structure and how genome structure can affect population dynamics (Chapters 10 and 14); (iii) how to measure pleiotropy and the structure of standing genetic variation for quantitative traits (Chapter 11), and other characters that are not multivariate normal (Chapter 12); (iv) how range expansion *per se* affects the frequency and distribution of detrimental alleles in natural populations, and how this process can establish and maintain range limits (Chapter 13); and (v) how large-effect loci, novel mutations, and cryptic genetic variation can contribute to rapid evolution during invasion (Chapter 14). These chapters represent significant innovations in the field of evolutionary genetics over the past 50 years, but given the largely theoretical nature of this section, it is worth returning to another insight from Waddington's (1965) opening remarks:

> It seems to me that what is needed, at the present stage of our understanding of evolution, is not so much a greater elaboration of formal theories of quantitative and population genetics of the kind we have been producing for the last 20 or 30 years, or even more analyses of wild populations in terms of genetics divorced from their ecology. What we need is more knowledge about the ways in which populations, in fact, meet evolutionary challenges: What intensities of natural selection can they put up with, how far and how fast can they modify their phenotype (including their habits)?

Guided by the theoretical insights presented in this section and rapid advancements in gene sequencing technology, environmental sensing, and the analysis of "big data," it should indeed be possible in the near future to comprehensively investigate the evolutionary genetics of invasive species at continental to global scales without "divorcing" wild populations from their ecology.

REFERENCES

Excoffier L, Foll M, Petit RJ (2009) Genetic consequences of range expansion. *Annual Review of Ecology, Evolution, and Systematics*, **40**, 481–501.

Jones FC, Grabherr MG, Chan YF *et al.* (2012) The genomic basis of adaptive evolution in threespine sticklebacks. *Nature*, **484**, 55–61.

Lowry DB, Willis JH (2010) A widespread chromosomal inversion polymorphism contributes to a major life-history transition, local adaptation, and reproductive isolation. *PLoS Biology*, **8**, e1000500.

Waddington CH (1942) The epigenotype. *Endeavour*, **1**, 18–20.

Waddington CH (1953) Genetic assimilation of an acquired character. *Evolution*, **7**, 118–126.

Waddington CH (1965) Introduction to the symposium. In: *The Genetics of Colonizing Species* (eds. Baker HG, Stebbins GL), pp. 1–6. Academic Press, New York.

Chapter 9

EVOLUTION OF PHENOTYPIC PLASTICITY IN COLONIZING SPECIES

Russell Lande

Division of Biology, Imperial College London, Silwood Park Campus, Ascot, Berkshire, SL5 7PY, UK

Abstract

I elaborate an hypothesis to explain inconsistent empirical findings comparing phenotypic plasticity in colonizing populations or species with plasticity from their native or ancestral range. Quantitative genetic theory on the evolution of plasticity reveals that colonization of a novel environment can cause a transient increase in plasticity: a rapid initial increase in plasticity accelerates evolution of a new optimal phenotype, followed by slow genetic assimilation of the new phenotype and reduction of plasticity. An association of colonization with increased plasticity depends on the difference in the optimal phenotype between ancestral and colonized environments, the difference in mean, variance and predictability of the environment, the cost of plasticity, and the time elapsed since colonization. The relative importance of these parameters depends on whether a phenotypic character develops by one-shot plasticity to a constant adult phenotype or by labile plasticity involving continuous and reversible development throughout adult life.

Previously published as an article in *Molecular Ecology* (2015) 24, 2038–2045, doi: 10.1111/mec.13037

INTRODUCTION

Many populations and species experience sudden shifts in environment following long-distance dispersal, host switches, invasion of their range by exotic species or rapid climate change. If the new environment differs substantially from the original environment, new phenotypic adaptations may be required to prevent local extinction. In view of the ubiquity of local genetic adaptation as a major basis for geographic variation within species (Schluter 2000), it is not surprising that adaptive evolution by natural selection has been detected in colonizing and invasive populations, though sometimes confounded with founder effects and random genetic drift (Dlugosch & Parker 2008; Prentis *et al.* 2008; Kilkenny & Galloway 2012; Colautti & Barrett 2013; Vandepitte *et al.* 2014).

Classical transplantation experiments with plants demonstrated that geographic variation of phenotypic adaptations within species is generally caused both by plasticity and genetic evolution (Turesson 1922; Clements 1929; Clausen *et al.* 1940). It is therefore plausible and has frequently been suggested that increased plasticity facilitates establishment and range expansion in a new environment. The role of phenotypic plasticity in ecological interactions, species coexistence, mating systems and speciation has also been emphasized (Stam 1983; Agrawal 2001; Miner *et al.* 2005; Levin 2010; Pfennig *et al.* 2010; Thilbert-Plante & Hendry 2011; Fitzpatrick 2012; Peterson & Kay 2014).

Baldwin (1896) first suggested that developmental plasticity in a changing environment promotes population persistence and orients the direction of future evolution. These concepts were subsequently applied and extended by Bateson (1963), Stam (1983), Sultan (1987), West-Eberhard (1989, 2003), Schlichting & Pigliucci (1998), Price *et al.* (2003) and others too numerous to list. Web of Science searches for key words plasticity AND (colonizing OR invasive), or plasticity AND 'climate change', each returned well over 1000 publications in ecology and evolution.

However, the role of phenotypic plasticity in colonization and invasion of novel environments by sexual species remains far from clear. Comparisons of plasticity in colonizing populations or species in novel environments vs. plasticity in their native or ancestral range are often inconsistent. Most notably, recent reviews on the subject reached conflicting conclusions. Davidson *et al.* (2011) found that invasive species showed increased plasticity compared with noninvasive congeners, while Palacio-López & Gianoli (2011) and Godoy *et al.* (2011) did not find such a difference. Here, I review quantitative genetic theory on the evolution of plasticity in a novel environment to suggest possible reasons for this inconsistency.

EVOLUTIONARY RESCUE AND THE BALDWIN EFFECT

Novel environments often represent demographic sinks where populations tend to decline. A requirement for evolutionary adaptation to the local environment may be one of the main reasons for the time lag often observed between colonization of a new habitat and subsequent population growth and range expansion. Another possible cause of time lags in invasions is that secondary colonies may become established in more favourable locations with increased local fitness and emigration; similarly, temporal change in environment can create conditions favourable for population growth and range expansion (Turchin 2003; Crooks 2005).

Evolutionary rescue of a population from extinction in a new environment, coupled to a model of population dynamics, was analysed by Gomulkiewicz & Holt (1995). Phenotypic plasticity was incorporated into this model by Chevin & Lande (2010) using norms of reaction. Among a discrete set of environments, plasticity is often measured as genotype–environment interaction (Via & Lande 1985), whereas in a continuous range of environments, plasticity is typically depicted by norms of reaction giving the expected phenotype as a function of the environment in which a given genotype develops (Schlichting & Pigliucci 1998). These alternate descriptions are in essence mathematically equivalent (Via *et al.* 1995).

Figure 1 illustrates evolutionary rescue with the Baldwin effect (constant plasticity). The environment is assumed to be constant, so the full effect of plasticity operates in the first generation, with subsequent changes in population growth rate due to Darwinian evolution. In this example, substantial adaptive plasticity is required for evolutionary rescue to prevent extinction in the new environment. The evolution of plasticity can further augment evolutionary rescue (Lande 2009; Chevin & Lande 2010; Chevin *et al.* 2010, 2013) as explained below.

Population bottlenecks during colonization and evolutionary rescue may cause considerable reduction

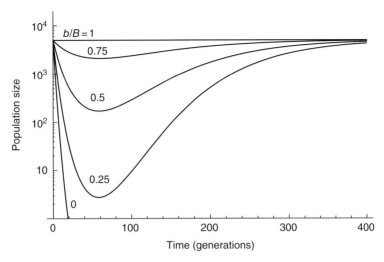

Fig. 1 Evolutionary rescue of a population by natural selection on a quantitative character with constant plasticity (Baldwin effect) in a constant environment. Population dynamics obey Gompertz density dependence (linear in log population size), with evolution of the mean phenotype increasing the mean Malthusian fitness under stabilizing selection towards a new optimum phenotype. After a sudden change of environment at time 0, the mean phenotype is initially far from the optimum. The norm of reaction of the population mean phenotype as a function of the environment is linear, with plasticity measured by its slope, b, assumed constant so that evolution only alters reaction norm elevation. The optimal phenotype is a linear function of the environment with slope B (modified from Chevin & Lande 2010).

of genetic variation and adaptive potential during periods of small population size. In the original Asilomar conference, Lewontin (1965) commented that Mayr greatly exaggerated the founder effect. Small population size can drastically reduce the number of genotypes and alleles, but a population of effective size N_e contains on average $1 - 1/(2N_e)$ of the heterozygosity or additive genetic variance in quantitative characters in the population from which it was sampled. A single outcrossed female therefore contains on average 3/4 of the additive genetic variance that existed in the population from which it was sampled (and larger fractions of dominance and epistatic variances). Multiple generations of small population size are therefore necessary to greatly reduce the expected genetic variance by random genetic drift. Prolonged small population size creates a substantial risk of extinction from demographic and environmental stochasticity (Lande *et al.* 2003) and from the fixation of slightly deleterious spontaneous mutations (Lande 1994, 1995).

After recovery to large population size, the reduction in heterozygosity measured from molecular polymorphisms may be long lasting due to low per-locus mutation rates. In contrast, the high mutability of polygenic quantitative characters implies a much more rapid

recovery of additive genetic variance and adaptive potential (Lande 1980, 1995). This largely explains the finding of Dlugosch & Parker (2008) that invasive populations often show reduced heterozygosity, without appreciable reduction in genetic variance of quantitative characters.

EVOLUTION OF PLASTICITY IN A NOVEL ENVIRONMENT

Most quantitative genetic theory on the evolution of phenotypic plasticity is based on a rather simple model of development, here termed *one-shot plasticity*. During a brief critical stage of early development, the environment experienced by an individual influences its subsequent development to a constant adult phenotype that is subject to selection. This provides a reasonable description of development for most morphological characters in species with deterministic growth.

All individuals in the sexual population in a given generation experience the same (macro)environment. Generations are discrete and nonoverlapping, but the environment is assumed to change continuously, with time t measured in generations. Assuming that norms of reaction are linear, the adult phenotype of an individual in generation t is

$$z_t = a + b\varepsilon_{t-\tau} + e. \tag{1}$$

Here, a is the elevation (or intercept) of the genotypic reaction norm, giving the additive genetic effect (breeding value) in the reference environment, $\varepsilon = 0$, and b is the slope of the plastic phenotypic response to the environment. Juveniles of generation t are exposed to environment $\varepsilon_{t-\tau}$ during a critical period of development a fraction of a generation τ before the adult phenotype is expressed and exposed to natural selection. An independent residual component of phenotypic variation, e, caused by genetic dominance, developmental noise and micro-environmental variation unrelated to the macro-environment, is assumed to be normally distributed with constant mean $\bar{e} = 0$ and variance σ_e^2 among individuals in every generation. The units of a and e are the same as for the character, z, whereas b is measured in character units per environmental unit. Reaction norm elevation in the reference environment a and slope b are assumed to have a bivariate normal distribution in the population, with additive genetic variances and covariance that remain constant during evolution of the means, $\bar{a}$ and $\bar{b}$. Thus, in any generation, the individual breeding values and adult phenotypes before selection also are normally distributed.

An important result of this theory is that the evolution of adaptive plasticity is limited by predictability of the environment between the critical stage of juvenile development and adult selection. Imperfect environmental predictability occurs because of temporal changes in the environment and/or juvenile dispersal between the times of development and selection.

Using this model, and assuming that the optimal phenotype is a linear function of the environment with slope B, Gavrilets & Scheiner (1993a) showed that the mean plasticity in the population evolves towards $\bar{b} = \rho_\tau B$. This is the optimal plasticity that maximizes the expected mean fitness in the population, although it is only partially adaptive (smaller than B) due to incomplete environmental predictability, ρ_τ, across the developmental time lag τ. This result relies on an implicit assumption that evolution of the mean reaction norm in the population is slow compared with fluctuations in the optimum phenotype. A cost of plasticity also can limit its evolution, but measurements of the cost suggest it is usually small (Van Buskirk & Steiner 2009).

After a sudden change in the environment, selection towards a new optimal phenotype favours the evolution of increased plasticity (Gavrilets & Scheiner 1993b; Via et al. 1995). This result depends on the simplifying assumption of no genetic correlation between reaction norm elevation and slope and also neglects the long-term evolutionary impact of temporal fluctuations in the new environment (Gavrilets & Scheiner 1993b).

Lande (2009) modelled the long-term evolution of plasticity during adaptation to a new optimal phenotype in a fluctuating environment following a sudden extreme change in the average environment. For simplicity, the predictability of background environmental fluctuations was assumed to be the same before and after the sudden change in average environment. An evolutionary justification for the assumption of no genetic correlation between reaction norm elevation and slope was provided by showing this to be a consequence of phenotypic canalization (or reduced phenotypic variance) resulting from a long history of stabilizing selection in the local average environment. The existence of canalization in a quantitative character can be tested by transforming the scale of measurement (Wright 1968, Chap. 11) such that the phenotypic mean and variance within environments are uncorrelated across environments, to determine whether a minimum phenotypic variance occurs in the average ancestral or native environment. Without such transformation, quantitative characters such as body size often display a positive correlation between phenotypic mean and variance among populations or environments that could obscure the patterns described below when selection in the new environment favours a substantial reduction in the mean phenotype.

If a sudden change in the average environment, δ, exceeds the usual range of background environmental fluctuations, $\delta^2 \gg \sigma_\varepsilon^2$, then the evolutionary dynamics unfold in two distinct phases (Lande 2009). During Phase 1, a rapid increase in plasticity allows the new optimal phenotype to be closely approached, and Phase 2 involves a relatively slow genetic assimilation of the new optimal phenotype, with gradual relaxation of plasticity to a level determined by the predictability of background environmental fluctuations and its replacement by gradually increased elevation of the reaction norm in the average ancestral environment. Figure 2 shows that following the sudden major shift in average environment, the evolution of plasticity produces a much more rapid recovery of mean fitness, and thus

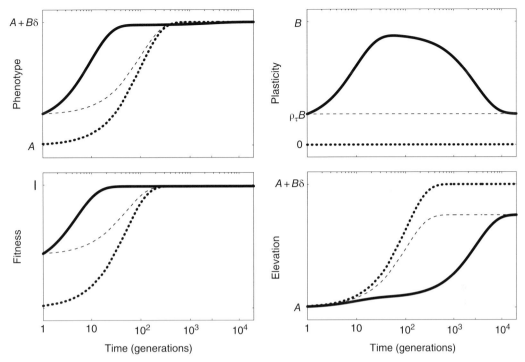

Fig. 2 Evolution of the expected mean phenotype and corresponding Malthusian fitness after colonization, determined by the expected plasticity and elevation of the mean norm of reaction in the population. When the sudden change in average environment exceeds the usual range of background environmental fluctuations, a rapid increase in plasticity allows the new optimal phenotype to be closely approached, followed by slow genetic assimilation of the new optimal phenotype with relaxation of plasticity to a level determined by the predictability of background environmental fluctuations. The two phases in the expected evolution of mean plasticity (slope) and elevation of the population reaction norm are shown with time on a log scale. *Dotted line*: Darwinian evolution with no plasticity. *Dashed line*: Darwinian evolution with constant plasticity (Baldwin effect). *Solid line*: Reaction norm evolution with genetic variance in plasticity (modified from Lande 2009).

a more powerful rescue effect, than by Darwinian evolution alone (with no plasticity) or the Baldwin effect (Darwinian evolution with constant plasticity). Figure 3 illustrates the pivoting of the average norm of reaction during the two phases of evolution.

The transient increase in plasticity described by this theory has important implications for understanding the role of plasticity in colonizing populations. In this scenario, the hypothesis of increased plasticity in colonizing populations would be confirmed only for a limited time following colonization of a new habitat, but not for sufficiently old events. Transient dynamics of plasticity may thus play a key role in inconsistent findings on comparisons of plasticity in colonizing populations with that in their native or ancestral range. This complication can be addressed using historical or

population genetic estimates of the age of colonization events. The timescale over which a transient increase in plasticity can be observed following colonization of a new extreme environment cannot be predicted without knowledge of inheritance of the quantitative characters being studied, the inheritance of plasticity in these characters and how selection acting on them changes with the environment in time and space (Lande 2009; Chevin *et al.* 2010, 2013). However, it seems likely that comparisons between invasive and noninvasive congeneric species usually involve divergence times too long to observe a transient increase in plasticity after colonization of a new extreme environment.

The magnitude of difference in optimal phenotypes between ancestral and colonized ranges is a crucial parameter in the theory. If the mean environment of

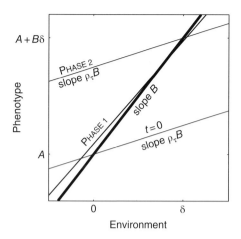

Fig. 3 Two phases in evolution of the mean norm of reaction for a quantitative trait after colonization of a new extreme environment. At time $t = 0$ the average environment suddenly shifts from 0 in the ancestral population to δ in the new environment. Background environmental fluctuations occur as a stationary time series with variance σ_ε^2 and autocorrelation across the developmental time lag ρ_τ in both the ancestral and new environments. Norms of reaction are linear, with plasticity measured by their slope, and the optimal phenotype is a linear function of the environment with slope B, so that the optimal phenotype shifts from A in the ancestral population to $A + \delta B$ in the new environment. The character is canalized in the ancestral environment, such that the genetic and phenotypic variances are minimized in the average ancestral environment (modified from Lande 2009).

colonization closely resembles that of the ancestral population, so that little difference exists in the optimal phenotypes, then little or no transient increase in plasticity is predicted. Predictability of the environment in time and space in the ancestral and colonized ranges is also important. A permanent increase (or decrease) in plasticity will evolve in the colonizing population if its environmental predictability is larger (or smaller) than that in the ancestral population, even if their average environments are similar. With a sudden change in both the average and predictability of the environment, a transient increase in plasticity can be followed by a permanent increase or decrease in plasticity. These complications can be addressed by measurements of phenotypes and natural selection acting on them in ancestral and colonized environments, and by studies of environmental mean, variance and predictability in space and time. Finally, a change in average environment may substantially alter the optimal phenotype for

some characters, but not others; and the evolution of trait plasticities may not occur independently because of genetic correlations among trait parameters in multivariate norms of reaction.

In a population continuously distributed in space, a permanent increase in plasticity is expected at the edge of the range (Chevin & Lande 2011). This occurs because maladaptive gene flow from relatively abundant more centrally located populations into less abundant peripheral populations creates a permanent deviation of the mean phenotype from the local optimum at the edge (Kirkpatrick & Barton 1997; Case & Taper 2000). Increased plasticity at a species boundary facilitates range expansion into a new habitat, with all the consequences and complications described above.

LABILE PLASTICITY AND ENVIRONMENTAL TOLERANCE

Physiological and behavioural characters typically display reversible development producing labile phenotypes that change continually during an individual lifetime, here termed *labile plasticity*. A theory for evolution of plasticity in labile characters undergoing continual development and selection through time requires an explicit model of developmental dynamics describing how the phenotype of an individual changes during its lifetime in response to environmental fluctuations. The phenotype of an individual at time t, denoted as z_t, can be represented as the sum of a constant additive genetic effect a, a constant microenvironmental contribution that varies among individuals (including nonadditive genetic effect and developmental noise), e, and a term describing labile changes in the individual phenotype in response to the macro-environment ε_t experienced by all individuals in the population at any given time.

Lande (2014) employed a model of individual developmental dynamics described by a first-order linear differential equation with the rate of development λ assumed to be a constant independent of the genotype and the environment. The (macro)environment is measured as a deviation from its average value, so that $\bar{\varepsilon} = 0$, and is assumed to have a stationary distribution with temporal variance σ_ε^2 and autocorrelation function $\rho_\tau = E[\varepsilon_t \varepsilon_{t-\tau}] / \sigma_\varepsilon^2$.

Evolutionary analysis was facilitated by assuming that the expected phenotype that a genotype develops in a constant controlled environment is a linear function of the environment, with slope b measuring the plasticity (as in eqn 1). In a fluctuating environment, labile developmental changes in the individual phenotype were assumed to occur on a timescale that is short compared with the population generation time, $1/\lambda \ll T$. This entails that juvenile growth and development are short compared with the average adult lifespan, so that selection acts primarily on labile adult phenotypes. The individual phenotype can be approximated as an integral over past environments with an exponentially diminishing contribution from longer time lags u,

$$z_t = a + b\lambda \int_0^\infty e^{-\lambda u} \varepsilon_{t-u} \mathrm{d}u + e \qquad (2)$$

An environmental tolerance curve describes fitness as a function of the environment for a genotype, population or species. Tolerance curves, or their limits expressed as climate envelopes, originally applied in physiological ecology and toxicology, are now frequently used to predict shifts in geographic ranges, or extinction, of species subject to secular climatic trends such as regional or global warming. Ecologists and conservation biologists using tolerance curves to predict species range shifts have, until recently, largely neglected the basic evolutionary and ecological processes of demography, adaptive evolution and plasticity (Morin & Thuiller 2009; Chevin *et al.* 2010; Normand *et al.* 2014). The lack of evolutionary mechanisms in these approaches is partly dictated by the absence of phenotypes in tolerance curves.

Chevin *et al.* (2010) pointed out with graphical examples that norms of reaction (giving the expected phenotype as a function of the environment) and tolerance curves (depicting fitness as a function of the environment) are each two-dimensional aspects of the three-dimensional relationship of fitness as a function of both phenotype and environment. This three-dimensional relationship, describing how the environment affects phenotypic selection, is a key ingredient in all evolutionary models of fluctuating selection on quantitative characters and the evolution of phenotypic plasticity.

Lande (2014) modelled the evolution of plasticity in a labile character undergoing continuous development and selection in a fluctuating environment. Notably, for both labile and one-shot plasticity, a given amount of plasticity corresponds to the same environmental tolerance curve, because norms of reaction and environmental tolerance curves are measured by experiments or observations in which the phenotypes

and fitnesses of replicate genotypes or population samples develop across a range of constant environments.

The evolution of labile plasticity, nevertheless, does differ significantly from the evolution of one-shot plasticity. In the absence of a cost of plasticity, the expected labile plasticity that evolves does not depend on environmental predictability, but always equals the slope of the optimum phenotype as a function of the environment, $\bar{b} = B$. In contrast to one-shot plasticity, labile plasticity may have a substantial cost because continual development requires extensive expenditure of materials and energy. With a cost of labile plasticity, the expected plasticity that evolves depends on both the environmental variance and predictability, and the cost of plasticity, $\bar{b} = B\sigma_\varepsilon^2 \bar{\rho} / \left(\sigma_\varepsilon^2 \bar{\rho} + \gamma_b / \gamma\right)$. Here, $\sigma_\varepsilon^2 \bar{\rho}$ is the product of environmental variance and the environmental predictability averaged over the continuous developmental time lag, $\bar{\rho} = \lambda \int_0^\infty e^{-\lambda \tau} \rho_\tau \mathrm{d}\tau$. The cost of plasticity is measured by the strength of stabilizing selection against plasticity relative to the strength of phenotypic stabilizing selection within environments, γ_b / γ. A sudden extreme change in average environment produces a transient increase in labile plasticity that separates into two distinct phases, although under a different condition than for one-shot plasticity, $\delta^2 \gg \sigma_\varepsilon^2 \bar{\rho} + \gamma_b / \gamma$.

Environmental tolerance curves derived from this theory support the prevailing concept in physiological ecology of a generalist-specialist trade-off between the width and height of tolerance curves (or niche width and maximum fitness) (Huey & Kingsolver 1989; Deutsch *et al.* 2008). The theory reveals that the magnitude of this trade-off depends on the cost of plasticity, as shown in Fig. 4.

CONCLUSION

Contrary to common conceptions in the literature, quantitative genetic theory on the evolution of phenotypic plasticity does not predict a general tendency for increased plasticity or environmental tolerance in colonizing and invasive populations and species. Whether or not this is observed in particular cases depends on several parameters including the optimal phenotype, and the mean, variance and predictability of the environment in the ancestral or native geographic range compared to those in the colonized

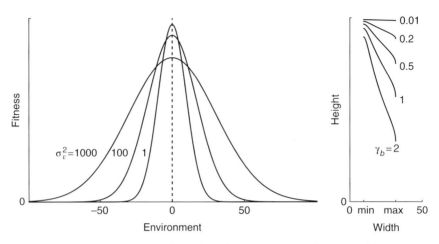

Fig. 4 *Left*: Environmental tolerance curves that evolve under different environmental variance σ_ε^2 for a given cost of plasticity measured by the strength of stabilizing selection against plasticity, $\gamma_b = 0.5$. Mean fitness is on an absolute scale. The average environment may differ among locations, but is depicted as standardized to 0 for comparison. *Right*: Trade-off between tolerance curve width and height, dependent on the cost of plasticity. Minimum and maximum widths of tolerance curves (min and max) occur, respectively, in constant and highly variable environments, determined by the strength of stabilizing selection on the character within environments and the genetic variance in plasticity (modified from Lande 2014).

range. Other critical parameters include the cost of plasticity and whether the phenotypic characters develop by one-shot plasticity to a final adult phenotype or by labile plasticity involving continuous and reversible development throughout adult life. Adaptation to a new extreme environment may often occur by a rapid transient increase in plasticity followed by slow genetic assimilation involving decreased plasticity, such that the time since colonization also determines whether increased plasticity will be observed in the new environment. The complexity of these processes and the variety of possible outcomes, in addition to heterogeneity among studies in times of observation since colonization, help to explain inconsistent findings in comparisons of plasticity in colonizing species vs. those in their ancestral or native range. Empirical advances on this subject will depend on measuring parameters of the environment, as well as development and inheritance of phenotypic characters and selection acting on them, in addition to reconstructing the history of colonization events.

ACKNOWLEDGEMENTS

I thank the conference participants and reviewers for discussion and comments. This work was supported by a Royal Society Research Professorship.

REFERENCES

Agrawal AA (2001) Phenotypic plasticity in the interactions and evolution of species. *Science*, **294**, 321–326.

Baldwin JM (1896) A new factor in evolution. *The American Naturalist*, **30**, 444–451.

Bateson G (1963) The role of somatic change in evolution. *Evolution*, **17**, 529–539.

Case TJ, Taper ML (2000) Interspecific competition, environmental gradients, gene flow, and the coevolution of species' borders. *The American Naturalist*, **155**, 583–605.

Chevin L-M, Lande R (2010) When do adaptive plasticity and genetic evolution prevent extinction of a density-regulated population? *Evolution*, **64**, 1143–1150.

Chevin L-M, Lande R (2011) Adaptation to marginal habitats by evolution of increased phenotypic plasticity. *Journal of Evolutionary Biology*, **24**, 1462–1476.

Chevin L-M, Lande R, Mace GM (2010) Adaptation, plasticity and extinction in a changing environment: towards a predictive theory. *PLoS Biology*, **8**, e1000357.

Chevin L-M, Collins S, Lefèvre F (2013) Phenotypic plasticity and evolutionary demographic responses to climate change: taking theory out to the field. *Functional Ecology*, **27**, 966–979.

Clausen J, Keck DD, Hiesey WM (1940) *Experimental Studies on the Nature of Species. I. Effect of Varied Environments on Western North American Plants*. Carnegie Institution of Washington, Washington, District of Columbia.

Clements FE (1929) Experimental methods in adaptation and morphogeny. *The Journal of Ecology*, **17**, 356–379.

Colautti RI, Barrett SCH (2013) Rapid adaptation to climate facilitates range expansion of an invasive plant. *Science*, **342**, 364–366.

Crooks JA (2005) Lag times and exotic species: the ecology and management of biological invasions in slow-motion. *Ecoscience*, **12**, 316–329.

Davidson AM, Jennions M, Nicotra AB (2011) Do invasive species show higher phenotypic plasticity than native species and, if so, is it adaptive? A meta-analysis. *Ecology Letters*, **14**, 419–431.

Deutsch CA, Tewksbury JJ, Huey RB *et al.* (2008) Impacts of climate warming on terrestrial ectotherms across latitude. *Proceedings of the National Academy of Sciences of the United States of America*, **105**, 6668–6672.

Dlugosch KL, Parker IM (2008) Founding events in species invasions: genetic variation, adaptive evolution, and the role of multiple introductions. *Molecular Ecology*, **17**, 431–449.

Fitzpatrick BM (2012) Underappreciated consequences of phenotypic plasticity for ecological speciation. *International Journal of Ecology*, **2012**, 256017.

Gavrilets S, Scheiner SM (1993a) The genetics of phenotypic plasticity. V. Evolution of reaction norm shape. *Journal of Evolutionary Biology*, **6**, 31–48.

Gavrilets S, Scheiner SM (1993b) The genetics of phenotypic plasticity. VI. Theoretical predictions for directional selection. *Journal of Evolutionary Biology*, **6**, 49–68.

Godoy O, Valladares F, Pilar C-D (2011) Multispecies comparison reveals that invasive and native plants differ in their traits but not in their plasticity. *Functional Ecology*, **25**, 1248–1259.

Gomulkiewicz R, Holt RD (1995) When does natural selection prevent extinction? *Evolution*, **49**, 201–207.

Huey RB, Kingsolver JG (1989) Evolution of thermal sensitivity of ectotherm performance. *Trends in Ecology & Evolution*, **4**, 131–135.

Kilkenny FF, Galloway LF (2012) Adaptive divergence at the margin of an invaded range. *Evolution*, **67**, 722–731.

Kirkpatrick M, Barton NH (1997) Evolution of a species' range. *The American Naturalist*, **150**, 1–23.

Lande R (1980) Genetic variation and phenotypic evolution during allopatric speciation. *The American Naturalist*, **116**, 463–479.

Lande R (1994) Risk of population extinction from fixation of new deleterious mutations. *Evolution*, **48**, 1460–1469.

Lande R (1995) Mutation and conservation. *Conservation Biology*, **9**, 782–791.

Lande R (2009) Adaptation to an extraordinary environment by evolution of phenotypic plasticity and genetic assimilation. *Journal of Evolutionary Biology*, **22**, 1435–1446.

Lande R (2014) Evolution of phenotypic plasticity and environmental tolerance of a labile quantitative character in a fluctuating environment. *Journal of Evolutionary Biology*, **27**, 866–875.

Lande R, Engen S, Sæther B-E (2003) *Stochastic Population Dynamics in Ecology and Conservation*. Oxford University Press, Oxford.

Levin DA (2010) Environment-enhanced self-fertilization: implications for niche shifts in adjacent populations. *The Journal of Ecology*, **98**, 1276–1283.

Lewontin RC (1965) Comment. In: *The Genetics of Colonizing Species* (eds Baker HG, Stebbins LG), pp. 481–484. Academic Press, New York, New York.

Miner BG, Sultan SE, Morgan SG, Padilla DK, Relyea RA (2005) Ecological consequences of phenotypic plasticity. *Trends in Ecology & Evolution*, **20**, 685–692.

Morin X, Thuiller W (2009) Comparing niche- and process-based models to reduce prediction uncertainty in species range shifts under climate change. *Ecology*, **90**, 1301–1313.

Normand S, Zimmermann NE, Schurr FM, Lischke H (2014) Demography as the basis for understanding and predicting range dynamics. *Ecography*, **37**, 1149–1154.

Palacio-López K, Gianoli E (2011) Invasive plants do not display greater phenotypic plasticity than their native or non-invasive counterparts: a meta-analysis. *Oikos*, **120**, 1393–1401.

Peterson ML, Kay KM (2014) Mating system plasticity promotes persistence and adaptation of colonizing populations of hermaphroditic angiosperms. *The American Naturalist*, **185**, 28–43.

Pfennig DW, Wund MA, Snell-Rood EC, Cruikshank T, Schlichting CD, Moczek AP (2010) Phenotypic plasticity's impacts on diversification and speciation. *Trends in Ecology & Evolution*, **25**, 459–467.

Prentis PJ, Wilson JRU, Dormontt EE, Richardson DM, Lowe AJ (2008) Adaptive evolution in invasive species. *Trends in Plant Science*, **13**, 288–294.

Price TD, Qvarnstrom A, Irwin DE (2003) The role of phenotypic plasticity in driving genetic evolution. *Proceedings of the Royal Society of London B: Biological Sciences*, **270**, 1433–1440.

Schlichting CD, Pigliucci M (1998) *Phenotypic Evolution. A Reaction Norm Perspective*. Sinauer, Sunderland, Massachusetts.

Schluter D (2000) *The Ecology of Adaptive Radiation*. Oxford University Press, Oxford.

Stam P (1983) The evolution of reproductive isolation in closely adjacent plant populations through differential flowering time. *Heredity*, **50**, 105–118.

Sultan SE (1987) Evolutionary implications of phenotypic plasticity in plants. *Evolutionary Biology*, **21**, 127–178.

Thibert-Plante X, Hendry AP (2011) The consequences of phenotypic plasticity for ecological speciation. *Journal of Evolutionary Biology*, **24**, 326–342.

Turchin P (2003) *Complex Population Dynamics. A Theoretical/Empirical Synthesis*. Princeton University Press, Princeton, New Jersey.

Turesson G (1922) The genotypical response of the plant species to the habitat. *Hereditas*, **3**, 211–350.

Van Buskirk J, Steiner UK (2009) The fitness costs of developmental canalization and plasticity. *Journal of Evolutionary Biology*, **22**, 852–860.

Vandepitte K, De Meyer T, Helsen K *et al.* (2014) Rapid genetic adaptation precedes the spread of an exotic plant species. *Molecular Ecology*, **23**, 2157–2164.

Via S, Lande R (1985) Genotype-environment interaction and the evolution of phenotypic plasticity. *Evolution*, **39**, 505–522.

Via S, Gomulkiewicz R, De Jong G, Scheiner SM, Schlichting CD, van Tienderen PH (1995) Adaptive phenotypic plasticity: consensus and controversy. *Trends in Ecology & Evolution*, **10**, 212–217.

West-Eberhard MJ (1989) Phenotypic plasticity and the origins of diversity. *Annual Review of Ecology and Systematics*, **20**, 249–278.

West-Eberhard MJ (2003) *Developmental Plasticity and Evolution.* Oxford University Press, Oxford.

Wright S (1968) *Evolution and the Genetics of Populations. Vol. 1. Genetic and Biometric Foundations.* University of Chicago Press, Chicago, Illinois.

Chapter 10

CHROMOSOME INVERSIONS, ADAPTIVE CASSETTES AND THE EVOLUTION OF SPECIES' RANGES

Mark Kirkpatrick and Brian Barrett

Department of Integrative Biology, University of Texas, Austin, TX 78712, USA

Abstract

A chromosome inversion can spread when it captures locally adapted alleles or when it is introduced into a species by hybridization with adapted alleles that were previously absent. We present a model that shows how both processes can cause a species range to expand. Introgression of an inversion that carries novel, locally adapted alleles is a particularly powerful mechanism for range expansion. The model supports the earlier proposal that introgression of an inversion triggered a large range expansion of a malaria mosquito. These results suggest a role for inversions as cassettes of genes that can accelerate adaptation by crossing species boundaries, rather than protecting genomes from introgression.

Previously published as an article in *Molecular Ecology* (2015) 24, 2046–2055, doi: 10.1111/mec.13074

INTRODUCTION

Why does not everything live everywhere? A mantra of Neo-Darwinism is that genetic variation is ubiquitous and there is heritability for virtually all phenotypes. If true, one would expect a species to adapt to the local conditions at the edge of its range and expand outward, ultimately spreading across the planet. But so far as we know, our own species is the only one in the history of the Earth to have ever done so.

Insects illustrate one explanation for range limits. This group comprises about one million described species, making up about two-thirds of all animal biodiversity (Zhang 2011). Despite this, they have not been able to colonize the world's open oceans, which make up two-thirds of its surface. The only arguable exceptions are five species in the genus *Halobates* (Cheng 1985). But these are water striders that live on the surface, and not even they are able to live an aquatic life in the marine environment. Insects originated some 479 million years ago (Misof *et al.* 2014), so it seems they have had enough time to adapt to salt water were they able. In short, an evolutionary constraint makes two-thirds of the Earth's surface a no-fly (or rather, no-swim) zone for two-thirds of all animal species.

A second explanation for range limits is that gene flow stymies adaptation at the limits of species' ranges. Haldane (1956) and Mayr (1963) suggested verbally that gene flow carries alleles from high-density populations at the core of the range out to its low-density periphery. Those alleles are adapted to ecological conditions at the core, but have low fitness in the conditions at the periphery, which suppresses population growth there and prevents range expansion. Since Haldane and Mayr, models have been developed that show that the tension between local adaptation and gene flow can indeed limit range expansion under some conditions (e.g. Kirkpatrick & Barton (1997); see the reviews of Gaston (2003), Sexton *et al.* (2009) and the Discussion).

Mosquitoes in the genus *Anopheles* present examples in which both genetic constraints and gene flow are thought to have played roles in first limiting the species' ranges and then enabling dramatic range expansions. These species are of particular interest as they include the most important vectors of malaria in the world – a disease responsible for one-sixth of childhood mortality in Africa each year (White *et al.* 2011). These mosquitoes segregate for several chromosome inversions (reviewed by Ayala *et al.* (2014)). The inversion

2La is strongly implicated in adaptation to arid habitats (Fouet *et al.* 2012). Remarkably, this inversion has been passed between two sibling species, *Anopheles gambiae* and *Anopheles arabiensis*, following a hybridization event between 3000 and 11 000 years ago (Besansky *et al.* 2003; Sharakhov *et al.* 2006; White *et al.* 2007, 2009). While the direction of introgression is still uncertain, it seems clear that this event allowed one of the species to expand its range into new habitats. This scenario suggests the species' range was limited by a genetic constraint that was broken by introduction of new genetic variation via hybridization (Ayala & Coluzzi 2005).

How did the *2La* inversion first become established in *Anopheles* mosquitoes? One hypothesis for how inversions evolve is consistent with the data on this inversion. When an inversion captures locally adapted alleles at two or more loci, it can spread because it prevents recombination from breaking them apart (Kirkpatrick & Barton (2006); see also the reviews by Hoffmann & Rieseberg (2008) and Kirkpatrick (2010)). Spread of the inversion is caused by the tension between local adaptation and gene flow. This suggests a plausible scenario for what triggered one of the most important biological invasions in human history. Inversion *2La* adapted to arid conditions, appeared and became established in either *A. arabiensis* or *A. gambiae*, perhaps through an interplay between local adaptation and gene flow. The inversion was then introduced into its sibling species by hybridization. That hybridization event enabled the recipient species to expand its ecological and geographical ranges, with devastating consequences for human health.

While examples such as *Anopheles* mosquitoes make a plausible case for the idea that inversions could be important to range expansions, there is as yet no population genetic theory showing the conditions in which that can happen or the possible outcomes. The most relevant theory that is available concerns local adaptation in populations that have fixed density. Gene swamping imposes a fundamental limit to the spatial resolution of adaptation. When a species range ends in a patch of habitat that is ecologically different, alleles that are favoured there cannot establish if the strength of selection is below a threshold set by the average dispersal distance relative to the size of the patch (Slatkin 1973). With two or more linked loci that segregate for locally adapted alleles, selection has reinforcing effects across the loci, making it easier for the alleles to establish at each locus (Yeaman & Whitlock 2011).

We would like to understand how these basic principles apply when there is feedback between local adaptation and population densities.

Here, we develop a very simple model that captures some key features of demography and evolution to learn how inversions can affect a species range. There are several questions we want to answer. When will locally favoured alleles be able to establish in the peripheral environment? When they do establish, what is the effect on population densities at the periphery? If an inversion captures locally adapted alleles that are already present, how will the species range evolve? How does that situation compare with the case in which an inversion carries novel, locally adapted alleles when it first appears?

We find that when an inversion establishes by capturing locally adapted alleles that are already present, the effect on population density and range size is often modest. But when an inversion appears carrying novel alleles, a dramatic biological invasion can result. We consider the consequences of direct selection against a new inversion, as when it reduces fertility in heterozygotes. This type of selection can easily prevent an inversion from establishing by capturing locally adapted alleles. An inversion can spread despite deleterious direct selection, however, if it carries novel alleles that are sufficiently beneficial.

The model

We consider the situation in which the periphery of a species range is ecologically distinct from its core. Alleles that are favoured in the core are locally maladapted to the peripheral habitat. Selection against those alleles can decrease population density in the periphery and limit the species range. We ask how the population responds to the introduction of alleles or an inversion that increases fitness in the peripheral habitat.

The species inhabits a spatially continuous one-dimensional habitat (Fig. 1). The species range is unbounded to the left, but has a hard limit (a reflecting boundary) at point x_{max} on the right, for example caused by a geographical barrier. The habitat limit x_{max} can be made arbitrarily large, so this model includes the case in which the habitat extends indefinitely both to the left and to the right. Time is continuous. Individuals disperse at random, and the dispersal variance is denoted as σ^2. We assume that densities are sufficiently high that random genetic drift can be neglected.

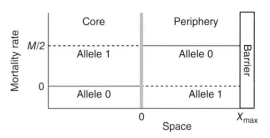

Fig. 1 Sketch of the assumptions about how selection varies in space.

The maximum intrinsic rate of increase is R. The actual rate of increase in a local population is less than R because of two factors. The first is density dependence: local populations have a carrying capacity, K. We assume the carrying capacity is constant everywhere so that the only factor constraining range expansion into empty habitat is the species' intrinsic lack of ability to adapt to a new environment. The second factor is additional mortality caused by selection against alleles that are not locally adapted. Our model therefore assumes hard selection (Wallace 1975): maladaptation of a population suppresses its growth.

Two loci, A and B, are carried on the same chromosome. The probability that they recombine is r. Alleles A_0 and B_0 are adapted to the range core, and alleles A_1 and B_1 are adapted to the periphery. For the sake of simplicity, we assume that selection acts in the haploid phase of the life cycle. The model also applies to diploids if fitness effects are multiplicative (that is, no dominance or epistasis). The mortality rate of genotype $A_i B_j$ at location x is written as $\mu_{ij}(x)$. Mating is random.

We begin by assuming that the fitness of an inversion is determined entirely by the alleles it carries at loci A and B. We return later to consider what happens when an inversion has direct fitness effects, for example when inversion heterozygotes suffer reduced fertility.

As we are interested in population dynamics, we follow the density rather than the frequency of the different types of chromosomes. We begin by considering chromosomes with the ancestral arrangement (that is, without the inversion). The density of chromosomes with genotype $A_i B_j$ at point x is denoted as $n_{ij}(x)$. The rate of change in time of that density is

$$\partial_t n_{ij}(x) = \frac{\sigma^2}{2}\partial_x^2 n_{ij}(x) + R\left[1 - \frac{n^*(x)}{K}\right]\left[n_{ij}(x) + \delta_{ij} r c(x)\right]$$
$$- \mu_{ij}(x)n_{ij}(x).$$

$$(1)$$

Here, n^* is the total density of all genotypes, $\delta_{ij} \overset{\text{def}}{=} -1$ if $i = j$, and $\delta_{ij} = 1$ if $i \neq j$. The quantity

$$c(x) \overset{\text{def}}{=} \frac{1}{2}[n_{00}(x)p_{11}(x) - n_{01}(x)p_{10}(x) - n_{10}(x)p_{01}(x) \\ + n_{11}(x)p_{00}(x)] \tag{2}$$

is related to the association (linkage disequilibrium) between alleles at the two loci; each genotype's density is weighted by the probability that mating produces recombinant genotypes. The derivation of eqn (1) is given in the Data S1 (Supporting information).

On the right side of eqn (1), the first of the three terms represents the effects of migration, which averages out differences in local densities. The second term corresponds to selection-independent birth and death. Inside the first set of square brackets is a term that represents density-dependent population regulation. Inside the second pair of square brackets, the term involving c reflects how recombination alters the densities of genotypes. The last of the three terms on the right side of (1) is the additional mortality incurred when a genotype is not adapted to the local habitat.

The dynamics of the inversion also follow eqn (1) with minor changes. We replace $n_{ij}(x)$ by $n_1(x)$, which is the density of inverted chromosomes at point x. We assume that genetic exchange between inverted and uninverted chromosomes (which can result either by double recombination events or gene conversion) is negligible, and so the term $[\delta_{ij}\, r\, c(x)]$ that appears in eqn (1) is dropped. We replace $\mu_{ij}(x)$ in eqn (1) by the mortality rate that corresponds to the alleles at loci A and B that the inversion captured.

The model simplifies substantially if we change the variables to nondimensional units. We measure densities relative to the carrying capacity so that $N_{ij} \overset{\text{def}}{=} n_{ij}/K$, rescale time by the maximum growth rate so that $T \overset{\text{def}}{=} Rt$ and rescale space in terms of dispersal and the growth rate so that $X \overset{\text{def}}{=} x\sqrt{2R}/\sigma$ (Kirkpatrick & Barton 1997). Equation (1) then becomes

$$\partial_T N_{ij} = \partial_X^2 N_{ij} + (1 - N^*)(N_{ij} + \delta_{ij}rC) - m_{ij}N_{ij}, \tag{3}$$

where N_{ij}, N^*, C and $m_{ij} \overset{\text{def}}{=} \mu_{ij}/R$ are implicitly functions of X. $C(X)$ is defined by eqn 2, but with N_{ij} replacing n_{ij} and X replacing x. This change of variables reduces the number of parameters by eliminating of σ^2, R and K. We can make a further simplification by assuming

that loci A and B have identical effects. That allows us further to assume that A_0B_1 and A_1B_0 chromosomes have equal densities at equilibrium, reducing the number of dynamic variables from five to four: N_{00}, N_{01}, N_{11} and N_1.

We now make specific assumptions about how selection varies in space (see Fig. 1). The species range crosses an ecotone that lies at $X = 0$. Alleles A_0 and B_0 are favoured to the left of this point, and alleles A_1 and B_1 are favoured to the right. X_{max} measures the size of the peripheral habitat to which the 1 allele is adapted.

Each allele carried by an individual that is not locally adapted increases its mortality rate by $M/2$. We assume that there is no epistasis, so the fitness effects of alleles are additive. The mortality rate of chromosomes with genotype $A_i B_j$ is then

$$m_{00}(X) = \begin{pmatrix} 0 \text{ if } X < 0 \\ M \text{ if } X \geq 0, \end{pmatrix}$$

$$m_{11}(X) = \begin{pmatrix} M \text{ if } X < 0 \\ 0 \text{ if } X \geq 0, \end{pmatrix} \tag{4}$$

$$m_{01}(X) = m_{10}(X) = M/2.$$

We will focus on cases where the inversion carries the alleles adapted to the peripheral habitat, so we assume that the inversion's mortality rate is $m_1(X) = m_{11}(X)$.

This final model has just three parameters: M, r and X_{max}. We analysed the model numerically by integrating the differential equations using *Mathematica* (Wolfram Research 2014). While it is possible to obtain analytic results, we hope that the qualitative patterns described below apply generally.

RESULTS

We begin by considering how establishment of alleles adapted to the periphery can affect population densities. Next, we study the effects of an inversion that spreads in the periphery because it captures those locally adapted alleles. We then contrast those results with the scenario in which an inversion first appears carrying locally adapted alleles that are not already present, as can happen by hybridization. Last, we extend the model to include cases in which the inversion itself has direct effects on fitness, as when inversion heterozygotes have reduced fertility.

CHAPTER 1

Fig. 1 Herbert G. Baker and G. Ledyard Stebbins in the field, Napa County, California 1973, on an excursion organized by the Bay Area Biosystematists.

Invasion Genetics: The Baker and Stebbins Legacy, First Edition. Edited by Spencer C. H. Barrett, Robert I. Colautti, Katrina M. Dlugosch, and Loren H. Rieseberg.
© 2017 John Wiley & Sons, Ltd. Published 2017 by John Wiley & Sons, Ltd.

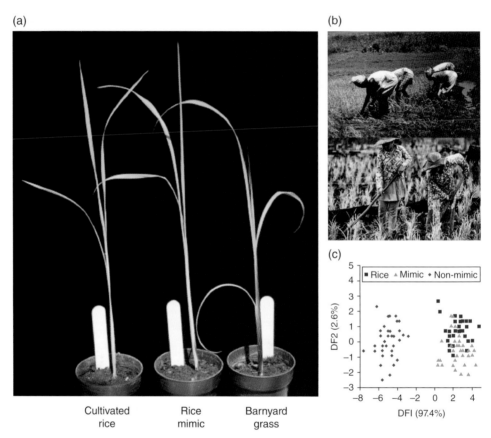

(a)

Cultivated rice   Rice mimic   Barnyard grass

(b)

(c)

Fig. 2 Generalist and specialist weeds in the barnyard grass complex: (a) from left to right—cultivated rice, the specialist rice mimic *Echinochloa phyllopogon*, and the generalist *Echinochloa crus-galli*; (b) weeding practices in rice exert selection pressures on the morphology of weed populations favouring variants of barnyard grass that resemble rice; (c) phenotypic resemblance between the generalist, the rice mimic and rice based on a discriminant functions analysis of nine quantitative characters. For further details, see Barrett (1983).

Fig. 3 Two invasive weeds identified by Baker (1965) as possessing general-purpose genotypes. (a) The sterile pentaploid short-styled morph of tristylous *Oxalis pes-caprae*, Tel Aviv, Israel (2013); (b) The clonal aquatic *Eichhornia crassipes* at Bacon Island Slough near Stockton, California (2014). The population of *E. crassipes* is composed of a single clone of the mid-styled morph and has persisted at this site for 40 years and during this time has dramatically increased in size as a result of clonal growth. Sexual reproduction is prevented at the site despite seed production because of unsuitable conditions for seed germination and seedling establishment (see Barrett 1980).

CHAPTER 3

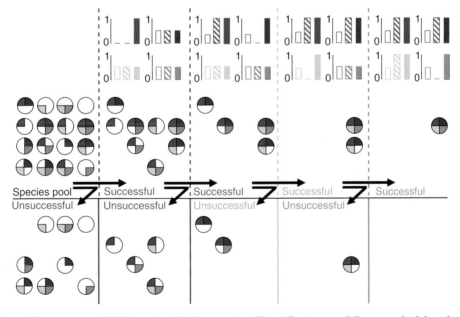

Fig. 2 Different characteristics might determine whether a species will or will not successfully pass each of the subsequent filters (questions), and consequently, the choice of the optimal comparator group for the successful species. In this example, each species (circle) can have four characteristics indicated by the four different colours, and the state of these characteristics determines whether a species can pass the filters of the same colours. For example, only species that possess the blue colour can pass the blue filter. The bar diagrams indicate the proportion of species that possess the particular colour; solid bars: species that successfully passed the respective filter, hatched bars: species that did not successfully pass the respective filter but successfully passed the preceding one(s), open bars: all the species from the initial species pool that did not pass the respective filter and also not the preceding ones. The difference in the proportion of species with each colour among successful species and unsuccessful species depends on the filter at which the colour is important and whether one also includes the unsuccessful species that failed already at one of the preceding filters.

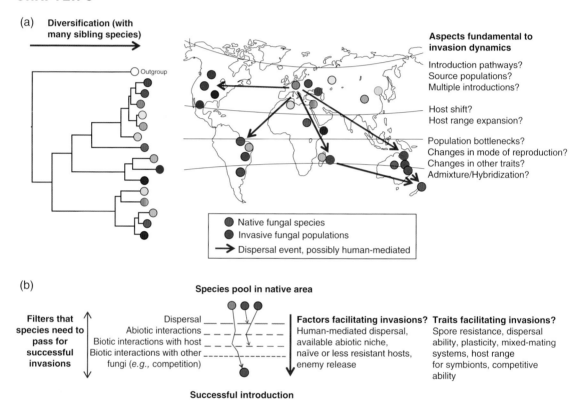

(a) Diversification (with many sibling species)

Outgroup

Aspects fundamental to invasion dynamics

Introduction pathways?
Source populations?
Multiple introductions?

Host shift?
Host range expansion?

Population bottlenecks?
Changes in mode of reproduction?
Changes in other traits?
Admixture/Hybridization?

● Native fungal species
● Invasive fungal populations
→ Dispersal event, possibly human-mediated

(b)

Species pool in native area

Filters that species need to pass for successful invasions

Dispersal
Abiotic interactions
Biotic interactions with host
Biotic interactions with other fungi (*e.g.*, competition)

Factors facilitating invasions?
Human-mediated dispersal, available abiotic niche, naïve or less resistant hosts, enemy release

Traits facilitating invasions?
Spore resistance, dispersal ability, plasticity, mixed-mating systems, host range for symbionts, competitive ability

Successful introduction

Fig. 1 An invasion scenario typical for fungi and essential questions to illuminate the process. (a) The figure shows the phylogenetic diversity and geographic distribution of a fungal group from which a species has invaded other regions and lists the questions fundamental to invasion dynamics: recognition of local species diversity in the native range, with the delimitation of sibling species, identification of the traits facilitating invasion, source populations, introduction pathways, occurrence of bottlenecks or of multiple introductions, possible evolutionary or ecological changes, such as host shift, changes in mode of reproduction or in other traits. The red circles represent introduced populations while the nonred circles represent native fungal species. (b) Successive filters that species must pass for invasions to be successful (dispersal, abiotic and biotic interactions) and possible mechanisms facilitating invasions (Human-mediated dispersal, available abiotic and biotic niche, enemy release, spore resistance, dispersal ability, plasticity, mixed mating systems, host range for symbionts or competitive ability).

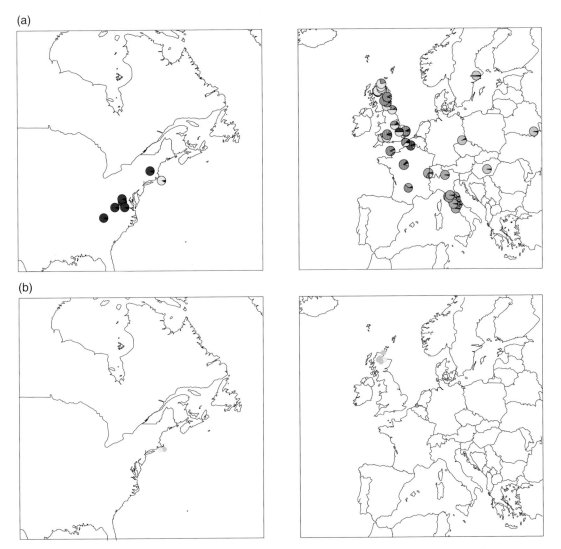

Fig. 3 Assignment of North American fungal *Microbotryum lychnidis-dioicae* samples, collected on *Silene latifolia*, to European populations. (a) Map of the samples collected in the United States and Europe (*N* = 328 individuals), with colours representing the mean membership proportions inferred by STRUCTURE for *K* = 6; higher *K* did not allow assigning more precisely American isolates to European populations, see Fig. S1. (b) Map of the samples assigned to the yellow cluster in (a) and with identical genotypes in the United States and in Scotland; a specific colour (yellow or brown) has been given for each of the two multilocus genotypes of the yellow cluster in the USA. Colours represent the inferred ancestry from *K* ancestral populations. The percentages indicated are the proportions of runs that found the main solution shown here.

CHAPTER 6

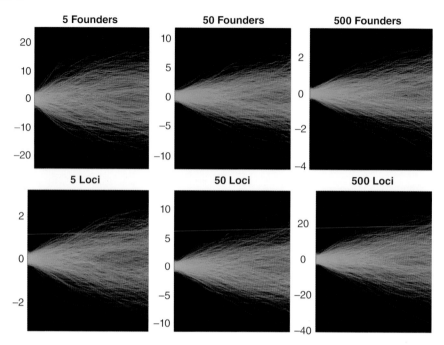

Box 2, Fig. 1 Rapid evolution of nonadaptive geographical clines in simulated invasions. The change in a population's mean phenotype for a quantitative trait (*y*-axis) in 20 populations distributed along a spatial gradient (*x*-axis) is shown for each of 1000 replicates in each panel. Highly significant clines ($P < 0.001$), based on linear regression, are shown in yellow (~50% of simulations in each panel), and simulations with non-significant clines shown are in blue.

CHAPTER 8

Box 1, Fig. 1 *Acmispon wrangelianus* growing in uninvaded sites (a) are much larger, much more fecund (produce over 300% more seeds), and receive much less herbivore damage compared to *A. wrangelianus* individuals in sites invaded by *Medicago polymorpha* (b).

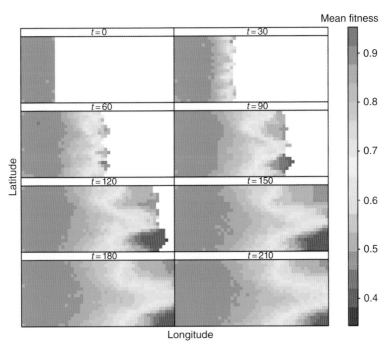

Fig. 3 Evolution of mean fitness during a range expansion. The simulated grid is 20 × 50 demes. Mutations are recessive and parameter values are as in Fig. 1.

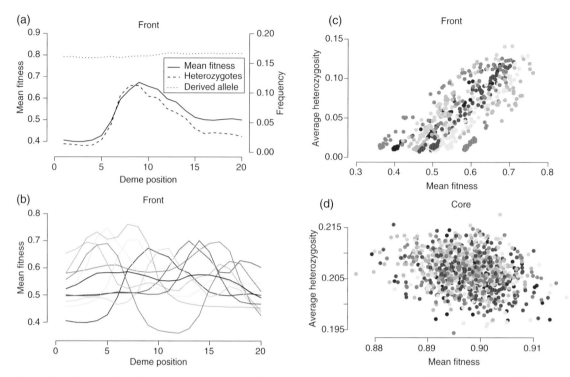

Fig. 4 Genetic properties of demes at the wave front and heterozygosity–fitness correlations (HFC) across demes in different parts of the species range. (a) Example of the mean fitness, the heterozygosity and the derived allele frequency at different latitudinal positions at the expansion front when the habitat has just been fully colonized ($t = 150$, simulation shown in Fig. 3). The deme mean fitness on the wave front correlates with heterozygosity, but not with derived allele frequency. (b) Mean fitness at the front of the expansion from 10 distinct simulation runs (coloured lines) and the average over all simulation runs (solid black line). (c) HFC on the expansion front at generation $t = 150$. (d) No significant HFC in core populations before the onset of the expansion ($t = 0$). In (c) and (d), each point represents the mean fitness and average heterozygosity of a single deme. Different colours in panels b–d correspond to 10 distinct simulation replicates. Parameter values are as in Fig. 3.

CHAPTER 18

Fig. 1 Native to forested areas in Central and South America, cane toads have colonized harsh arid landscapes within Australia. The first panel shows an artificial water body in the semi-desert habitat near Longreach in central Queensland, and the second panel shows a high density of adult toads around such a water body at night. Photographs by M. Greenlees.

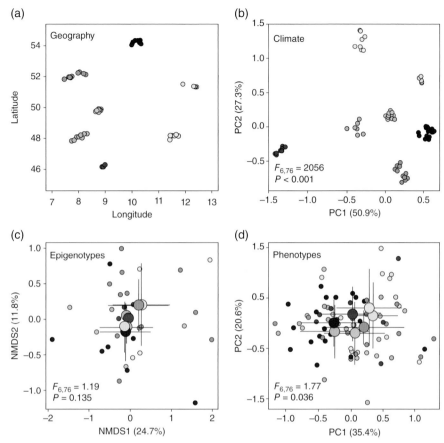

Fig. 1 Geographic (a), climatic (b), epigenetic (c), and phenotypic (d) distances among Japanese knotweed clones. Small symbols represent 83 different Central European origins of invasive Japanese knotweed (*Fallopia japonica*), with different colors for each of the seven geographic regions (see Table 1). The large symbols in the last two panels represent the regional means (±SD). Climate and phenotype plots are based on principal component analysis (PCA) of nine bioclimatic variables and seven phenotypic traits, respectively, whereas the epigenetic plot represents a non-metric multidimensional scaling (NMDS) analysis of 19 polymorphic methylation-sensitive amplified polymorphism (MSAP) markers. The test statistics are the results of multivariate tests for regional differentiation (see section "Materials and Methods" for details).

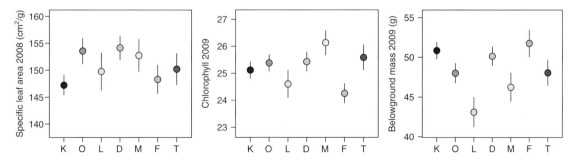

Fig. 2 Regional variation in phenotype among invasive Japanese knotweed clones. The data (means ± SD) are from common-garden progeny of 83 Central European origins of *Fallopia japonica* distributed across seven geographic regions: K = Kiel, O = Osnabrück, L = Leipzig, D = Darmstadt, M = München, F = Freiburg, and T = Ticino. The colors are as in Fig. 1.

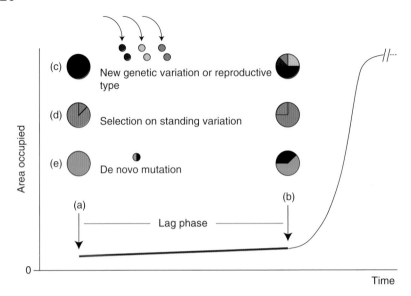

Box 1, Fig. 1 Example of the lag phase and potential genetic causes, showing hypothetical patterns of genetic variation at a single locus at (a) initial establishment and before (b) the onset of accelerated expansion. (c) shows an increase in genetic variation following immigration of new genotypes or new sexes. (d) illustrates selection on standing variation. (e) represents the origin of a *de novo* adaptive mutation.

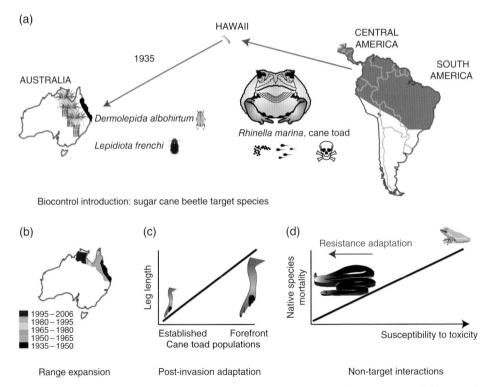

Box 3, Fig. 1 The cane toad (*Rhinella marina*) was introduced (a) to control cane beetles in sugar cane fields in northeastern Queensland in 1935 after successful use in Hawaii. Since then, cane toads have (b) expanded across tropical and subtropical Australia (Urban *et al.* 2008), increasing their rate of spread through, among others (Phillips & Shine 2006; Shine 2010), (c) evolution of longer legs (Phillips *et al.* 2006). Cane toads eat a wide variety of nontarget invertebrates reducing their population sizes. (d) Most native predators have declined as well due to lethal toxic ingestion of the toads, tadpoles and/or eggs with one known exception: the Australian black snake that has evolved physiological resistance to cane toad toxins (Phillips & Shine 2006; Shine 2010).

Establishing locally adapted alleles at the periphery

In our model, if alleles adapted to the peripheral habitat are able to establish, they cause population densities there to increase. Figure 2 shows that the outcome depends strongly on the size of the peripheral habitat. If it is large, then before the locally adapted alleles arrive, the population density will decline to very low densities at large distances from the ecotone boundary (Fig. 2, top panel). Introducing locally adapted alleles in this case can trigger a dramatic range expansion. When the peripheral region to the right of the ecotone is smaller, before the locally adapted alleles appear, spillover from the core population maintains moderate densities in the periphery even though it is a demographic sink (Fig. 2, bottom left panel). If alleles that are adapted to the periphery establish in this case, the density increase is more modest. The last case is when the peripheral habitat is smaller than a threshold size needed for the locally adapted alleles to establish (Fig. 2, bottom right panel). In our model, the minimum size of

the peripheral habitat required to establish locally adapted alleles decreases with increasing selection against maladapted alleles (M) and decreasing recombination rates (r).

The inversion captures locally adapted alleles

Imagine that local selection is sufficiently strong, the peripheral habitat sufficiently large, and the recombination rate sufficiently low that locally adapted alleles are able to establish. We now ask what happens if an inversion arises in the peripheral habitat and captures those alleles. For brevity, we refer to this situation as the 'capture' scenario.

An example of an inversion that establishes by the capture scenario is shown in Fig. 3. The inversion drives to extinction the uninverted chromosomes that carry alleles adapted to the periphery (the A_0B_1, A_1B_0 and A_1B_1 genotypes). At equilibrium, all that remains are the uninverted chromosomes with the A_0B_0

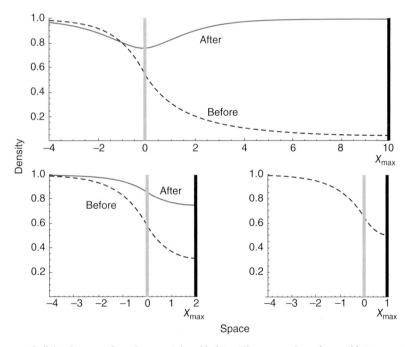

Fig. 2 Establishment of alleles that are adapted to a peripheral habitat. The curves show the equilibrium population densities before and after the alleles A_1 and B_1 establish. The three panels differ in the size of the peripheral region ($X_{max} = 10, 2, 1$). With the smallest peripheral region (bottom right), alleles A_1 and B_1 are swamped by gene flow from the core habitat and cannot establish. Other parameter values are $M = 1$ and $r = 0.1$.

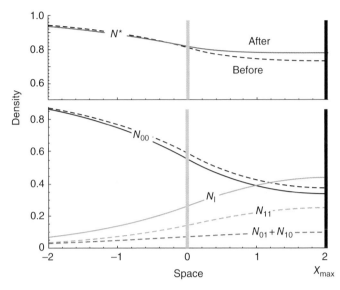

Fig. 3 The inversion spreads by capturing locally adapted alleles that are already established in the periphery. Dashed curves show densities before the inversion appears; solid curves show the equilibrium with the inversion. *Upper panel*: Total population densities. *Lower panel*: Densities of individual genotypes. Parameter values are $M = 0.5$, $r = 0.1$, $X_{max} = 2$.

genotype at the core of the range and the inversion with A_1B_1 genotype at the periphery. Mixing of the two genotypes happens around the ecotone boundary at $X = 0$.

While the inversion causes large changes in genotype frequencies, it has relatively little effect on the total population density. Examples are shown in Fig. 4. The total densities of all four genotypes before the inversion appears are shown by the curves labelled N^*_{all}, and the total densities after the inversion establishes are labelled N^*_I. The difference between the curves is small and decreases as the size of the peripheral habitat grows.

We note that quite different outcomes are possible under alternative assumptions about selection (eqn (4)). In particular, we have found cases in which an inversion that establishes by the capture scenario can cause large increases in density and range size.

Adaptive cassettes: the inversion carries new adaptive alleles

Recall that our model was inspired by the history of *Anopheles* mosquitoes, in which the introgression of a chromosomal inversion between sibling species may

have triggered a major range expansion. We now ask how the model behaves when an inversion that is introduced to a peripheral population carries locally adapted alleles that did not previously occur in the species.

These results are easy to anticipate in the light of the earlier discussion. The inversion is a nonrecombining unit, and so it evolves exactly as a single allele that has the fitness of both allele A_1 and B_1 combined. If that advantage is sufficiently strong to avoid swamping from the A_0B_0 genotype, then the inversion establishes and can cause a large range expansion. The inversion enjoys two benefits: it introduces the locally adapted alleles, and it suppresses recombination with the ancestral alleles.

Figure 4 shows examples. The only genotype present before the inversion appears is A_0B_0, and its densities are shown by the curves labelled N^*_{00}. After the inversion establishes, the densities reach the same equilibrium as in the capture scenario (the curves labelled N^*_I). The increases in density can be very large and are particularly dramatic when the peripheral habitat is big.

This scenario corresponds to the history hypothesized for inversion *2La* in *Anopheles*. Following its introduction to a species by hybridization, the inversion introduced new alleles for aridity tolerance that

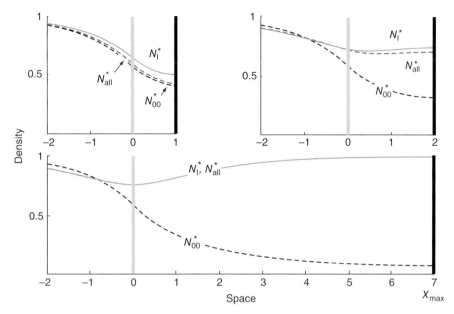

Fig. 4 Comparison of density changes when an inversion establishes via the capture vs the adaptive cassette scenarios. The curves labelled N_{all}^* show the equilibrium total densities before the inversion appears in the capture scenario (all genotypes are already present). The curves labelled N_{00}^* show the total densities before the inversion appears in the adaptive cassette scenario (only the A_0B_0 genotype is present). The curves labelled N_I^* are the equilibrium densities after the inversion establishes under both scenarios. In the bottom panel, the curves for N_I^* and N_{all}^* are indistinguishable. Parameter values are $M = 0.5$ and $r = 0.1$.

allowed that mosquito to vastly expand its range. We can think of an inversion in this situation as an 'adaptive cassette' of genes: it replaces a piece of chromosome with an alternate form that might carry several, or even many, alleles adapted to a new environment.

Because the inversion in our model carries two locally adapted alleles, it has greater immunity to swamping than those alleles do in the ancestral population. That means there are situations in which locally adapted alleles will not spread when introduced by mutation, but can invade when they are carried by an inversion. An example of parameters that lead to this situation is $M = 3$, $r = 0.25$, $X_{max} = 2$.

Direct fitness effects of an inversion

To this point, we have assumed that the only effects that an inversion has on fitness result from the alleles it carries at loci A and B. Inversions can, however, have direct fitness effects in several ways. The break points where the chromosome is inverted can disrupt a gene or alter gene expression. Alternatively, inversion

heterozygotes can suffer reduced fertility because recombination inside the inverted region generates aneuploid gametes. While some taxa of animals have mechanisms that partly or entirely suppress this cost, inversion heterozygotes have reduce fitness in other animals and most plants (Lande 1979, 1985). Because our model assumes there is no dominance, it is not suited to a full analysis of this situation.

We can, however, modify our model to understand when underdominant fitness effects will prevent inversions from spreading when they first appear. To do that, we simply add a term to the equation for the dynamics of the inversion that reflects direct selection on the inversion (that is, beyond the fitness effects of the alleles it carries):

$$\partial_T N_I = \partial_X^2 N_I + (1 - N^*)N_I - m_{11}N_I - s_I N_I \qquad (5)$$

(where N_I, N^* and m_{11} are functions of X and T). Here, s_I is the additional fitness cost that the inversion incurs from direct selection. This can represent simple forms of direct selection (e.g. deleterious effects of the breakpoints). It can approximate the effects of selection against heterozygotes when the inversion first

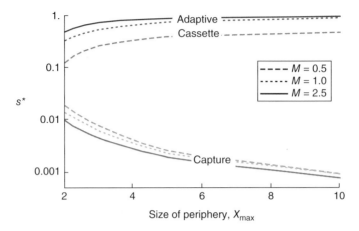

Fig. 5 The minimum direct fitness cost of an inversion that prevents it from establishing (s^*) as a function of the genic mortality rate (M) and the size of the peripheral habitat (X_{max}) under the adaptive cassette and the capture scenarios. The recombination rate is $r = 0.1$. Note the Y-axis is logarithmic.

arises: although the model assumes selection in the haploid phase, its dynamics are very similar to selection in diploids when the inversion is rare. That allows us to study when direct selection against the inversion will prevent it from spreading when rare. (Note that our model is in continuous time, and so s_i can take values <1, unlike a selection coefficient in discrete time.)

We used this modified model to determine numerically the minimum direct fitness cost that will prevent an inversion from establishing, which we denote as s^*. There are again two cases to consider: the capture scenario and the adaptive cassette scenario.

Figure 5 shows results for how s^* depends on the allelic mortality rate (M) and the size of the peripheral habitat (X_{max}). Even weak direct selection against the inversion will generally prevent it from establishing under the capture scenario. The size of s^* can be two or more orders of magnitude smaller than the strength of selection acting on the locally adapted alleles. That result can be understood intuitively. The inversion invades because its fitness is higher than the mean fitness of the resident population, which in turn results from the presence of chromosomes that have either one or two alleles that are locally maladapted. But those genotypes are typically rare. Consequently, the fitness advantage that drives the inversion to invade is typically weak.

There are two situations in which an underdominant inversion might establish by the capture scenario. One is when the inbreeding rate is high. Then, the inversion will occur frequently as a homozygote and so evade the fitness cost suffered by heterozygotes. A second situation is when drift is strong relative to selection against heterozygotes. In that case, the inversion can drift to a frequency high enough that selection can cause it to increase further (Charlesworth *et al.* 1987).

The adaptive cassette scenario leads to quite different results (Fig. 5). If the peripheral habitat is sufficiently large that the ancestral $A_0 B_0$ is essentially absent far from the ecotone boundary, the inversion will invade so long as the deleterious effects of direct selection do not depress its growth rate below 0 (which happens in our model when $s_i > 1$). In sum, an inversion can establish under much broader conditions under the adaptive cassette scenario than under the capture scenario.

DISCUSSION

When an inversion arises in a peripheral population, it can capture locally adapted alleles at two or more loci that are already present (the 'capture' scenario). The inversion can then spread because it suppresses recombination between the loci, binding together the favourable alleles. In some cases, the inversion can avoid swamping and become established even when the alleles that contribute towards its fitness advantage are unable to avoid swamping individually. We find, however, that spread of the inversion in these situations typically leads to only modest increases in the density of peripheral populations and expansion of the range.

An inversion can have much greater impact when it arrives carrying locally adapted alleles that were not previously found in the species. This seems most likely to occur via hybridization, as apparently happened in *Anopheles*. Here, the inversion acts as an 'adaptive cassette' that swamps out several (or perhaps many) alleles that have low fitness in the peripheral habitat for adapted alleles. This basic idea has been long appreciated in microbial genetics, where there are many examples of horizontal gene transfer that have enabled large expansions of species' niches (Ochman *et al.* 2000; Darmon & Leach 2014). If it introduces new adaptive genetic variation, an inversion can trigger a large range expansion.

An inversion introduced by hybridization may have had a substantial prior evolutionary history in the donor species. That history could provide opportunities for it to accumulate additional mutations on its genetic background, perhaps amplifying benefits of local adaptation with positive epistatic effects.

In some animals and in many plants, inversion heterozygotes suffer reduced fertility because recombination inside the inversion generates aneuploid gametes (reviewed in Lande (1979, 1985)). Under the adaptive cassette scenario, the inversion can establish if the novel alleles it carries have a sufficiently large advantage in the peripheral habitat. Under the capture scenario, however, even weak direct selection against the inversion can prevent it from establishing. Two caveats apply to that conclusion: strong inbreeding and strong drift might allow an underdominant inversion to establish by the capture mechanism. Both factors could be important in the establishment of inversions and translocations in some taxa (such as plants).

Our model was built to be simple, and we sacrificed both realism and generality in hopes of capturing simple conclusions about the role that inversions might have for mediating range evolution. In many situations, however, the outcomes might be mitigated by factors that the model does not consider. For example, we assume that there is vacant habitat with the same carrying capacity as the species has in the range core. That would not be the case if a competitor (for example, the species from which a new inversion is introduced by hybridization) occupies the habitat beyond the current edge of the range.

Our findings build on earlier theory that shows how three factors can limit a species' range. First, ranges

can be constrained by lack of genetic variation to more extreme conditions. Genetic drift can play an important role by eliminating adaptive genetic variation in small peripheral populations (Alleaume-Benharira *et al.* 2006; Bridle *et al.* 2010). Theory suggests that when range expansion is triggered by new mutations, they may typically have intermediate phenotypic effects and originate in the range periphery (Turner & Wong 2010; Behrman & Kirkpatrick 2011; Kirkpatrick & Peischl 2013). Second, range expansion can be prevented by the flow into peripheral populations of genes that are not adapted to the local ecological conditions (Kirkpatrick & Barton 1997; Gomulkiewicz *et al.* 1999; Ronce & Kirkpatrick 2001; Garcia-Ramos & Rodriguez 2002). That effect can be magnified by biotic interactions including competition, predation, parasitism and hybridization (Case & Taper 2000; Nuismer & Kirkpatrick 2003; Goldberg & Lande 2006, 2007; Holt *et al.* 2011). The outcome is sensitive to details about the underlying genetics (Barton 2001) and demography (Filin *et al.* 2008). A third way in which range expansion can be thwarted is when the environment changes in time as well as space. Species can lag behind a shifting environmental gradient, and the resulting maladaptation can limit range size (Pease *et al.* 1989; Duputie *et al.* 2012). This effect can be exacerbated by drift in small populations (Atkins & Travis 2010; Bourne *et al.* 2014). Environmental change also makes it more difficult for new beneficial mutations to establish (Kirkpatrick & Peischl 2013).

An interesting perspective is that our model includes roles for both the genetic constraint and the gene flow mechanisms of range limits. When an inversion triggers a range expansion by introducing new alleles to a species, it is breaking a genetic constraint. An inversion can also cause density to increase in peripheral populations by capturing locally adapted alleles that are already present. In this case, the spread of the inversion is driven by its fitness advantage over maladapted genotypes that are present because of gene flow.

Do locally adapted inversions often play a role in range expansions? There is strong evidence for local adaption of some inversions, notably in Diptera (Hoffmann & Rieseberg 2008). The most direct evidence comes from estimates of how the fitness of an inversion in *Anopheles funestus* varies spatially (Ayala *et al.* 2012). Those studies, however, do not tell us whether the inversions contributed to range expansions. More direct evidence on that point comes from other systems. In threespine stickleback (*Gasterosteus aculeatus*), three inversions have evolved to high frequency in many independent invasions of freshwater from marine populations (Jones *et al.* 2012). In the yellow monkey-flower (*Mimulus guttatus*), an inversion that has major effects on growth and life history has been repeatedly involved in the colonization of a novel habitat (Lowry & Willis 2010). In the butterfly *Heliconius numata*, an inversion controls a polymorphism in wing colour that allows the species to mimic several unrelated species of toxic model butterflies (Joron *et al.* 2011). *H. numata* is widespread in the Amazon Basin, and it is plausible that the inversion allowed it to expand into new regions where it mimics a new model. In the red imported fire ant (*Solenopsis invicta*), a large inversion that alters behaviour, morphology and life history has been a key to its invasion of the southern United States (Tsutsui & Suarez 2003; Wang *et al.* 2013). In Drummond's rockcress (*Boechera stricta*), a new inversion that carries at least two locally adapted loci appears to have spread recently as the species colonized new habitat (C.-R. Lee and T. Mitchell-Olds, personal communication).

Some of these examples share another interesting feature. We noted earlier that a major range expansion of an *Anopheles* mosquito seems to have been enabled by inversions that introgressed by hybridization. Recent unpublished results suggest that the locally adapted inversions in *Heliconius numata* may also have appeared by hybridization (M. Joron, personal communication). The history of the locally adapted inversion in *Mimulus guttatus* is ambiguous: it seems to be present in congeners, but it is unclear whether that is the result of introgression or an ancestral polymorphism that predates divergence of those species (Oneal *et al.* 2014).

If confirmed, these systems would highlight the potential for inversions to play the role of adaptive cassettes. This is an example of a more general but underappreciated point that hybridization can provide alleles that have already survived the test of natural selection (Lewontin & Birch 1966). In closing, we note that the adaptive cassette scenario inverts the evolutionary role of inversions relative to the more common view in which they protect the genome from introgression (Noor *et al.* 2001; Rieseberg 2001; Navarro & Barton 2003; Fishman *et al.* 2013; Nosil & Feder 2013). Given the genetic diversity of inversion systems and the ecological diversity of the species in which they are found, it seems likely that inversions play both of these roles in different evolutionary contexts.

ACKNOWLEDGEMENTS

We are grateful to D. Ayala, M. Joron, C.-R. Lee, D. Lowry and T. Mitchell-Olds for discussions. We thank D. Ayala, C. Cheng, D. Hooper, D. Humphreys, L. Redding, L. Rieseberg and three anonymous reviewers for comments on the manuscript. This work was supported by NSF grant DEB-0819901.

REFERENCES

Alleaume-Benharira M, Pen IR, Ronce O (2006) Geographical patterns of adaptation within a species' range: interactions between drift and gene flow. *Journal of Evolutionary Biology*, **19**, 203–215.

Atkins KE, Travis JMJ (2010) Local adaptation and the evolution of species' ranges under climate change. *Journal of Theoretical Biology*, **266**, 449–457.

Ayala FJ, Coluzzi M (2005) Chromosome speciation: humans, Drosophila, and mosquitoes. *Proceedings of the National Academy of Sciences of the United States of America*, **102**, 6535–6542.

Ayala D, Guerrero RF, Kirkpatrick M (2012) Reproductive isolation and local adaptation quantified for a chromosome inversion in a malaria mosquito. *Evolution*, **67**, 946–958.

Ayala D, Ullastres A, González J (2014) Adaptation through chromosomal inversions in Anopheles. *Frontiers in Genetics*, **5**, 129.

Barton NH (2001) Adaptation at the edge of a species' range. In: *Integrating Ecology and Evolution in a Spatial Context* (eds Silvertown J, Antonovics J), pp. 365–392. Blackwell, Oxford.

Behrman KD, Kirkpatrick M (2011) Species range expansion by beneficial mutations. *Journal of Evolutionary Biology*, **24**, 665–675.

Besansky NJ, Krzywinski J, Lehmann T *et al.* (2003) Semipermeable species boundaries between *Anopheles gambiae* and *Anopheles arabiensis*: evidence from multilocus DNA sequence variation. *Proceedings of the National Academy of Sciences of the United States of America*, **100**, 10818–10823.

Bourne EC, Bocedi G, Travis JMJ *et al.* (2014) Between migration load and evolutionary rescue: dispersal, adaptation and the response of spatially structured populations to environmental change. *Proceedings of the Royal Society B-Biological Sciences*, **281**, 20132795.

Bridle JR, Polechova J, Kawata M, Butlin RK (2010) Why is adaptation prevented at ecological margins? New insights from individual-based simulations. *Ecology Letters*, **13**, 485–494.

Case TJ, Taper ML (2000) Interspecific competition, environmental gradients, gene flow, and the coevolution of species' borders. *American Naturalist*, **155**, 583–605.

Charlesworth B, Coyne JA, Barton NH (1987) The relative rates of evolution of sex-chromosomes and autosomes. *American Naturalist*, **130**, 113–146.

Cheng L (1985) Biology of Halobates (Heteroptera: Gerridae). *Annual Review of Entomology*, **30**, 111–135.

Darmon E, Leach DRF (2014) Bacterial genome instability. *Microbiology and Molecular Biology Reviews*, **78**, 1–39.

Duputie A, Massol F, Chuine I, Kirkpatrick M, Ronce O (2012) How do genetic correlations affect species range shifts in a changing environment? *Ecology Letters*, **15**, 251–259.

Filin I, Holt RD, Barfield M (2008) The relation of density regulation to habitat specialization, evolution of a species' range, and the dynamics of biological invasions. *American Naturalist*, **172**, 233–247.

Fishman L, Stathos A, Beardsley PM, Williams CF, Hill JP (2013) Chromosomal rearrangements and the genetics of reproductive barriers in Mimulus (monkeyflowers). *Evolution*, **67**, 2547–2560.

Fouet C, Gray E, Besansky NJ, Costantini C (2012) Adaptation to aridity in the malaria mosquito *Anopheles gambiae*: chromosomal inversion polymorphism and body size influence resistance to desiccation. *PLoS ONE*, **7**, e34841.

Garcia-Ramos G, Rodriguez D (2002) Evolutionary speed of species invasions. *Evolution*, **56**, 661–668.

Gaston KJ (2003) *The Structure and Dynamics of Geographic Ranges*. Oxford University Press, Oxford.

Goldberg EE, Lande R (2006) Ecological and reproductive character displacement on an environmental gradient. *Evolution*, **60**, 1344–1357.

Goldberg EE, Lande R (2007) Species' borders and dispersal barriers. *American Naturalist*, **170**, 297–304.

Gomulkiewicz R, Holt RD, Barfield M (1999) The effects of density dependence and immigration on local adaptation and niche evolution in a black-hole sink environment. *Theoretical Population Biology*, **55**, 283–296.

Haldane JBS (1956) The relation between density regulation and natural selection. *Proceedings of the Royal Society Series B-Biological Sciences*, **145**, 306–308.

Hoffmann AA, Rieseberg LH (2008) Revisiting the impact of inversions in evolution: from population genetic markers to drivers of adaptive shifts and speciation? *Annual Review of Ecology Evolution and Systematics*, **39**, 21–42.

Holt RD, Barfield M, Filin I, Forde S (2011) Predation and the evolutionary dynamics of species ranges. *American Naturalist*, **178**, 488–500.

Jones FC, Grabherr MG, Chan YF et al. (2012) The genomic basis of adaptive evolution in threespine sticklebacks. *Nature*, **484**, 55–61.

Joron M, Frezal L, Jones RT et al. (2011) Chromosomal rearrangements maintain a polymorphic supergene controlling butterfly mimicry. *Nature*, **477**, 203–U102.

Kirkpatrick M (2010) How and why chromosome inversions evolve. *Plos Biology*, **8**, e1000501.

Kirkpatrick M, Barton NH (1997) Evolution of a species' range. *American Naturalist*, **150**, 1–23.

Kirkpatrick M, Barton N (2006) Chromosome inversions, local adaptation, and speciation. *Genetics*, **173**, 419–434.

Kirkpatrick M, Peischl S (2013) Evolutionary rescue by beneficial mutations in environments that change in space and time. *Philosophical Transactions of The Royal Society B-Biological Sciences*, **368**, 20120082.

Lande R (1979) Effective deme sizes during long-term evolution estimated from rates of chromosomal rearrangement. *Evolution*, **33**, 234–251.

Lande R (1985) The fixation of chromosomal rearrangements in a subdivided population with local extinction and colonization. *Heredity*, **54**, 323–332.

Lewontin RC, Birch LC (1966) Hybridization as a source of variation for adaptation to new environments. *Evolution*, **20**, 315–336.

Lowry DB, Willis JH (2010) A widespread chromosomal inversion polymorphism contributes to a major life-history transition, local adaptation, and reproductive isolation. *Plos Biology*, **8**, e1000500.

Mayr E (1963) *Animal Species and Evolution*. Harvard University Press, Cambridge.

Misof B, Liu S, Meusemann K et al. (2014) Phylogenomics resolves the timing and pattern of insect evolution. *Science*, **346**, 763–767.

Navarro A, Barton NH (2003) Accumulating postzygotic isolation genes in parapatry: a new twist on chromosomal speciation. *Evolution*, **57**, 447–459.

Noor MAF, Grams KL, Bertucci LA, Reiland J (2001) Chromosomal inversions and the reproductive isolation of species. *Proceedings of the National Academy of Sciences of the United States of America*, **98**, 12084–12088.

Nosil P, Feder JL (2013) Genome evolution and speciation: toward quantitative descriptions of pattern and process. *Evolution*, **67**, 2461–2467.

Nuismer SL, Kirkpatrick M (2003) Gene flow and the coevolution of parasite range. *Evolution*, **57**, 746–754.

Ochman H, Lawrence JG, Groisman EA (2000) Lateral gene transfer and the nature of bacterial innovation. *Nature*, **405**, 299–304.

Oneal E, Lowry DB, Wright KM, Zhu ZR, Willis JH (2014) Divergent population structure and climate associations of a chromosomal inversion polymorphism across the *Mimulus guttatus* species complex. *Molecular Ecology*, **23**, 2844–2860.

Pease CP, Lande R, Bull JJ (1989) A model of population growth, dispersal and evolution in a changing environment. *Ecology*, **70**, 1657–1664.

Rieseberg LH (2001) Chromosomal rearrangements and speciation. *Trends in Ecology & Evolution*, **16**, 351–358.

Ronce O, Kirkpatrick M (2001) When sources become sinks: migrational meltdown in heterogeneous habitats. *Evolution*, **55**, 1520–1531.

Sexton JP, McIntyre PJ, Angert AL, Rice KJ (2009) Evolution and ecology of species range limits. *Annual Review of Ecology Evolution and Systematics*, **40**, 415–436.

Sharakhov IV, White BJ, Sharakhova MV *et al.* (2006) Breakpoint structure reveals the unique origin of an inter-specific chromosomal inversion (2La) in the *Anopheles gambiae* complex. *Proceedings of the National Academy of Sciences of the United States of America*, **103**, 6258–6262.

Slatkin M (1973) Gene flow and selection in a cline. *Genetics*, **75**, 733–756.

Tsutsui ND, Suarez AV (2003) The colony structure and population biology of invasive ants. *Conservation Biology*, **17**, 48–58.

Turner JRG, Wong HY (2010) Why do species have a skin? Investigating mutational constraint with a fundamental population model. *Biological Journal of the Linnean Society*, **101**, 213–227.

Wallace B (1975) Hard and soft selection revisited. *Evolution*, **29**, 465–473.

Wang J, Wurm Y, Nipitwattanaphon M *et al.* (2013) A Y-like social chromosome causes alternative colony organization in fire ants. *Nature*, **493**, 664–668.

White BJ, Hahn MW, Pombi M *et al.* (2007) Localization of candidate regions maintaining a common polymorphic inversion (2La) in *Anopheles gambiae*. *Plos Genetics*, **3**, 2404–2414.

White BJ, Cheng CD, Sangare D *et al.* (2009) The population genomics of trans-specific inversion polymorphisms in *Anopheles gambiae*. *Genetics*, **183**, 275–288.

White BJ, Collins FH, Besansky NJ (2011) Evolution of *Anopheles gambiae* in relation to humans and malaria. *Annual Review of Ecology, Evolution, and Systematics*, **42**, 111–132.

Wolfram_Research (2014) *Mathematica 10.0*, 10.0 edn. Wolfram Research, Champaign, Illinois.

Yeaman S, Whitlock MC (2011) The genetic architecture of adaptation under migration-selection balance. *Evolution*, **65**, 1897–1911.

Zhang Z-Q (2011) Animal biodiversity: an introduction to higher-level classification and taxonomic richness. *Zootaxa*, **3148**, 7–12.

SUPPORTING INFORMATION

Additional supporting information can be found in the online version of the *Molecular Ecology* article.

Data S1 Supporting information.

Chapter 11

THE DISTRIBUTION OF GENETIC VARIANCE ACROSS PHENOTYPIC SPACE AND THE RESPONSE TO SELECTION

Mark W. Blows and Katrina McGuigan

School of Biological Sciences, University of Queensland, St Lucia, QLD 4072, Australia

Abstract

The role of adaptation in biological invasions will depend on the availability of genetic variation for traits under selection in the new environment. Although genetic variation is present for most traits in most populations, selection is expected to act on combinations of traits, not individual traits in isolation. The distribution of genetic variance across trait combinations can be characterized by the empirical spectral distribution of the genetic variance–covariance (**G**) matrix. Empirical spectral distributions of **G** from a range of trait types and taxa all exhibit a characteristic shape; some trait combinations have large levels of genetic variance, while others have very little genetic variance. In this study, we review what is known about the empirical spectral distribution of **G** and show how it predicts the response to selection across phenotypic space. In particular, trait combinations that form a nearly null genetic subspace with little genetic variance respond only inconsistently to selection. We go on to set out a framework for understanding how the empirical spectral distribution of **G** may differ from the random expectations that have been developed under random matrix theory (RMT). Using a data set containing a large number of gene expression traits, we illustrate how hypotheses concerning the distribution of multivariate genetic variance can be tested using RMT methods. We suggest that the relative alignment between novel selection pressures during invasion and the nearly null genetic subspace is likely to be an important component of the success or failure of invasion, and for the likelihood of rapid adaptation in small populations in general.

Previously published as an article in *Molecular Ecology* (2015) 24, 2056–2072, doi: 10.1111/mec.13023

INTRODUCTION

Biological invasions have long been of interest due to their impact on biodiversity and endemism. Initially, work focused on understanding the ecological factors contributing to invasion success, but recognition that rapid adaptation is a feature of some successful invasions (Blossey & Notzold 1995; Felker-Quinn *et al.* 2013) has led to increasing attention being paid to the genetic characteristics of successful invaders (Lee 2002; Dlugosch & Parker 2008). While the study of rapid evolution as a consequence of other human-induced environmental perturbations, such as the evolution of insecticide resistance (Scott *et al.* 2000; Ffrench-Constant *et al.* 2004) or antibiotic resistance (Baquero & Blazquez 1997; Palmer & Kishony 2013), is aided by knowing the primary selective agent, and therefore what traits might initially respond to selection, invasions represent a greater challenge. The selective forces acting on the initial invading population are generally unknown. Furthermore, invasive species are taxonomically diverse, and many are not amendable to common garden or other experimental designs typically used to identify evolutionary events.

Understanding evolutionary responses during invasions is therefore closely aligned with understanding how rapid adaptation occurs in general. At its most basic level, the response to selection will be directly proportional to the genetic variation for the trait under selection (Falconer 1981). Theoretical investigations suggest that rapid adaptation is more likely to be based on standing genetic variation than on new mutations (Barrett & Schluter 2008; Orr & Unckless 2008). In support of this view, studies in several taxa have identified alleles associated with the adaptation to novel environments that were present in low frequency in known or putative colonizers (Colosimo *et al.* 2005; Vandepitte *et al.* 2014). It might therefore be assumed that taxa with abundant standing genetic variation should be able to adapt more rapidly than taxa with lower levels of variation. However, quantitative genetic variation is found in almost any individual trait (Blows & Hoffmann 2005) and in all but the smallest populations (Willi *et al.* 2006). This suggests that genetic variation per se is not predictive of invasion success. Molecular genetic studies further suggest that invasion success might be independent of levels of neutral genetic variation in populations (Dlugosch & Parker 2008; Rollins *et al.* 2013). Taken together, these observations indicate the need for a more nuanced approach

to the investigation of the availability of genetic variation for rapid adaptation.

As any response to selection in nature is likely to involve more than a single quantitative trait (Blows 2007), an appreciation of the nature of multivariate genetic variation is an obvious place to start. The potential for genetic covariation among traits to affect the response to selection is captured by the multivariate breeder's equation (Lande 1979):

$$\Delta z = \mathbf{G}\beta$$

where β is a vector selection gradients, and $\mathbf{G}$ is the additive genetic variance–covariance matrix. The role of genetic covariation in promoting adaptation under altered environmental conditions has been addressed in several theoretical (Gomulkiewicz & Houle 2009; Duputie *et al.* 2012; Villmoare 2013; Kopp & Matuszewski 2014) and empirical (Agrawal & Stinchcombe 2009; Selz *et al.* 2014; Turner *et al.* 2014) studies. In general, the distribution of genetic variation across trait space suggests that adaptation might be very rapid if selection favours multivariate phenotypes associated with high levels of genetic variation. Conversely, if selection favours trait combinations associated with very low levels of genetic variation, in small populations with low absolute fitness, extinction might occur prior to adaptation (Gomulkiewicz & Houle 2009; Kopp & Matuszewski 2014).

In this study, we outline a framework for studying the distribution of genetic variance in multivariate sets of traits. We begin by describing how the empirical spectral distribution of $\mathbf{G}$ governs the evolutionary potential of a set of traits. We review what is known about the empirical spectral distribution in real data sets, show that phenotypic trait combinations with very low levels of genetic variance are common and detail how these are expected to impact on the response to selection. Finally, we highlight the importance of considering the correct null distribution for the empirical spectral distribution, and how advances made in random matrix theory can be used to determine the extent of genetic covariance among traits and therefore the prevalence of genetic constraint.

THE EMPIRICAL SPECTRAL DISTRIBUTION OF G AND ITS EVOLUTIONARY IMPLICATIONS

Rather than focusing on bivariate genetic correlations that comprise it, the G-matrix is most usefully interpreted as characterizing the genetic variance in all possible combinations of the traits contained in the matrix (Blows & Hoffmann 2005; Blows 2007; Aguirre *et al.* 2014). The empirical spectral distribution, represented by the ranked eigenvalues of **G**, provides a summary of how the genetic variance in a set of traits is distributed across phenotypic space, which can often be in dramatic contrast to levels of genetic variance in individual traits (Dickerson 1955; Blows & Hoffmann 2005). The empirical spectral distribution has been formally described in only a handful of studies, with much of the genetic variance being found in one or a few dimensions, and consequently, genetic variation is often low for a substantial proportion of the phenotypic space (Hine & Blows 2006; Kirkpatrick 2009; Walsh & Blows 2009). In Fig. 1, we take advantage of a data set of published G-matrices, recently complied by Pitchers *et al.* (2014b). Across the three categories of trait types identified by Pitchers *et al.* (2014b), the same pattern of high genetic variance in a few dimensions (eigenvectors), contrasting with very low levels of genetic variance for much of phenotypic space prevails (Fig. 1).

When some eigenvalues of **G** are very small, the associated eigenvectors may form part of what has been termed a nearly null genetic subspace (Gomulkiewicz & Houle 2009; Houle & Fierst 2013; Hine *et al.* 2014). The nearly null genetic subspace plays a key role in our understanding of genetic constraints on the response to selection for two reasons. First, although an eigenvalue of **G** of zero would reveal the presence of an absolute genetic constraint (Mezey & Houle 2005), a null subspace in which evolution is prohibited, it is not feasible to use standard quantitative genetic experiments to prove the presence of null spaces. This is because rejecting the null hypothesis of a zero genetic variance will always be dependent on the statistical power of a given experiment. Consequently, defining the extent of the nearly null subspace, that may or may not include a true null subspace, is likely to be the most practical approach to determining the extent of genetic constraint. Second, while the nearly null space can be considered to represent quantitative genetic constraints where evolution will be slow, in the context of limited population size in real populations, failure to adapt to selection in these regions of phenotypic space is a real possibility (Gomulkiewicz & Houle 2009).

The empirical spectral distribution describes the relative sizes of the eigenvalues of **G** and governs how the population will respond to a given direction of selection, β. When both β and **G** are known,

the evolutionary implications of the empirical spectral distribution of **G** are straightforward to determine. A spectral decomposition of the breeder's equation (Walsh & Blows 2009; Aguirre *et al.* 2014) displays how the eigenvalues of **G** influence the direction of response to selection:

$$\Delta z = \mathbf{G}\beta = \sum_i \lambda_i \boldsymbol{g}_i \boldsymbol{g}_i^T \beta$$

where λ_i is the *i*th eigenvalue of **G**, and $\boldsymbol{g}_i \boldsymbol{g}_i^T \beta$ is the projection of β along the *i*th eigenvector of **G**, $\boldsymbol{g}_i$. The size of a given eigenvalue, in combination with how closely β is associated with the corresponding eigenvector, determines the contribution to the response from that part of the space. Note that the spectral decomposition does not imply that selection occurs along eigenvectors of **G**; rather, it facilitates the determination of whether a specific direction of selection occurs in the region of phenotypic spaces associated with the most (higher eigenvalues of **G**) or least (lower eigenvalues of **G**) genetic variance.

When β lies in a direction that is contained in the nearly null genetic subspace, represented by the lower eigenvectors of **G**, two general consequences for the response to selection are likely. First, the response will become increasingly biased away from the direction of selection as **G** becomes increasingly ill-conditioned, where the leading eigenvalue is much larger than the lower eigenvalues (Lande 1979; Walsh & Blows 2009). There is some empirical evidence to indicate that natural populations respond to selection in directions that are biased towards greater genetic variance (Schluter 1996), particularly if β is most closely associated with the nearly null subspace (Chenoweth *et al.* 2010). Second, individual traits are likely to respond in a direction opposite to their selection gradients as the direct response in these individual traits is overwhelmed by the indirect response generated by the larger eigenvalues (e.g. Hine *et al.* 2011, 2014; see Fig. 2).

PLEIOTROPY AND THE FREQUENCY OF NEARLY NULL GENETIC SUBSPACES

Genetic covariance is generated among traits through both linkage disequilibrium and pleiotropy. Pleiotropy is likely to underlie most of the genetic covariance

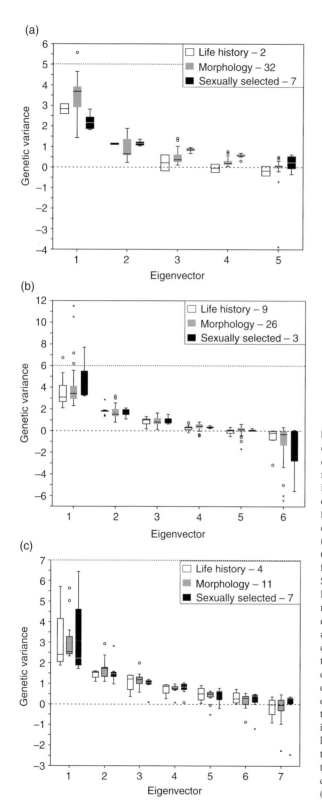

Fig. 1 Empirical spectral distributions of published genetic correlation matrices, derived from a subset of the Pitchers *et al.* (2014a,b) data set, presented as box plots (boxes represent the 1st–3rd interquartile range with bar indicating the median; whiskers represent the range of the data, with values 1.5–3 times the interquartile range (IQR) from the end of the box indicated by circles, and more extreme cases shown as crosses). (a) 41 five-trait matrices, (b) 38 six-trait matrices, (c) 22 seven-trait matrices. Correlation rather than covariance matrices were chosen for presentation to avoid the confounding issue of scale. Several matrices of fewer or more traits were included by Pitchers *et al.* (2014b), but are not presented here. One matrix (file Ell1998.195) was excluded from the five-trait data set as it was not symmetrical. Pitchers *et al.* (2014b) allocated the set of traits in each matrix to one of three trait classes: life-history traits, morphology and sexually selected traits. No obvious differences were found between the trait classes in their spectral distributions. The typical exponential decay in the ranked eigenvalues reported in earlier studies (Kirkpatrick 2009) is apparent here. Many of these matrices are subject to substantial sampling error, as indicated by the frequency of negative eigenvalues for the lower eigenvectors (falling below the dashed line), and by the presence of some leading eigenvalues that exceed the trace of the matrix, which for correlation matrices is constrained to equal the number of traits in the matrix (dotted line).

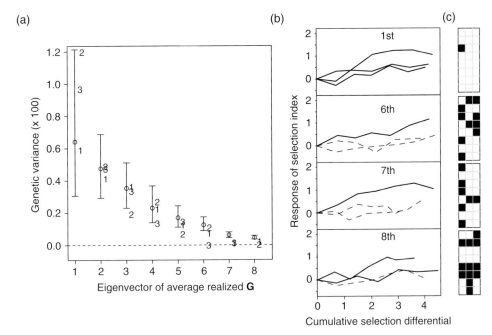

Fig. 2 The relationship between standing genetic variance and responses to selection determined through high-dimensional artificial selection on cuticular hydrocarbons in *Drosophila serrata*. (a) The distribution of genetic variance determined in the base population and from the responses to selection in each experimental replicate, presented on the common scale of the eight eigenvectors of the average realized **G**. Genetic variance in the base population was determined through variance component analysis of a paternal half-sib breeding design before selection commenced (open circles, with 95% credible intervals from 1000 samples of the posterior distributions estimated from a Bayesian analysis). The realized genetic variance was determined for each of three temporal replicates of the total selection experiment (numerical values 1, 2 and 3). (b) Phenotypic responses in four of the eight selection treatments, corresponding to the 1st, 6th, 7th and 8th eigenvectors of the base population **G**. Each of the eight selection treatments was applied to three replicate populations, with 50% truncation selection for six generations. The responses to artificial selection (in units of phenotypic standard deviations of the selection vector), scaled by their respective selection differentials, provide an estimate of the genetic variance in the trait in the base population, or the realized heritability. Solid lines and dashed lines, respectively, represent significant realized heritability (indicating response to selection) versus no significant response. As rates of evolution are on the scale of standard deviations, the responses to selection shown do not reflect the extent of change observed in individual traits (see Hine *et al.* 2014). (c) The response of each individual CHC (eight rows) in each replicate selection line (three columns) plotted for the four treatments in b, where the response was either in the direction of the selection gradient (open) or in the direction opposite to the selection gradient applied (black). The other four treatments (2nd–5th) are not presented here, but were qualitatively similar to the 1st, with significant realized heritability of all replicate lines and relatively fewer traits responding in directions opposing their selection gradient (Hine *et al.* 2014). Redrawn from data in Hine *et al.* (2014).

observed among quantitative traits (Falconer 1981), and so the extent of pleiotropy will be a key determinant of the frequency of nearly null genetic subspaces. Although there are numerous examples of individual loci and alleles with pleiotropic effects, the extent of pleiotropy remains poorly characterized and contentious (Wagner & Zhang 2011; Paaby & Rockman 2013). Gene knockout (or knockdown)

studies in several model organisms indicate that most genes in such assays affect few traits, suggesting the extent of pleiotropy is typically limited (Wang *et al.* 2010; Wagner & Zhang 2011). Similarly, QTL mapping studies of intertaxa crosses have also typically supported limited, rather than extensive, pleiotropy (or close physical linkage) (Albert *et al.* 2008; Wagner *et al.* 2008).

Much of the recent discussion of the extent of pleiotropy has been stimulated by the work of Gunter Wagner and colleagues (Wagner *et al.* 2008; Wagner & Zhang 2011), taking advantage of recent methodological advances to test a hypothesis of universal pleiotropy, whereby all traits are affected by each gene. Within this context, interpretation of limited pleiotropy can still encompass loci with pleiotropic effects on tens to hundreds of phenotypic traits (see figure 3 in Wagner & Zhang 2011). Furthermore, pleiotropic effects might be restricted to sets of traits, generating variational modules (co-varying traits) through modular pleiotropy (Wang *et al.* 2010; Wagner & Zhang 2011). Given that most estimates of **G** consider sets of potentially functionally related traits (e.g. different measures of a body region), this might be precisely the context in which we expect pleiotropy to be pervasive, and evolutionary outcomes affected.

In contrast to this emerging view that most genes have pleiotropic effects on few traits, there is a large body of work that indicates quantitative genetic covariance among pairs of traits (Roff 1996) or multiple traits (Kirkpatrick 2009; Walsh & Blows 2009; Figs 1 and 2) is common, that correlated responses to selection frequently occur (Bohren *et al.* 1966; Weber 1992; Beldade *et al.* 2002), and therefore that the effects of pleiotropy tend to be seen broadly across the phenotype. As argued by Johnson & Barton (2005) on the basis of what is known about per trait and genomic mutation rates, relatively few traits can have a unique genetic basis.

Recent analyses of pleiotropy in mutation accumulation lines have suggested that pleiotropy is both common and widespread and that pleiotropic mutation might contribute to the presence of nearly null spaces. Extensive investigation of mutational variance in wing shape in *Drosophila melanogaster* indicated the consistent presence of regions of the 24-trait wing shape phenotypic space associated with very low (statistically undetectable) mutational variance (Houle & Fierst 2013). Further evidence that pleiotropic mutation might contribute to the presence of nearly null genetic subspaces comes from analyses of mutational covariance in gene expression phenotypes of *Drosophila serrata*. Over 20% of 677 sets of five randomly combined expression traits were determined to contain co-varying traits (McGuigan *et al.* 2014b). As traits were combined into sets without reference to their function, the relatively high level of covariance detected under robust statistical testing suggests that pleiotropic mutational effects

are widespread among expression traits. In that study (McGuigan *et al.* 2014b), only the major axis of each mutation (genetic) covariance matrix was considered; as we detail below, the presence of a subspace associated with high levels of genetic variance implies the presence of a nearly null subspace.

There is a clear need for further work to clarify how the pleiotropic nature of alleles might limit the genetically accessible dimensions of phenotypic space. Genome-wide association studies in humans are one area that is likely to generate insights into these relationships, with ongoing identification of pleiotropic effects on multiple diseases (Solovieff *et al.* 2013). As pointed out by Hill & Zhang (2012), the statistical framework for testing for pleiotropic effects of genes is likely to prevent identification when phenotypic effects are small; the extent to which the structure of **G** is determined by very many small-effect pleiotropic loci is unknown.

THE RESPONSE TO SELECTION IN NEARLY NULL GENETIC SUBSPACES

We know remarkably little about the evolutionary responses of these newly established nearly null subspaces of the multivariate phenotype. In particular, it is surprising that little systematic attention has been given to using manipulative experiments to determine the availability of genetic variation in complex phenotypes, beyond selection on two traits (e.g. Weber 1992; Beldade *et al.* 2002; Conner 2003). Experiments of this type have consistently shown that a selection response in the direction of both the major and minor axis of the two-dimensional phenotype is achieved. In these cases however, a substantial proportion of the genetic variance remained in the direction of the minor axis (Walsh & Blows 2009), and so these experiments have not been informative about how a response to selection would occur in a nearly null genetic subspace.

In a recent evolutionary manipulation using a higher-dimensional artificial selection experiment, Hine *et al.* (2014) applied artificial selection to a suite of eight cuticular hydrocarbons in *Drosophila serrata*. Cuticular hydrocarbons are involved in species recognition and mate choice (Blows & Allan 1998; Higgie *et al.* 2000) as well as providing a barrier to desiccation (Gibbs 1998; Foley & Telonis-Scott 2011), a major physiological challenge in terrestrial invertebrates (Chown *et al.* 2011). By applying directional selection along eight genetic eigenvectors of the eight-trait phenotypic space, the evolutionary manipulation spanned the entire phenotypic space. This experiment demonstrated that the estimate of multivariate standing genetic variance in a population, calculated from a standard breeding design, successfully predicted the genetic variance available for a response to selection across the phenotypic space (Fig. 2a). Importantly, the distribution of high versus low genetic variance across phenotypic space resulted in qualitatively different selection responses. The selection treatments associated with low genetic variance (the putative nearly null space) differed from high genetic variance treatments in two ways. First, the selection response was stochastic when the genetic variance in the base population was low, with less than half of the selection lines responding over these three treatments (Fig. 2b). Second, more individual traits responded to selection in the direction opposite to their selection gradient when genetic variance was low (Fig. 2c).

The increase in the stochastic nature of the response for the same set of traits as the genetic variance decreased indicated that sampling effects during the establishment of these small replicate experimental populations had an important effect on the response. It therefore seems likely that the response in the last three eigenvectors was based on alleles segregating at low frequency in the base population, and so prone to be subject to sampling (Hine *et al.* 2014). In an earlier artificial selection experiment on the same eight traits, Hine *et al.* (2011) established that for a trait combination that was closely associated with the nearly null subspace, the selection response was in large part based on a single putative allele that was at low frequency in the base population.

The Hine *et al.* (2014) experiment provided manipulative evidence that observations of genetic variance in individual traits have limited relevance to an evolutionary context in which combinations of traits will be under selection; in general, neither the rate nor direction of evolutionary response of single traits will be predicable from observations of their genetic variance in isolation when trait combinations are under selection (Walsh & Blows 2009). The prevalence of inconsistent responses to selection for traits with low genetic variance extends the theoretical concept of a nearly null subspace, as originally envisaged by Gomulkiewicz & Houle (2009) who emphasized the situation when

adaptation is slow relative to population decline, to encompass a lack of response in some populations as a consequence of sampling variation associated with low frequency alleles (i.e. drift).

Artificial selection experiments of the kind employed by Hine *et al.* (2014) will not be possible in many species, and therefore, more indirect approaches are needed to help understand how likely it will be that evolutionary responses will be governed by a nearly null genetic subspace. Taken at face value, the prevalence of low levels of genetic variance in some parts of the phenotypic space (Fig. 1) implies that the response to selection is likely to be subject to inconsistent responses for many sets of morphological and life-history traits in small invading or colonizing populations. Pitchers *et al.* (2014b) investigated various metrics that have been used to describe the size and shape of **G** to assess the potential impact of **G** on the selection response. In their meta-analysis of published data, these authors found little evidence that such metrics explained any variation in evolutionary rates among traits, suggesting that perhaps the nature of multivariate genetic variance does not impact on the long-term efficacy of the response. In the remainder of this paper, we outline why a more detailed consideration of the distribution of genetic variance is required before we are in a position to fully understand its potential impact on the response.

THE SPECTRAL DISTRIBUTION OF RANDOM MATRICES

If we are to understand how genetic variance is distributed for multiple traits, we first need to consider how the spectral distribution of random matrices behaves. While the metrics summarized by Pitchers *et al.* (2014b) will all capture some aspect of the empirical spectral distribution of **G**, they share two limitations. First, such metrics lack a formal statistical framework for determining whether an observed pattern significantly deviates from a defined null hypothesis. Second, they do not attempt to describe the distribution of eigenvalues in a comprehensive manner, relying often on either the relative size of the lead eigenvalue (Kirkpatrick 2009) or the variance of the eigenvalues (Pavlicev *et al.* 2009). Here, we detail how these problems can be resolved through a formal comparison of the full empirical spectral distribution of **G** with the spectral distribution expected from a matrix of the same size (number of traits and samples) in which covariance is generated only by sampling variance.

The importance of considering the behaviour of the spectral distribution as a consequence of sampling is shown in Box 1 for the simple case of the sample covariance matrix **P**. For a P-matrix of size $p = 10$ traits and $n = 250$ individuals, the magnitude of the leading eigenvalue is larger than any of the individual trait phenotypic variances and on average about twice the size of the smallest eigenvalue of **P**. If fewer individuals are measured for each trait, say $n = p = 10$, then the eigenvalues will vary by over three orders of magnitude, simply as a consequence of sampling (Johnstone 2007). Therefore, considerable variance among the eigenvalues of a P-matrix is expected just from sampling. From random matrix theory (RMT), this highly predictable behaviour of the eigenvalues, where the bulk distribution of the eigenvalues follows a Marchenko–Pastur distribution, is known as the quarter-circle law (Box 1, Fig. 1b).

The empirical spectral distribution of **G** tends to display an exponential decay, suggesting that pleiotropy generates considerable covariance among traits (Fig. 1). It is however clear from RMT that the presence of an exponential decline in the eigenvalues of a G-matrix cannot be taken as evidence for the presence of pleiotropy without further analysis; such a decline is an expected outcome of random sampling and the behaviour of the eigenvalues of such matrices. Unfortunately, an appropriate statistical null hypothesis to test the empirical spectral decomposition of **G** against has not yet been established in quantitative genetics. A sampling bias in the eigenvalues of **G** was first noted by Hill & Thompson (1978), where they established the probability of **G** being nonpositive

---

Box 1 The spectral distribution of random P- and G-matrices

The expected spectral distribution of random symmetrical matrices has been the subject of intense interest in mathematical physics since the discovery of the semi-circle law by Wigner (1955). In a series of further results in what is now known since random matrix theory, the spectral distributions of a number of different classes (termed ensembles) of random symmetrical matrix have been shown to conform to a set of related distributions (Bai & Silverstein 2010). For example, consider the phenotypic variance–covariance (**P**) matrix as an example of a sample covariance matrix:

$$P = \frac{1}{n} X^T X,$$

where **X** is a square matrix of n rows (individuals) and p columns (traits), containing the observations. Such sample covariance matrices form the basis of principle components analysis, which will be familiar to many biologists.

Figure 1a shows the spectral distribution of a sample covariance matrix among random traits for 200 P-matrices estimated from simulated data sets with $p = 10$ traits and $n = 250$ individuals. Observations of individuals were drawn from the distribution N(0,1) for each of the 10 traits, and the identity matrix was used as the covariance matrix among them so that covariances among traits appear only as a consequence of sampling.

When all the eigenvalues are displayed as a histogram, the distribution follows the Marchenko–Pastur (MP) distribution, which is given by (Johnstone 2007):

$$f(x) = \frac{\sqrt{(b-x)(x-a)}}{2\pi\gamma x},$$

where, $a = (1 - \sqrt{\gamma})^2, b = (1 + \sqrt{\gamma})^2, \gamma = \frac{p}{n}$ and the function is plotted in the interval $a < x < b$. For our simulated example, $\gamma = 0.04$ and the fit of the MP distribution to the eigenvalues of **P** is shown in Fig. 1b. The predictable behaviour of the bulk of the eigenvalues of a sample covariance matrix in this fashion is sometimes referred to as the quarter-circle law (Johnstone 2007).

We took the same 200 simulated data sets of $n = 250$ observations, allocated five rows to each of 50 groups (representing, for example, five individuals sampled from each of 50 inbred lines) and estimated the between-group (**G**) covariance matrix using a multivariate linear model that fitted an unstructured covariance matrix at the between-group and residual (within-group) levels, using restricted maximum likelihood. The spectral distributions of these G-matrices are given in Fig. 2. Once again, the leading eigenvalue of **G** is substantial (Fig. 2a), averaging over 0.2 of the phenotypic variance in an individual trait, equating to a heritability of around 20% generated through random covariance

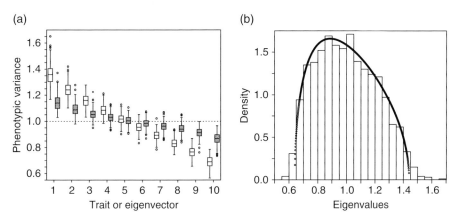

Box 1, Fig. 1 Bulk distribution of the eigenvalues of 200 10-trait **P**-matrices estimated from simulated data. (a) scree plot presented as a box plot (as in Fig. 1). Open (unshaded) boxes correspond to the eigenvalues of **P**, while shaded boxes are the phenotypic variances of the individual 10 traits contained in **P**, ordered from largest to smallest variance within each 10-trait set. The dashed line (at phenotypic variance = 1) indicates the variance that was simulated for all traits. (b) Distribution of the observed eigenvalues and the theoretical prediction. Histogram (total area normalized to 1) of the 2000 eigenvalues from the 200 **P** constructed from the simulated data. The dotted line is the Marchenko–Pasteur distribution for = 0.04.

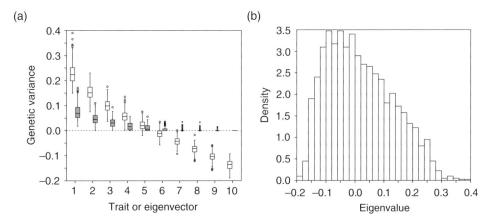

Box 1, Fig. 2 Bulk distribution of the eigenvalues of the 200 10-trait **G**-matrices estimated from the simulated data. (a) scree plot presented as a box plot, as in Fig. 1a. The dashed line indicates zero; note that the restricted maximum likelihood estimates of the individual trait among-group variance components (shaded grey as in Fig. 1a) were constrained to be positive, but the covariance matrix among them was not similarly constrained to be positive semi-definite. (b) Normalized (unit area) histogram of eigenvalues of the 2000 eigenvalues from the 200 **G** estimated from the simulated data.

among traits. Note that the fitting of an unstructured covariance matrix at the among-line level allows negative eigenvalues, although individual trait variances were constrained to be positive. Often, estimation procedures will force the among-line covariance matrix to be positive semi-definite (Meyer & Kirkpatrick 2008), which then results in the typical exponential decay in eigenvalues seen for real data (main text Figs 1 and 2).

Allowing negative eigenvalues in the current context is important as it allows the spectral distribution of random data to be explored. Although the histogram of genetic eigenvalues appears similar to that seen for the phenotypic eigenvalues (Fig. 1b), no theory currently exists for establishing the eigenvalue limiting distribution of variance component matrices (I. Johnstone, personal communication).

definite for a given experimental design. In particular, Hill and Thompson noted the tendency for the leading eigenvalue of **G** to be biased upwards and the smallest eigenvalues to be biased downwards. Box 1 (Fig. 2a) shows the spectral distribution of a 10-trait between-group covariance matrix estimated from a linear model in a fashion typical of quantitative genetic analyses. It can be seen how the sampling bias commented upon by Hill & Thompson (1978) is a component of the as-yet unresolved Marchenko–Pastur-like behaviour of the eigenvalues of a variance-component covariance matrix.

TESTING OBSERVED SPECTRAL DISTRIBUTIONS OF G AGAINST THE TRACY–WIDOM DISTRIBUTION

Interrogating an observed empirical spectral distribution of **G** to determine whether the eigenvalues deviate from the random expectation could take three general approaches. First, the bulk distribution of eigenvalues could be compared against the theoretical distribution for the random expectation, an approach that has been applied in the characterization of the modular nature of protein (Luo *et al.* 2006) and gene expression (Luo *et al.* 2007) networks. Such an approach is not yet available for studying G-matrices, as the theoretical distribution for variance component matrices is currently unknown (Box 1). Second, as nearly null spaces are of particular interest, determining whether the smallest eigenvalues of **G** are significantly smaller than random expectation might be valuable. However, estimation procedures for variance component matrices often force the resulting G-matrix to be positive semi-definite (Meyer & Kirkpatrick 2008), and therefore, the sampling properties of the smallest eigenvalues are unlikely to follow established theory (Feldheim & Sodin 2010) at this edge of the distribution.

Third, much is known in RMT concerning the behaviour of the largest eigenvalues of real symmetrical matrices, and this can be used to test how the genetic variance is restricted to fewer dimensions in empirical estimates of **G** (Box 2). For simple sample covariance matrices, the leading eigenvalue will follow the Tracy–Widom (TW) distribution with centring and scaling parameters of the distribution that have been well established (Johnstone 2001). For example, Patterson *et al.* (2006) showed how the leading eigenvalue from a sample covariance matrix generated from

biallelic marker loci could be tested against the TW distribution to establish the presence of population structure. In a quantitative genetics context, as G-matrices are usually estimated from a multivariate linear model specific in form to a particular experiment, determining the appropriately centred and scaled TW distribution is more complicated than in the case of a sample covariance matrix because exact expressions for the centring and scaling constants are unknown. In the absence of such expressions, the appropriate centring and scaling of the TW distribution can be approximated through an empirical approach outlined by Saccenti *et al.* (2011) (Box 2).

It is important at this point to consider what hypothesis is being tested against the *TW* distribution and how this relates to established analyses of multivariate genetic variance. In the case of a P-matrix, where all traits have been standardized to have the same phenotypic variance, the presence of a significant leading eigenvalue tested against the TW distribution unambiguously demonstrates the presence of significant correlation among the traits. Unfortunately, in the case of G-matrices, the genetic variances are expected to vary among traits even when the phenotypic variances are the same, and so such a definitive statement concerning the presence of covariance cannot be made. For example, when a G-matrix is modelled using an unstructured covariance matrix or a factor-analytic structure without specific variances (Hine & Blows 2006; Meyer 2007), the leading eigenvalue of **G** could represent significant genetic variance entirely as a consequence of the genetic variance in a single trait. In contrast, factor-analytic modelling that includes specific variances, where the multivariate genetic variance is modelled as common genetic covariance captured by up to *p* genetic factors, and as a vector of *p* specific genetic variances that accounts for genetic variance restricted to each individual trait, is an effective way of determining the extent of genetic covariance generated by pleiotropy because it partitions out from **G** any variation that is unique to each individual trait, and not shared through covariance with another trait(s) (Meyer 2009; McGuigan & Blows 2010; McGuigan *et al.* 2014a,b).

Although it is an effective tool for detecting genetic covariance, the factor-analytic approach requires mixed-model convergence, which is often difficult to obtain once *p* exceeds about 10 traits. Furthermore, reduced rank estimates of the eigenvalues of **G** can become biased when too few genetic factors are

Box 2 The leading eigenvalue of G and the Tracy–Widom distribution

The leading eigenvalues of certain classes of random matrices are known to converge to the Tracy–Widom (TW) distribution as p becomes large. First demonstrated for Wigner matrices (Tracy & Widom 1996), subsequent work has shown how the leading (or edge) eigenvalues of other classes of random matrices follow the TW distribution after appropriate scaling and centring. In particular, the leading eigenvalue of sample covariance matrices (Johnstone 2001) and the F-matrix in the context of hypothesis testing of fixed effects in MANOVA (Johnstone 2009) have been shown to converge to this distribution.

Unfortunately, as in the case of the eigenvalue bulk distribution of variance component matrices, the appropriate scaling and centring parameters for the TW distribution for the leading eigenvalue of a variance-component matrix are unknown (I. Johnstone, personal communication). We have therefore adopted an empirical fitting approach (Saccenti et al. 2011) that rescales and centres an empirical distribution of the leading eigenvalues from G-matrices to closely approximate the appropriate TW distribution. For the 200 leading eigenvalues ($\lambda_{g_{max}}$) from the G-matrices from Box 1, the TW statistic was calculated using the expression for Wigner matrices (Bai & Silverstein 2010, Theorem 5.20):

$$TW_w = p^{2/3}(\lambda_{g_{max}} - 2).$$

These values were then rescaled following the approach of Saccenti et al. (2011, equation 17):

$$TW_G = -1.206 + \frac{1.268}{0.197}(TW_w + 8.225),$$

where −1.206 and 1.268 are the mean and standard deviation of the TW distribution, respectively, and −8.225 and 0.197 are the mean and standard deviation of the distribution of the unscaled TW_w statistic. The approximate scaling and centring of the TW distribution established in this way recovered the random expectation of 10 of the 200 (5%) leading eigenvalues of the G estimated from the simulated data (Box 1) exceeding the 5% significance threshold (Fig. 1).

A similar procedure can be followed for the 2nd, 3rd, 4th eigenvalues and so on, where the first three moments of the TW distribution are changed accordingly (Tracy & Widom 2009). This sequential approach would be very useful in data sets where p is large. Similar to factor-analytic modelling, sequential comparison to the TW distribution would allow the proportion of genetic variance that is significantly captured by the leading eigenvalues to be determined. However, when p is small, these distributions are unlikely to remain relevant, and further work is required to determine under what conditions such an approach might be useful for real G-matrices.

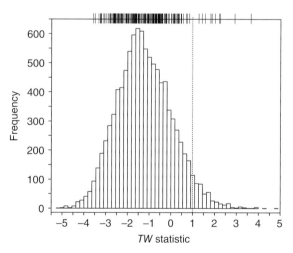

Box 2, Fig. 1 Testing the size of the leading eigenvalue of a G-matrix against the random expectation under the Tracy–Widom distribution. The histogram represents 10000 values drawn from the scaled and centred gamma distribution ($k = 46.446$, $\theta = 0.186054$, $\alpha = 9.84801$) known to accurately approximate the TW distribution (Chiani 2014). The leading eigenvalues of 200 G, estimated from simulated random data as described in Box 1, are shown as vertical lines across the top of the figure. The dotted vertical line is the 5% significance threshold for the TW distribution (0.979); 10 observed eigenvalues fall above (to the right) of this value.

estimated, potentially making hypothesis testing of nonzero leading eigenvalues unreliable (Meyer & Kirkpatrick 2008). Application of the TW distribution is therefore likely to be particularly useful for data sets with large p, or when **G** is estimated outside of a mixed-model framework where partitioning trait-specific and shared variance is not available. When p is large, the likelihood of a single-trait genetic variance exceeding the size of the leading eigenvalue capturing genetic covariance among traits will be small (see Box 1 Fig. 2a). As we show here (Box 3), the factor-analytic specific variance modelling approach can be used in conjunction with the TW distribution, with the factor-analytic modelling providing full-rank matrices in which trait-specific variances have been partitioned out and the TW distribution providing a means of hypothesis testing that does not rely on repeated analysis of the observed data (from 0 to p potential nonzero eigenvalues) and is insensitive to the known bias that can occur through fitting too few factors.

In an example of an application of the TW distribution to real data, we determined the frequency of genetic covariance among a very large number (8780) of gene expression traits that had been measured in a simple quantitative genetic experimental design (Box 3). We assessed how often the leading eigenvalue of a G-matrix was larger than expected by chance when gene expression traits were randomly assigned to five-trait sets (Box 3). Of the 1756 **G**, 95% had a leading eigenvalue that exceeded the 5% threshold for significance, indicating that pleiotropy is widespread among these gene expression traits. This is consistent with the previous analysis of this data using factor-analytic modelling (M. W. Blows, S. L. Allen, J. M. Collet, S. F. Chenoweth & K. McGuigan, in review), and factor-analytic modelling of the same traits in a set of mutation accumulation lines, where widespread pleiotropic mutation was found (McGuigan *et al.* 2014b).

Box 3 Genetic covariance among sets of gene expression traits

Taking the approach outlined in Box 2 and applying it to G-matrices estimated from real data requires a two-step process. First, the size (p) and nature (covariance or correlation matrix) of the real **G** need to be simulated using randomly generated data that match the characteristics of the observed data. This allows the mean and variance of the TW_w distribution of the leading eigenvalue for that particular case to be established. Second, the mean and variance of the simulated TW_w distribution are then used to centre and scale this distribution to result in TW_G, and finally to centre and scale the observed leading eigenvalues to TW_G using the Saccenti *et al.* (2011) approach; significant departures from the random expectation can then be determined. It should be noted that this approach will be problematic to implement in empirical situations where the mixed model used to the estimate **G** takes many minutes to converge, and hence, running the same model a large number of times to generate the simulated distribution may not be feasible.

To illustrate the approach on real data, we take advantage of a large number of **G** (1756) estimated among sets of five randomly combined gene expression traits. Briefly, the experimental design comprised 30 inbred lines established from a natural population of *Drosophila serrata*, with two independent RNA

extractions from each line (Allen *et al.* 2013). Univariate mixed-model analysis indicated that 8780 gene expression traits displayed significant between-line variance at a false discovery rate of 5% (McGuigan *et al.* 2014b). The heritabilities of these 8780 traits are shown in Fig. 1a and should be interpreted as broad-sense heritability given the inbred experimental design. Traits were randomly assigned to one of 1756 five-trait sets, which were then subjected to multivariate mixed linear modelling. For each five-trait set, at the between-line (genetic) level, specific trait variances were estimated for each individual trait, along with a covariance matrix that was constrained to contain all positive eigenvalues (i.e. a factor-analytic, full-rank covariance structure as in Meyer 2009; McGuigan & Blows 2010; McGuigan *et al.* 2014b). The G-matrices analysed here represent the common genetic covariance among the five expression traits in each set. To generate the simulated G-matrices, random data drawn from the distribution N(0,1) replaced the real observations. The observed data were standardized to N(0,1) prior to factor-analytic modelling, and therefore, the random data were on the same scale as the original data. The random data were then subjected to the same modelling procedures as the observed data, and 1756 random G-matrices were

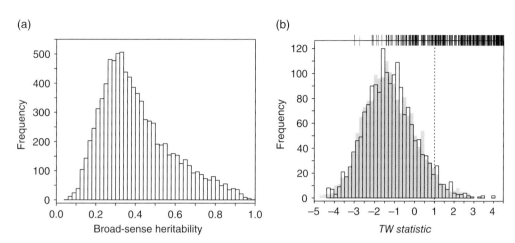

Box 3, Fig. 1 Genetic analysis of 8780 gene expression traits. (a) Histogram of the broad-sense heritability of 8780 of 11 604 gene expression traits that displayed significant between-line variance at a false discovery rate of 5% (redrawn from data in McGuigan *et al.* 2014b). (b) Applying a TW statistic test of covariance among the 8780 gene expression traits. Histograms are of values drawn from the theoretical TW distribution (grey shaded bars), and values of the TW$_G$ statistic for the leading eigenvalues from five-trait **G** estimated from the simulated random data (open bars). From the simulated data, −5.633 and 0.088 were the mean and standard deviation of the distribution of the unscaled TW$_w$ statistic that we used to result in TW$_G$. Only 1740 rather than 1756 values of $\lambda_{g_{max}}$ were retained for the TW$_w$ distribution as 16 simulated matrices had leading eigenvalues that fell well outside the bulk of the distribution and had a detrimental effect on scaling the TW$_G$ to fit the theoretical TW distribution. Shown as solid lines at the top of the graph are the observed leading eigenvalues ($\lambda_{g_{max}}$) from 1756 **G** estimated from the randomly combined sets of five gene expression traits. There were 1451 observed eigenvalues with larger TW statistics (>4.5), which are not plotted to aid visualization of the significance threshold and the match between theoretical and simulated TW distributions.

estimated. In this example, the time taken for model convergence is relatively short, and although many models are being run (two sets of 1756), the total number of models is less than we would typically run under a factor-analytic model testing framework to determine the statistical support for nonzero eigenvalues (e.g. for a five-trait matrix, six models varying in the number of factors are required).

The scaled TW$_G$ distribution closely matched the theoretical TW distribution (Fig. 1b), indicating that the Saccenti *et al.* (2011) scaling procedure (Box 2) performed well for these data. Of the 1756 leading eigenvalues of the observed **G**, 1667 exceeded the 5% critical threshold for significance, suggesting that 95% of the five-trait sets displayed significant genetic covariance among the randomly allocated traits.

PHENOME-WIDE DISTRIBUTION OF GENETIC VARIANCE

Houle (2010) has mounted a persuasive argument for the need to develop phenome-wide characterizations of the phenotype–genotype map and how selection acts across the phenome if evolutionary predictions are to become meaningful. As we outlined above, the extent of pleiotropic effects among traits, and factors determining this, are not well understood. The high prevalence of genetic covariance among small random

sets of gene expression traits (Box 3) suggests pleiotropic effects are common. Given that expression traits were combined at random, without a priori information on gene function, the high incidence of genetic covariance suggests the possibility that, by randomly assigning traits, we were repeatedly seeing the effects of the same pleiotropic allele(s) across many traits. Under such a scenario, the empirical spectral distribution of a phenome-wide **G** might display a similar pattern to that for much smaller sets of traits (Fig. 1). If so, there may be a very large number of trait combinations across the phenome that could fall within a nearly null genetic subspace.

The genetic analysis of 1000s of traits at once is beyond current multivariate statistical methodologies, with p as small as 10-20 being problematic for multivariate mixed-model approaches. The so-called curse of dimensionality hits exponentially hard as the number of traits increases; a G-matrix for the 8780 gene expression traits requires the estimation of over 38 million individual genetic variances and covariances. While such a G-matrix has not been estimated directly, it has been approximated using the established genetic relationships within each of the 1756 sets of 5 traits. The leading eigenvalue of this extreme-dimension **G** was substantially larger than the null expectation and represented an underlying genetic factor that had a common effect on a very large number of expression phenotypes (M. W. Blows, S. L. Allen, J. M. Collet, S. F. Chenoweth & K. McGuigan, in review). Although the methods underlying this approach to approximating high-dimensional **G** require further development, these approaches might ultimately allow us to characterize the nearly null subspace of the phenome.

WIDER IMPLICATIONS OF THE EMPIRICAL SPECTRAL DISTRIBUTION OF G

There are a number of implications of the empirical spectral distribution of **G** that impinge on how the evolutionary potential of populations is assessed and quantified. First, perhaps the most obvious consequence of the common pattern to the empirical spectral distribution (Fig. 1) is that the distribution of genetic variance may be very uneven among traits combinations of functionally related morphological or life-history traits. If this pattern is verified using the RMT approaches we develop here, it is important to note that such spectral distributions could also be generated by poor initial selection of traits; multicollinearity at the level of phenotypic traits could result in an individual trait being included that does not explain any independent phenotypic variance (Lewontin 1978; Fristrup 2001). This possibility highlights the fact that identifying regions of phenotypic space that might be evolutionarily inaccessible depends not just on analysis of **G**, but also of **P** (McGuigan & Blows 2007).

Second, the details of what trait combinations are under selection matter; different trait combinations will have very different evolutionary potentials arising from standing genetic variation. The observation that genetic variance might typically be unevenly distributed through phenotypic space indicates that simple measures of standing genetic variance will not be informative of the potential evolutionary responses of populations. In particular, it is becoming increasingly recognized that estimates of additive genetic variance (and either heritability or coefficients of variance derived from them) have limited applicability for understanding evolutionary potential outside of simple artificial selection (Hansen & Houle 2008; Walsh & Blows 2009). There is also evidence that levels of neutral molecular genetic diversity do not correlate with population persistence and adaptation (Dlugosch & Parker 2008; Rollins *et al.* 2013).

Unfortunately, we have only the experiment of Hine *et al.* (2014) that directly tests the evolutionary response of trait combinations from across the phenotypic space in a comprehensive manner to allow us to predict how variation in the alignment of selection with genetic variance might typically influence responses. From this experiment, it was clear that the response to selection in the nearly null subspace was inconsistent, which in combination with other information we have for cuticular hydrocarbons in *Drosophila serrata* suggests that rare alleles are likely to underlie the nearly null subspace. This has two general implications. First, the potential contribution of rare alleles to adaptive responses matches with what is generally understood about persistence and adaptation in small populations, such as might colonize novel habitats. Both short-term bottlenecks and more persistent small population sizes will result in genetic drift, with random sampling of alleles in each generation potentially altering the genetic diversity of the population. If adaptation depends on sampling of rare, high-fitness alleles, then adaptation might occur or not simply as a consequence of sampling. Furthermore, the reduced efficacy of selection in small populations ensures that adaptation will be slower for a given level of genetic variance, and extinction might occur before fitness increases (Gomulkiewicz & Houle 2009). This interpretation is also consistent with the apparent contribution of repeated introduction to colonization success (Roman & Darling 2007; Dlugosch & Parker 2008; Facon *et al.* 2008): chance loss of rare alleles would be reduced under such a scenario.

Second, the nature of the genetic details underlying the potential response to selection in trait combinations that occupy the nearly null subspace is likely to make identifying the alleles difficult. The contribution of low frequency alleles to the variation in quantitative traits has been notoriously difficult to establish. Exploiting the power of genome-wide association mapping data sets in humans, Yang *et al.* (2010) have demonstrated that rare alleles, and alleles of small effect, are likely to contribute substantially to the variation. Theoretical models of invasion success under different specific genetic models have suggested that adaptation will become increasingly unlikely as the number of contributing loci increases, but perhaps of more interest here, alleles with small effect will reduce extinction risk (Gomulkiewicz *et al.* 2010). Therefore, developing predictive models concerning the role of specific loci in invasion success is likely to be very challenging.

To gain a more complete understanding of nearly null genetic subspaces and their potential impact on the response to selection, we will need to determine why they exist; why do some trait combinations have vastly more genetic variance than others? There are two forces that could potentially generate the uneven distribution of genetic variance: mutation and stabilizing selection. While mutation and selection will often act jointly on the genetic variance of a particular phenotype, their relative importance in determining why a substantial proportion of high-dimensional phenotypes have very little genetic variance is now the key to understanding the nature of evolutionary limits. If some parts of the phenotype have low genetic variance because they experience very little or no mutation, the ability to respond to selection will be restricted under all conditions a population experiences. We have very little information on the input of new mutational variance to multivariate phenotypes. Recently, Houle & Fierst (2013) reported the presence of nearly null genetic subspace in the wing shape of populations of *D. melanogaster* diverging through accumulation of new mutations. Further data are required to assess whether mutational variance mighty typically be unevenly distributed across phenotypic space and whether nearly null mutational spaces exist. There is some evidence from neutral molecular markers that successful colonizers might have high mutation rates (Kanuch *et al.* 2014) and thus replace variation lost through selection at high rates. Interestingly, one of the (few) common genetic features of successful invasions appears to be repeated colonization, or admixture during colonization of multiple source populations (Roman & Darling 2007; Dlugosch & Parker 2008;

but see Rius & Darling 2014). This might indicate that stabilizing selection operates within source populations to reduce variance, rather than certain parts of the phenotypic space being inaccessible through mutation.

Alternatively, if the input of mutational variance of all phenotypes is similar, but strong selection on some parts of the phenotype results in low genetic variance, then evolutionary responses in these directions will be possible once the regime of selection changes. As well as altering the alignment of selection to genetic variance, changes in the selection regime could result in the breakdown of linkage among alleles that might be contributing to the presence of a nearly null subspace. Theoretically, we expect populations existing in stable conditions to be evolving in the vicinity of their adaptive optimum and thus to be under stabilizing selection, but studies of real populations rarely uncover evidence of stabilizing selection (Kingsolver *et al.* 2012). Genetic analyses have demonstrated that phenotypic covariances between traits and fitness (selection) are not necessarily underlain by genetic covariances, as would be required for evolution (Rausher 1992; Stinchcombe *et al.* 2002; Pemberton 2010). Further, as with **G** itself, selection analysis of one or few traits might result in misunderstanding of selection when other traits under selection are not included in the analysis (Morrissey *et al.* 2010, 2012; McGuigan *et al.* 2011). We have previously demonstrated variance-reducing (stabilizing) selection operating on genetic variances of wing shape and cuticular hydrocarbons in *D. serrata* (McGuigan & Blows 2009; McGuigan *et al.* 2011). However, much more work is required to determine whether, and how strongly, selection acts to reduce genetic variance in populations, and how the strength of this stabilizing selection relates to the uneven distribution in genetic variance across phenotypic space.

In conclusion, the question of how much genetic variation is available to selection is far more complicated than an estimate of a univariate heritability or neutral marker diversity. To understand the distribution of genetic variance across the phenotype, we need new developments in both our biological experiments and statistical tools. From a biological perspective, we need a comprehensive understanding of the processes of mutation and (stabilizing) selection acting on sets of functionally related traits, so that we can determine the causes of the observed empirical spectral distribution of G-matrices. A greater challenge will then be to be able to extend this understanding to a phenome-wide level, which is dependent on phenotyping organisms in ways, and on a scale, that we have yet to devise (Houle 2010). From a statistical perspective, new theory is required for the spectral distribution of variance-component covariance matrices, and again we need principled ways of characterizing and testing the distribution of genetic variance across the phenome, including methods for estimating very high-dimensional G-matrices.

ACKNOWLEDGEMENTS

We thank Iain Johnstone for many helpful discussions on the application of random matrix theory to variance component matrices, and Scott Allen and Steve Chenoweth for access to the gene expression data in Box 3.

REFERENCES

Agrawal AF, Stinchcombe JR (2009) How much do genetic covariances alter the rate of adaptation? *Proceedings of the Royal Society B-Biological Sciences*, **276**, 1183–1191.

Aguirre JD, Hine E, McGuigan K, Blows MW (2014) Comparing G: multivariate analysis of genetic variation in multiple populations. *Heredity*, **112**, 21–29.

Albert AYK, Sawaya S, Vines TH *et al.* (2008) The genetics of adaptive shape shift in stickleback: pleiotropy and effect size. *Evolution*, **62**, 76–85.

Allen SL, Bonduriansky R, Chenoweth SF (2013) The genomic distribution of sex-biased genes in *Drosophila serrata*: X-chromosome demasculinization, feminization, and hyper-expression in both sexes. *Genome Biology and Evolution*, **5**, 1986–1994.

Bai Z, Silverstein JW (2010) *Spectral Analysis of Large Dimensional Random Matrices*, 2nd edn. Springer, New York.

Baquero F, Blazquez J (1997) Evolution of antibiotic resistance. *Trends in Ecology & Evolution*, **12**, 482–487.

Barrett RDH, Schluter D (2008) Adaptation from standing genetic variation. *Trends in Ecology & Evolution*, **23**, 38–44.

Beldade P, Koops K, Brakefield PM (2002) Developmental constraints versus flexibility in morphological evolution. *Nature*, **416**, 844–847.

Blossey B, Notzold R (1995) Evolution of increased competitive ability in invasive nonindigenous plants: a hypothesis. *Journal of Ecology*, **83**, 887–889.

Blows MW (2007) A tale of two matrices: multivariate approaches in evolutionary biology. *Journal of Evolutionary Biology*, **20**, 1–8.

Blows MW, Allan RA (1998) Levels of mate recognition within and between two *Drosophila* species and their hybrids. *American Naturalist*, **152**, 826–837.

Blows MW, Hoffmann AA (2005) A reassessment of genetic limits to evolutionary change. *Ecology*, **86**, 1371–1384.

Bohren BB, Hill WG, Robertson A (1966) Some observations on asymmetrical correlated responses to selection. *Genetical Research*, **7**, 44–57.

Chenoweth SF, Rundle HD, Blows MW (2010) The contribution of selection and genetic constraints to phenotypic divergence. *American Naturalist*, **175**, 186–196.

Chiani M (2014) Distribution of the largest eigenvalue for real Wishart and Gaussian random matrices and a simple approximation for the Tracy-Widom distribution. *Journal of Multivariate Analysis*, **129**, 69–81.

Chown SL, Sorensen JG, Terblanche JS (2011) Water loss in insects: an environmental change perspective. *Journal of Insect Physiology*, **57**, 1070–1084.

Colosimo PF, Hosemann KE, Balabhadra S *et al.* (2005) Widespread parallel evolution in sticklebacks by repeated fixation of ectodysplasin alleles. *Science*, **307**, 1928–1933.

Conner JK (2003) Artificial selection: a powerful tool for ecologists. *Ecology*, **84**, 1650–1660.

Dickerson GE (1955) Genetic slippage in response to selection for multiple objectives. *Cold Spring Harbor Symposia on Quantitative Biology*, **20**, 213–224.

Dlugosch KM, Parker IM (2008) Founding events in species invasions: genetic variation, adaptive evolution, and the role of multiple introductions. *Molecular Ecology*, **17**, 431–449.

Duputie A, Massol F, Chuine I, Kirkpatrick M, Ronce O (2012) How do genetic correlations affect species range shifts in a changing environment? *Ecology Letters*, **15**, 251–259.

Facon B, Pointier JP, Jarne P, Sarda V, David P (2008) High genetic variance in life-history strategies within invasive populations by way of multiple introductions. *Current Biology*, **18**, 363–367.

Falconer DS (1981) *Introduction to Quantitative Genetics*, 2nd edn. Longmans Green, London/New York.

Feldheim ON, Sodin S (2010) A universality result for the smallest eigenvalues of certain sample covariance matrices. *Geometric and Functional Analysis*, **20**, 88–123.

Felker-Quinn E, Schweitzer JA, Bailey JK (2013) Meta-analysis reveals evolution in invasive plant species but little support for Evolution of Increased Competitive Ability (EICA). *Ecology and Evolution*, **3**, 739–751.

Ffrench-Constant RH, Daborn PJ, Le Goff G (2004) The genetics and genomics of insecticide resistance. *Trends in Genetics*, **20**, 163–170.

Foley BR, Telonis-Scott M (2011) Quantitative genetic analysis suggests causal association between cuticular hydrocarbon composition and desiccation survival in *Drosophila melanogaster*. *Heredity*, **106**, 68–77.

Fristrup KM (2001) A history of character concepts in evolutionary biology. In: *The Character Concept in Evolutionary Biology* (ed. Wagner GP), pp. 13–35. Academic Press, San Diego, California, USA.

Gibbs AG (1998) Water-proofing properties of cuticular lipids. *American Zoologist*, **38**, 471–482.

Gomulkiewicz R, Houle D (2009) Demographic and genetic constraints on evolution. *American Naturalist*, **174**, E218–E229.

Gomulkiewicz R, Holt RD, Barfield M, Nuismer SL (2010) Genetics, adaptation, and invasion in harsh environments. *Evolutionary Applications*, **3**, 97–108.

Hansen TF, Houle D (2008) Measuring and comparing evolvability and constraint in multivariate characters. *Journal of Evolutionary Biology*, **21**, 1201–1219.

Higgie M, Chenoweth S, Blows MW (2000) Natural selection and the reinforcement of mate recognition. *Science*, **290**, 519–521.

Hill WG, Thompson R (1978) Probabilities of non-positive definite between-group or genetic covariance matrices. *Biometrics*, **34**, 429–439.

Hill WG, Zhang XS (2012) On the pleiotropic structure of the genotype-phenotype map and the evolvability of complex organisms. *Genetics*, **190**, 1131–1137.

Hine E, Blows MW (2006) Determining the effective dimensionality of the genetic variance–covariance matrix. *Genetics*, **173**, 1135–1144.

Hine E, McGuigan K, Blows MW (2011) Natural selection stops the evolution of male attractiveness. *Proceedings of the National Academy of Sciences of the United States of America*, **108**, 3659–3664.

Hine E, McGuigan K, Blows MW (2014) Evolutionary constraints in high-dimensional trait sets. *American Naturalist*, **184**, 119–131.

Houle D (2010) Numbering the hairs on our heads: the shared challenge and promise of phenomics. *Proceedings of the National Academy of Sciences of the United States of America*, **107**, 1793–1799.

Houle D, Fierst J (2013) Properties of spontaneous mutational variance and covariance for wing size and shape in *Drosophila melanogaster*. *Evolution*, **67**, 1116–1130.

Johnson T, Barton N (2005) Theoretical models of selection and mutation on quantitative traits. *Philosophical Transactions of the Royal Society B-Biological Sciences*, **360**, 1411–1425.

Johnstone IM (2001) On the distribution of the largest eigenvalue in principal components analysis. *Annals of Statistics*, **29**, 295–327.

Johnstone IM (2007) High dimensional statistical inference and random matrices. *Proceedings of the International Congress of Mathematics*, **00**, 307–333.

Johnstone IM (2009) Approximate null distribution of the largest root in multivariate analysis. *Annals of Applied Statistics*, **3**, 1616–1633.

Kanuch P, Berggren A, Cassel-Lundhagen A (2014) Genetic diversity of a successful colonizer: isolated populations of *Metrioptera roeselii* regain variation at an unusually rapid rate. *Ecology and Evolution*, **4**, 1117–1126.

Kingsolver JG, Diamond SE, Siepielski AM, Carlson SM (2012) Synthetic analyses of phenotypic selection in natural populations: lessons, limitations and future directions. *Evolutionary Ecology*, **26**, 1101–1118.

Kirkpatrick M (2009) Patterns of quantitative genetic variation in multiple dimensions. *Genetica*, **136**, 271–284.

Kopp M, Matuszewski S (2014) Rapid evolution of quantitative traits: theoretical perspectives. *Evolutionary Applications*, **7**, 169–191.

Lande R (1979) Quantitative genetic analysis of multivariate evolution, applied to brain:body size allometry. *Evolution*, **33**, 402–416.

Lee CE (2002) Evolutionary genetics of invasive species. *Trends in Ecology & Evolution*, **17**, 386–391.

Lewontin RC (1978) Adaptation. *Scientific American*, **239**, 157–169.

Luo F, Zhong JX, Yang YF, Scheuermann RH, Zhou JZ (2006) Application of random matrix theory to biological networks. *Physics Letters A*, **357**, 420–423.

Luo F, Yang Y, Zhong J et al. (2007) Constructing gene co-expression networks and predicting functions of unknown genes by random matrix theory. *BMC Bioinformatics*, **8**, 299.

McGuigan K, Blows MW (2007) The phenotypic and genetic covariance structure of Drosophilid wings. *Evolution*, **61**, 902–911.

McGuigan K, Blows MW (2009) Asymmetry of genetic variation in fitness-related traits: apparent stabilizing selection on g_{max}. *Evolution*, **63**, 2838–2847.

McGuigan K, Blows MW (2010) Evolvability of individual traits in a multivariate context: partitioning the additive genetic variance into common and specific components. *Evolution*, **64**, 1899–1911.

McGuigan K, Rowe L, Blows MW (2011) Pleiotropy, apparent stabilizing selection and uncovering fitness optima. *Trends in Ecology & Evolution*, **26**, 22–29.

McGuigan K, Collet JM, Allen SL, Chenoweth SF, Blows MW (2014a) Pleiotropic mutations are subject to strong stabilizing selection. *Genetics*, **197**, 1051–1062.

McGuigan K, Collet JM, McGraw EA et al. (2014b) The nature and extent of mutational pleiotropy in gene expression of male *Drosophila serrata*. *Genetics*, **196**, 911–921.

Meyer K (2007) Multivariate analyses of carcass traits for Angus cattle fitting reduced rank and factor analytic models. *Journal of Animal Breeding and Genetics*, **124**, 50–64.

Meyer K (2009) Factor-analytic models for genotype x environment type problems and structured covariance matrices. *Genetics Selection Evolution*, **41**, 21.

Meyer K, Kirkpatrick M (2008) Perils of parsimony: properties of reduced rank estimates of genetic covariance matrices. *Genetics*, **180**, 1153–1166.

Mezey JG, Houle D (2005) The dimensionality of genetic variation for wing shape in *Drosophila melanogaster*. *Evolution*, **59**, 1027–1038.

Morrissey MB, Kruuk LEB, Wilson AJ (2010) The danger of applying the breeder's equation in observational studies of natural populations. *Journal of Evolutionary Biology*, **23**, 2277–2288.

Morrissey MB, Parker DJ, Korsten P *et al.* (2012) The prediction of adaptive evolution: empirical applications of the secondary theorem of selection and comparison to the breeder's equation. *Evolution*, **66**, 2399–2410.

Orr HA, Unckless RL (2008) Population extinction and the genetics of adaptation. *American Naturalist*, **172**, 160–169.

Paaby AB, Rockman MV (2013) The many faces of pleiotropy. *Trends in Genetics*, **29**, 66–73.

Palmer AC, Kishony R (2013) Understanding, predicting and manipulating the genotypic evolution of antibiotic resistance. *Nature Reviews Genetics*, **14**, 243–248.

Patterson N, Price AL, Reich D (2006) Population structure and eigenanalysis. *PLoS Genetics*, **2**, e190.

Pavlicev M, Cheverud JM, Wagner GP (2009) Measuring morphological integration using eigenvalue variance. *Evolutionary Biology*, **36**, 157–170.

Pemberton JM (2010) Evolution of quantitative traits in the wild: mind the ecology. *Philosophical Transactions of the Royal Society B-Biological Sciences*, **365**, 2431–2438.

Pitchers W, Wolf J, Tregenza T, Hunt J, Dworkin I (2014a) Data from: Evolutionary rates for multivariate traits: the role of selection and genetic variation, Dryad Digital Repository, doi: 10.5061/dryad.g4t8c.

Pitchers W, Wolf JB, Tregenza T, Hunt J, Dworkin I (2014b) Evolutionary rates for multivariate traits: the role of selection and genetic variation. *Philosophical Transactions of the Royal Society B-Biological Sciences*, **369**, 20130252 doi: 10.1098/rstb.2013.0252.

Rausher MD (1992) The measurement of selection on quantitative traits: biases due to environmental covariances between traits and fitness. *Evolution*, **46**, 616–626.

Rius M, Darling JA (2014) How important is intraspecific genetic admixture to the success of colonising populations? *Trends in Ecology & Evolution*, **29**, 233–242.

Roff DA (1996) The evolution of genetic correlations: an analysis of patterns. *Evolution*, **50**, 1392–1403.

Rollins LA, Moles AT, Lam S *et al.* (2013) High genetic diversity is not essential for successful introduction. *Ecology and Evolution*, **3**, 4501–4517.

Roman J, Darling JA (2007) Paradox lost: genetic diversity and the success of aquatic invasions. *Trends in Ecology & Evolution*, **22**, 454–464.

Saccenti E, Smilde AK, Westerhuis JA, Hendriks M (2011) Tracy-Widom statistic for the largest eigenvalue of autoscaled real matrices. *Journal of Chemometrics*, **25**, 644–652.

Schluter D (1996) Adaptive radiation along genetic lines of least resistance. *Evolution*, **50**, 1766–1774.

Scott M, Diwell K, McKenzie JA (2000) Dieldrin resistance in *Lucilia cuprina* (the Australian sheep blowfly): chance, selection and response. *Heredity*, **84**, 599–604.

Selz OM, Lucek K, Young KA, Seehausen O (2014) Relaxed trait covariance in interspecific cichlid hybrids predicts morphological diversity in adaptive radiations. *Journal of Evolutionary Biology*, **27**, 11–24.

Solovieff N, Cotsapas C, Lee PH, Purcell SM, Smoller JW (2013) Pleiotropy in complex traits: challenges and strategies. *Nature Reviews Genetics*, **14**, 483–495.

Stinchcombe JR, Rutter MT, Burdick DS *et al.* (2002) Testing for environmentally induced bias in phenotypic estimates of natural selection: theory and practice. *American Naturalist*, **160**, 511–523.

Tracy CA, Widom H (1996) On orthogonal and symplectic matrix ensembles. *Communications in Mathematical Physics*, **177**, 727–754.

Tracy CA, Widom H (2009) The distributions of random matrix theory and their applications. In: *New trends in Mathe-matical Physics* (ed. Sidoravicius V), pp. 753–765. Springer, the Netherlands.

Turner KG, Hufbauer RA, Rieseberg LH (2014) Rapid evolution of an invasive weed. *New Phytologist*, **202**, 309–321.

Vandepitte K, De Meyer T, Helsen K *et al.* (2014) Rapid genetic adaptation precedes the spread of an exotic plant species. *Molecular Ecology*, **23**, 2157–2164.

Villmoare B (2013) Morphological integration, evolutionary constraints, and extinction: a computer simulation-based study. *Evolutionary Biology*, **40**, 76–83.

Wagner GP, Zhang JZ (2011) The pleiotropic structure of the genotype-phenotype map: the evolvability of complex organisms. *Nature Reviews Genetics*, **12**, 204–213.

Wagner GP, Kenney-Hunt JP, Pavlicev M *et al.* (2008) Pleiotropic scaling of gene effects and the 'cost of complexity'. *Nature*, **452**, 470–473.

Walsh B, Blows MW (2009) Abundant genetic variation + strong selection = multivariate genetic constraints: a geometric view of adaptation. *Annual Review of Ecology Evolution and Systematics*, **40**, 41–59.

Wang Z, Liao BY, Zhang JZ (2010) Genomic patterns of pleiotropy and the evolution of complexity. *Proceedings of the National Academy of Sciences of the United States of America*, **107**, 18034–18039.

Weber KE (1992) How small are the smallest selectable domains of form. *Genetics*, **130**, 345–353.

Wigner EP (1955) Characteristic vectors of bordered matrices with infinite dimensions. *Annals of Mathematics*, **62**, 548–564.

Willi Y, Van Buskirk J, Hoffmann AA (2006) Limits to the adaptive potential of small populations. *Annual Review of Ecology Evolution and Systematics*, **37**, 433–458.

Yang JA, Benyamin B, McEvoy BP *et al.* (2010) Common SNPs explain a large proportion of the heritability for human height. *Nature Genetics*, **42**, 565–569.

DATA ACCESSIBILITY

Gene expression data can be found in Gene Expression Omnibus (GEO) database under accession nos. GSE45801. SAS IML code for the simulations of the random P- and G-matrices is available as online Supporting Information.

SUPPORTING INFORMATION

Additional supporting information can be found in the online version of the *Molecular Ecology* article.

Data S1 Simulation of 200 P and G matrices with p=10 traits and n=250 individuals.

Chapter 12

INFORMATION ENTROPY AS A MEASURE OF GENETIC DIVERSITY AND EVOLVABILITY IN COLONIZATION

Troy Day

Department of Mathematics and Statistics, Jeffery Hall, Queen's University, Kingston, ON K7L 3N6, Canada
Department of Biology, Queen's University, Kingston, ON K7L 3N6, Canada

Abstract

In recent years, several studies have examined the relationship between genetic diversity and establishment success in colonizing species. Many of these studies have shown that genetic diversity enhances establishment success. There are several hypotheses that might explain this pattern, and here I focus on the possibility that greater genetic diversity results in greater evolvability during colonization. Evaluating the importance of this mechanism first requires that we quantify evolvability. Currently, most measures of evolvability have been developed for quantitative traits whereas many studies of colonization success deal with discrete molecular markers or phenotypes. The purpose of this study is to derive a suitable measure of evolvability for such discrete data. I show that under certain assumptions, Shannon's information entropy of the allelic distribution provides a natural measure of evolvability. This helps to alleviate previous concerns about the interpretation of information entropy for genetic data. I also suggest that information entropy provides a natural generalization to previous measures of evolvability for quantitative traits when the trait distributions are not necessarily multivariate normal.

Previously published as an article in *Molecular Ecology* (2015) 24, 2073–2083, doi: 10.1111/mec.13082

INTRODUCTION

The publication of the symposium volume *The Genetics of Colonizing Species* by Baker & Stebbins in 1965 was a landmark in the study of species invasions and the colonization of new habitats. A great deal of the published discussion of the symposium and the accompanying publications themselves centred on understanding those characteristics of species that make for good colonizers. For example, rapid growth, plasticity, short generation times and the ability to self have all been suggested as traits that increase colonization ability.

In addition to examining the attributes of individuals that make for good colonizers, one might also consider population-level characteristics that increase the likelihood of colonization success. Several of the contributions in Baker & Stebbins (1965) follow up ideas along this theme as well, but since that time (perhaps because of the influence of that volume), there has been an ever increasing interest in studies of this sort.

One particular area of focus has been the hypothesis that increased trait variation (e.g. genetic diversity) leads to increased colonization success (Hughes *et al.* 2008; Lee & Gelembiuk 2008; Forsman 2014). For example, Forsman (2014) recently surveyed the literature and identified 18 experimental studies on animals and plants that manipulated the genetic (and/or phenotypic) diversity of founder groups and then assessed the effect on establishment success. Of the 18 studies, all but one reported a significant positive relationship between establishment success and genetic diversity. González-Suárez *et al.* (2015) compiled observational data for 511 invasion events involving 97 different mammalian species. Interestingly, they found that establishment success was positively associated with intraspecific variation in adult body mass but negatively associated with variation in neonate body mass. Along similar lines, González-Suárez & Revilla (2013) used a large mammalian data set to show that the risk of extinction (for example due to habitat change) tends to decrease as intraspecific variation in adult body mass, litter size and age at sexual maturity each increases.

The above studies suggest that intraspecific diversity in some traits might enhance colonization success or reduce the likelihood of extinction if the environment changes. As has been noted previously, however, there are many possible explanations for this pattern (González-Suárez & Revilla 2013; Forsman 2014; González-Suárez *et al.* 2015). For example, high diversity might result in a high probability that preadapted types are already common among the colonizing individuals. Similarly, if some form of niche complimentary plays a role in colonization success, then high diversity might result in a high probability that the appropriate set of types is present among the colonizers (see also Loreau & Hector 2001).

Another explanation is that some form of evolutionary adaptation is required for successful establishment and that high (genetic) diversity results in a greater likelihood of such adaptation as opposed to extinction (Willi *et al.* 2006; Lee & Gelembiuk 2008). Dlugosch & Parker (2008) have shown that, across many species, colonizing populations have reduced genetic diversity when measured using discrete molecular markers. This would therefore lead one to suspect that the evolvability of such populations might often be compromised. When measured for quantitative traits, however, this pattern was less apparent. Furthermore, they suggest that many quantitative traits have evolved rapidly in colonizing populations despite the fact that diversity is sometimes reduced. Thus, the importance of this evolutionary hypothesis remains unclear.

Given the variety of hypotheses for diversity-colonization success relationships, studies of such patterns have unsurprisingly employed a wide variety of measures of diversity. For example, the focus of most of the 18 studies examined by Forsman (2014) was on genetic diversity, and most quantified genetic diversity as the number of genotypes present in the colonizers (e.g. Reusch *et al.* 2005 with *Zostera marina*, Wang *et al.* 2012 with *Spartina alterniflora*, Agashe 2009 with *Tribolium castaneum*, Ellers *et al.* 2011 with *Orchesella cincta*, Hovick *et al.* 2012 and Crawford & Whitney 2010 with *Arabidopsis thaliana*, Robinson *et al.* 2013 with *Daphnia magna* and Drummond & Vellend 2012 with *Taraxacum officinale*). However, several studies used proxies for genetic diversity such as effective population size (Newman & Pilson 1997 with *Clarkia pulchella*), level of inbreeding or relatedness (Leberg 1990 with *Gambusia holbrooki*, and Gamfeldt *et al.* 2005 with *Balanus improvisus*), degree of multiple mating in the parents of founders (Mattila & Seeley 2007 with honey bees) or phenotypic polymorphism (Wennersten *et al.* 2012 with *Tetrix subulata*). Finally, some studies directly measured heterozygosity or levels of polymorphism (Porcaccini & Piazzi 2001 with *Posidonia oceanica*, Markert *et al.* 2010 with *Americamysis bahia* and Martins & Jain 1979 with *Trifolium hirtum*).

The most suitable measure of diversity to use is presumably determined, in part, by the mechanisms that translate diversity into increased colonization success. After all, it will be these mechanisms that determine which aspects of genetic diversity are most important. Therefore, it might prove useful to examine how best to measure diversity in the context of different hypotheses. That is the purpose of this study. I focus on the hypothesis that greater genetic diversity leads to a greater evolvability in the colonizing population, and I ask how should evolvability be measured in the context of colonization studies such as those mentioned above? I will show using some relatively simple evolutionary considerations that Shannon's measure of the information entropy of the standing genetic variation is a natural measure of evolvability.

QUANTIFYING EVOLVABILITY

Although measures of evolvability has been derived in previous studies (Houle 1992; Hansen 2003; Hansen & Houle 2008), these focus on quantitative traits and restrict attention to cases where the trait follows a normal (or a multivariate normal) distribution. As can be seen from the above list of experimental studies, much of the work on the diversity-colonization success relationship has focused on molecular genetic data or discrete phenotypic traits. It is not clear how previously proposed measures of evolvability might be adapted to this situation, and therefore, I will derive a measure of evolvability by essentially 'starting from scratch'. Interestingly, the measure of evolvability obtained has some features that make it a potential generalization of previous measures of evolvability used for quantitative traits, but for situations where multivariate normality no longer necessarily holds (see Box 1).

I begin by considering the colonization of a new habitat. I use the terms 'native population' and 'native habitat' in reference to those of the source of the colonists. Similarly, I use the term 'novel population' and 'novel habitat' in reference to the colonists. Although this terminology is motivated by the study of colonizing species, the considerations below apply equally to other instances of adaptation to novel environments. For example, this includes instances of lab adaptation as well as adaptation to changes in the environment.

When a species colonizes a new area, often the selective conditions differ from those of the native population. For instance, the new area might have a different

temperature profile, salinity or photoperiod, in addition to a different set of biotic factors such as predation, competition and parasitism. Consequently, natural selection will favour different alleles in the novel habitat as compared with the native habitat. For a newly colonizing population to be successful, it must therefore not only overcome the demographic challenges associated with a small population size, but also potentially evolve adaptations to its new habitat if it is to avoid extinction.

To make things concrete, consider a single locus with m potentially segregating alleles (extensions to multiple loci will be considered briefly in the discussion). The probability distribution of alleles in the native population will be referred to as the 'native distribution'. The novel habitat will typically select for a different distribution of alleles, and I refer to this new distribution as the 'target distribution'. I will define the target distribution more precisely later, but for the time being, it should simply be viewed as the allelic distribution that is favoured by selection in the new habitat.

For successful adaption to the new habitat (as opposed to extinction), it might be necessary for the native distribution to evolve into the target distribution. We would therefore like a way of quantifying how easy it is for this to happen. I will restrict attention to the simplest case where the alleles favoured in the novel habitat are present in the standing genetic variation of the native population. This is effectively the same type of assumption that is made in the analyses of evolvability for quantitative traits (Hansen & Houle 2008). Extensions to situations where the favoured alleles must first arise through mutation in the novel habitat are considered briefly in the discussion.

Now consider a native distribution and the target distribution to which it must evolve. How easy is it for this to occur through selection? If the native distribution can achieve target distributions that are very different from itself, then, all else equal, we would say that the native distribution is highly evolvable. At the same time, however, if the native distribution can only achieve such very different target distributions if selection is extremely strong, then we might say that its evolvability is quite low. Thus, I seek a measure that quantifies the magnitude of the change in distribution that can be achieved *per unit strength of selection*, as has been used in other analyses (Hansen & Houle 2008).

The first step is to quantify how much change occurs in going from the native to the target distribution. Suppose p_i is the native distribution of allele frequencies and p_i^* is the target distribution. How can we measure

Box 1 The relationship between measures of evolvability

There are two useful ways to compare the measure of evolvability developed here with that proposed for quantitative traits by Houle (1992) and Hansen & Houle (2008).

The most direct comparison can be made by using the general measure of evolvability in eqn (4). This measure makes no assumptions about the strength or form of selection or about the form of the target distribution. Furthermore, in the case of a continuous random variable X representing the breeding value of a quantitative trait, and target and native probability density functions given by $p^*(x)$ and $p(x)$, respectively, the corresponding quantity D can be computed as

$$D(p^* \| p) = \int p^*(x) \log_b \left(\frac{p^*(x)}{p(x)} \right) dx.$$

Therefore, we can compute the evolvability in eqn (4) for any distribution of interest.

In the context of quantitative traits, we restrict attention to those specific types of selection (and mutation) that are compatible with both the native and the target distributions being Gaussian and having a common variance. Under these conditions, the evolutionary change that occurs in going from the native distribution $p(x)$ to the target distribution $p^*(x)$ is completely described by the difference in the means of these two distributions. In this case, $D(p^* \| p)$ can be calculated as

$$D(p^* \| p) = \frac{1}{2} \frac{(\Delta \bar{x})^2}{v}$$

where $\Delta \bar{x}$ is the difference in the means and v is the (common) variance.

Now if we use the mean-standardized nondimensional breeder's eqn from Hansen & Houle (2008), we have $\Delta \bar{x} = I_A \beta$ where I_A is the mean-standardized variance of the breeding value distribution and β is the mean-standardized selection gradient. Thus, we have

$$D(p^* \| p) = I_A \frac{\beta^2}{2}$$

Hansen and Houle proposed $\sqrt{I_A}$, or equivalently I_A, as a measure of evolvability for such traits. The above eqn shows that this is a special case of eqn (4) where the strength of selection is measured by $S(p_i^*, p_i) = \beta^2/2$.

A second type of comparison with previous measures of evolvability can be made by considering how evolvability as measured by information entropy in eqn (7) (which assumes truncation selection and no knowledge of the selective regime in the novel habitat) compares with previous measures if the native distribution is a continuous distribution of breeding values. Although it is well-known that there is no equivalent measure of information entropy for continuous random variables, Shannon himself defined the entropy in such cases by *analogy* with the discrete case. In particular, he used

$$-\int p(x) \log_b p(x) dx$$

as the information entropy for a continuous random variable X (Shannon & Weaver 1949). The above expression is usually referred to as differential entropy to distinguish it from the right-hand side of eqn (7).

Interestingly, for a mean-standardized normal distribution of breeding values, the differential entropy can be calculated as $\ln(\sqrt{I_A} \sqrt{2\pi e})$ where e is the base of the natural logarithm. Therefore, for normally distributed quantitative traits, information entropy reduces to a simple transformation of the coefficient of variation $\sqrt{I_A}$. Thus, in this special case, we again obtain a measure of evolvability that is effectively the same as that proposed for quantitative traits. This lends support to the idea that information entropy provides a type of generalization of other measures of evolvability for cases where the distribution of breeding values is not Gaussian.

the change achieved when p_i evolves into the distribution p_i^*, from the standpoint of selection and evolution? If the target distribution can be obtained simply by randomly sampling the native distribution, then, in a sense, no change is necessary to go from p_i to p_i^* because the two distributions are effectively the same from the standpoint of selection. This is because no selection (i.e. no biased sampling) is required to obtain the target distribution. On the other hand, if it is very unlikely to

obtain the target distribution by randomly sampling the native distribution, then a great deal of change is required to go from p_i to p_i^* because the two distributions would then be very different from the standpoint of selection. Thus, we can use the probability of obtaining the target distribution from the native distribution via random sampling as an evolutionarily relevant measure of the change achieved in evolving from one into the other through selection.

To formalize this idea, we need to calculate the probability of obtaining the target distribution when sampling from the native distribution. I begin by first considering a sample of size n (shortly I will take the limit as $n \to \infty$). To work with a sample of size n, we first need to characterize the target distribution for a sample of size n. Let A_i be random variables denoting the number of alleles of type i that are obtained when drawing a sample of n alleles from the target distribution (with $\sum A_i = n$). Now, for any sample obtained from the target distribution, we can calculate the probability of obtaining this same collection of alleles when sampling n alleles from the native distribution. Denoting this probability by Z_n, it is given by the multinomial probability:

$$Z_n = \frac{n!}{\prod_i A_i!} \prod_i p_i^{A_i} \tag{1}$$

The probability Z_n in eqn (1) is, itself, a random variable because the A_i are random variables. Different samples from the target distribution will, by chance, result in different numbers of each type of allele, A_i. Consequently, different samples will result in different probabilities Z_n of obtaining the sample when drawing from the native distribution. To obtain a 'typical' value for Z_n, we can calculate an average of Z_n over the draws in the sample for a large sample size n. The multiplicative nature of (1) means that a 'typical' value of Z_n is best characterized by its geometric average, $(Z_n)^{1/n}$. This represents the (geometric) average probability across all draws in a sample, and it takes a value between 0 (the distributions are very different) and 1 (the distributions are effectively the same).

The quantity $(Z_n)^{1/n}$ is also a random variable, but we expect that, from the law of large numbers, it will not vary much if the sample size n is large. Before formally considering this limit, however, I first take the negative logarithm of $(Z_n)^{1/n}$. Denoting the resulting quantity by L_n gives

$$L_n = -\log_b \left(\frac{n!}{\prod_i A_i!} \prod_i p_i^{A_i} \right)^{1/n}$$
$$= -\frac{1}{n}\log_b n! + \frac{1}{n}\sum_i \log_b A_i! - \frac{1}{n}\sum_i A_i \log_b p_i \tag{2}$$

where I have left the base of the logarithm unspecified. L_n is a non-negative random variable, and as the native and target distributions become increasingly different, L_n takes on increasingly large positive values.

The use of the logarithm in (2) has two advantages. First, as will be seen below, the strength of selection is often naturally measured on a additive scale and using the logarithm here means that we are then measuring the change in distribution on an additive scale as well.

Second, using the logarithm provides a convenient way to interpret the change achieved in evolving from the native to the target distribution. The base b is arbitrary and so it can be viewed as setting the scale of measurement. In particular, a value of $L_n = 0$ means that there is a (average) probability of 1 of obtaining the target distribution by randomly sampling the native distribution. A value of $L_n = 1$ means that there is a (average) probability of $1/b$ of obtaining the target distribution. Similarly, a value of $L_n = 2$ means that there is a (average) probability of $1/b^2$ of obtaining the target distribution, and so forth.

Relative values of L_n also have meaningful interpretations. For example, suppose two populations differ by one unit in the amount of change, L_n, that occurs when evolving from the native into the target distribution, and consider working with the base $b = 10$. Then, the population with the larger value of L_n will have evolved ten times as much as the other population in the sense that the probability of it giving rise to the target distribution simply through random sampling will have increased ten times more than that of the other population.

With these preliminaries, we can now take the limit $n \to \infty$. In this limit, the random variable L_n approaches a fixed (i.e. deterministic) value, and defining $D = \lim_{n \to \infty} L_n$, we obtain

$$D(p_i^* \| p_i) = \sum_i p_i^* \log_b \left(\frac{p_i^*}{p_i} \right) \tag{3}$$

where I have used the notation $D(p_i^* \| p_i)$ to indicate that D depends on both the target and native distributions (the notation '$\|$' in the argument of D is a convention borrowed from information theory). Formally, the above limit holds 'almost surely' and makes use of the strong law of large numbers, Stirling's approximation $\log_b n! = n\log_b n - n + O(\log_b n)$ and the continuous mapping theorem. Analogous calculations have been used in the context of statistical mechanics and Bayesian statistics (for example, see Jaynes 2003). The quantity $D(p_i^* \| p_i)$ in (3) that characterizes the amount of change that occurs in evolving from the native into the target distribution is known in information theory as the Kullback-Leibler divergence between the distributions p_i^* and p_i.

To quantify the evolvability of the native population, we also need to know how strong selection must be in order to evolve the target distribution from the native distribution (Hansen & Houle 2008). Many measures of the strength of selection have been proposed but, as far as I am aware, there is no single measure that is suitable under all possible forms of selection. Therefore, although there is a very general way to measure the amount of evolution that occurs through one generation of selection, there does not appear to be a similarly general way to measure the strength of selection required to cause this evolution. Thus, for the moment, I simply denote the strength of selection required to evolve the distribution p_i^* from p_i in a single generation as $S(p_i^*, p_i)$.

With the above two ingredients in hand, we can now define the evolvability $\mathcal{E}$ of the target distribution from the native distribution, as the amount of change in allelic distribution that occurs in going from one to the other, per unit strength of selection. We have

$$\mathcal{E} = \frac{D(p_i^* \| p_i)}{S(p_i^*, p_i)}. \tag{4}$$

Box 1 discusses the relationship between this general formulation and the specific measures of evolvability for quantitative traits proposed by Hansen & Houle (2008). The next task is to simplify the general measure of evolvability given by eqn (4) for populations that colonize novel habitats.

INFORMATION ENTROPY AS EVOLVABILITY WHEN ADAPTING TO NOVEL ENVIRONMENTS

Equation (4) is a very general measure of evolvability that can be applied to any distribution of alleles at a single locus and to any target distribution (it can also be extended to account for a continuum of alleles; Box 1). It also highlights an important point – the evolvability of a population depends not only on the allelic distribution of that population p_i, but also on the form of selection as quantified by the difference between p_i and p_i^*. A similar dependence occurs with previous measures of evolvability for quantitative traits as well although selection in such cases is necessarily restricted to specific functional forms (e.g. Hansen & Houle 2008; Chevin 2012).

In many situations, we do not know exactly what the form of selection will be, and therefore, we cannot specify a particular target distribution. For example, this is often the case for populations that colonize novel habitats. In such cases, the best we can do is to calculate an expected evolvability over the different forms of selection that might occur (e.g. see Kirkpatrick 2009; Chevin 2012). To do so, we first specify the set of possible alleles. We then specify a class of target distributions for this set of alleles that captures the different forms of selection of interest. Finally, we calculate the expected evolvability $\mathbb{E}[\mathcal{E}]$ by averaging (4) over these different target distributions, where each target distribution is weighted by its probability of occurrence.

There are two equivalent ways that we can proceed in this direction. The first starts by specifying a class of fitness functions. These can then be used to obtain a suitable measure of the strength of selection for each target, $S(p_i^*, p_i)$. They can also be used to derive the target distributions by determining the distribution of alleles that is produced by each fitness function. From there, one can compute the quantity $D(p_i^* \| p_i)$. The second approach starts by specifying a class of target distributions. These will directly determine the quantity $D(p_i^* \| p_i)$ for each target. We then derive a fitness function for each target distribution by determining the function required to produce the target through selection. Finally, these fitness functions can be used to obtain a suitable measure of the strength of selection for each target, $S(p_i^*, p_i)$. I follow the second approach below as the general measure of evolvability in eqn (4) has been developed by fixing attention on the native and target distributions themselves.

From the standpoint of evolvability during colonization, some types of target distributions are perhaps of more interest than others. To the best of my knowledge, all previous measures of evolvability for quantitative traits are based on an assumption that, whatever the nature of selection might be, it favours a particular phenotypic or genotypic value. This is embodied by the assumption that selection is directional in these studies (Hansen & Houle 2008). This also seems like a reasonable assumption for adaptation to a novel habitat, and so I employ a form of directional selection as well.

There are many different ways to model directional selection. There is an 'obvious' choice for which of these to use in the context of quantitative traits, however, because the form that is chosen must be compatible with the assumption that all distributions remain multivariate normal. Thus, most such studies assume (sometimes implicitly) that the fitness of different types is specified by a function from the exponential family.

In the context of eqn (4), there is no similar constraint that dictates our choice for the form of fitness because we are allowing for arbitrary target distributions. Thus to model directional selection for a particular allele k, we are free to choose any target distribution subject only to the constraint that the frequency of allele k increases through selection and that of other alleles decreases (i.e. $p_k^* > p_k$ and $p_i^* < p_i$ for all $i \neq k$). This freedom is both a luxury and a curse as we must restrict things more in order to make further progress, but any such restriction is necessarily somewhat arbitrary.

Perhaps the most obvious further restriction for modelling directional selection is to assume truncation selection. This is the most extreme form of directional selection for a particular allele. It is also the simplest form of directional selection in the sense of having the fewest parameters. Furthermore, it is a natural form of selection that one might impose artificially if attempting to quantify the evolvability of a population by selecting it in different directions.

Another convenient feature of truncation selection is that it permits an unambiguous measure of the strength of selection. If we denote the frequency of any favoured type under truncation selection by q, then under random mating the evolutionary change in q as a result of one generation of selection is given by (Wright 1935)

$$\Delta q = \frac{q(1-q)}{2} \frac{\partial \ln \bar{W}}{\partial q} \tag{5}$$

where Δq is the change in q and $\bar{W}$ is the population mean fitness. This is an example of Wright's adaptive topography (Wright 1935; Lande 1976; Barton & Turelli 1987), and the magnitude of the quantity $\partial \ln \bar{W} / \partial q$ (which is sometimes referred to as the selection gradient) provides a widely used normalized measure of the strength of selection.

For these reasons, I proceed with an assumption of truncation selection for a particular allele k. In this case, the target distribution is $p_k^* = 1$ and $p_i^* = 0$ for all $i \neq k$, and the associated fitness function that produces this target distribution is $W_k > 0$ and $W_i = 0$ for all $i \neq k$, where W_i is the marginal fitness of allele i. It is worth noting though that this form of target distribution applies more generally if, instead of quantifying evolvability over a single generation of selection, we were to quantify it over multiple generations. For example, suppose we assume a more general form of directional selection for allele k where we require only that $W_k > W_i$

for all $i \neq k$. This is no longer truncation selection as all alleles might have nonzero fitness. Under these conditions, if no processes other than selection are occurring, then over time the population will evolve to a distribution concentrated entirely at allele k. In other words, the allelic distribution would converge to $p_i = 1$ if $i = k$ and $p_i = 0$ otherwise. Thus, the same target distribution applies to this more general form of directional selection if we quantify evolvability over a large number of generations.

Under truncation selection, we have $\bar{W} = p_k W_k$ and therefore $S(p_i^*, p_i) = |\partial \ln \bar{W} / \partial p_k| = 1/p_k$. Substituting this and $p_k^* = 1$ and $p_i^* = 0$ for all $i \neq k$ into (4) gives

$$\mathcal{E} = -p_k \log_b p_k \tag{6}$$

Equation (6) gives the evolvability of the target distribution from the native distribution when there is truncation selection for allele k. As we do not know *a priori* which allele will have the highest fitness during colonization, the final step is to calculate the expected evolvability $\mathbb{E}[\mathcal{E}]$ over all possible favoured alleles. To do so, we need to specify the probability that allele k will be the favoured allele. If we have no *a priori* knowledge of which allele will be favoured, the only reasonable assumption is that each allele has equal chance of being the favoured allele. Thus, taking the expectation of (6) over a uniform distribution for k gives

$$\mathbb{E}[\mathcal{E}] = -\sum_k p_k \log_{\hat{b}} p_k \tag{7}$$

where the normalization constant $1/m$ has been absorbed into the new base of the logarithm, $\hat{b}$. Equation (7) is Shannon's measure of the information entropy of the native allelic distribution (Shannon 1948). Thus, the information entropy of the native allelic distribution provides a natural measure of the evolvability of a population under truncation directional selection when colonizing a novel habitat (Box 2 provides some intuition for thinking about information entropy).

DISCUSSION AND CONCLUDING REMARKS

The results derived here suggest that Shannon's information entropy is a sensible measure of genetic diversity in the context of evolvability in novel habitats. Shannon information is frequently used as a measure

Box 2 Gaining an intuition for information entropy

Although information entropy is used frequently in ecology, it is less common in evolutionary biology (although see Frank 2012 for uses of information theory in evolutionary biology that are quite distinct from that explored here). There are two simple ways to begin developing an intuition for what information entropy represents.

The first is rooted in the information-theoretic origins of entropy. Suppose we know the native allelic distribution for some population and imagine that a randomly chosen individual disperses to a novel habitat. For simplicity suppose the organism is haploid. The allelic identity of this single colonist can be viewed as a random variable drawn from the known native allelic distribution. We can then ask the qualitative question, how much new information do we gain about the allelic identity of this colonist if we were to actually measure it?

Although the above question is vague in that we have not specified what is meant by 'information', we can make some qualitative progress without being more precise. For example, if the native population contained only a single allelic type, then clearly we would gain no new information by measuring the colonist. This is because we already know with certainty what its identity must be. On the other hand, if there are m potential alleles, and if each of them is equally frequent in the native population, then we would gain a great deal of information by measuring the colonist. This is because its allelic identity prior to measurement is maximally uncertain. And it seems reasonable that we would gain an intermediate degree of new information if the native allelic distribution was somewhere between these two extremes.

Information entropy captures the above qualitative intuition in a precise way. Roughly speaking, it is a measure of the uncertainty of a random variable. Low information entropy corresponds to a low degree of uncertainty in the outcome of a random variable. In such cases, we gain very little new information about a realization of the random variable by seeing its value because there is not much uncertainty in its outcome. On the other hand, high information entropy corresponds to a high degree of uncertainty in the outcome of a random variable. In this case, we gain a great deal of information about a realization of the random variable by seeing its value.

Another useful way to think about the information entropy of a distribution is as a measure of variability. Studies of variance are common in evolutionary biology because, under certain assumptions, genetic variance plays an important role in evolutionary change through natural selection. More generally *genetic variability* is perhaps a more suitable quantity from the standpoint of selection as there needs to be variation for selection to act. Importantly, variability and variance are not always the same thing.

As an example, consider a distribution of discrete allelic values where each allele specifies the value of a quantitative trait like body size. Figure 1 presents distributions from two hypothetical populations that, in an important sense, have the same variability. Both populations have

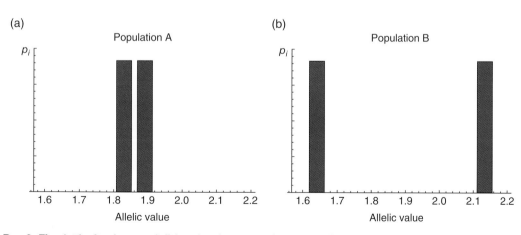

Box 2, Fig. 1 The distributions of allelic values for two populations, A and B. Both populations contain two alleles with equal frequency, and therefore, they have equal variability as measured by information entropy. However, the variance in allelic value of population B is larger than that of population A.

only two alleles and each is equally frequent. As a result, their information entropy is the same. However, the variance in body size in the two populations is very different, with population B having a higher variance. From an evolutionary standpoint, however, there is an important sense in which the two populations are equally evolvable. Therefore variation, as measured by information entropy, can be a more suitable measure of evolvability. Along similar lines, Figure 2 presents distri-butions from two populations, population A with high variance but low variability (as measured by information entropy) and populations B with low variance but high variability (again measured by information entropy). In this case, it seems reasonable that we would want to classify population B as being more evolvable than population A, again suggesting that variability as meas-ured by information entropy is a more suitable measure of evolvability than variance.

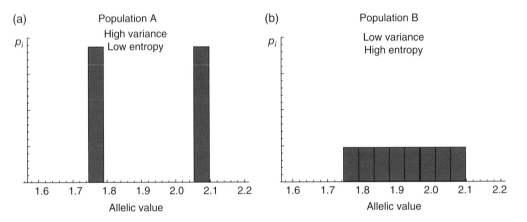

Box 2, Fig. 2 The distributions of allelic values for two populations, A and B. The populations have differing variances and variability as measured by information entropy. Population A has a high variance in allelic value but low variability, while population B has the opposite.

of species diversity in the ecological literature, and there have also been several instances of its use in pop-ulation genetics as a measure of genetic diversity (Sherwin *et al.* 2006; Kosman & Leonard 2007; Sherwin 2010). The earliest such instance that I am aware of is Lewontin's (1972) use of information entropy for quantifying patterns of allelic diversity in humans. Nevertheless, this measure has not seen widespread use in population genetics despite the fact that many software packages can routinely compute this quantity.

One possible reason for a general reluctance to use Shannon information in population genetics has to do with its interpretation (Nei 1975; Hennick & Zeven 1991). In ecology, several different measures of diver-sity are often used and debated, and the discussions sometimes centre around the phenomenological prop-erties of each measure. For example, discussions often

consider questions like whether a measure adequately captures species evenness versus richness, or whether it places 'too much' weight on rare species.

In population genetics, the tradition has been to focus more on the mechanistic interpretation of diver-sity measures. For example, measures like heterozygo-sity or percentage polymorphic loci have clear population-genetic interpretations. And as Nei (1975) remarked, Shannon information was 'designed to measure the amount of information in information engineering and is not related to any genetic entity. As such it is not clear what the value of this quantity means in terms of genetic materials'.

The derivation presented here partially addresses this issue of interpretation. It shows that information entropy is a natural measure of evolvability during colonization. As an example, suppose we chose the base $\hat{b}=10$ and we wish to compare the evolvability of two native populations, A and B, and suppose further that population A has value of $\mathbb{E}[\mathcal{E}]$ that is one unit larger than that of population B. Then, under truncation directional selection, population A is 10 times more evolvable than population B in the sense that, per unit strength of selection, population A can evolve a 10-fold greater likelihood of giving rise to the target distribution through simple random sampling.

Interestingly, of the 18 studies on colonization described in the introduction, the earliest of these (Martins & Jain 1979) was also the only study to employ information entropy as a measure of genetic diversity. Coincidentally, Martins & Jain's (1979) study also appears to have been motivated, in large part, by the symposium volume of Baker & Stebbins (1965). They studied rose clover, *Trifolium hirtum*, and examined the effect of the information entropy of the allelic distribution of colonists over two years. In the first year of the study, they found no effect, but in the second year, they found a rather strong positive relationship between entropy and the establishment success of new roadside colonies.

The measure of information entropy in eqn (7) leaves the base of the logarithm, $\hat{b}$, unspecified. One natural choice is $\hat{b} = m$ where m is the number of alleles potentially segregating. Information entropy is always maximized when all alleles are present at equal frequency, and therefore with this choice of $\hat{b}$, the maximum possible information entropy is 1. This choice is analogous to the use of base 2 logarithms in information theory where the random variables of interest are often binary, taking one of two possible values. Thus, while information entropy is measured in *bits* when using base 2, it is measured in so-called *m-ary* units when using base m.

It should also be emphasized that, when using any measure to compare the evolvability of two populations, one should ensure that the *potential* allelic types in each population are the same. Otherwise the meaning of any comparison is unclear. This does not mean that the same alleles need to be segregating in all populations, but only that the alleles segregating in each population are a subset of the same set of m possible alleles. Furthermore, all else equal $\mathbb{E}[\mathcal{E}]$ will tend to increase as the number of alleles included in a sample increases, and therefore, it is also important to control for sampling effort when making comparisons.

The derivation presented here has focused entirely on quantifying evolvability in terms of the allelic distribution of a single locus. Often, however, data are available for multiple loci. In such cases, there are two ways that one might make use of such data. The first is simply to average the value of $\mathbb{E}[\mathcal{E}]$ over all loci. In fact this was exactly the approach taken by Martins & Jain (1979), and it is analogous to the calculation of average heterozygosity across multiple loci Nei (1975).

The second approach is instead to calculate the *joint* information entropy across all loci. To do so, one would use the distribution of all genotypes of interest and calculate the information entropy of this distribution. In other words, each locus would be viewed as a random variable, and the genotype distribution would be regarded as the joint distribution of alleles across all loci. The joint information entropy would then be calculated as in eqn (7), but where the summation takes place over all elements of the joint probability distribution. This second approach is preferable in some ways as it gives a total measure of evolvability, accounting for any possible association of alleles through linkage disequilibrium, whereas the first approach gives the average evolvability of a single locus.

As with previous measures of evolvability the measure derived here focuses only on standing genetic variation. While this is likely an important component of adaptation in colonization, novel mutations are likely also important (Schluter & Barrett 2008). Although accounting for this in measures of evolvability is difficult without knowing more about mutational pathways and fitness relationships, we might still make some progress by viewing the target distribution as the distribution obtained from the standing variation that is as close as possible to the distribution favoured by selection. The rationale would be that the closer the population is to the real distribution favoured by selection, perhaps the longer the population can persist before going extinct. As a result, the greater will be the likelihood that the appropriate mutations arise before extinction occurs.

Finally, it is important to emphasize the limitations of information entropy as a measure of evolvability. As eqn (4) shows, the evolvability of any population depends on the form of selection. Consequently, there is no single measure that is appropriate under all conditions. The derivation of information entropy in eqn (7) from eqn (4) rests on two important assumptions: (i) that there is truncation selection in favour of a particular allele and (ii) that all alleles are equally likely to be the favoured allele. If either of these assumptions is relaxed, then a different measure of evolvability might be obtained. For example, if we have reason to believe that certain alleles are more likely to be favoured during colonization than others, then the expected evolvability can be written more generally as $-\mathbb{E}[\,p_k \log_b p_k\,]$ where the expectation is taken over an appropriate, nonuniform, distribution. Likewise, relaxing the assumption of truncation selection will typically produce still different measures. Thus, it is important to choose a measure of evolvability that appropriately captures the situation of interest.

ACKNOWLEDGEMENTS

I thank Mark Blows and David Houle for discussion and David Houle, Steve Frank and two anonymous referees for very helpful feedback on the manuscript. This work was supported by a research grant from the Natural Sciences and Engineering Research Council of Canada. I also thank Wiley publishing for financial support to attend the symposium.

REFERENCES

Agashe D (2009) The stabilizing effect of intraspecific genetic variation on population dynamics in novel and ancestral habitats. *The American Naturalist*, **174**, 255–267.

Baker HG, Stebbins GL (1965) *The Genetics of Colonizing Species*. Academic Press, Waltham, Massachusetts.

Barton NH, Turelli M (1987) Adaptive landscapes, genetic distance, and the evolution of quantitative characters. *Genetical Research*, **49**, 157–173.

Chevin L-M (2012) Genetic constraints on adaptation to a changing environment. *Evolution*, **67**, 708–721.

Crawford KM, Whitney KD (2010) Population genetic diversity influences colonization success. *Molecular Ecology*, **19**, 1253–1263.

Dlugosch KM, Parker IM (2008) Founding events in species invasions: genetic variation, adaptive evolution, and the role of multiple introductions. *Molecular Ecology*, **17**, 431–449.

Drummond EBM, Vellend M (2012) Genotypic diversity effects on the performance of *Taraxacum officinale* populations increase with time and environmental favorability. *PLoS ONE*, **7**, e30314.

Ellers J, Rog S, Braam C, Berg MP (2011) Genotypic richness and phenotypic dissimilarity enhance population performance. *Ecology*, **92**, 1605–1615.

Forsman A (2014) Effects of genotypic and phenotypic variation on establishment are important for conservation, invasion, and infection biology. *Proceedings of the National Academy of Science*, **111**, 302–307.

Frank SA (2012) Natural selection. V. How to read the fundamental equations of evolutionary change in terms of information theory. *Journal of Evolutionary Biology*, **25**, 2377–2396.

Gamfedlt L, Wallen J, Jonsson PR, Berntsson KM, Havenhand JN (2005) Increasing intraspecific diversity enhances settling success in a marine invertebrate. *Ecology*, **86**, 3219–3224.

González-Suárez M, Revilla E (2013) Variability in life-history and ecological traits is a buffer against extinction in mammals. *Ecology Letters*, **16**, 242–251.

González-Suárez M, Bacher S, Jeschke JM (2015) Intraspecific trait variation is correlated with establishment success of alien mammals. *The American Naturalist*. in press.

Hansen TF (2003) Is modularity necessary for evolvability? Remarks on the relationship between pleiotropy and evolvability. *Molecular BioSystems*, **69**, 83–94.

Hansen TF, Houle D (2008) Measuring and comparing evolvability and constraint in multivariate characters. *Journal of Evolutionary Biology*, **21**, 1201–1291.

Hennick S, Zeven AC (1991) The interpretation of Nei and Shannon-Weaver within population variation indices. *Euphytica*, **51**, 235–240.

Houle D (1992) Comparing evolvability and variability in quantitative traits. *Genetics*, **130**, 195–204.

Hovick SM, Gumuser ED, Whitney KD (2012) Community dominance patters, not colonizer genetic diversity, drive colonization success in a test using grassland species. *Plant Ecology*, **213**, 1365–1380.

Hughes AR, Inouye BD, Johnson MTJ, Underwood N, Vellend M (2008) Ecological consequences of genetic diversity. *Ecology Letters*, **11**, 609–623.

Jaynes ET (2003). *Probability theory: the logic of science*. Cambridge University Press, Cambridge, UK.

Kirkpatrick M (2009) Patterns of quantitative genetic variation in multiple dimensions. *Genetica*, **136**, 271–284.

Kosman E, Leonard KJ (2007) Conceptual analysis of methods applied to assessment of diversity within and distance between populations with asexual and mixed mode of reproduction. *New Phytologist*, **174**, 683–696.

Lande R (1976) Natural selection and random genetic drift in phenotypic evolution. *Evolution*, **30**, 314–334.

Leberg PL (1990) Influence of genetic variability on population growth: implications for conservation. *Journal of Fish Biology*, **37** (supplement A), 193–195.

Lee CE, Gelembiuk GW (2008) Evolutionary origins of invasive populations. *Evolutionary Applications*, **1**, 427–448.

Lewontin R (1972) The apportionment of human diversity. *BMC Evolutionary Biology*, **6**, 381–398.

Loreau M, Hector A (2001) Partitioning selection and complementarity in biodiversity experiments. *Nature*, **412**, 72–76.

Markert JA, Champlin DM, Gutjahr-Gobell R *et al.* (2010) Population genetic diversity and fitness in multiple environments. *BMC Evolutionary Biology*, **10**, 205.

Martins PS, Jain SK (1979) Role of genetic variation in the colonizing ability of Rose Clover (*Trifolium hirtum* All.). *The American Naturalist*, **114**, 591–595.

Mattila HR, Seeley TD (2007) Genetic diversity in honey bee colonies enhances productivity and fitness. *Science*, **317**, 362–364.

Nei M (1975) *Molecular population genetic and evolution*. North-Holland Publishing, Amsterdam.

Newman D, Pilson D (1997) Increased probability of extinction due to decreased genetic effective population size: experimental populations of *Clarkia pulchella*. *Evolution*, **51**, 354–362.

Porcaccini G, Piazzi L (2001) Genetic polymorphism and transplantation success in the mediterranean seagrass *Posidonia oceanica*. *Restoration Ecology*, **9**, 332–338.

Reusch TBH, Ehlers A, Hammerli A, Worm B (2005) Ecosystem recovery after climatic extremes enhanced by genetic diversity. *Proceedings of the National Academy of Science*, **102**, 2826–2831.

Robinson JD, Wares JP, Drake JM (2013) Extinction hazards in experimental *Daphnia magna* populations: effects of genotype diversity and environmental variation. *Ecology and Evolution*, **3**, 233–243.

Schluter D, Barrett R (2008) Adaptation from standing genetic variation. *Trends in Ecology and Evolution*, **23**, 38–44.

Shannon CE (1948) A mathematical theory of communications. *The Bell System Technical Journal*, **27**, 379–423.

Shannon CE, Weaver WW (1949) *The mathematical theory of communication*. University of Illinois Press, Urbana, IL.

Sherwin WB (2010) Entropy and information approaches to genetic diversity and its expression: genomic geography. *Entropy*, **12**, 1765–1798.

Sherwin WB, Jabot F, Rush R, Rossetto M (2006) Measurement of biological information with applications from genes to landscapes. *Molecular Ecology*, **15**, 2857–2869.

Wang XY, Shen DW, Jiao J (2012) Genotypic diversity enhances invasive ability of *Spartina alterniflora*. *Molecular Ecology*, **21**, 2542–2551.

Wennersten L, Johansson J, Karpestam E, Forsman A (2012) Higher establishment success in more diverse groups of pygmy grasshoppers under seminatural conditions. *Ecology*, **93**, 2519–2525.

Willi Y, Van Buskirk J, Hoffman AA (2006) Limits to the adaptive potential of small populations. *Annual Review of Ecology Evolution and Systematics*, **37**, 433–458.

Wright S (1935) Evolution in populations in approximate equilibrium. *Journal of Genetics*, **30**, 257–266.

Chapter 13

EXPANSION LOAD: RECESSIVE MUTATIONS AND THE ROLE OF STANDING GENETIC VARIATION

Stephan Peischl[*][†] *and Laurent Excoffier*[*][†]

[*]Institute of Ecology and Evolution, University of Berne, Berne 3012, Switzerland
[†]Swiss Institute of Bioinformatics, Lausanne 1015, Switzerland

Abstract

Expanding populations incur a mutation burden – the so-called expansion load. Previous studies of expansion load have focused on codominant mutations. An important consequence of this assumption is that expansion load stems exclusively from the accumulation of new mutations occurring in individuals living at the wave front. Using individual-based simulations, we study here the dynamics of standing genetic variation at the front of expansions, and its consequences on mean fitness if mutations are recessive. We find that deleterious genetic diversity is quickly lost at the front of the expansion, but the loss of deleterious mutations at some loci is compensated by an increase of their frequencies at other loci. The frequency of deleterious homozygotes therefore increases along the expansion axis, whereas the average number of deleterious mutations per individual remains nearly constant across the species range. This reveals two important differences to codominant models: (i) mean fitness at the front of the expansion drops much faster if mutations are recessive, and (ii) mutation load can increase during the expansion even if the total number of deleterious mutations per individual remains constant. We use our model to make predictions about the shape of the site frequency spectrum at the front of range expansion, and about correlations between heterozygosity and fitness in different parts of the species range. Importantly, these predictions provide opportunities to empirically validate our theoretical results. We discuss our findings in the light of recent results on the distribution of deleterious genetic variation across human populations and link them to empirical results on the correlation of heterozygosity and fitness found in many natural range expansions.

Previously published as an article in *Molecular Ecology* (2015) 24, 2084–2094, doi: 10.1111/mec.13154

INTRODUCTION

Identifying and understanding the ecological and evolutionary processes that cause range expansions, range shifts or contractions has a long tradition in evolutionary biology (Darwin 1859; MacArthur 1972; Sexton *et al.* 2009). More recently, the growing appreciation of the consequences of dynamic range margins on the ecology, population genetics and behaviour of species has changed our views about several evolutionary processes, such as the evolution of dispersal (Phillips *et al.* 2006; Shine *et al.* 2011; Lindström *et al.* 2013), life history traits (Phillips *et al.* 2010) and species range limits (Peischl *et al.* 2015).

Strong genetic drift at the margins of expanding populations allows some neutral genetic variants that are on the wave front to strongly increase in frequencies and spread over large territories in newly colonized habitats (Edmonds *et al.* 2004), a phenomenon called 'gene surfing' (Klopfstein *et al.* 2006). Gene surfing of neutral variation has been investigated both theoretically (Hallatschek & Nelson 2008; Excoffier *et al.* 2009; Slatkin & Excoffier 2012) and empirically (Hallatschek & Nelson 2008; Moreau *et al.* 2011; Graciá *et al.* 2013). Gene surfing can also affect the spread of selected variants (Travis *et al.* 2007; Burton & Travis 2008; Lehe *et al.* 2012; Peischl *et al.* 2013, 2015). Population genetics models of range expansions predict that expanding populations incur a mutation burden – the 'expansion load' (Peischl *et al.* 2013). Expansion load is a transient phenomenon, but it can persist for several hundreds to thousands of generations, and may limit the ability of a species to colonize new habitats (Peischl *et al.* 2015).

Previous studies of expansion load assumed that mutations were codominant. An important consequence of this assumption is that standing genetic variation has no effect on the dynamics of mean fitness at the front of expanding populations (Peischl *et al.* 2013). In particular, the total number of mutations per individual, and hence the individual's fitness, remains approximately constant if new mutations are ignored (Peischl *et al.* 2013, 2015). In additive models, expansion load thus stems exclusively from the accumulation of new mutations that occur in individuals living at the front of the expansion.

Empirical evidence for expansion load may come from humans, where a proportional excess of deleterious mutations in non-African populations has been found (Lohmueller *et al.* 2008; Subramanian 2012; Torkamani *et al.* 2012; Peischl *et al.* 2013; Fu *et al.* 2014; Lohmueller 2014). Importantly, when focusing on mutations that occurred during or after the out-of-Africa expansion, the excess of deleterious variants is not restricted to rare variants (Peischl *et al.* 2013). This suggests that proportionally more deleterious mutations have risen to high frequencies in human populations located in newly settled habitats. In contrast to what would be expected from expansion-load theory, recent analyses found no significant differences in the average allele frequency of predicted deleterious alleles (Fu *et al.* 2014; Simons *et al.* 2014; Do *et al.* 2015). The average number of predicted deleterious mutations carried by an individual is, however, slightly but significantly larger in non-Africans (Fu *et al.* 2014). In addition, non-African individuals have significantly more loci homozygous for predicted deleterious alleles than African individuals (Lohmueller *et al.* 2008; Subramanian 2012; Fu *et al.* 2014). Whether and how human past demography affected the efficacy of selection and the spatial distribution of mutation load is thus still ongoing, and the interested readers are referred to the recent review of Lohmueller (2014) who provides a constructive attempt at reconciling views on this subject.

It is an old observation that deleterious mutations tend to be (partially) recessive (Morton *et al.* 1956; Mukai *et al.* 1972). More recently, it has been shown that most deleterious mutations have small effects and that their effect size correlates negatively with recessiveness (Garcia-Dorado & Caballero 2000; Peters *et al.* 2003; Eyre-Walker & Keightley 2007; Agrawal & Whitlock 2011). Importantly, if mutations are recessive, the number of deleterious mutations per individual alone is not informative about the mutation load (Kimura *et al.* 1963). For instance, if deleterious mutations are completely recessive, mutation load is determined by sites that are homozygous for deleterious alleles. Thus, if mutations are even partially recessive, the genotypic partitioning of deleterious variation is more important than the total number of deleterious mutations carried by an individual.

Past demographic events have been shown to affect the genotypic composition of standing genetic variation and therefore the mutation load, to an extent that is still debated (Kirkpatrick & Jarne 2000; Lohmueller *et al.* 2008; Simons *et al.* 2014; Do *et al.* 2015). Kirkpatrick & Jarne (2000) studied analytically the effect of a single-generation bottleneck on the mutation load and showed that a severe bottleneck would always increase the load relative to a population at mutation–selection equilibrium for partially recessive variants and that effect is stronger for more recessive alleles. In their model, most deleterious variants are lost, but others sharply increase in frequency and contribute proportionally more to the load. Recently, these results were confirmed and extended to more complex bottleneck scenarios that were estimated from human genomic data (Gravel 2014). Interestingly, it seems that changes in the distribution of the number of deleterious variants carried by individuals after a bottleneck can be used to infer whether selection is predominantly recessive or additive (Balick *et al.* 2014).

Similar to bottlenecks, range expansions are also known to affect the genotypic composition of neutral standing genetic variation (Excoffier *et al.* 2009), but seem to have a larger effect than bottlenecks on the mutation load (Peischl *et al.* 2013). The role of standing genetic variation in models of expansion load remains however unclear when mutations are recessive. We investigate here the effect of recessive mutations on the dynamics of expansion load. In particular, we use individual-based simulations to investigate the role of standing genetic variation, the width of the habitat and the composition of expansion load with respect to allele frequencies and mutational effects.

MODEL AND RESULTS

Model

We model a population of diploid monoecious individuals that occupy discrete demes located on a one- or two-dimensional grid (Kimura & Weiss 1964). Generations are discrete and nonoverlapping, and mating within each deme is random. Adult individuals migrate to adjacent demes with probability m per generation. Migration is homogeneous and isotropic, except that the boundaries of the habitat are reflecting, that is individuals cannot migrate out of the habitat.

Population size grows logistically within demes. The expected number of offspring in the next generation produced by the N_j adults in deme j is

$$N_j^* = \frac{R_0}{1+(R_0-1)N_j/K}N_j,$$

where R_0 is the fundamental (geometric) growth rate and K is the deme's carrying capacity (Beverton & Holt 1957). To model demographic stochasticity, the actual number of offspring, N'_j is then drawn from a Poisson distribution with mean N^*_j. Mating pairs are formed by randomly drawing individuals (with replacement) according to their relative fitness, and each mating pair produces a single offspring. The process is repeated N'_j

times, leading to approximately Poisson-distributed numbers of offspring per individual.

The relative fitness of individuals is determined by n independently segregating biallelic loci. The alleles at locus i are denoted a_i (wild type) and A_i (derived). Mutations occur in both directions and the genome-wide mutation rate is u; in each new gamete k, randomly chosen sites change their allelic state, where k is drawn from a Poisson distribution with mean u. The fitness contributions of the genotypes $a_i a_i$, $a_i A_i$ and $A_i A_i$ at locus i are 1, $1-hs_i$ and $1-s_i$, respectively. Here, s_i denotes the strength of selection at locus i and h is the dominance coefficient. Fitness effects are multiplicative across loci, such that the fitness of an individual is given by $w = \prod_i w_i$, where w_i is the fitness effect of the *i*th locus of the focus individual, that is there is no epistasis. In contrast to absolute fitness, relative fitness is density independent in our model. This assumption seems conservative, because if the fitness of individuals was density dependent (i.e. individuals would have more similar fitness at low densities), neutral processes at the expansion front would become even more important. In the following, we will focus on codominant ($h = 0.5$) or fully recessive ($h = 0$) mutations and refer to the Fig. S6 (Supporting information) for results obtained for intermediate degrees of recessiveness. We assume that mutation effects are drawn from the same distribution of fitness effects (DFE) for all individuals (independently from their current fitness).

We perform individual-based simulations of the above-described model in 1D or 2D habitats. Our simulations start from ancestral populations located in 10 leftmost (rows of) demes of the range. After a burn-in phase that ensures that the ancestral populations are at mutation–selection–drift balance, the population expands from left to right until the habitat is filled. Because we are mainly interested in the role of standing genetic variation, we focus on relatively short expansions, that is colonization of a 1×50 (1D) or a 20×50 (2D) deme habitat. The long-term dynamics of expansion load have been studied elsewhere (Peischl *et al.* 2013, 2015).

Impact of standing genetic variation on expansion load

For simplicity, we first consider expansions along a one-dimensional habitat and assume that all mutations have the same effect, that is we set $s_i = s$, and investigate 2D habitats and more complex distributions of fitness effects in later sections. We mainly focus here on mildly deleterious mutations with effects on the order of $Ns = 1$, because these mutations contribute most to mutation load from standing genetic variation, and they have been shown to behave essentially like neutral mutations on the wave front during range expansions (Peischl *et al.* 2013). In Peischl *et al.* (2013), we derived an analytical approximation for the rate of change of mean fitness at the expansion front due to the establishment of new mutations. This approximation can be modified in a straightforward way to account for recessive mutations, and the numerical evaluation of these results is shown in Fig. 1.

If mutations are codominant ($h = 0.5$), expansion load is caused exclusively by the establishment of new mutations occurring during the expansion, and standing genetic variation has a negligible effect on the dynamics of mean fitness (Peischl *et al.* 2013). Mean fitness at the wave front decreases at a constant rate over time (Fig. 1), and the rate at which mean fitness decreases per generation is proportional to the number of new mutations entering the population per generation (Peischl *et al.* 2013).

The dynamics of expansion load changes dramatically if mutations are recessive (Fig. 1). The analytical approximation obtained in Peischl *et al.* (2013), which ignores standing genetic variation, is a poor fit to the observed dynamics of mean fitness (Fig. 1). In the first few generations, mean fitness decreases much faster than predicted by analytical theory for the accumulation of new mutations (cf. solid and dashed black lines in Fig. 1). Over the course of the expansion, the rate at which expansion load is created slows down and gradually approaches the analytical prediction. Then, changes in expected mean fitness arise exclusively from new mutations (cf. solid and dashed black lines for $t > 50$ in Fig. 1, see also Fig. S1, Supporting information). This shows that standing genetic variation plays an important role in the establishment of expansion load if mutations are recessive, especially during early phases of expansions. This result is qualitatively similar to that of Kirkpatrick & Jarne (2000), who showed that load was always increasing after a single-generation bottleneck. The main novelty here is to consider the effects of recurrent bottlenecks such as those occurring during range expansions. For additional examples for the evolution of mean fitness during range expansions, including different migration rates (Fig. S2, Supporting information), carrying capacities (Fig. S3, Supporting information) and distribution of fitness effects (Fig. S4, Supporting information), we refer to the Supporting Information.

We next investigate the evolution of the genotypic composition of standing genetic variation on the

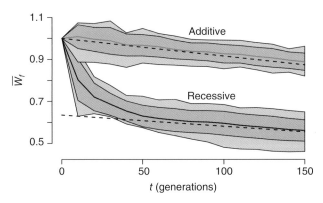

Fig. 1 Evolution of mean fitness at the wave front. Dashed lines show analytical predictions for the evolution of the mean fitness due to de novo mutations initially derived in Peischl *et al.* (2013), and adjusted here to account for the dominance coefficient of deleterious mutations by simply changing the selection coefficients of heterozygotes from $1-s/2$ to $1-hs$. In the codominant case, the analytical prediction is adjusted to match the mean fitness at the onset of the expansion, whereas in the recessive case it is adjusted to the mean fitness observed at generation $t = 150$. Simulations show results for the combination of standing and new genetic variation. Top shaded areas and lines show results for additive mutations (h = 0), and bottom shaded areas and lines show results for recessive codominant mutations (h = 0.5). Solid lines indicate the average mean fitness from 50 simulations, and dark and light shaded areas indicate ± one standard deviation and the minimum and maximum of mean fitness, respectively. Other parameter values are $n = 1000$, $K = 100$, $u = 0.1$, $m = 0.1$, $s = 0.01$, $R = 2$.

front of an expansion along a single dimension. In general, we find that the change in levels of diversity is very similar in the neutral, codominant and recessive case. In all cases, strong drift at the expansion front leads to increased inbreeding, a reduction of heterozygosity and higher homozygosity (Fig. 2). Indeed, the average number of heterozygous loci per individual decreases during the expansion (Fig. 2a), whereas the number of loci that are homozygous for the derived allele increases (Fig. 2b). Because we simulated a fixed number of loci, the derived allele frequency shown in Fig. 2c is proportional to the average number of mutations carried by an individual. Note that for a given selection coefficient, the initial allele frequency in the core depends on the mode of selection, and is generally higher if mutations are recessive. However, Fig. 2c shows that the total number of mutations per individual remains nearly constant during the expansion. Therefore, range expansions (and bottlenecks) should have a relatively weak effect on the individual fitness component that is due to codominant mutations, in agreement with several recent observations (Fu *et al.* 2014; Simons *et al.* 2014; Do *et al.* 2015). Also, at any given locus, mutations are either lost or fixed over the course of the expansion, and the probability of fixation of a given mutation is close to its initial frequency (Peischl *et al.* 2013), suggesting that (mildly and moderately) deleterious mutations are behaving like neutral mutations on the wave front (dashed lines in Fig. 2a–c). In 2D expansions, the dynamics of genotype frequencies are qualitatively very similar to 1D expansions (Fig. S5, Supporting information). Strong genetic drift is therefore the major force driving the evolution of genotype frequencies at the wave front.

The nearly neutral evolution of allele frequencies on the expansion front reveals a critical role of the degree of dominance on the build-up of the expansion load. If mutations are codominant, the fitness of an individual is determined by the total number of mutations it carries (Wright 1930). Thus, Fig. 2c shows that in the codominant case, standing genetic variation would have a negligible impact on fitness. In contrast, if mutations are recessive, the fitness of an individual is determined by its number of loci homozygous for the derived allele. Because the number of derived homozygous loci per individual rapidly increases at the front of the expansions, standing genetic variation has a severe effect on fitness if mutations are recessive (Figs 1 and 2b).

Gene flow on the wave front of 2D expansions restores diversity and fitness

In the following section, we focus on completely recessive mutations ($h = 0$). Figure 3 shows an example of the evolution of the mean fitness during an expansion in a 2D habitat (20×50 demes). As in 1D expansions, the mean fitness drops to low levels on the expansion front within the first few (≈ 30) generations and then continues to gradually decreases at a slower rate. There is however a considerable variation in fitness across the wave front of 2D expansions (fitness differences of more than 40%, Figs 3 and 4). At the end of the expansion (Fig. 3, $t = 150$), we find a high-fitness ridge along the expansion axis in the central part of the newly settled species range, surrounded by sectors of low fitness on the lateral edges of the species range (see also Fig. 4a). This is partially caused by the lack of immigrants at the lateral edge of the species range (boundary effect). However, the location of the high-fitness ridge varies across simulation runs, suggesting that a boundary effect alone cannot explain the observed patterns (Fig. 4b).

Figure 4a shows the variation in fitness, heterozygosity and derived allele frequency across the wave front at the end of the expansion shown in Fig. 3. We find that the average number of mutations per individual is uniform across the expansion front, which means that the variation in fitness across the expansion front is not driven by a differential accumulation of mutations. Contrastingly, variation in heterozygosity across demes is substantial, ranging from demes with almost zero heterozygosity to demes with heterozygosity as high as before the onset of the expansion (cf. Figs 2a and 4a,c). Genetic variation is quickly lost along the expansion axis during the expansion, but gene flow between nearby demes having established different mutations at high frequency typically restores heterozygosity, especially after the expansion. Interestingly, diversity is lower in populations close to the lateral edges of the expansions due to a border effect translating into reduced gene flow, which explains the occurrence of lateral regions of low heterozygosity and mean fitness (Fig. 4b). Because the deleterious effects of recessive mutations are masked in heterozygotes, heterozygosity correlates strongly with mean fitness across the wave front (cf. solid and dashed line in Fig. 4a, and see Fig. 4c).

Using a generalized linear mixed model, we estimate the effects of several key quantities that can

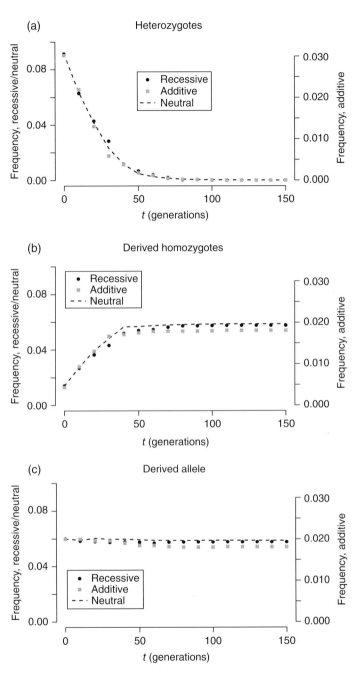

Fig. 2 Evolution of standing genetic variation on the wave front of a range expansion in one dimension. Each panel shows results for codominant, recessive and neutral mutations. For better comparison, neutral mutations were assumed recessive deleterious during the burn-in phase of the simulations, but neutral (in all demes) after the onset of the expansion. Parameter values are as in Fig. 1. Note the different scales on the y-axis for recessive (or neutral) and additive mutations in each plot.

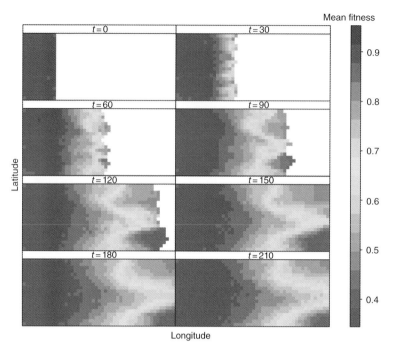

Fig. 3 Evolution of mean fitness during a range expansion. The simulated grid is 20 × 50 demes. Mutations are recessive and parameter values are as in Fig. 1. (*See insert for color representation of the figure.*)

be measured in experimental set-ups (deme coordinates, simulation number (random effect) and average heterozygosity as independent variables, and mean fitness as dependent variable), to identify whether we find different heterozygosity–fitness correlations (HFC) in core and in front populations. We indeed find a strongly positive HFC at the front of the expansion (Fig. 4c, regression slope ≈0.4, $P \approx 10^{-5}$, see Table S1, Supporting information), but not in the ancestral population (Fig. 4d, $P \approx 0.7$, see Table S2, Supporting information). The contributions of all other parameters to mean fitness are not significantly different from zero (see Table S1 and S2, Supporting information). Interestingly, weaker but similar correlations are found at the individual level within demes, suggesting that HFC created after range expansions could be detected in samples from single populations. We performed linear regression of heterozygosity and fitness for 50 individuals sampled from the same deme. Repeating this across demes at the wave front (at generation $t = 150$) and across simulation replicates, we found an average regression slope of ≈0.1 ($P < 0.05$ in 78% of demes,

Figs 5 and S8, Supporting information). Furthermore, we found that latitudinal position of demes has no significant effect on the strength of within deme HFC (Table S3, Supporting information).

Expansion load is driven by a few mutations occurring at high frequency

So far we assumed that all mutations had the same effect s. To investigate the composition of expansion load with respect to mutation fitness effects, we now consider the case where mutation fitness effects are drawn from an exponential distribution with mean s. Figure 6A,B shows the site frequency spectrum (SFS) observed in core and front populations, respectively. In core populations, the SFS shows the pattern expected for sites under negative selection (Bustamante *et al.* 2001), with a large excess of low-frequency variants. On the wave front, the total number of segregating sites is reduced in marginal populations (cf. Fig. 6a,b). More interestingly, as compared to core populations, we see a markedly different SFS on the front, with a

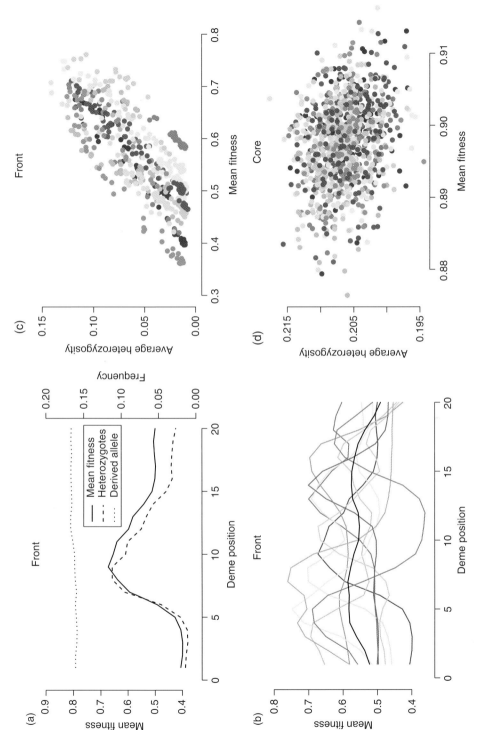

Fig. 4 Genetic properties of demes at the wave front and heterozygosity–fitness correlations (HFC) across demes in different parts of the species range. (a) Example of the mean fitness, the heterozygosity and the derived allele frequency at different latitudinal positions at the expansion front when the habitat has just been fully colonized (t = 150, simulation shown in Fig. 3). The deme mean fitness on the wave front correlates with heterozygosity, but not with derived allele frequency. (b) Mean fitness at the front of the expansion from 10 distinct simulation runs (coloured lines) and the average over all simulation runs (solid black line). (c) HFC on the expansion front at generation t = 150. (d) No significant HFC in core populations before the onset of the expansion (t = 0). In (c) and (d), each point represents the mean fitness and average heterozygosity of a single deme. Different colours in panels b–d correspond to 10 distinct simulation replicates. Parameter values are as in Fig. 3. (*See insert for color representation of the figure.*)

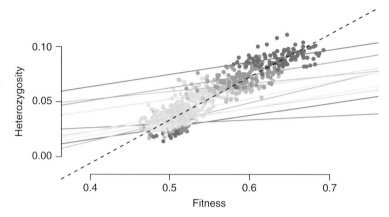

Fig. 5 Within deme heterozygosity–fitness correlations (HFC). Each point corresponds to the fitness and heterozygosity of a single individual, as inferred from a single simulation on a 20 by 50 grid. Grayscales correspond to different demes at the wave front (at generation $t = 150$). Gray lines show the linear regression lines of heterozygosity and fitness within demes. The dashed black line shows the overall (between deme) HFC for comparison.

clear deficit of rare and intermediate-frequency variants and an increase in high-frequency variants (Fig. 6B). Thus, even though fewer polymorphic sites with deleterious variants are found in more recently colonized areas than in the ancestral region, the alleles at polymorphic sites tend to be at higher frequency in more recently colonized populations.

Figure 6c,d shows the distribution of polymorphic loci stratified according to their mutation effect sizes. The eight mutation effect classes have been defined such that they represent the eight quantiles of the DFE, that is the rate at which mutations of a given category enter the population are equal for all categories. As expected, we find that the number of polymorphic loci generally decreases with increasing mutation effect size and that large-effect mutations tend to be present at lower frequencies than low-effect mutations (see Fig. 6c,d). Compared to core populations, the allele frequencies at polymorphic sites on the wave front tend to be larger across all mutational effect categories (cf. yellow and red coloured areas in Fig. 6c,d). Furthermore, the increase in allele frequency is most pronounced for small effect mutations. Thus, expansion load is driven mainly by standing deleterious mutations of small to moderate effect (i.e. up to $N_s < 2$ for the parameter values used in Fig. 6, see also Fig. S7 (Supporting information) which shows analogous results for larger mean s) that rise to high frequency during the expansion.

DISCUSSION

We have investigated here the dynamics of an expansion load caused by recessive mutations. Using individual-based simulations, we have shown that shifts in the genotypic composition of standing genetic variation can lead to a rapid drop of mean fitness at the onset of an expansion (see Figs 1 and 2, and Figs S2–S4, Supporting information) without necessarily affecting the total number of deleterious alleles per individuals (see Figs 2, 4, and S5, Supporting information). Figure 2 shows that genotype frequencies evolve almost neutrally at the expansion front and that strong genetic drift at the expansion front increases the number of derived homozygote sites per individual. The derived homozygote frequency at the expansion front approaches the initial frequency of the derived allele over the course of the expansion (see Fig. 2b). The total expansion load due to standing genetic variation, which is proportional to the number of derived homozygous sites, is therefore limited by the initial frequency of deleterious mutations. Thus, if many loci are polymorphic for deleterious variants at the onset of the expansion, the (recessive) expansion load from standing genetic variation can dominate the total mutation load (see Fig. 1). A similar phenomenon, although of lesser magnitude, occurs if only some of the mutations would be fully or if they would be partially recessive (see Fig. S6, Supporting information).

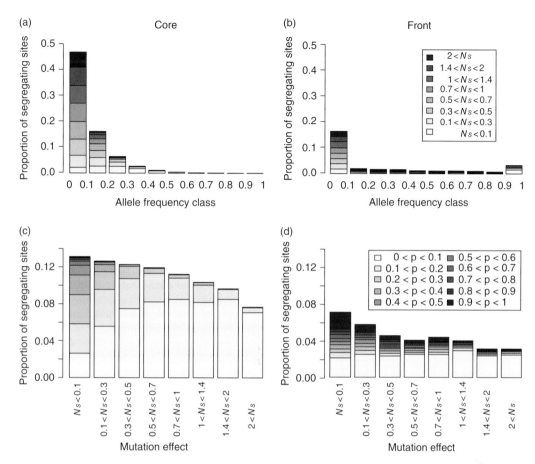

Fig. 6 Distributions of segregating sites in core and front populations after 2D range expansions. The distributions are stratified according to allele frequencies (site frequency spectrum, top row) and mutation effects (bottom row). The plots are normalized with respect to the total number of sites that segregating in any of the colonized demes. In the top row, allele frequencies (p) are binned in decimal intervals, and the boundaries of these intervals are indicated on the x-axis. Results were recorded 150 generations after the onset of the expansion, which is shortly after the habitat was colonized completely (mean time to colonization ≈130 generations, see also Fig. 3). Panels (a) and (c) show results for a core population [XY coordinates (5, 10)], (b) and (d) for a front population [XY *coordinates* (45, 10)]. Mutations are recessive and their effects are drawn from an exponential distribution with mean s = 0.01. Other parameter values are as in Fig. 1.

Even though we find that our results are robust with respect to changes in the migration rate (Fig. S2, Supporting information) and carrying capacity (Fig. S3, Supporting information), it would be interesting to further explore the parameter space. Unfortunately, individual-based simulations are computationally intensive, especially, if both N and m are large, which prevents an exhaustive exploration of the parameter space. Theoretical results suggest, however, that the effective population size at the front of expanding populations depends only weakly on the local carrying capacities (Hallatschek & Nelson 2008). Similarly, migration rates seem to have a weak effect on the effective population size at the expansion front (Peischl *et al.* 2013; see also Fig. S2, Supporting information). Allee effects, that is a reduction in fitness when conspecific density is low, strongly decrease the strength of drift at expansion fronts (Hallatschek & Nelson 2008; Roques *et al.* 2012) and could therefore mitigate expansion load. Importantly, Allee effects appear to be important in many examples of range expansions or invasive species (Green 1997; Taylor & Hastings 2005).

The effect of range expansions on deleterious genetic diversity is also reflected in the site frequency spectrum (SFS, see Fig. 6). As compared to stationary populations in the core of the species range, populations from more recently colonized areas have fewer segregating sites, but proportionally more high- and low-frequency variants (cf. Fig. 6a,b). These differences in the SFS of core and front populations should provide an opportunity to evidence expansion load from sequence data and to infer important quantities such as the distribution of fitness effects (Keightley & Eyre-Walker 2007; Boyko *et al.* 2008; Racimo & Schraiber 2014). The development of statistical and computational methods able to infer parameters under spatially explicit models including range expansions and selection remains, however, a major challenge (Sousa *et al.* 2014).

Interestingly, human genomic data are consistent with our predictions for genomic signatures of expansion load. In particular, the number of segregating sites is higher in African populations than in non-African populations (Lohmueller *et al.* 2008), non-African populations show an excess of low-frequency and high-frequency deleterious alleles (Lohmueller *et al.* 2008; Fu *et al.* 2014), the average number of sites that are homozygous for predicted deleterious variants sites is larger in non-African individuals (Fu *et al.* 2014), and the average number of predicted deleterious mutations per individual is slightly, but significantly, larger in non-Africans (Fu *et al.* 2014). Determining mutation load (or, alternatively, fitness) from genomic variation data is, however, an intrinsically difficult problem because mutation load depends on many unknown parameters (selection coefficients that may vary over space and time, epistatic interactions, dominance relationships, etc.), and the relevance of comparing a population with deleterious mutations to a theoretical population free of such mutations is questionable (Lesecque *et al.* 2012). Testing theoretical predictions of the effect of a range expansion, or other demographic scenarios (e.g. Simons *et al.* 2014; Do *et al.* 2015), on functional diversity with human genomic data might nevertheless be extremely useful to substantially increase our understanding of the complex interactions of demography and selection.

We assumed here that fitness is density independent, but one could imagine that fitness differences between individuals could be decreased at low densities, making drift relatively stronger than selection on range margins, promoting expansion load even further than in the cases studied here. We also assumed here that

selection is soft, that is demographic parameters are independent of fitness (Wallace 1975), but it would be interesting to extend our results to models of hard selection, where mutation load on the front can stop an expansion and even drive parts of the species range to extinction (Peischl *et al.* 2015). Our results suggest that admixture during range expansions, or secondary contact between expanding lineages, could mitigate expansion load and prevent marginal populations from collapsing. A previous study of range expansions under an additive model with hard selection has shown that suppressing recombination at the wave front can have beneficial effects for the spread of high-fitness lineages (Peischl *et al.* 2015). Recombination modifiers, such as inversions, could have a similar effect if mutations are recessive and facilitate the spread of admixed lineages. An interesting example for studying the potentially beneficial role of admixture and suppressed recombination during range expansions is from the clam genus *Corbicula*, which includes both sexual and asexual (androgenetic diploid) lineages. Sexual populations are restricted to their native Asian areas, but the androgenetic lineages are widely distributed and extend as far as in America and Europe where they are invasive (Pigneur *et al.* 2014). Intriguingly, the invasive lineages also show an excess of heterozygosity, which is preserved through clonal reproduction. No such excess of heterozygosity is found in the native range, suggesting that the combination of asexual reproduction and high heterozygosity may have been key drivers of the invasion.

An interesting prediction of our model is that if a given proportion of deleterious mutations are recessive, then heterozygosity–fitness correlations (HFC) should naturally occur in populations that have recently expanded their range (see Fig. 4a,c). Importantly, the positive correlation between heterozygosity and fitness in recently colonized areas can be observed at both the individual level and the population level (see Figs 4 and 5). Even though our simulations modelled a single expansion in a 2D habitat, we would expect similar HFCs if there was a secondary contact between expanding populations from different areas (e.g. from different last glacial maximum refuge areas). The HFC should be even stronger in the case of a secondary contact, because the isolation between expanding lineages should be larger and different recessive alleles could have fixed in different refugia or during the expansion from these refugia. HFC have been observed in many cases of natural range expansions and invasive species

(Chapman *et al.* 2009), but their underlying mechanisms and their role during range expansions and invasions are still unclear (Szulkin *et al.* 2010; Rius & Darling 2014). A particular interesting example of HFC is found in the invasive weed *Silene vulgaris*, where, as predicted by our model (see Figs 4 and 5), HFC correlations are observed in the recently invaded North American range, but not in their native European range (Keller *et al.* 2014). It remains, however, unclear whether admixture between divergent lineages has indeed a causal role in range expansions. A combination of transplantation experiments and genomic data analyses could certainly be used to test the predictions of our model.

In summary, we have investigated here the evolution of standing genetic variation during range expansions, the dynamics of mean fitness on the expansion front if mutations are recessive, and the genomic signature of range expansions. Importantly, our results make predictions that can be tested in natural populations. Empirical validation of our results would increase our understanding of the interactions of demography and selection (Lohmueller 2014) and could help us identifying key drivers of range expansions and biological invasions (Rius & Darling 2014).

ACKNOWLEDGEMENTS

SP was supported by a Swiss NSF Grant No. 31003A-143393 to LE. We thank Vitor Sousa and Isabelle Duperret for stimulating discussions on this topic. We are grateful for the helpful comments from four anonymous reviewers.

REFERENCES

Agrawal AF, Whitlock MC (2011) Inferences about the distribution of dominance drawn from yeast gene knockout data. *Genetics*, **187**, 553–566.

Balick DJ, Do R, Reich D, Sunyaev SR (2014) Response to a population bottleneck can be used to infer recessive selection. *bioRxiv*, 003491.

Beverton R, Holt S (1957) On the dynamics of exploited fish populations. Fisheries Investigation Series 2 (19). London: Ministry of Agriculture, Fisheries and Food.

Boyko AR, Williamson SH, Indap AR *et al.* (2008) Assessing the evolutionary impact of amino acid mutations in the human genome. *PLoS Genetics*, **4**, e1000083.

Burton OJ, Travis JMJ (2008) The frequency of fitness peak shifts is increased at expanding range margins due to mutation surfing. *Genetics*, **179**, 941–950.

Bustamante CD, Wakeley J, Sawyer S, Hartl DL (2001) Directional selection and the site-frequency spectrum. *Genetics*, **159**, 1779–1788.

Chapman J, Nakagawa S, Coltman D, Slate J, Sheldon B (2009) A quantitative review of heterozygosity–fitness correlations in animal populations. *Molecular Ecology*, **18**, 2746–2765.

Darwin C (1859) *On the Origins of Species by Means of Natural Selection*. Murray, London.

Do R, Balick D, Li H *et al.* (2015) No evidence that selection has been less effective at removing deleterious mutations in Europeans than in Africans. *Nature Genetics*, **47**, 126–131.

Edmonds CA, Lillie AS, Cavalli-Sforza LL (2004) Mutations arising in the wave front of an expanding population. *Proceedings of the National Academy of Sciences of the United States of America*, **101**, 975–979.

Excoffier L, Foll M, Petit RJ (2009) Genetic consequences of range expansions. *Annual Review of Ecology, Evolution, and Systematics*, **40**, 481–501.

Eyre-Walker A, Keightley PD (2007) The distribution of fitness effects of new mutations. *Nature Reviews Genetics*, **8**, 610–618.

Fu W, Gittelman RM, Bamshad MJ, Akey JM (2014) Characteristics of neutral and deleterious protein-coding variation among individuals and populations. *The American Journal of Human Genetics*, **95**, 421–436.

Garcia-Dorado A, Caballero A (2000) On the average coefficient of dominance of deleterious spontaneous mutations. *Genetics*, **155**, 1991–2001.

Graciá E, Botella F, Anadón JD *et al.* (2013) Surfing in tortoises? Empirical signs of genetic structuring owing to range expansion. *Biology Letters*, **9**, 20121091.

Gravel S (2014) When is selection effective? bioRxiv, 010934.

Green RE (1997) The influence of numbers released on the outcome of attempts to introduce exotic bird species to New Zealand. *Journal of Animal Ecology*, **66**, 25–35.

Hallatschek O, Nelson DR (2008) Gene surfing in expanding populations. *Theoretical Population Biology*, **73**, 158–170.

Keightley PD, Eyre-Walker A (2007) Joint inference of the distribution of fitness effects of deleterious mutations and population demography based on nucleotide polymorphism frequencies. *Genetics*, **177**, 2251–2261.

Keller S, Fields P, Berardi A, Taylor D (2014) Recent admixture generates heterozygosity–fitness correlations during the range expansion of an invading species. *Journal of Evolutionary Biology*, **27**, 616–627.

Kimura M, Weiss GH (1964) The stepping stone model of population structure and the decrease of genetic correlation with distance. *Genetics*, **49**, 561–576.

Kimura M, Maruyama T, Crow JF (1963) The mutation load in small populations. *Genetics*, **48**, 1303.

Kirkpatrick M, Jarne P (2000) The effects of a bottleneck on inbreeding depression and the genetic load. *American Naturalist*, **155**, 154–167.

Klopfstein S, Currat M, Excoffier L (2006) The fate of mutations surfing on the wave of a range expansion. *Molecular Biology and Evolution*, **23**, 482–490.

Lehe R, Hallatschek O, Peliti L (2012) The rate of beneficial mutations surfing on the wave of a range expansion. *PLoS Computational Biology*, **8**, e1002447.

Lesecque Y, Keightley PD, Eyre-Walker A (2012) A resolution of the mutation load paradox in humans. *Genetics*, **191**, 1321–1330.

Lindström T, Brown GP, Sisson SA, Phillips BL, Shine R (2013) Rapid shifts in dispersal behavior on an expanding range edge. *Proceedings of the National Academy of Sciences*, **110**, 13452–13456.

Lohmueller KE (2014) The distribution of deleterious genetic variation in human populations. bioRxiv.

Lohmueller KE, Indap AR, Schmidt S *et al.* (2008) Proportionally more deleterious genetic variation in European than in African populations. *Nature*, **451**, U994–U995.

MacArthur RH (1972) *Geographical Ecology: Patterns in the Distribution of Species*. Princeton University Press, Oxfordshire, UK.

Moreau C, Bherer C, Vezina H *et al.* (2011) Deep human genealogies reveal a selective advantage to be on an expanding wave front. *Science*, **334**, 1148–1150.

Morton NE, Crow JF, Muller HJ (1956) An estimate of the mutational damage in man from data on consanguineous marriages. *Proceedings of the National Academy of Sciences of the United States of America*, **42**, 855.

Mukai T, Chigusa SI, Mettler L, Crow JF (1972) Mutation rate and dominance of genes affecting viability in *Drosophila melanogaster*. *Genetics*, **72**, 335–355.

Peischl S, Dupanloup I, Kirkpatrick M, Excoffier L (2013) On the accumulation of deleterious mutations during range expansions. *Molecular Ecology*, **22**, 5972–5982.

Peischl S, Kirkpatrick M, Excoffier L (2015) Expansion load and the evolutionary dynamics of a species range. *American Naturalist*, **185**, 1–13.

Peters A, Halligan D, Whitlock M, Keightley P (2003) Dominance and overdominance of mildly deleterious induced mutations for fitness traits in *Caenorhabditis elegans*. *Genetics*, **165**, 589–599.

Phillips BL, Brown GP, Webb JK, Shine R (2006) Invasion and the evolution of speed in toads. *Nature*, **439**, 803.

Phillips BL, Brown GP, Shine R (2010) Life-history evolution in range-shifting populations. *Ecology*, **91**, 1617–1627.

Pigneur LM, Etoundi E, Aldridge DC, Marescaux J, Yasuda N, Van Doninck K (2014) Genetic uniformity and long-distance clonal dispersal in the invasive androgenetic Corbicula clams. *Molecular Ecology*, **23**, 5102–5116.

Racimo F, Schraiber JG (2014) Approximation to the distribution of fitness effects across functional categories in human segregating polymorphisms. *PLoS Genetics*, **10**, e1004697.

Rius M, Darling JA (2014) How important is intraspecific genetic admixture to the success of colonising populations? *Trends in Ecology and Evolution*, **29**, 233–242.

Roques L, Garnier J, Hamel F, Klein EK (2012) Allee effect promotes diversity in traveling waves of colonization. *Proceedings of the National Academy of Sciences*, **109**, 8828–8833.

Sexton JP, McIntyre PJ, Angert AL, Rice KJ (2009) Evolution and ecology of species range limits. *Annual Review of Ecology Evolution and Systematics*, **40**, 415–436.

Shine R, Brown GP, Phillips BL (2011) An evolutionary process that assembles phenotypes through space rather than through time. *Proceedings of the National Academy of Sciences*, **108**, 5708.

Simons YB, Turchin MC, Pritchard JK, Sella G (2014) The deleterious mutation load is insensitive to recent population history. *Nature Genetics*, **46**, 220–224.

Slatkin M, Excoffier L (2012) Serial founder effects during range expansion: a spatial analog of genetic drift. *Genetics*, **191**, 171–181.

Sousa V, Peischl S, Excoffier L (2014) Impact of range expansions on current human genomic diversity. *Current Opinion in Genetics and Development*, **29**, 22–30.

Subramanian S (2012) The abundance of deleterious polymorphisms in humans. *Genetics*, **190**, 1579–1583.

Szulkin M, Bierne N, David P (2010) Heterozygosity-fitness correlations: a time for reappraisal. *Evolution*, **64**, 1202–1217.

Taylor CM, Hastings A (2005) Allee effects in biological invasions. *Ecology Letters*, **8**, 895–908.

Torkamani A, Pham P, Libiger O *et al.* (2012) Clinical implications of human population differences in genome-wide rates of functional genotypes. *Frontiers in Genetics*, **3**, 1–19.

Travis JMJ, Munkemuller T, Burton OJ *et al.* (2007) Deleterious mutations can surf to high densities on the wave front of an expanding population. *Molecular Biology and Evolution*, **24**, 2334–2343.

Wallace B (1975) Hard and soft selection revisited. *Evolution*, **29**, 465–473.

Wright S (1930) The genetical theory of natural selection. A review. *Journal of Heredity*, **21**, 349–356.

DATA ACCESSIBILITY

The code used for the simulations is available on GitHub: https://github.com/CMPG/ADMRE.

SUPPORTING INFORMATION

Additional supporting information can be found in the online version of the *Molecular Ecology* article.

Fig. S1 Evolution of mean fitness on the front of a 1D expansion.

Fig. S2 Evolution of mean fitness at the front of a 2D expansion, in a habitat of 10×50 demes.

Fig. S3 Evolution of mean fitness at the front of a 2D expansion.

Fig. S4 Evolution of mean fitness at the front of a 2D expansion.

Fig. S5 Evolution of genotype frequencies at the front of a two-dimensional expansion, in a habitat of 20×50 demes.

Fig. S6 Evolution of mean fitness at the front of the expansion for varying degrees of dominance.

Fig. S7 Distributions of segregating sites in core and front population after a 2D range expansion.

Fig. S8 Slope and *P*-values of linear regression of individual heterozygosity and fitness.

Table S1 HFC on the expansion front.

Table S2 HFC in core populations.

Table S3 Within-deme HFC.

Chapter 14

THE DEVIL IS IN THE DETAILS: GENETIC VARIATION IN INTRODUCED POPULATIONS AND ITS CONTRIBUTIONS TO INVASION

Katrina M. Dlugosch, Samantha R. Anderson, Joseph Braasch, F. Alice Cang, and Heather D. Gillette

Department of Ecology and Evolutionary Biology, University of Arizona, PO Box 210088, Tucson, AZ 85721, USA

Abstract

The influence of genetic variation on invasion success has captivated researchers since the start of the field of invasion genetics 50 years ago. We review the history of work on this question and conclude that genetic variation—as surveyed with molecular markers—appears to shape invasion rarely. Instead, there is a significant disconnect between marker assays and ecologically relevant genetic variation in introductions. We argue that the potential for adaptation to facilitate invasion will be shaped by the details of genotypes affecting phenotypes, and we highlight three areas in which we see opportunities to make powerful new insights. (i) The genetic architecture of adaptive variation. Traits shaped by large-effect alleles may be strongly impacted by founder events yet more likely to respond to selection when genetic drift is strong. Large-effect loci may be especially relevant for traits involved in biotic interactions. (ii) Cryptic genetic variation exposed during invasion. Introductions have strong potential to uncover masked variation due to alterations in genetic and ecological environments. (iii) Genetic interactions during admixture of multiple source populations. As divergence among sources increases, positive followed by increasingly negative effects of admixture should be expected. Although generally hypothesized to be beneficial during invasion, admixture is most often reported among sources of intermediate divergence, supporting the possibility that incompatibilities among divergent source populations might be limiting their introgression. Finally, we note that these details of invasion genetics can be coupled with comparative demographic analyses to link genetic changes to the evolution of invasiveness itself.

Previously published as an article in *Molecular Ecology* (2015) 24, 2095–2111, doi: 10.1111/mec.13183

INTRODUCTION

At a symposium on The Genetics of Colonizing Species in 1964, some of the best minds in evolutionary biology, genetics, ecology and applied biology came together to ponder questions about how the founding of new populations might fundamentally alter the genetics and colonization success of species in novel environments (Baker & Stebbins 1965). Baker & Stebbins (1965: p.vii) noted in their preface to the proceedings of the symposium that 'When the approximately thirty biologists who attended started to exchange facts and ideas, all of them realized at once that each had things to say which were of great value to the others, and which were new to them. [...] We hope that some of the spirit of adventure which many of the participants experienced at the symposium will find its way to the readers of this volume'. Indeed, the symposium became the beginning of a vigorous field of enquiry marked by its goal of bridging genetics, evolution and ecology to understand colonization, with particular insights provided by contemporary species invasions (Barrett 2015; Bock *et al.* 2015).

The symposium was naturally dominated by a discussion of how founding events might diminish genetic diversity within populations. Deceptively simple questions of how much genetic variation establishes in colonizing populations and to what degree the quantity of variation affects the subsequent success of new populations captivated and continue to captivate invasion biologists (Lockwood *et al.* 2005; Dlugosch & Parker 2008a; Uller & Leimu 2011; Blackburn *et al.* 2015). An emphasis on the role of genetic variation in invasion has only grown as evidence has mounted that invaders frequently show evolutionary changes in traits putatively related to fitness and/or the propensity to invade new environments (Hendry *et al.* 1999; Cox 2004; Bossdorf *et al.* 2005; Colautti & Barrett 2013; Colautti & Lau 2015). Thus, it is clear that evolution is happening, seemingly in response to natural selection, but the extent to which adaptation during colonization might be constrained by genetic variation remains largely unknown, despite its fundamental importance.

Here, we trace the history of thought about the role of genetic variation in invasions, and we argue that the nature of genetic variation (the 'details') will be more relevant to facilitating invasion than its total quantity *per se*. We take a closer look at specific attributes of adaptive genetic variation in founding populations, including its genetic architecture, its expression under different environments and its interaction among divergent source populations. All of these aspects of variation show strong potential to influence adaptation during invasion, and open promising avenues for further investigation. We conclude by noting that these details of invasion genetics can allow us to connect the impacts of specific evolutionary changes to population growth and spread, generating a more complete understanding of the importance of genetic variants for the process of invasion itself.

GENETIC VARIATION IN INVASIONS: A HISTORY

One of the most universal features of invasions is that founding populations will experience demographic bottlenecks of some magnitude after introduction and/or range expansion. Baker (1955) made a powerful case for the potential severity of demographic bottlenecks at founding well before the 1964 symposium. Baker argued that successful long-distance dispersal was strongly associated with species' ability to reproduce without a mate (termed 'reproductive assurance'), using strategies such as self-fertilization or asexual propagation (Baker 1955, 1965). Stebbins (1957) referred to this association of colonization with reproductive strategy as Baker's Law, and Baker argued that it results from the frequent lack of mates and/or mating opportunities during population establishment (Pannell 2015). In addition to the immediate losses of genetic diversity expected from small numbers of founders, methods of reproductive assurance are generally associated with further declines in genetic variation, all else being equal (Ellstrand & Roose 1987; Hamrick & Godt 1996). For these reasons, Baker predicted that most successful colonizers would be ones to thrive without genetic variation, relying instead upon a single best 'general-purpose-genotype' capable of colonizing a wide variety of environments (Baker 1955, 1974; Ferrero *et al.* 2015).

Despite Baker's assertions, there was much debate at the symposium about the extent to which demographic bottlenecks and self-fertilization/asexuality would actually reduce variation in real founding populations. Many participants assumed that this was so, but some geneticists pointed out important theoretical considerations to the contrary. Confusion on this point culminated in an exposition by Lewontin (Baker & Stebbins 1965: p. 481), wherein he clarified that while demographic bottlenecks will likely lead to the loss of some rare alleles, they will generally not substantially reduce genetic variance in quantitative traits (determined by common variants at multiple loci; Fig. 1). Mayr (1965a) further argued that new mutations will arise reasonably quickly in founding populations as opportunities for mutants to occur expand with the population (see Box 1). Empirically, Allard (1965) presented data showing a high degree of mating system lability, phenotypic variation and local adaptation in species that are predominantly self-fertilizing, creating a disconnect between traits conferring reproductive assurance in colonizers and a significant lack of genetic variation. Ultimately, this debate could not be settled at the meeting, as data on genetic variation in colonizing populations were almost entirely lacking at the time, outside of chromosomal inversion polymorphisms which showed variable patterns of loss and maintenance of diversity in *Drosophila* (Carson 1965; Dobzhansky 1965; Mayr 1965b).

A separate debate was had at the symposium about whether any lost variation would meaningfully influence the evolutionary and ecological success of founding populations. In line with his hypothesis of the general-purpose-genotype, Baker (1965) suggested that while genetic variation and adaptability could be beneficial or even necessary in the long run, founding populations might have a reduced need for adaptation due to a relatively low-competition environment. Alternatively, Fraser's (1965) discussion of work on the genetics of *D. melanogaster* bristle number highlighted the possibility that lost diversity at some loci could free additive variation at other epistatically interacting loci, potentially increasing the genetic variation available for adaptation after a bottleneck (i.e. conversion of epistatic to additive variance; see Bock *et al.* 2015). Discussion after Fraser's study further emphasized the possible benefits of genetic drift during colonization, including the potential for higher level selection among divergent founding demes and shifting balance in adaptive landscapes (Fraser 1965). In a note of caution, however, F. Wilson (1965) conveyed that biocontrol introductions often seemed to show frustratingly low adaptability or ecological amplitude. Biocontrol introductions can be particularly low in diversity—owing to the difficulty identifying, collecting and propagating diversity during deployment—implying that extreme bottlenecks could indeed ultimately limit colonization success (Wilson 1965).

Decades later, molecular data from protein markers began to shed some light on genetic changes after

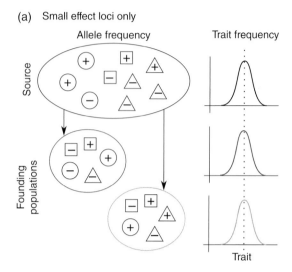

(a) Small effect loci only

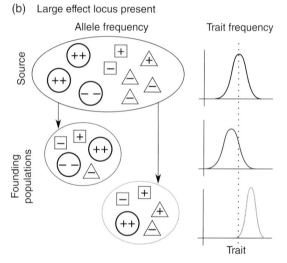

(b) Large effect locus present

Fig. 1 Genetic architecture will shape how population bottlenecks during colonization events impact quantitative trait variation. Panels show founding populations sampling allelic variation (acting to increase [+] or decrease [−] the trait value) across loci (shapes). Allelic variants are shown in proportion to their frequency in a population. (a) Traits governed by many loci of small effect are expected to change little in mean or variance, as founding populations sample common allelic variants, and fixation at some loci has little effect on the trait. (b) In contrast, traits that are shaped at least in part by a locus of large effect may shift in both mean and variance in response to either fixation or frequency shifts at these influential loci.

Box 1 New mutations in colonizing populations

It is reasonable to expect that rapid adaptation which contributes directly to the success of introduced and invasive species will be derived largely from standing genetic variation in the native range, given the short time frame available for response to selection before extinction in struggling founder populations. Nevertheless, recent estimates show that a variety of mutation types occur frequently (Lynch & Conery 2000; Denver *et al.* 2004; Ossowski *et al.* 2010; Stapley *et al.* 2015). Founding populations that are able to increase from small numbers will quickly produce many opportunities for new mutations to appear. Growing populations also provide better opportunities

for fixation of rare beneficial variants, which will be the case for new mutations when they arise (Otto & Whitlock 1997).

For example, mutation rates in the model plant *Arabidopsis thaliana* have been estimated conservatively at 6×10^{-9} substitutions/site/generation for single nucleotide polymorphisms ('SNPs', i.e. point mutations; Ossowski *et al.* 2010), 1.3 copies/individual for copy number changes ('CNVs', i.e. gene duplications and deletions; DeBolt 2010; Ossowski *et al.* 2010) and 1.0 mutations/individual for non-microsatellite insertion/deletion changes ('indels', including frameshift mutations; Ossowski *et al.* 2010). Assuming

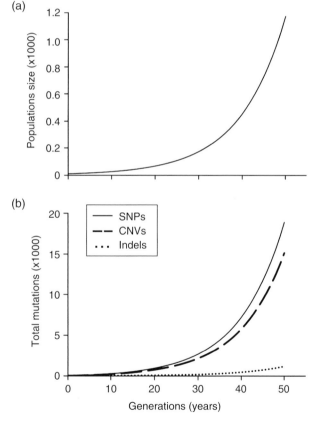

Box 1, Fig. 1 New mutations appearing in a theoretical population of *A. thaliana* growing exponentially from 10 founders without selection. (a) Over 50 years, the population size grows to 1174 individuals at a 10% annual rate of increase. (b) During that time, the cumulative number of mutations arising in the population rises sharply according to average rates, particularly for SNPs and CNVs.

a population of 10 founders of an annual plant growing exponentially at 10% a year, nearly 20 000 different SNP mutations and over 15 000 CNV variants will have occurred in the population within 50 years (Fig. 1). Even if only a small percentage of these mutations have fitness effects (on the order of 20% in *A. thaliana*; Ossowski *et al.* 2010), there will be many opportunities for new mutations to respond to natural selection in the new environment during the early years of population establishment.

Importantly, unlike random SNPs, which should most often be neutral if silent or deleterious if nonsynonymous, copy number changes of intact genes have a much greater potential to result in beneficial phenotypic effects, particularly through changes in gene expression (Kondrashov 2012; Hirase *et al.* 2014; Żmieńko *et al.* 2014). The role of copy number changes in rapid adaptation is largely unknown (Kondrashov 2012) and seems likely to be a promising area of study in invasion genetics.

founding events, and it was clear that real demographic bottlenecks were reducing variation at individual loci in many cases (Barrett & Richardson 1986). In a classic study, Nei *et al.* (1975) published theoretical expectations for the quantity of allelic variation persisting after bottlenecks, showing that founding population size and the rate at which the new population grows both critically determine the loss of variation at markers due to drift. They noted that allelic richness (allele number) will be more sensitive to founding population size and less sensitive to rate of increase, while the reverse is true of heterozygosity, meaning that different metrics of genetic variation will reveal different parts of the story of population history. Nei *et al.* (1975) demonstrated these inferences with an example from invasion biology: the introduction of *D. pseudo-obscura* into Bagota showed a strong loss of allelic diversity consistent with rapid population growth from a small founder number (in line with historical records).

Invasion biologists increasingly questioned how invasions progress despite potentially costly genetic bottlenecks (e.g. Briskie & Mackintosh 2004). An idea developed that invaders must somehow be resolving a 'genetic paradox of invasion' (Allendorf & Lundquist 2003; Frankham 2005), wherein introduced species somehow thrived in new environments when they should be suffering from deleterious losses of genetic diversity. Qualitative reviews of the accumulating empirical data showed little in the way of a consistent pattern of high or low bottleneck severity within successful invasions (Brown & Marshall 1981; Barrett & Richardson 1986; Gray *et al.* 1986; Barrett & Husband 1990; Lee 2002; Cox 2004; Lambrinos 2004; Bossdorf *et al.* 2005), but several reviews did emphasize that invasions were often the result of multiple introductions (Ellstrand & Schierenbeck

2000; Allendorf & Lundquist 2003; Lockwood *et al.* 2005; Novak & Mack 2005; Wares *et al.* 2005; Roman & Darling 2007; Suarez & Tsutsui 2008). Increasing attention was paid to cases of elevated or rising genetic diversity due to multiple introductions, as demonstrations of a potential genetic advantage for invaders and a general resolution to the 'genetic paradox' (e.g. Novak & Mack 1993; Kolbe *et al.* 2004; Frankham 2005; Lavergne & Molofsky 2007; Roman & Darling 2007; Facon *et al.* 2008; Hufbauer 2008), although arguments lingered that bottlenecks might not be fundamentally problematic for invaders in the first place (e.g. Koskinen *et al.* 2002; Dlugosch & Parker 2008b).

More recently, Dlugosch & Parker (2008a) quantitatively summarized the genetic diversity data available at the time for 80 species of animals, plants and fungi. This analysis showed that changes in intrapopulation genetic variation were generally normally distributed around a modest loss of variation (approx. 15–20%, depending on molecular marker and diversity metric), with larger losses apparent in allelic diversity than in heterozygosity for studies that measured both, as expected. Large increases in diversity were extremely rare, and multiple introductions had only small positive effects on diversity on average. Uller & Leimu (2011) revisited this question with a meta-analysis of the effect size of introduction on heterozygosity (change relative to variance among marker and population) in 85 species of animals and plants. Their results again showed that large changes in diversity have been uncommon. The magnitude of diversity change was unrelated to a metric of invasiveness, and for cases in which founder number was known, large losses of variation (effect size > |2|) were not observed unless founder numbers were extremely small (<15 individuals). These quantitative reviews have largely put to rest the argument that multiple introductions have been critical in providing genetic rescue from severe and deleterious founder effects in most cases.

Taken together, the accumulated data make a strong case that successful invaders frequently experience genetic bottlenecks, but they are neither dramatically depauperate in, nor especially well endowed with, genetic variation relative to native populations of the same species. Certainly, very strong demographic bottlenecks will limit the success of introductions for a variety of reasons (Wilson 1965; Lockwood *et al.* 2005; Agashe *et al.* 2011; Hufbauer *et al.* 2013; Szűcs *et al.* 2014), but demographic barriers to establishment (e.g. stochastic extinction, Allee effects) would seem to be the more important determinants of introduction failure at extremely low founder population sizes (Blackburn *et al.* 2015). Instead, founder populations that are large enough to overcome demographic constraints should consequently retain significant amounts of genetic variation, and indeed this appears to be the case.

THE DEVIL IN THE DETAILS

If changes in genetic variation are not a central determinant of introduction success, are we to conclude that genetic diversity is not important to invasions? On the contrary, many studies of introduced species indicate that genetics might play an integral role in the progress of an invasion, but they suggest that role is determined by *what* genetic variation is introduced, rather than *how much*. Certainly, total genetic variation has often been intended to serve as a proxy for the likelihood that an adaptive genetic variant is present when the selective environment shifts, but it has become increasingly clear that the disconnect between total and adaptive variation can be large (e.g. Merilä & Crnokrak 2001; McKay & Latta 2002; Leinonen *et al.* 2008). Below we examine three aspects of the nature of genetic variation which will have a particularly significant influence on the adaptive potential of invading populations, all of which have come into focus largely since the formative start of the field of invasion genetics 50 years ago: (i) individual loci whose variants have large phenotypic effects, (ii) cryptic genetic variation that is exposed in the introduced range, and (iii) genetic interactions during admixture of previously isolated alleles that have been brought together by multiple introductions.

Loci of large effect

A major and open question regarding the nature of adaptive genetic variation is the degree to which it is dominated by a small number of genes whose alleles have large effects on the phenotype. Large-effect loci include both those that might contribute to quantitative trait variation and (more obviously) those controlling discrete Mendelian traits. The genetic architecture of quantitative traits is not easily observed, but it will shape patterns of adaptation, and has attracted significant debate (Orr 2005). Fisher's early work on this topic asserted that quantitative traits must be overwhelmingly governed by many loci of small effect (the infinitesimal model; Fisher 1930), predicting that adaptation would proceed continuously towards a fitness optimum at a rate governed by available genetic variance and the strength of selection. Much later, Kimura (1985) pointed out that beneficial large-effect mutations were much less likely than small-effect mutations to be lost to genetic drift because they experience stronger selection. Orr (1998) further argued that mutations involved in the process of climbing to an adaptive optimum should first involve a few large-effect loci, followed by loci with a constant decrease in effect size, a reinvigoration of the geometric model of adaptation originally explored by Fisher (1930). Additional theoretical support for the importance of large-effect loci comes from studies of adaptive landscapes, where interactions among loci create multiple adaptive optima separated by valleys of low fitness (Wright 1932). Complex fitness landscapes appear to be common in nature, and adaptation under these conditions is much more likely if large-effect loci can facilitate jumps across low-fitness trait space (Whitlock *et al.* 1995; Whitlock 1997; De Visser & Krug 2014).

The architecture of trait variation should be particularly important for adaptive evolution in invading populations for several reasons. First, as articulated so well by Lewontin at the Baker & Stebbins symposium (Baker & Stebbins 1965: p. 481), traits governed by many small-effect loci should lose little standing variation during founding events, because demographic bottlenecks during colonization will affect genetic variation through the loss of individual allelic variants, particularly rare variants. In contrast, traits that are affected by loci of large effect may experience significant changes in mean and variance purely via sampling effects at these loci (Fig. 1). We expect that the impact of this sampling on traits and fitness would be more negative than positive on average, because large-effect alleles that are *favourable* in a novel environment where adaptation is needed would have been either neutral or selected *against* in the native range (a different environment). Alleles that were deleterious in the native range should be rare and most susceptible to loss during founding. In this way, a large-effect allele can amplify the impacts of founder/bottleneck effects on the phenotype and may therefore have more potential to alter invader establishment and/or spread in often unpredictable ways.

Second, while large-effect loci might have negative impacts on adaptive variation during founder events, these loci may also enhance the response of traits to selection during range expansion. With a greater impact on fitness, beneficial alleles of large effect should respond more strongly to selection and more effectively avoid loss due to drift than individual small-effect alleles (Kimura 1985). Adaptation can and does often proceed successfully via the collective action of many loci of small effect (Olson-Manning *et al.* 2012); however, large-effect alleles may be particularly important to invasions. Invading populations are subject to occasional very low effective population sizes (i.e. strong genetic drift) both during initial founding events and at the invasion front during the process of spatial expansion ('allele surfing'; Hallatschek *et al.* 2007; Excoffier *et al.* 2009; Peischl & Excoffier 2015). Effective population size at the invasion front may be extremely low even if the invading population is large. Strong genetic drift will be particularly problematic for new mutations that might contribute to novel variation in founding populations (Box 1), as these must rise from extreme rarity to contribute to adaptation. Therefore, regardless of their importance to adaptation in stable populations, we predict that the presence or absence of large-effect loci will influence which traits are able to adapt within invasions.

Recent reviews find that large-effect loci do appear to be an important genetic basis of trait variation. A meta-analysis of QTL effects in plants found that while almost 90% of QTL identified were of small effect, the remaining loci each explained a large proportion of phenotypic variation (estimates >20%) in a study (Louthan & Kay 2011). Certainly, large-effect loci are easier to detect in a QTL analysis and the magnitude of their effects can be inflated by experimental artefacts (Beavis 1994; Otto & Jones 2000), but even in well-studied model organisms such as *Drosophila* and humans, large-effect loci appear to play a significant

role in adaptation alongside the many detectable loci of small effect (Olson-Manning *et al.* 2012). Genomic studies have also greatly expanded the types of mutations that we know can have large effects on phenotype, particularly structural mutations such as copy number variation (CNVs, i.e. gene duplications; Lynch & Conery 2000) and inversions (Kirkpatrick 2010), as well as regulatory mutations (Hoekstra & Coyne 2007; Wray 2007). Indeed, copy number variants, largely unappreciated before genome-scale sequencing, are now known to be one of the major forms of mutation differentiating closely related species—such as humans and other apes—and individuals of the same species (Lynch & Conery 2000; Freeman *et al.* 2006). Copy number changes occur at nearly the same rate as point mutations and seem much more likely than other types of mutations to have beneficial effects (Kondrashov 2012; Hirase *et al.* 2014; Żmieńko *et al.* 2014). Thus, it would appear that many types of major mutations are frequent and able to contribute both standing and *de novo* large-effect variants to small founder populations (Box 1).

Our first window into the genetic basis of adaptive variation in invasion came from patterns of chromosomal inversions (easily observed in *Drosophila* salivary glands as mentioned above; Carson 1965; Dobzhansky 1965). It is now clear that such structural changes can be associated with large and potentially adaptive phenotypic effects (Hoffmann *et al.* 2004; Kirkpatrick 2010). Through physical rearrangement of loci and/or suppression of recombination, inversions can retain associations between complimentary alleles at multiple loci, preserving coadapted gene complexes and/or making a larger effect locus out of multiple loci of smaller effect (Hoffmann & Rieseberg 2008; Yeaman 2013). The potential for inversions to allow the rapid spread of advantageous loci during invasions seems strong in principle (Kirkpatrick & Barrett 2015). Both the invasion of Australia by *D. melanogaster* (Hoffmann & Weeks 2007) and the invasion of the Americas by *D. subobscura* (Prevosti *et al.* 1988) show evidence of rapid adaptation in chromosomal inversion frequencies. This is particularly impressive in South America, where *D. subobscura* is inferred to have invaded via just a few founders, yet the species rapidly re-evolved clines in inversion frequency characteristic of the native range (Pascual *et al.* 2007). Inversions are now increasingly identified in studies of the genetic basis of adaptation across taxa; for example, in the post-Pleistocene invasion of freshwater lakes by threespine sticklebacks, an inversion is among several large-effect loci that control the repeated evolution of freshwater and benthic forms (Jones *et al.* 2012).

Studies that map the genetic basis of phenotypic variation in contemporary invasions have been slow to appear, but these clearly indicate an important role for loci of large effect (Bock *et al.* 2015). Paterson *et al.* (1995) were among the first to map QTL in a colonizer, showing that a small number of QTL controlled the propensity of johnsongrass (*Sorghum halepense*) to produce asexually via rhizomes. Linde (2001) similarly found three major QTL controlling flowering time differences among ecotypes of the invasive plant shepherd's purse (*Capsella bursa-pastoris*). Most recently, Whitney *et al.* (2015) have identified three major QTL associated with range expansion in the sunflower *Helianthus annuus texanus*.

As we accumulate information about the genetic basis of invader phenotypes, we can begin to ask what types of traits are most likely to have standing variation in large-effect loci, or to gain it more easily through mutations. One class of traits for which large-effect loci are already well known are those under frequency-dependent selection in the native range. For example, variation at social recognition loci in both the fire ant (*Solenopsis invicta*) and the Argentine ant (*Linepithema humile*) has been lost during founder events, resulting in decreased conspecific aggression and increased invasiveness (Tsutsui *et al.* 2000, 2003; Krieger & Ross 2002), giving us our most famous cases of *positive* effects of genetic bottlenecks on invasion. Another major class of loci under frequency-dependent selection are self-incompatibility (SI) loci in plants, at which loss of alleles can clearly be detrimental to reproduction in founding populations (e.g. Elam *et al.* 2007). These cases demonstrate the important phenotypic effects of the loss of large-effect alleles, through the positive or negative nature of the consequences, are less predictable. There are also many classic examples of invaders circumventing loss of SI diversity by evolving self-compatibility in invading populations (Baker 1965; Barrett 2015; Ferrero *et al.* 2015), emphasizing the impressive adaptability of introduced species under seemingly unlikely conditions.

Interestingly, a recent review of QTL studies in plants concludes that large-effect loci are more commonly found in traits governing biotic interactions than in traits associated with adaptation to abiotic conditions (Louthan & Kay 2011). This pattern may be due to greater spatial variation in selection for biotic interactions (i.e. a rougher adaptive landscape) favouring fixa-tion of large-effect alleles, as opposed to the presence of more available large-effect variation in these traits *per se*. Regardless, the maintenance of variation in large-effect loci across populations under spatially and temporally varying selection in native populations may provide more opportunities for adaptation in these traits during invasion. Given that one of the major hypotheses for invasion success is the escape from negative biotic interactions in the native range (Keane & Crawley 2002), the potential for especially high adaptability in traits underlying precisely these interactions should be of great interest to invasion geneticists.

Linking phenotypic variation to its genetic basis is a major undertaking in any study system, but invaders may be especially well suited to these approaches among nonmodel organisms (Box 2). The contemporary nature of their evolutionary changes means that adaptive variation exists within and among current populations across the range, facilitating genetic mapping, identification of current targets of selection, and observation of the impacts of genetic variants on populations in native and invaded environments. An exciting window into the future of these opportunities is provided by Hamilton *et al.* (2015) in this volume, who compare the phenotypic effects of many loci associated with adaptation in the native range of the model *Arabidopsis thaliana* (Fournier-Level *et al.* 2011; Hancock *et al.* 2011) between the native and introduced ranges (Hamilton *et al.* 2015). Their results reveal a large number of loci with conditional effects, altering fitness in only one environment, but just a handful of loci that have significant effects across both ranges, with opposing consequences for fitness in each region (antagonistic pleiotropy). This type of study is powerful for identifying the genetic basis of adaptation to novel ranges, the effect size and number of loci involved, and whether alleles at individual loci can shift populations from one phenotypic optimum in the native range to a new optimum in an invasion. Clearly, important and exciting insights into the connections between genetic architecture and invasiveness await the accumulation of this type of information.

Cryptic genetic variation

To establish adaptive genetic variants in a founding population, it is ideal if these alleles are segregating at high frequency in source population(s). As noted above, we expect that many loci will not contain such

Box 2 Identifying the genetic basis of adaptive variation

The architecture of adaptive genetic diversity is vastly more observable than ever before (Prentis *et al*. 2008; Stinchcombe & Hoekstra 2008; Stewart *et al*. 2009). When traits are known to have evolved within invasions, or traits are inferred to be under selection via 'reverse' genetics (see below), their genetic basis can be identified using classical 'forward' genetics (Fig. 1). Genome wide markers can be screened across hundreds of individuals in systems with no prior genetic information, using genotype-by-sequencing approaches such as RADseq (Davey *et al*. 2011; Narum *et al*. 2013). Genetic maps can be obtained from sequences of individuals with known genetic relationships (i.e. via controlled crosses or pedigrees), or using genome wide association (GWAS) approaches in natural populations (Narum *et al*. 2013). The final step of associating trait variation with genotypes should work particularly well for invasive species, because trait variation among genotypes is likely to be segregating within extant populations or easily obtained via intraspecific controlled crosses (opportunities not

available for most studies of species-level divergence, for example).

In a complimentary fashion, 'reverse' genetics tools that screen loci for evidence of natural selection (e.g. gene expression comparisons, scans for sweeps or differentiation of marker variation, and correlations of allele frequencies with environmental variables; Ekblom & Galindo 2011; Manel & Holderegger 2013) now allow genomic information itself to suggest which loci might be involved in adaptation (Bock *et al*. 2015). Using alignments of genes in candidate regions to loci in model organisms, phenotypic effects can be hypothesized and explored further. Invasions pose particular analytical challenges for reverse genetics, due to their nonequilibrium and often complex demographic history, which may generate allele frequency shifts that mimic signatures of selection (Excoffier *et al*. 2009). Putative loci under selection must be evaluated directly for their phenotypic and fitness effects. For example, Vandepitte *et al*. (2014) recently identified several regions of the Pyrenean rocket (*Sisymbrium austriacum*)

GOAL (FORWARD/TOP-DOWN GENETICS):
Associate traits with genetic markers and/or candidate loci

APPROACH	SOURCE OF TRAIT VARIATION
QTL map ◄—	cross between invading and native genotype
GWAS ◄—	trait variation within native source region

KNOWLEDGE TRANSFER
Loci with known phenotypic effect suggest regions of the genome of interest

Use reverse genetics to ask:
Is there evidence for selection on genetic variants?

KNOWLEDGE TRANSFER
Candidate loci suggest traits of interest, based on alignment to model organisms

Use forward genetics to ask:
What are phenotypic effects of alleles?

GOAL (REVERSE/BOTTOM-UP GENETICS):
Identify markers and/or candidate genes under selection

APPROACH
Identify significant patterns in:
- gene expression differences
- allele diversity (sweeps)
- allele frequency divergence
- allele frequency correlation with environment

SOURCE OF TRAIT VARIATION
Comparisons between invading and native populations

Box 2, Fig. 1 A combination of forward, reverse and candidate gene approaches can resolve the genetic basis of variation that has been involved in the evolution of invaders.

genome that have differentiated during invasion, and some of these are located within genes that control flowering time in the closely related model *A. thaliana*. Flowering time can now be evaluated further in this system using 'forward' genetics and field studies of selection on this trait to validate its role in adaptation.

A persistent challenge is to firmly link individual mutations to their phenotypic effects, as mapping of quantitative trait loci or regions under selection can encompass multiple loci and variant sites. For example, multiple mutations in just one gene independently control different aspects of adaptive coat colour variation involved in the colonization of light sand substrate by deer mice (Linnen *et al.* 2013). Nevertheless, by combining 'forward' and 'reverse' genetics, and investigating candidate genes in these regions based on information from model organisms, it is possible to work through these links and build an understanding of the details of the genetics of adaptations (Stinchcombe & Hoekstra 2008; Fig. 1). Indeed, invasions may prove to offer outstanding opportunities to study the rapid evolution of loci in close physical linkage.

variation, given that source environments differ from the invasion and will have selected for different trait optima. This situation might be avoided if adaptive variation in the new environment is not under selection in the old, as will be the case for 'cryptic' genetic variation—that is variation only observed in the invasion. Cryptic genetic variation of a trait is revealed whenever the phenotypic effects of genetic variants differ depending on the ecological environment (G × E interactions) or differ depending on the genetic environment (allele frequencies at other loci; G × G interactions), such that there is increased genetic variation of the trait within some environments (Hermisson & Wagner 2004; McGuigan & Sgrò 2009; Paaby & Rockman 2014). These previously hidden sources of variation would seem to be particularly important for rapid evolutionary change, and contemporary species introductions should be outstanding places to look for evidence of adaptation via cryptic variants (Lee & Gelembiuk 2008).

There is a long history of interest in the potential for exposure of a population to a stressful environment to increase heritable variation in this way (Waddington 1956; Hoffmann & Merilä 1999; Badyaev 2005). This field has grown out of the well-known pioneering work of Waddington (1956), who argued that selection will act to create buffering mechanisms to stabilize optimal phenotypes under typical conditions (masking genetic variation), but that these buffering mechanisms may fail under atypical/stressful conditions, as he demonstrated in *D. melanogaster*. At the Baker & Stebbins symposium, Waddington (1965) himself raised questions about the stability of phenotypes and the expression of genetic variation during the colonization of new environments.

Evidence is now accumulating that this phenomenon is common under natural environmental variation, in both individual loci as well as quantitative traits (Dworkin *et al.* 2003; Latta *et al.* 2015). Cryptic allelic variants may be strictly neutral in the historical environment, having no phenotypic effects. For example, a large body of work has shown that the heat-shock chaperone protein Hsp90 buffers the effects of mis-folded proteins in a wide variety of taxa, such that genetic variation in those proteins lacks phenotypic effect. When Hsp90 expression is reduced through environmental effects or its own mutations, genetic variation in the associated proteins is exposed and can reveal adaptive variants (Paaby & Rockman 2014). Alternatively, alleles may be under selection for their phenotypic effects, but have new pleiotropic phenotypes expressed in the new environment. For example, Duveau & Félix (2012) showed that a locus under selection for life history variation in *Caenorhabditis elegans* produces novel morphological effects under environmental variation. These observations suggest that cryptic variation may often be present in ecologically relevant traits. Indeed, conditionally neutral allelic variants appear to be a common genetic basis for local adaptations in natural populations (Colautti *et al.* 2012; Olson-Manning *et al.* 2012; Hamilton *et al.* 2015), which means that cryptic variation should have many opportunities to accumulate in traits related to adaptation to a range of environmental variation that is not uncommon within species ranges (Paaby & Rockman 2014).

The exposure of cryptic variation might be especially relevant for species introductions, given that we expect shifts in the genetic background due to founder effects and/or admixture, as well as shifts in the biotic and abiotic ecological environment in the new range. Compelling cases have been made that cryptic variation has played a role in historical cases of adaptation in body size in sticklebacks invading freshwater lakes (McGuigan *et al.* 2011) and in adaptive loss of eyes in cavefish invading caves (Rohner *et al.* 2013). To date, there is no study that we are aware of which tests for evidence of cryptic variation surfacing in a contemporary invasion, although some studies of invader plasticity seem to suggest this possibility (e.g. Purchase & Moreau 2012). A recent demonstration of dominance reversal in allelic effects (where a formally recessive allele becomes dominant) during invasion of the copepod *Eurytemora affinis* highlights the potential for the genetic basis of adaptive variation to change fundamentally in an invader's new environmental context (Posavi *et al.* 2014). In a theoretical analysis, Masel (2006) showed that cryptic variation might often be biased towards adaptive variation (i.e. biased against deleterious mutations), if [source] populations occasionally experience conditions similar to the new environment during their evolutionary history, exposing cryptic variation to selection. Given that invasions often seem to occur in niche space that is similar to native environments (Petitpierre *et al.* 2012; Strubbe *et al.* 2013), it is not unreasonable to imagine that native populations might have experienced relevant conditions in the past.

To understand whether cryptic variation is important to invasions, we must disentangle it from a set of nonmutually exclusive factors that can influence adaptive variation in these populations (Fig. 2). Admixture or hybridization (and specifically the transgressive segregation that these foster) can also clearly increase genetic variance during invasion, particularly in traits that are divergent in source populations/species, but this increase should be robust to environmental context. Conversion of epistatic to additive variance has also been of long-standing interest as a way in which additive variation (but not total genetic variation) can increase during invasion (Goodnight 1988), although the conditions under which conversion is expected are relatively narrow and there is little evidence for its role to date (Turelli & Barton 2006; Van Heerwaarden *et al.* 2008). In contrast, cryptic variation should increase total genetic variance only in the novel environment (Fig. 2). This context-dependent variation highlights an important point: our null expectation for variation in a founder population should be that expressed by founder genotypes in their *source* environment, something which is rarely examined. As loci underlying adaptations in invaders are identified, a key area of

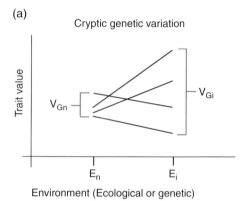

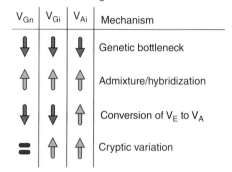

Fig. 2 Cryptic genetic variation can be revealed whenever genetic variance increases in a new ecological or genetic environment. (a) If genetic variance increases in the invaded environment (subscript 'i'), relative to the native environment (subscript 'n'), cryptic variation might contribute adaptive variation to an invasion. (b) Cryptic variation is one of several potential mechanisms underlying changes in total (V_G), additive (V_A) and epistatic (V_E) genetic variance of a trait in an invading population, relative to its source population(s).

interest should be their effects under different environmental conditions and genetic backgrounds, particularly the comparison of native vs. invading contexts.

Admixture

Thus far, we have considered the details of genetic diversity in founding events derived from a single source, but we know that multiple introductions are very common features of invasions (Dlugosch & Parker 2008a; Uller & Leimu 2011). As discussed above, multiple introductions appear to have modest effects on the total amount of genetic variation in most introduced populations, as measured by molecular markers. In contrast, there is a very strong potential for multiple introductions to have significant impacts on adaptive variation in invading populations if the introductions come from different areas of the native range, resulting in admixture of divergent source populations (Ellstrand & Schierenbeck 2000; Verhoeven *et al.* 2011; Rius & Darling 2014). Admixture can infuse the invasion with novel alleles, which can have particularly strong effects on phenotype for loci of large-effect and/or cryptic variants arising from a history of spatially varying selection across native sources. Perhaps most importantly, admixture creates unique opportunities for genetic interactions among previously isolated alleles and/or loci, which can dramatically affect phenotypes and fitness in admixed genotypes (Waser & Price 1989; Lynch 1991; Edmands 1999; Keller & Waller 2002).

Genetic interactions are well known from studies of inbreeding depression, outbreeding depression, and heterosis observed through genetic crosses. These observations are part of a general set of expectations for the fitness effects of crosses between parents of varying genetic distances, based on the mechanisms underlying the genetic interactions (Lynch 1991; Fig. 3). Outcrossing is expected to be beneficial when a focal population has some fixed recessive deleterious alleles—genetic load—that can be rescued through dominance interactions with more fit alleles, generating diminishing returns as genetic distance increases and genetic load is relieved. In contrast, the effects of epistatic interactions among alleles at different loci rise or even accelerate as genetic distance between parents increases. Epistatic interactions are expected to build up in relatively isolated populations due to selection for locally coadapted gene complexes or to genetic drift (creating Bateson-Dobzhansky-Muller incompatibilities), acting separately in each population (Lynch & Walsh 1998). Epistatic effects in the F1 generation may sometimes be positive (heterotic) but are expected to become increasingly negative with genetic distance, particularly in the F2 and later generations when co-evolved multilocus genotypes are broken apart by recombination (i.e. 'hybrid breakdown'; Fig. 4; Orr & Turelli 2001; Bomblies *et al.* 2007).

Although both positive and negative genetic interactions are expected, admixture has generally been hypothesized to be beneficial to invasions. Signatures of

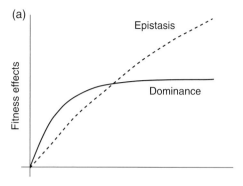

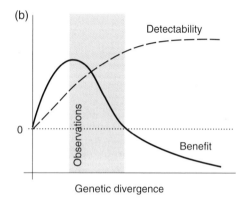

Fig. 3 Genetic divergence between potential source regions will shape both their genetic interactions and our ability to observe their admixture when present. (a) Dominance interactions (positive effects from the rescue of loci homozygous for deleterious recessive alleles) are maximized at low levels of divergence, while the fitness effects of epistatic interactions (often negative due to incompatible allele combinations, particularly in F2 and later generations) continue to increase with divergence among loci. (b) The power to detect admixture in a data set increases with genetic divergence, as native subpopulations become more identifiable, but net fitness benefits from their genetic interactions should become increasingly negative. As a result, observations of admixture are predicted to concentrate at intermediate levels of genetic divergence (after Lynch 1991).

positive genetic interactions have been sought by looking for correlations between heterozygosity and fitness traits (Heterozygosity-Fitness Correlations, 'HFCs'; Rius & Darling 2014). It is important to note that these analyses typically use molecular marker heterozygosity under the assumption that it is in linkage disequilibrium with sites that affect fitness. Linkage disequilibrium with neutral markers is expected to decay quickly in most

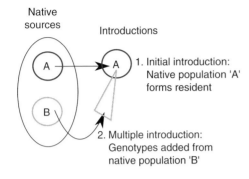

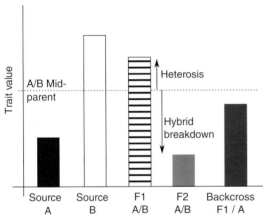

Fig. 4 A realistic scenario for the effects of admixture on fitness-related traits. Multiple introductions are likely to occur via input of new genotypes ('B') into an existing introduced population of genotypes ('A'). If fitness-related traits (e.g. body size, seed number) in the source populations differ genetically, then the expected result of the F1 cross is the mid-parent value. F1 progeny may experience heterosis or 'hybrid vigour' due to genetic interactions between sources A and B, but these often break down in the F2. Hybrid breakdown might often be ameliorated by backcrosses to the resident population, a likely scenario for introgression of new genotypes during invasion.

cases, and so there is a limited window in which this type of analysis is expected to reveal positive effects of admixture when they are present. Moreover, if presumed fitness-rated traits are also varying due to local adaptation in the native source range (Colautti *et al.* 2009), and/or if epistatic genetic interactions are generating nonlinear fitness effects of heterozygosity, then HFCs are no longer expected (Lynch & Walsh 1998; Chapman *et al.* 2009; Szulkin *et al.* 2010). Nevertheless, significant HFCs have shown apparent benefits of admixture during invasion.

For example, Keller *et al.* (2014) elegantly showed a positive HFC for reproduction in admixed introductions of the plant *Silene vulgaris*, a relationship that was not present in a zone of historical admixture in the native range. This approach may be best suited to detecting fitness effects of very recent admixture (relative to recombination rate), particularly among sources with a low degree of divergence, such that epistatic interactions are expected to be minimal.

In place of HFC searches, it might be ideal to combine genetic surveys of invasion history with experimental crosses of known or potential sources, looking directly for the fitness effects of crossing at varying levels of divergence. Experimental crosses have the particular benefit of allowing progeny performance to be evaluated against a mid-parent expected value (Fig. 4). Mid-parent comparisons are important because source populations will often have divergent life history and/or morphological traits due to local adaptation, and these are often the same traits that are used to quantify fitness effects (e.g. body size, offspring size or number). For example, an invading population with an offspring number that is intermediate between values observed in source populations may in fact be experiencing 'hybrid vigour', if reproduction exceeds the mid-parent expectation. Traits of experimental progeny can be evaluated at the F1, F2 and backcross generations and compared to invading genotypes that are potentially the products of these histories (Facon *et al.* 2008; Turgeon *et al.* 2011).

Identifying admixture and/or its potential source regions for further study introduces its own challenges. While historical records can indicate multiple introductions, ultimately it is essential to observe genetic mixing directly to be certain admixture is present. A typical approach is to survey marker variation in the native and invaded ranges as broadly as possible, delineate any distinct genetic subpopulations in the native source region and identify signatures of these subpopulations mixing in the introduced range (Cristescu 2015). There are two underappreciated challenges associated with these steps. First, sampling in the native range must be sufficient for a reasonable argument to be made that invaders are truly admixed, and not simply derived from an unsampled region in the native range (including a zone of historical admixture). Second, sampling in the native range and marker variability must be high enough to resolve different source populations as unique. This latter concern is particularly serious, because admixture may have

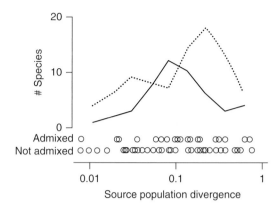

Fig. 5 A survey of studies reporting admixture or no admixture in introduced species (circles), as a function of genetic differentiation among sites in the native range (F_{ST} and related metrics, based on nuclear genetic markers; $N = 70$ species). Lines show the number of species for which admixture was reported (heavy line) or not reported (dashed line) in a sliding window of width $1.0 \ln(F_{ST})$. Data are available at Dryad, doi:10.5061/dryad.s2948.

important phenotypic effects at low levels of divergence among sources (Fig. 3).

Genetic reconstructions of invasion history are accumulating, affording opportunities to examine the conditions under which admixture is detectable and potentially affecting invader fitness. Reports of admixture to date are reasonably common (37% of 70 invaders studied with nuclear markers; Fig. 5), but they appear to be underrepresented at both low and high levels of source population divergence (Fig. 5). This pattern matches our expectation that admixture should be harder to detect at low levels of source divergence, and is consistent with our prediction that admixture might be unfavourable at higher levels of source divergence. We can imagine a variety of alternative factors that will also affect these patterns (e.g. sampling design, mating system, generation time, bottleneck severity, number of introductions and introduction vector), but the idea that there might be conditions under which genetic variation itself resists multiple introductions is in direct opposition to the long-held view that admixture should be beneficial during invasions. There is clearly a need for direct examination of the fitness consequences of admixture among potential source populations, particularly across both low and high levels of divergence.

CONTRIBUTIONS OF GENETIC DIVERSITY TO INVASION

In the event that we are able to identify genetic changes that have been adaptive in invading populations, we will have outstanding opportunities to assess the contribution of adaptation to the process of invasion itself. We typically do not know whether adaptations that we observe in well-established invaders reflect higher fitness in the novel environment post-invasion without contribution to the process of invasion itself. A potential example of adaptation unrelated to invasiveness is provided by the highly invasive plant garlic mustard (*Alliaria petiolata*) in North America, which is thought to have invaded largely due to its allelopathic inhibition of competing species. Recent studies suggest some populations may be evolving reduced allelopathic activity in the face of competition with resistant conspecifics and increasingly resistant natives (Lankau *et al.* 2009), a putative shift away from the traits that facilitated the invasion itself.

In general, we have little knowledge of the population-level consequences of individual genetic changes (Kinnison & Hairston 2007; Gaston 2009). We routinely assume that invading genotypes with higher reproduction and/or survival relative to other lineages must be invading with increased population growth, density and/or spread. While selection will favour the evolution of the invading population towards such a high-fitness genotype, there may be extrinsic factors limiting the consequences of adaptation for population growth itself. One of the primary barriers to connecting the fitness of individual genotypes with population-level performance is density-dependent changes in vital rates that occur as a population grows (Antonovics & Levin 1980; Ozgul *et al.* 2009), as will often be the case when invaders reach the high densities that can be a hallmark of 'invasiveness'. Density dependence is an extremely common property of populations, but in general we know almost nothing about how fitness differences among genotypes change under competition with density of conspecifics, within invasions or any other system. Fisher (1930) went so far as to suggest that fitness increases could even contribute to 'environmental degradation', imposing limits on population growth as high-fitness individuals consumed more resources. Invasions offer particularly outstanding opportunities to study density dependence in genotype performance, as many invasions have active expansion fronts at which density is currently varying under relevant conditions.

Whether adaptation or beneficial admixture ultimately matter to population establishment, persistence, spread and 'invasiveness' merits careful consideration (Molofsky *et al.* 2014). A promising way forward is to integrate comparative demography into the study of the evolution of invaders. The best example of this that we are aware of to date is in work seeking to understand the contribution of evolutionary increases in leg length/individual dispersal speed of cane toads (*Bufo marinus*) to the rate of expansion of its invasion into Australia (Perkins *et al.* 2013). In this case, it seems likely that adaptation for dispersal has changed the rate of spread at least in part, although adaptation may not have played much role in the successful establishment and severe ecological impact of this species once established.

FUTURE DIRECTIONS

The issues that we have considered above suggest many opportunities for new insights into the importance of genetic variation in invasion, including the following:
• The genetic basis of variation in ecologically relevant traits. Invasions provide outstanding opportunities to understand the contributions of different classes of mutations—such as chromosomal inversions and copy number variants—to rapid adaptive evolution. We predict that evolutionary changes that are important to invasion by introduced species will more often occur in traits affected by large-effect loci, through unpredictable founder/bottleneck effects at these loci, through the arrival of new mutations that are able to rise quickly in frequency, and through response to selection at low effective population size during range expansion.
• The relative contribution of large-effect loci in abiotic vs. biotic trait variation. Where the genetic basis of adaptive variation in invaders can be identified, it could reveal important connections between evolutionary and ecological opportunities for introduced species. Nonnative species are particularly likely to experience alterations in biotic interactions; if these same interactions are more evolutionarily labile due to underlying trait architecture, then invaders may be predisposed to benefit from adaptation during invasion.
• The genetic variation within founder genotypes in the invaded vs. the native environment. Disentangling potential bottleneck, admixture and cryptic variation effects will enable a deeper understanding of the realized evolutionary constraints and opportunities experienced by founding populations.

• The fitness effects of genetic admixture between source populations with different levels of genetic divergence. In particular, experimental crosses and comparisons to mid-parent values should reveal the potential fitness benefits of multiple introductions of modest genetic differentiation, as well as the potential resistance to admixture from negative genetic interactions among more divergent sources.
• The demographic impacts of evolutionary change. Comparative demography of genetic variants can be used to connect individual evolutionary changes to their consequences for aspects of invasiveness (population growth, density and spread) across all stages of invasion.

Baker often referred to his study of the genetics of colonizers as 'genecology' (Baker 1965), and this connection between genes and ecology remains a worthy goal 50 years later. By understanding the details of invader genetics, we can move towards understanding how their ecology might realistically evolve.

ACKNOWLEDGEMENTS

We would like to thank the organizers and participants of Invasion Genetics: The Baker and Stebbins Legacy Symposium held at Asilomar, California, USA, in 2014, which celebrated 50 years of invasion genetics and inspired this synthesis. Comments from RI Colautti, R Baucom and an anonymous reviewer substantially improved this manuscript.

REFERENCES

Agashe D, Falk JJ, Bolnick DI (2011) Effects of founding genetic variation on adaptation to a novel resource. *Evolution*, **65**, 2481–2491.

Allard R (1965) Genetic systems associated with colonizing ability in predominantly self-pollinated species. In: *The Genetics of Colonizing Species* (eds Baker H, Stebbins G), pp. 49–76. Academic Press, New York.

Allendorf FW, Lundquist LL (2003) Introduction: population biology, evolution, and control of invasive species. *Conservation Biology*, **17**, 24–30.

Antonovics J, Levin D (1980) The ecological and genetic consequences of density-dependent regulation in plants. *Annual Review of Ecology and Systematics*, **11**, 411–452.

Badyaev A (2005) Stress-induced variation in evolution: from behavioural plasticity to genetic assimilation. *Proceedings of the Royal Society B-Biological Sciences*, **272**, 877–886.

Baker H (1955) Self-compatibility and establishment after "long-distance" dispersal. *Evolution*, **9**, 347–349.

Baker H (1965) Characteristics and modes of origin of weeds. In: *The Genetics of Colonizing Species* (eds Baker H, Stebbins G), pp. 147–172. Academic Press, New York.

Baker H (1974) The evolution of weeds. *Annual Review of Ecology and Systematics*, **5**, 1–24.

Baker H, Stebbins G (eds) (1965) *The Genetics of Colonizing Species*. Academic Press, New York.

Barrett SCH (2015) Foundations of invasion genetics: the legacy of Baker and Stebbins. *Molecular Ecology*, **24**, 1927–1941.

Barrett SCH, Husband BC (1990) The genetics of plant migration and colonization. In: *Plant Population Genetics, Breeding, and Genetic Resources* (eds Brown AHD, Clegg MT, Kahler AL, Weir BS), pp. 254–277. Sinauer Associates Inc., Sunderland.

Barrett SCH, Richardson BJ (1986) Genetic attributes of invading species. In: *Ecology of Biological Invasions, an Australian Perspective* (eds Groves R, Burdon JJ), pp. 21–33. Australian Academy of Sciences, Canberra.

Beavis WD (1994) The power and deceit of QTL experiments: lessons from comparative QTL studies. *Proceedings of the Forty-ninth Annual Corn and Sorghum Industry Research Conference*, **49**, 250–266.

Blackburn T, Lockwood JL, Cassey P (2015) The influence of numbers on invasion success. *Molecular Ecology*, **24**, 1942–1953.

Bock D, Caseys C, Cousens R *et al.* (2015) What we still don't know about invasion genetics. *Molecular Ecology*, **24**, 2277–2298.

Bomblies K, Lempe J, Epple P *et al.* (2007) Autoimmune response as a mechanism for a Dobzhansky-Muller-type incompatibility syndrome in plants. *PLoS Biology*, **5**, e236.

Bossdorf O, Auge H, Lafuma L *et al.* (2005) Phenotypic and genetic differentiation between native and introduced plant populations. *Oecologia*, **144**, 1–11.

Briskie JV, Mackintosh M (2004) Hatching failure increases with severity of population bottlenecks in birds. *Proceedings of the National Academy of Sciences of the USA*, **101**, 558–561.

Brown A, Marshall D (1981) Evolutionary changes accompanying colonization in plants. In: *Evolution Today, Proceedings of the Second International Congress of Systematic and Evolutionary Biology* (eds Scudder G, Reveal J), pp. 351–353. Hunt Institute, Pittsburgh, Pennsylvania.

Carson H (1965) Chromosomal morphism in geographically widespread species of Drosophila. In: *The Genetics of Colonizing Species* (eds Baker H, Stebbins G), pp. 503–531. Academic Press, New York.

Chapman JR, Nakagawa S, Coltman DW, Slate J, Sheldon BC (2009) A quantitative review of heterozygosity-fitness correlations in animal populations. *Molecular Ecology*, **18**, 2746–2765.

Colautti RI, Barrett SCH (2013) Rapid adaptation to climate facilitates range expansion of an invasive plant. *Science*, **342**, 364–366.

Colautti RI, Lau JA (2015) Contemporary evolution during invasion: evidence for differentiation, natural selection, and local adaptation. *Molecular Ecology*, **24**, 1999–2017.

Colautti RI, Maron JL, Barrett SCH (2009) Common garden comparisons of native and introduced plant populations: latitudinal clines can obscure evolutionary inferences. *Evolutionary Applications*, **2**, 187–199.

Colautti RI, Lee C-R, Mitchell-Olds T (2012) Origin, fate, and architecture of ecologically relevant genetic variation. *Current Opinion in Plant Biology*, **15**, 199–204.

Cox GW (2004) *Alien Species and Evolution: The Evolutionary Ecology of Exotic Plants, Animals, Microbes, and Interacting Native Species*. Island Press, Washington.

Cristescu ME (2015) Genetic reconstructions of invasion history. *Molecular Ecology*, **24**, 2212–2225.

Davey JW, Hohenlohe PA, Etter PD *et al.* (2011) Genome-wide genetic marker discovery and genotyping using next-generation sequencing. *Nature Reviews Genetics*, **12**, 499–510.

De Visser JAGM, Krug J (2014) Empirical fitness landscapes and the predictability of evolution. *Nature Reviews Genetics*, **15**, 480–490.

DeBolt S (2010) Copy number variation shapes genome diversity in *Arabidopsis* over immediate family generational scales. *Genome Biology and Evolution*, **2**, 441–453.

Denver DR, Morris K, Lynch M, Thomas WK (2004) High mutation rate and predominance of insertions in the *Caenorhabditis elegans* nuclear genome. *Nature*, **430**, 679–682.

Dlugosch KM, Parker IM (2008a) Founding events in species invasions: genetic variation, adaptive evolution, and the role of multiple introductions. *Molecular Ecology*, **17**, 431–449.

Dlugosch KM, Parker IM (2008b) Invading populations of an ornamental shrub show rapid life history evolution despite genetic bottlenecks. *Ecology Letters*, **11**, 701–709.

Dobzhansky T (1965) "Wild" and "domestic" species of Drosophila. In: *The Genetics of Colonizing Species* (eds Baker H, Stebbins G), pp. 533–551. Academic Press, New York.

Duveau F, Félix M-A (2012) Role of pleiotropy in the evolution of a cryptic developmental variation in *Caenorhabditis elegans*. *PLoS Biology*, **10**, e1001230.

Dworkin I, Palsson A, Birdsall K, Gibson G (2003) Evidence that Egfr contributes to cryptic genetic variation for photoreceptor determination in natural populations of *Drosophila melanogaster*. *Current Biology*, **13**, 1888–1893.

Edmands S (1999) Heterosis and outbreeding depression in interpopulation crosses spanning a wide range of divergence. *Evolution*, **53**, 1757–1768.

Ekblom R, Galindo J (2011) Applications of next generation sequencing in molecular ecology of non-model organisms. *Heredity*, **107**, 1–15.

Elam DR, Ridley CE, Goodell K, Ellstrand NC (2007) Population size and relatedness affect fitness of a self-incompatible invasive plant. *Proceedings of the National Academy of Sciences of the USA*, **104**, 549–552.

Ellstrand NC, Roose ML (1987) Patterns of genotypic diversity in clonal plant species. *American Journal of Botany*, **74**, 123–131.

Ellstrand NC, Schierenbeck KA (2000) Hybridization as a stimulus for the evolution of invasiveness in plants? *Proceedings of the National Academy of Sciences of the USA*, **97**, 7043–7050.

Excoffier L, Foll M, Petit RJ (2009) Genetic consequences of range expansions. *Annual Review of Ecology, Evolution, and Systematics*, **40**, 481–501.

Facon B, Pointier J-P, Jarne P, Sarda V, David P (2008) High genetic variance in life-history strategies within invasive populations by way of multiple introductions. *Current Biology*, **18**, 363–367.

Ferrero V, Barrett SCH, Castro S *et al.* (2015) Invasion genetics of the Bermuda buttercup (*Oxalis pes-caprae*): complex intercontinental patterns of genetic diversity, polyploidy and heterostyly characterize both native and introduced populations. *Molecular Ecology*, **24**, 2143–2155.

Fisher RA (1930) *The Genetical Theory of Natural Selection*. Clarendon Press, Oxford.

Fournier-Level A, Korte A, Cooper MD *et al.* (2011) A map of local adaptation in *Arabidopsis thaliana*. *Science*, **334**, 86–89.

Frankham R (2005) Resolving the genetic paradox in invasive species. *Heredity*, **94**, 385.

Fraser A (1965) Colonization and genetic drift. In: *The Genetics of Colonizing Species* (eds Baker H, Stebbins G), pp. 117–125. Academic Press, New York.

Freeman JL, Perry GH, Feuk L *et al.* (2006) Copy number variation: new insights in genome diversity. *Genome Research*, **16**, 949–961.

Gaston KJ (2009) Geographic range limits of species. *Proceedings of the Royal Society B-Biological Sciences*, **276**, 1391–1393.

Goodnight CJ (1988) Epistasis and the effect of founder events on the additive genetic variance. *Evolution*, **42**, 441–454.

Gray AJ, Mack RN, Harper JL *et al.* (1986) Do invading species have definable genetic characteristics? *Philosophical Transactions of the Royal Society B: Biological Sciences*, **314**, 655–674.

Hallatschek O, Hersen P, Ramanathan S, Nelson DR (2007) Genetic drift at expanding frontiers promotes gene segregation. *Proceedings of the National Academy of Sciences of the USA*, **104**, 19926–19930.

Hamilton J, Okada M, Korves T, Schmitt J (2015) The role of climate adaptation in colonization success in *Arabidopsis thaliana*. *Molecular Ecology*, **24**, 2253–2263.

Hamrick JL, Godt MJW (1996) Effects of life history traits on genetic diversity in plant species. *Philosophical Transactions of the Royal Society B: Biological Sciences*, **351**, 1291–1298.

Hancock AM, Brachi B, Faure N *et al.* (2011) Adaptation to climate across the *Arabidopsis thaliana* genome. *Science*, **334**, 83–86.

Hendry AP, Kinnison MT, Dec N (1999) The pace of modern life: measuring rates of contemporary microevolution. *Evolution*, **53**, 1637–1653.

Hermisson J, Wagner GP (2004) The population genetic theory of hidden variation and genetic robustness. *Genetics*, **168**, 2271–2284.

Hirase S, Ozaki H, Iwasaki W (2014) Parallel selection on gene copy number variations through evolution of three-spined stickleback genomes. *BMC Genomics*, **15**, 735.

Hoekstra HE, Coyne JA (2007) The locus of evolution: evo devo and the genetics of adaptation. *Evolution*, **61**, 995–1016.

Hoffmann AA, Merilä J (1999) Heritable variation and evolution under favourable and unfavourable conditions. *Trends in Ecology & Evolution*, **14**, 96–101.

Hoffmann AA, Rieseberg LH (2008) Revisiting the impact of inversions in evolution: from population genetic markers to drivers of adaptive shifts and speciation? *Annual Review of Ecology and Systematics*, **39**, 21–42.

Hoffmann AA, Weeks AR (2007) Climatic selection on genes and traits after a 100 year-old invasion: a critical look at the temperate-tropical clines in *Drosophila melanogaster* from eastern Australia. *Genetica*, **129**, 133–147.

Hoffmann AA, Sgrò CM, Weeks AR (2004) Chromosomal inversion polymorphisms and adaptation. *Trends in Ecology & Evolution*, **19**, 482–488.

Hufbauer RA (2008) Biological invasions: paradox lost and paradise gained. *Current Biology*, **18**, 246–247.

Hufbauer RA, Rutschmann A, Serrate B, Vermeil de Conchard H, Facon B (2013) Role of propagule pressure in colonization success: disentangling the relative importance of demographic, genetic and habitat effects. *Journal of Evolutionary Biology*, **26**, 1691–1699.

Jones FC, Grabherr MG, Chan YF *et al.* (2012) The genomic basis of adaptive evolution in threespine sticklebacks. *Nature*, **484**, 55–61.

Keane R, Crawley M (2002) Exotic plant invasions and the enemy release hypothesis. *Trends in Ecology & Evolution*, **17**, 164–170.

Keller L, Waller D (2002) Inbreeding effects in wild populations. *Trends in Ecology & Evolution*, **17**, 230–241.

Keller SR, Fields PD, Berardi AE, Taylor DR (2014) Recent admixture generates heterozygosity-fitness correlations during the range expansion of an invading species. *Journal of Evolutionary Biology*, **27**, 616–627.

Kimura M (1985) *The Neutral Theory of Molecular Evolution*. Cambridge University Press, Cambridge.

Kinnison MT, Hairston NG (2007) Eco-evolutionary conservation biology: contemporary evolution and the dynamics of persistence. *Functional Ecology*, **21**, 444–454.

Kirkpatrick M (2010) How and why chromosome inversions evolve. *PLoS Biology*, **8**, e1000501.

Kirkpatrick M, Barrett B (2015) Chromosome inversions, adaptive cassettes, and the evolution of species' ranges. *Molecular Ecology*, **24**, 2046–2055.

Kolbe JJ, Glor RE, Rodríguez Schettino L *et al.* (2004) Genetic variation increases during biological invasion by a Cuban lizard. *Nature*, **431**, 177–181.

Kondrashov FA (2012) Gene duplication as a mechanism of genomic adaptation to a changing environment. *Proceedings of the Royal Society B-Biological Sciences*, **279**, 5048–5057.

Koskinen MT, Haugen TO, Primmer CR (2002) Contemporary fisherian life-history evolution in small salmonid populations. *Nature*, **419**, 826–830.

Krieger MJB, Ross KG (2002) Identification of a major gene regulating complex social behavior. *Science*, **295**, 328–332.

Lambrinos JG (2004) How interactions between ecology and evolution influence contemporary invasion dynamics. *Ecology*, **85**, 2061–2070.

Lankau RA, Nuzzo V, Spyreas G, Davis AS (2009) Evolutionary limits ameliorate the negative impact of an invasive plant. *Proceedings of the National Academy of Sciences of the USA*, **106**, 15362–15367.

Latta LC, Peacock M, Civitello DJ *et al.* (2015) The phenotypic effects of spontaneous mutations in different environments. *The American Naturalist*, **185**, 243–252.

Lavergne S, Molofsky J (2007) Increased genetic variation and evolutionary potential drive the success of an invasive grass. *Proceedings of the National Academy of Sciences of the USA*, **104**, 3883–3888.

Lee CE (2002) Evolutionary genetics of invasive species. *Trends in Ecology & Evolution*, **17**, 386–391.

Lee CE, Gelembiuk GW (2008) Evolutionary origins of invasive populations. *Evolutionary Applications*, **1**, 427–448.

Leinonen T, O'Hara RB, Cano JM, Merilä J (2008) Comparative studies of quantitative trait and neutral marker divergence: a meta-analysis. *Journal of Evolutionary Biology*, **21**, 1–17.

Linde M (2001) Flowering ecotypes of *Capsella bursa-pastoris* (L.) Medik. (Brassicaceae) analysed by a cosegregation of phenotypic characters (QTL) and molecular markers. *Annals of Botany*, **87**, 91–99.

Linnen CR, Poh Y-P, Peterson BK *et al.* (2013) Adaptive evolution of multiple traits through multiple mutations at a single gene. *Science*, **339**, 1312–1316.

Lockwood JL, Cassey P, Blackburn T (2005) The role of propagule pressure in explaining species invasions. *Trends in Ecology & Evolution*, **20**, 223–228.

Louthan AM, Kay KM (2011) Comparing the adaptive landscape across trait types: larger QTL effect size in traits under biotic selection. *BMC Evolutionary Biology*, **11**, 60.

Lynch M (1991) The genetic interpretation of inbreeding depression and outbreeding depression. *Evolution*, **45**, 622–629.

Lynch M, Conery JS (2000) The evolutionary fate and consequences of duplicate genes. *Science*, **290**, 1151–1155.

Lynch M, Walsh B (1998) *Genetics and Analysis of Quantitative Traits*. Sinauer Associates Inc., Sunderland.

Manel S, Holderegger R (2013) Ten years of landscape genetics. *Trends in Ecology & Evolution*, **28**, 614–621.

Masel J (2006) Cryptic genetic variation is enriched for potential adaptations. *Genetics*, **172**, 1985–1991.

Mayr E (1965a) The nature of colonization in birds. In: *The Genetics of Colonizing Species* (eds Baker HG, Stebbins GL), pp. 29–47. Academic Press, New York.

Mayr E (1965b) Summary. In: *The Genetics of Colonizing Species* (eds Baker H, Stebbins G), pp. 553–562. Academic Press, New York.

McGuigan K, Sgrò CM (2009) Evolutionary consequences of cryptic genetic variation. *Trends in Ecology & Evolution*, **24**, 305–311.

McGuigan K, Nishimura N, Currey M, Hurwit D, Cresko WA (2011) Cryptic genetic variation and body size evolution in threespine stickleback. *Evolution*, **65**, 1203–1211.

McKay JK, Latta RG (2002) Adaptive population divergence: markers, QTL and traits. *Trends in Ecology & Evolution*, **17**, 285–291.

Merilä J, Crnokrak P (2001) Comparison of genetic differentiation at marker loci and quantitative traits. *Journal of Evolutionary Biology*, **14**, 892–903.

Molofsky J, Keller SR, Lavergne S, Kaproth MA, Eppinga MB (2014) Human-aided admixture may fuel ecosystem transformation during biological invasions: theoretical and experimental evidence. *Ecology and Evolution*, **4**, 899–910.

Narum SR, Buerkle CA, Davey JW, Miller MR, Hohenlohe PA (2013) Genotyping-by-sequencing in ecological and conservation genomics. *Molecular Ecology*, **22**, 2841–2847.

Nei M, Maruyama T, Chakraborty R (1975) The bottleneck effect and genetic variability in populations. *Evolution*, **29**, 1–10.

Novak SJ, Mack RN (1993) Genetic variation in *Bromus tectorum* (Poaceae): comparison between native and introduced populations. *Heredity*, **71**, 167–176.

Novak SJ, Mack RN (2005) Genetic bottlenecks in alien plant species: the influence of mating systems and introduction dynamics. In: *Species Invasions: Insights into Ecology, Evolution, and Biogeography* (eds Sax D, Stachowicz J, Gaines S), pp. 201–228. Sinauer Associates Inc., Sunderland, Massachusetts.

Olson-Manning CF, Wagner MR, Mitchell-Olds T (2012) Adaptive evolution: evaluating empirical support for theoretical predictions. *Nature Reviews Genetics*, **13**, 867–877.

Orr HA (1998) The population genetics of adaptation: the distribution of factors fixed during adaptive evolution. *Evolution*, **52**, 935–949.

Orr HA (2005) The genetic theory of adaptation: a brief history. *Nature Reviews Genetics*, **6**, 119–127.

Orr HA, Turelli M (2001) The evolution of postzygotic isolation: accumulating Dobzhansky-Muller incompatibilities. *Evolution*, **55**, 1085–1094.

Ossowski S, Schneeberger K, Lucas-Lledó JI *et al.* (2010) The rate and molecular spectrum of spontaneous mutations in *Arabidopsis thaliana. Science*, **327**, 92–94.

Otto SP, Jones CD (2000) Detecting the undetected: estimating the total number of loci underlying a quantitative trait. *Genetics*, **156**, 2093–2107.

Otto SP, Whitlock MC (1997) The probability of fixation in populations of changing size. *Genetics*, **146**, 723–733.

Ozgul A, Tuljapurkar S, Benton TG *et al.* (2009) The dynamics of phenotypic change and the shrinking sheep of St. Kilda. *Science*, **325**, 464–467.

Paaby AB, Rockman MV (2014) Cryptic genetic variation: evolution's hidden substrate. *Nature Reviews Genetics*, **15**, 247–258.

Pannell JR (2015) Evolution of the mating system and ability to self-fertilize in colonizing plants. Molecular Ecology.

Pascual M, Chapuis MP, Mestres F *et al.* (2007) Introduction history of *Drosophila subobscura* in the New World: a microsatellite-based survey using ABC methods. *Molecular Ecology*, **16**, 3069–3083.

Paterson AH, Schertz KF, Lin YR, Liu SC, Chang YL (1995) The weediness of wild plants: molecular analysis of genes influencing dispersal and persistence of johnsongrass, *Sorghum halepense* (L.) Pers. *Proceedings of the National Academy of Sciences of the USA*, **92**, 6127–6131.

Peischl S, Excoffier L (2015) Expansion load: recessive mutations and the role of standing genetic variation. *Molecular Ecology*, **24**, 2084–2094.

Perkins TA, Phillips BL, Baskett ML, Hastings A (2013) Evolution of dispersal and life history interact to drive accelerating spread of an invasive species. *Ecology Letters*, **16**, 1079–1087.

Petitpierre B, Kueffer C, Broennimann O *et al.* (2012) Climatic niche shifts are rare among terrestrial plant invaders. *Science*, **335**, 1344–1348.

Posavi M, Gelembiuk GW, Larget B, Lee CE (2014) Testing for beneficial reversal of dominance during salinity shifts in the invasive copepod *Eurytemora affinis*, and implications for the maintenance of genetic variation. *Evolution*, **68**, 3166–3183.

Prentis PJ, Wilson JRU, Dormontt EE, Richardson DM, Lowe AJ (2008) Adaptive evolution in invasive species. *Trends in Plant Science*, **13**, 288–294.

Prevosti A, Ribo G, Serra L *et al.* (1988) Colonization of America by *Drosophila subobscura*: experiment in natural populations that supports the adaptive role of chromosomal-inversion polymorphism. *Proceedings of the National Academy of Sciences of the USA*, **85**, 5597–5600.

Purchase CF, Moreau DTR (2012) Stressful environments induce novel phenotypic variation: hierarchical reaction norms for sperm performance of a pervasive invader. *Ecology and Evolution*, **2**, 2567–2576.

Rius M, Darling JA (2014) How important is intraspecific genetic admixture to the success of colonising populations? *Trends in Ecology & Evolution*, **29**, 233–242.

Rohner N, Jarosz DF, Kowalko JE *et al.* (2013) Cryptic variation in morphological evolution: HSP90 as a capacitor for loss of eyes in cavefish. *Science*, **342**, 1372–1375.

Roman J, Darling JA (2007) Paradox lost: genetic diversity and the success of aquatic invasions. *Trends in Ecology & Evolution*, **22**, 454–464.

Stapley J, Santure A, Dennis S (2015) Transposable elements as agents of rapid adaptation may explain the genetic paradox of invasive species. *Molecular Ecology*, **24**, 2241–2252.

Stebbins G (1957) Self fertilization and population variability in the higher plants. *The American Naturalist*, **91**, 337–354.

Stewart CN, Tranel PJ, Horvath DP *et al.* (2009) Evolution of weediness and invasiveness: charting the course for weed genomics. *Weed Science*, **57**, 451–462.

Stinchcombe JR, Hoekstra HE (2008) Combining population genomics and quantitative genetics: finding the genes underlying ecologically important traits. *Heredity*, **100**, 158–170.

Strubbe D, Broennimann O, Chiron F, Matthysen E (2013) Niche conservatism in non-native birds in Europe: niche unfilling rather than niche expansion. *Global Ecology and Biogeography*, **22**, 962–970.

Suarez A, Tsutsui N (2008) The evolutionary consequences of biological invasions. *Molecular Ecology*, **17**, 351–360.

Szűcs M, Melbourne BA, Tuff T, Hufbauer RA (2014) The roles of demography and genetics in the early stages of colonization. *Proceedings of the Royal Society B-Biological Sciences*, **281**, 20141073.

Szulkin M, Bierne N, David P (2010) Heterozygosity-fitness correlations: a time for reappraisal. *Evolution*, **64**, 1202–1217.

Tsutsui N, Suarez A, Holway D, Case T (2000) Reduced genetic variation and the success of an invasive species. *Proceedings of the National Academy of Sciences of the USA*, **97**, 5948–5953.

Tsutsui ND, Suarez A, Grosberg RK (2003) Genetic diversity, asymmetrical aggression, and recognition in a widespread invasive species. *Proceedings of the National Academy of Sciences of the USA*, **100**, 1078–1083.

Turelli M, Barton NH (2006) Will population bottlenecks and multilocus epistasis increase additive genetic variance? *Evolution*, **60**, 1763–1776.

Turgeon J, Tayeh A, Facon B *et al.* (2011) Experimental evidence for the phenotypic impact of admixture between wild and biocontrol Asian ladybird (*Harmonia axyridis*) involved in the European invasion. *Journal of Evolutionary Biology*, **24**, 1044–1052.

Uller T, Leimu R (2011) Founder events predict changes in genetic diversity during human-mediated range expansions. *Global Change Biology*, **17**, 3478–3485.

Van Heerwaarden B, Willi Y, Kristensen TN, Hoffmann AA (2008) Population bottlenecks increase additive genetic variance but do not break a selection limit in rain forest Drosophila. *Genetics*, **179**, 2135–2146.

Vandepitte K, de Meyer T, Helsen K *et al.* (2014) Rapid genetic adaptation precedes the spread of an exotic plant species. *Molecular Ecology*, **23**, 2157–2164.

Verhoeven KJF, Macel M, Wolfe LM, Biere A (2011) Population admixture, biological invasions and the balance between local adaptation and inbreeding depression. *Proceedings of the Royal Society B-Biological Sciences*, **278**, 2–8.

Waddington C (1956) Genetic assimilation of the bithorax phenotype. *Evolution*, **10**, 1–13.

Waddington C (1965) Introduction to the symposium. In: *The Genetics of Colonizing Species* (eds Baker H, Stebbins G), pp. 1–6. Academic Press, New York.

Wares J, Hughes A, Grosberg K (2005) Mechanisms that drive evolutionary change: insights from species introductions and invasions. In: *Species Invasions: Insights into Ecology, Evolution, and Biogeography* (eds Sax D, Stachowicz J, Gaines S), pp. 229–257. Sinauer Associates Inc, Sunderland, Massachusetts.

Waser NM, Price MV (1989) Optimal outcrossing in *Ipomopsis aggregata*: seed set and offspring fitness. *Evolution*, **43**, 1097–1109.

Whitlock MC (1997) Founder effects and peak shifts without genetic drift: adaptive peak shifts occur easily when environments fluctuate slightly. *Evolution*, **51**, 1044–1048.

Whitlock MC, Phillips PC, Moore FB-G, Tonsor SJ (1995) Multiple fitness peaks and epistasis. *Annual Review of Ecology and Systematics*, **26**, 601–629.

Whitney KD, Broman K, Kane N *et al.* (2015) QTL mapping identifies candidate alleles involved in adaptive introgression and range expansion in a wild sunflower. Molecular Ecology.

Wilson F (1965) Biological control and the genetics of colonizing species. In: *The Genetics of Colonizing Species* (eds Baker H, Stebbins G), 307–329. Academic Press, New York.

Wray GA (2007) The evolutionary significance of cis-regulatory mutations. *Nature Reviews Genetics*, **8**, 206–216.

Wright S (1932) The roles of mutation, inbreeding, cross-breeding, and selection in evolution. In: *Proceedings of the Sixth International Congress on Genetics*, 1, pp. 356–366.

Yeaman S (2013) Genomic rearrangements and the evolution of clusters of locally adaptive loci. *Proceedings of the National Academy of Sciences of the USA*, **110**, E1743–E1751.

Żmieńko A, Samelak A, Kozłowski P, Figlerowicz M (2014) Copy number polymorphism in plant genomes. *TAG. Theoretical and Applied Genetics*, **127**, 1–18.

DATA ACCESSIBILITY

Molecular genetic diversity data from literature review are available through Dryad; Dryad doi:10.5061/dryad.s2948.

DISCUSSION

RUSSELL LANDE

KATRINA DLUGOSCH – I am interested in your point that the type of plasticity in invaders may depend on timing. We know that very rapid evolution is possible in invasive populations. Could you say more about situations where higher plasticity is a kind of transient state prior to local adaptation?

RUSSELL LANDE – Yes, that is exactly right. In this type of model, plasticity will increase in a transient and rapid way to the new optimum, assuming that after a sudden change in the environment that predictability is similar to what it was in the native range. Then the predictability aspect of the environment takes over again. So this may differ between one-shot plasticity and labile plasticity as well, and this will also depend on the type of characters studied, and the timescale since the invasion.

ACER VAN WALLENDAEL (FORDHAM UNIVERSITY) – You talked about phenotypic plasticity, but I was wondering if you were referring to that on the species scale, on the gene scale or different individuals?

RUSSELL LANDE – I mostly think about phenotypic plasticity on the individual or genotype level, as well as the population scale. Plasticity has a genotypic component and is really defined with respect to a given genotype

that is replicated across environments and allowed to develop phenotypes across the environments. That is how norms of reaction are measured empirically. The average of those is the population-level norm of reaction composed of individual variation. As with any sort of trait, it can evolve in space and time and vary within a species at the population level. All the usual levels of variation apply, although phenotypic plasticity is not a character *per se* of an individual, unless you are talking about a labile character, but if an individual develops a final phenotype, you cannot measure its plasticity. It has to be measured in a more complicated way.

MARK KIRKPATRICK

BEN PHILLIPS – Presumably, there has to be the evolution of some pre-zygotic reproductive isolation to allow the spread you describe to occur; otherwise, there would just be a hybrid zone, right?

MARK KIRKPATRICK – We are on a slippery slope when we get into species definitions! But you can certainly have a situation where there is incomplete but still very strong reproductive isolation between two species. Once every 1000 generations, a pair hybridizes,

Invasion Genetics: The Baker and Stebbins Legacy, First Edition. Edited by Spencer C. H. Barrett, Robert I. Colautti, Katrina M. Dlugosch, and Loren H. Rieseberg.

253

and 90% of their hybrid offspring die. But every once in a while, one survives and carries some genetic material that introgresses. Obviously, I am making up this scenario, but it is very easy to imagine almost completely reproductively isolated species that would allow this process to happen.

LOREN RIESEBERG – Regarding your model on the genetic limits to the range of species, it suggests that gene flow from the center of the range can pull back populations at the edge, or make them maladapted. The alternative hypothesis would be mutation limitation of populations on the edge of the range and thus limited genetic variation. Is there evidence for this, and do you think this model is likely to be valid?

MARK KIRKPATRICK – On some level, that must happen. There are many examples of species that are simply limited by physiological challenges. For example, a palm tree in principle could get a mutation that would make it freeze resistant, but it just has not got it yet. There are also situations in which it is impossible to imagine how any mutation could possibly do the trick. There are trivial examples, like mutations that would allow life in outer space. More interesting situations occur when a species range is limited by competition with, for example, a congener. In that case, there may not be any kind of genetic variation that will allow it to become more efficient at using a resource that its competitor is already exploiting at maximum efficiency. So I think that there really are genetic constraints, and that some of them are unbreakable. I don't know what it is about insect physiology that prevents them from living in salt water, but that seems like a real constraint.

LOREN RIESEBERG – With some plant species, sunflowers, for example, the northern limit of their distribution is southern Canada, and I have often wondered which model actually explains why they do not move further north. Obviously, it is getting colder and there are shorter seasons, but it is unclear if their distributions are limited by gene flow or by mutation because they have continuous ranges.

MARK KIRKPATRICK – There are generalizations that have been claimed in the literature for terrestrial species on continents: their northern limits are often limited by physiological constraints and their southern limits by biotic interactions. The same is said about the intertidal zone, with the upper limits set by abiotic conditions and the lower limits by biotic interactions.

KATRINA DLUGOSCH – In recent years, the role of admixture among multiple introductions in promoting invasion has been of considerable interest. Do you think admixture could be another situation where inversions play an important role in introgression? Or do you think that whether or not an inversion is present will be more important to adaptation within a single introduction versus with admixture with a different introduction from another part of the range?

MARK KIRKPATRICK – Inversions may just be a version "writ large" of what can happen at the level of a single locus. Inversions have a couple of things going for them. They capture multiple genes, and can get the selective benefit of locally adapted alleles at all of those loci. Inversions can avoid being lost by swamping that might otherwise happen to the individual alleles that they capture. The idea that hybridization can introduce adaptive genetic material goes back a long way. Richard Lewontin and L.C. Birch wrote a paper (*Evolution*, 1966, **20**: 315–336) on this, and there are probably earlier mentions. There is also the controversy over hybrid speciation. There are many examples, of course, but it is not clear that it is a very frequent phenomenon. What are likely to be more frequent are occasional hybridization events that introduce new alleles into a species that have not arisen by endogenous mutation.

TIM WRIGHT (NEW MEXICO STATE UNIVERSITY) – You seemed to suggest that a lot of inversions come into species by hybridization. Is there any reason to believe that this is a more likely route than being generated *de novo* within a species?

MARK KIRKPATRICK – I would love to know more about how frequent inversions of different sizes are, and how often they do or do not have deleterious side effects that prevent them from spreading. There are many genetic and biological features that go into determining whether an inversion will be successful, and it may be a very rare combination of those factors that conspire to generate something that is adaptive. It is easy to imagine (again in the absence of any data!) that if a beneficial inversion appeared in one species, it might not occur by mutation in a sympatric species, even if it was favored there as well.

RUSSELL LANDE – You probably already know all of this background, but inversions do have a fairly high rate of spontaneous occurrence; the smaller they are, the less selective disadvantage they have due to recombination within the inversion. And in Diptera, which is the main group that you talked about, big inversions are a couple of orders of magnitude more common, especially as polymorphisms within a population. Diptera have a special meiotic mechanism that suppresses recombination within the inversion. In most other species, you tend to see much smaller inversions. Some of these may not actually be adaptive, although the small ones are more likely to get fixed if they encompass some sort of adaptive variation. However, that is not always the case because the smaller they are, the more nearly neutral they are, because they have fewer genes that could be adaptive because they have less recombination within them. I was wondering, the introgressions you were talking about, you did not make it clear, were they, even in the mosquitoes, actually polymorphic from the population that they were introgressed from? Or were they fixed? Do you know about that? I am just worried, for example, with regard to the fish you discussed: are they actually adaptive? You do not want to interpret every inversion that gets into another species as being a preadaptation in the other species. It may be a post-adaptation or it may even be neutral in some cases.

Mark Kirkpatrick – Regarding the size of an inversion and what kind of effect it might have, I was talking about just one of the different ideas people have about how inversions evolve. Another important hypothesis is that, no matter how small an inversion is, it breaks the chromosome in two points. An inversion can therefore break one or more genes, which can have a direct fitness effect independent of whatever loci are inside the inversion.

Russell Lande – But usually breakpoints do not have any phenotypic effects.

Mark Kirkpatrick – Usually? I don't know how you can say that as I am not sure those data exist. There are definitely examples of inversions where the breakpoints do influence gene expression. Regarding your second question, I am not sure it is known whether the alleles are polymorphic in the source populations because typically the source populations are poorly studied. That is an important question for the future. The third point you made touched on the sticklebacks. The situation there is that these inversions form clines that have evolved independently and repeatedly, which is very strong evidence for local adaptation. Multiple invasions of freshwater have involved the same inversions that have gone to

high frequency (F.C. Jones *et al.*, 2012, *Nature*, **484**: 55–61), so that is not drift. Inversions are an extremely diverse, heterogeneous type of mutation, and I am sure some of them are neutral, some of them do break genes and are directly selected, and some of them are locally adapted. I think that there is every possible scenario and no one story fits all.

MARK BLOWS

Mark Kirkpatrick – I have a couple of hypotheses for why your **G**-Max (principal component 1 of a **G**-matrix) might be so large. One is boring; the other might be interesting. The boring one is that there is systematic bias in the way we estimate genetic variances and covariances, and the way those errors feed into the eigenvalues will make the estimate of the biggest one too big and the estimate of the smallest one too small. The possibly more interesting one is that there are some kinds of selection and genetic architectures that actually make **G**-Max line up with the direction of selection. For example, with a trait affected by a major locus with two alleles, if directional selection on the trait causes the rare allele to spread, the genetic variance for the trait will increase. What are your thoughts on these hypotheses?

Mark Blows – On the first hypothesis: yes, estimation procedures do tend, under some conditions, to bias the lead eigenvalue. However, from what I have observed, it is the bias that comes from random matrix theory that is likely to be far larger than the estimation bias generated through restricted maximum likelihood (REML) under certain circumstances. That would be my take. On the orientation question: I do not know, because unfortunately we do not have that kind of data. I am really interested in how these matrices evolve in response to both selection and the input of mutation, and I think that is where we need to go in the next few years.

Johanna Schmitt – What is really a trait and what should we empiricists be measuring? The traits that are evolutionarily important are combinations along the different axes that you are identifying in the **G**-Matrix. So how should empiricists change what we measure, and how should we be thinking about genetic architecture?

Mark Blows – In terms of what we measure, I do not think we need to change anything at all. It is just letting go of what we measure being the thing that is probably important! The notion that as biologists we are really good at picking the things that are important for the organisms, I don't believe that is true at all and selection analyses demonstrate this. You find far greater selection on trait combinations than on anything we measure in a univariate sense. And I think that makes sense because many of these traits act in concert to confer fitness, and it is just formalizing that. So I don't think we need to change a lot about what we measure, but it does change the way we deal with genetic analysis and getting used to the fact that there is a vast difference in the amount of genetic variance in these trait combinations. This contrasts with when I did my PhD, when the mantra was "everything is heritable, it is about 0.3, and just get on with it." It changes the way that you think in that way. It is not meant to be a depressing story at all. It is meant to be a way of then moving on to things such as why do those trait combinations have such vastly different genetic variances? Can we get information on mutation–selection balance, for instance, that shows us that the ones with virtually no variance are under heavy selection or do not have a lot of mutational input. So there is the potential for a lot of explanatory power if we go about it in the right way.

JENNIFER LAU – This is a follow-up question about what empiricists should do. Do we really need to make sure we estimate the **G**-Matrix (**G**) instead of just doing small sample size, phenotypic selection analysis? Also, I was wondering how variability in the magnitude of genetic variance across different trait combinations could potentially complicate some of the other challenges like unmeasured traits. Does multivariate genetic variation exacerbate these problems?

MARK BLOWS – The unmeasured traits problem is not solved by this approach and it is always present. In terms of what we should be measuring, there are a number of avenues that might provide good information. If we accept that pleiotropy is going to be fairly prevalent, then the trait-by-trait view of the natural world is not going to work and the models indicate this. I think the phenotypic covariance matrix (**P**) deserves far more attention than it is currently getting. It is a construct that has really good statistical properties. At least in the data sets that I see, **P** matches **G**. I shouldn't say this as a quantitative geneticist—I'll go out of business! But **P** matches **G** fairly well. So those trait combinations with small genetic variances have small phenotypic variances,

and therefore there are ways of thinking about finding signal that we are currently not using.

MARK VAN KLEUNEN – The **G**-Matrix is dependent on the environment in which it is measured and that might be very relevant for colonizing organisms that encounter novel environments. And I know the idea by Carl Schlichting on the "hidden reaction norm" (*Annals of the New York Academy of Sciences*, 2008, **1133**: 187–203), particularly the suggestion that the genetic variance component might increase when you put an organism in a novel environment. My question is: do you think this is true? And could you explain why we find these unexpected high rates of adaptive evolution in invasive organisms? And the second question is: do you have an expectation for what might happen to the covariance components? Do they increase or decrease?

MARK BLOWS – I am going to hedge my bets on this. Asking whether the **G**-Matrix changes in a different environment, or whether it evolves after it has been under selection are very relevant questions. The tools are there to determine exactly how those two-dimensional random variables, which are the matrices, change. I personally would not focus on whether the covariances change, I think of them as variances: Do the variances change in any particular direction? A lot of the analyses available—the genetic covariance tensor is one—will tell you exactly what trait combinations have changed in variance. Will moving to a harsher environment increase genetic variance? There is some comparative data that suggest it might. I have always been a little agnostic on that. I personally do not see any expectation as to what might occur under a harsher or more benign environment. So that is where I am hedging. I have not thought about that for a long time. I do not see any *a priori* reason why you would expect to find much more variance under a particular environment given standard morphological traits, for example.

RUSSELL LANDE – I think that we may be conflating the claim of a high rate of adaptive evolution in invasive species because most of the data are really about change in the average phenotype, which I do not think is really being talked about in most of the contexts about increased genetic variation. It is a possible mechanism for increased rate of adaptive phenotypic evolution, but most of the data that have been used to make the claim of increased rates of adaptive evolution are simply phenotypic data, and much of that will be simply due to plasticity. It is not adaptive genetic evolution at all, except perhaps in the sense I was talking about earlier of evolution of plasticity *per se*. Most of those data sets are just phenotypic observations. Nobody has really done common garden experiments to separate out genetic and environmental effects even on the change in the mean phenotype, much less looked at genetic variability.

The data are also pretty limited on whether extreme environments cause increased genetic variability or not, but that is actually the ingredient in the models that I was talking about. If a population is adapted to a certain range of environments that are fluctuating around most of the time, then, according to C.H. Waddington, there is some sort of canalization. If the environment changes a lot, or if you introduce experimentally a major mutation, then sometimes you do see increased genetic variance, but that is getting out of the zone of canalization. But many of the earlier experiments were with threshold characters, and that is a very different situation than with continuously varying characters, which is what Mark Blows was talking about. There is much less genetic information on the degree to which those are canalized and how wide the zone of canalization may be. Now, it is known that if you do artificial stabilizing selection, you do not see a lot of change in the genetic variance or the environmental variance, which is partly development noise and partly phenotypic plasticity with respect to environmental variation you may not even be measuring. But that is probably because there is some history of stabilizing selection or fluctuating stabilizing selection in most populations already. But if you do artificial disruptive selection, you see not only large increases in genetic variance but also big increases in the environmental variance, the developmental noise, and probably also to some extent plasticity. So this is a complicated subject that has not been very well investigated empirically for continuously varying characters.

TROY DAY

RUSSELL LANDE – The last analysis you undertook seemed to assume no sampling effects in the matrix that you started with. Do you have ideas on how to deal with that?

TROY DAY – You are right, that will introduce another source of variation in the calculation. What we are doing right now is an attempt to deal with that kind of variation in an entirely different way to calculate the entropy that does not involve binning the data. It involves measuring the distances between the data points and using that as a reflection of the density of the breeding value distribution locally.

RUSSELL LANDE – It struck me at the beginning of your talk, that to finally get to classical information as the measure from which everything else followed was almost circular. In particular, I did not see how taking an arbitrary width of your bin is valid as so many things could cancel out in the end. Does that only happen when you have a very tiny width? I was glad to hear what you said at the end in response to my question about the sampling, that you are considering getting rid of the binning altogether. That seems to be a much more natural way of doing things than being forced into using bins to begin with. I do not think that all the material you covered at the beginning added much, except that it helped you make the statement that you were avoiding "lunatic information theory."

TROY DAY – Where to start? The specific question about bin width—it doesn't depend on bin width. Information entropy falls out regardless of bin width. Bin width is an arbitrary thing that you can choose however you want. But the bigger question is about *justifying* that particular metric of evolvability rather than just starting with it at the outset. You could simply start with it and people do use things like mutual information already. Really, it is simply a tool to try and find dependencies among variables. And in the context of invasions that is one of the things you might want to know—you want to know if there are holes in genotype space, such as population states that are difficult to evolve. If there are such holes, then there are dependencies among variables. And so you could just ask this question using mutual information, but I think it is helpful to have some evolutionary justification for why mutual information might be a natural thing, or information entropy might be a natural thing to look at aside from just saying, "well people use this in stats, maybe I'll use it here." You need to have some evolutionary interpretation for what it means. So really we need two things to quantify evolvability. We need a measure of how much evolution happens and a measure of the strength of selection. And there are lots of choices for either of these. I might choose other things, like quantifying moments of the distribution, but the two I chose lead in a natural way to something that I think is useful.

RUSSELL LANDE – Back at the beginning of your derivation using bins, you said "previous theory assumes directional selection and a Gaussian fitness function but I'm just going to assume there is one thing selection is going for and that is a bin." That seems to be equally arbitrary from the point of view of how natural selection really acts. I would suggest you just start with information for that reason alone.

TROY DAY – Again that is a fair enough comment. The reason I chose that was to reflect the idea that if there are holes in genotype space, then that means there are particular things in genotype space that cannot be selected. The way to quantify that is to select for a particular thing, defined to be a bin, and then allow the bin you are selecting for to vary over whatever space of genotypes you are interested in.

MARK KIRKPATRICK – I want to ask a slightly different question, but first I want to say that I am very happy that you are looking at non-normal distributions, despite the pressure from Russ! Not everything is Gaussian, so that is great. But to turn it around, is it true that if you really do have a normally distributed trait, then by your measure it has equal evolvability no matter the variance or what the mean effect distribution; is that correct?

TROY DAY – The bigger the variance, the bigger the evolvability, but it is independent of the mean.

MARK KIRKPATRICK – In many situations, evolutionary biologists would think the coefficient of variation is a relevant measure of evolvability, because it doesn't really matter what the variance is so much as what the variance is relative to the mean. A variance of 1 mm^2 is big for an ant but not very big for an elephant. Could you comment?

TROY DAY – I have a technical comment, which is that the entropy of a Gaussian is essentially the log of the variance, but it still depends on units. The coefficient of variation that you raised does not have units. And this particular measure has units. And you can scale it so that it has units that are relative to the bin width. This gives it a meaningful interpretation because it is then measured relative to the resolution of selection. So you have a measure of variability that is going to tell you about evolvability, but how evolvable something is will depend on the size of the resolution of selection. If we have really big bins, then I will be selecting in a way for several things at the same time, all lumped into one bin. If I have very

narrow bins, then the evolvability would be different because selection is then differentiating more finely, between say 1 mm of this and then the next millimeter over. The reason it is not unit-less is because it is dependent on the resolution of selection that is used for the bins.

I should say that I think part of the discomfort that Russ has about the use of bins is related to this. You *can* do the same kind of analysis with continuous distributions. What I have done is discretized things and used the discretization as a measure of the resolution of selection. You could do it entirely with continuous distributions. In fact, those results for the normal distribution are the entropy for the continuous distribution. There are some subtleties to doing this, but I think at the end of the day that might be a better way to do it.

STEPHAN PEISCHL

RUSSELL LANDE – What do you think would happen if you relaxed the assumptions about how selection works on new mutations? I am always skeptical of models that have unconditionally advantageous mutations as you then get the mean fitness going to infinity in the center of the range. I guess if you had some sort of optimum phenotype that would only augment your results, but it probably is not going to have a big effect on the front. But I am more concerned about your assumptions about dispersal at a fine-grain of space. Allowing only limited dispersal between discrete demes will definitely exaggerate the results you obtained. I was wondering if you have explored models that would more closely approximate continuous space, and what influence that might have?

STEPHEN PEISCHL – With reference to your first question, you are right that the assumptions on selection are important, but we find very similar results. The results do not really change if we have the distribution of fitness effects that change also with the fitness of the organism in various ways. The important thing is really that at any given point in time, the number of deleterious mutations is higher than the number of beneficial ones. With reference to your second question on dispersal, I did not show it, but we investigated several modes of long-distance dispersal, either bimodal distribution (all these results were based on nearest-neighbor migration), but we also looked at Gaussian dispersal curves, or where there is either nearest-neighbor migration, or migration involving some long distance among demes. We observed that these different patterns of dispersal affect the spatial scale of the process because of faster expansion, but the general patterns do not change at all. I think the reason is that as long as there is some kind of dispersal that leads to something like a Fisher wave, there will always be a small effective population size on the wave front. I was surprised by how robust these results are with respect to changes in migration rates or patterns of migration.

LOREN RIESEBERG – I was wondering if you have started to model cases in which invasion fronts that have loads come together and mix. Have you modeled the mixture in this context?

STEPHEN PEISCHL. – Yes, we have looked at this, and there are several interesting findings. With recessive mutations we observed immediate heterosis because of different deleterious mutations in different populations. Also, if you have additive mutations, you see similar effects with some delay because of increased genetic variance that selection can act on. So our results suggest that hybridization may be really important for invasive success.

BEN PHILLIPS – There seems to be a slight discrepancy between expansion load with an accumulation of deleterious mutations on the invasion front, but also the finding that inbreeding depression is evolving downward, which sets up a situation for selfing to evolve. Is there any way to reconcile these two results?

STEPHAN PEISCHL – I don't think there is a conflict. Inbreeding depression can be reduced in either of two ways: by the loss of deleterious mutations or by fixation of deleterious mutations. If these mutations are at high frequency, it does not matter if a population is a selfer or an outcrosser.

RUSSELL LANDE – Since there are so many botanists here, I thought it might be worthwhile trying to clarify or elaborate on the apparent contradiction of lower inbreeding depression at an advancing front but lower fitness due to accumulation of deleterious mutations, as discussed by John Pannell. The way that botanists usually measure inbreeding depression is by comparing the fitness of selfing and outcrossing *within* populations. And I am fairly sure that is what John Pannell was doing in local populations at the advancing front. So the small population size leads to the fixation of deleterious mutations, which would normally be segregating in the larger populations behind the front. When del-

eterious mutations are segregating, they contribute to inbreeding depression; but when they are fixed they do not. However, if populations that have fixed deleterious mutations at the front were crossed with populations in the center, then you would see significant hybrid vigour, and then that would contribute to the inbreeding depression. So it is all a matter of the populations in which selfing and outcrossing are conducted.

STEPHAN PEISCHL – I agree with that explanation. I asked John Pannell whether they had conducted the crosses you mention, but they have not. We were also interested to see what happens when you take individuals of *Mercurialis annua* from populations representing the two invasion fronts that occur along different sides of the Iberian peninsula and cross them to see if you have hybrid vigour, which is what we would expect.

RUTH HUFBAUER – Over how many generations does expansion load start to appear, and what kind of population sizes were you modeling? To what degree does expansion load depend on population size?

STEPHAN PEISCHL – With regard to timescale and expansion load, one of the most important factors is migration—how much dispersal occurs. For example, with island-like migration expansion load disappears quickly. But with nearest-neighbor migration and a very structured population expansion, load will last for a long time. It also depends on how long the expansion continues. In terms of the population sizes that we used in the simulations, the results I presented were mainly for population sizes of 100 diploid individuals per deme, so relatively low. But, we also did simulations of up to a 1000 diploids per deme. Population size has some effect since when we used large local population sizes in the simulations, we obtained a wave profile that was more shallow, with population sizes at the wave front staying really small even when carrying capacities are large. At least in our simulations, the results we obtained are really robust with regards to population size.

KATRINA DLUGOSCH – Empirically, I think there are a lot of invasions now for which people have been tracking patterns of expansion at various time scales. Therefore, I think in the future it will be really fascinating to incorporate empirical data into some of these models.

SEAN HOBAN (NATIONAL INSTITUTE FOR MATHEMATICAL AND BIOLOGICAL SYNTHESIS) – Your simulations are mostly on the scale of a few thousand generations. Does the surfing phenomenon you describe happen in really rapid expansions such as 10 to a 100 generations?

STEPHAN PEISCHL – It depends, but there are cases where surfing happened in as little as 16 generations. There are examples from human expansion, for example, French colonization of Quebec in Canada, and we are currently investigating how strong the effect is, especially with respect to deleterious mutations. But it can happen quickly, especially in rapidly expanding populations because you have lots of founder effects in a very short time.

SEAN HOBAN – In terms of the effect size that each locus has on fitness, if there is variation, would genes of large effect or small effect be more likely to surf?

STEPHAN PEISCHL – We found that with beneficial mutations, you tend to observe more large effect mutations compared to a stationary population. And with deleterious mutations, it is those with small to moderate effect sizes rather than large effect mutations that occur.

KATRINA DLUGOSCH

MARK KIRKPATRICK – There are several plants with inversions that look like they are locally adapted, but which have high rates of selfing. This would appear to take away the recombination suppression advantage, which is a bit of a mystery. Regarding your work on rapid evolution in yellow starthistle (*Centaurea solstitialis*; see K.M. Dlugosch *et al.*, 2015, *Nature Plants*, **1**: 15066), I wonder what the rates of selfing or inbreeding are.

KATRINA DLUGOSCH – Yellow starthistle is an obligate outcrosser with high diversity and high population size. It seems like a situation where inversions could be really important. I think it is a study system where recombination is going to be frequent, and so having a cassette of alleles that can be moved together might allow local adaptation to move quickly.

ANDREW VEALE (UNIVERSITY OF BRITISH COLUMBIA) – We have had some excellent talks on theory and the population genetics of colonizing and expanding populations; however, with regard to management, what do people see as the future of invasion genetics where we have to justify funding received from management agencies?

RIMA LUCARDI – I am an employee of the US Forest Service, and one of the things that I am concerned with is that my stakeholders want rapid assessment of what the problems are with specific invasions. So we want to address this as quickly and as affordably as possible, and then try to give them prescriptions that allow them to use the funding that can be allocated in the most effective manner possible. How can invasion genetics help?

KATRINA DLUGOSCH – There are at least three ways in which invasion genetics can potentially inform management strategies. First, there is now the potential for predicting where invasive genotypes will perform well, which could say something about where invaders will move and where the limits of their range might be. Climatic niche modeling has become popular for addressing similar questions, but genomic analyses can also inform these questions. Second, genomic work is also relevant for biocontrol: it is worthwhile to identify source regions of the native range of invaders as areas from which biocontrol agents are obtained, as they are most likely to be effective. It costs millions of dollars to bring a biocontrol agent to field release, and so there is willingness by agencies to invest some money in population genomics to improve the accuracy of geographical sampling. Finally, the last topic concerns genome scans and the identification of candidate loci associated with traits causing invasiveness. These approaches may be able to inform us about possible "Achilles heels" we can find in particular species that could be exploited to assist future control measures. These are early days, but I believe there is every reason to be optimistic about the potential for invasion genetics to inform future management strategies.

REFERENCES

Dlugosch KM, Cang FA, Barker BS *et al.* (2015) Evolution of invasiveness through increased resource use in a vacant niche. *Nature Plants*, **1**, 15066.

Jones FC, Grabherr MG, Chan YF *et al.* (2012) The genomic basis of adaptive evolution in threespine sticklebacks. *Nature*, **484**, 55–61.

Lewontin RC, Birch LC (1966) Hybridization as a source of variation for adaptation to new environments. *Evolution*, **20**, 315–336.

Schlichting CD (2008) Hidden reaction norms, cryptic genetic variation, and evolvability. *Annals of the New York Academy of Sciences*, **1133**, 187–203.

Part 3

Invasion Genomics

INTRODUCTION

Loren H. Rieseberg*† and Kathryn A. Hodgins‡

* Department of Botany, University of British Columbia, 1316-6270 University Blvd., Vancouver, BC V6T 1Z4, Canada
† Department of Biology, Indiana University, Bloomington, IN 47405, USA
‡ School of Biological Sciences, Monash University, Clayton, VIC 3800, Australia

This section of the book concerns the genomics of colonizing species, an area of study that only became possible in the past 15 years due to advances in high-throughput sequencing, functional genomics, and computational technologies.

Despite early recognition of the potential utility of genomic approaches for studying invasion biology, the development of genomic tools and resources for invasive species lagged behind that of model species, as well as domesticated plants and animals (Basu *et al.* 2004). To catch up with other species, proposals were put forward to focus community efforts on one or a handful of "model" invaders (Stewart *et al.* 2009). However, the rapid decline in sequencing costs has made it possible to apply genomic approaches to essentially any organism. As a consequence, various invasive species are targets of ongoing genomic studies, and results are beginning to appear in the literature. For example, the first reference genome of an agricultural weed was published in 2014 (Peng *et al.* 2014), and microarray and population genomic studies have identified molecular changes associated with invasion success in several plant and animal invaders (Lockwood & Somero 2011; Hodgins *et al.* 2013; Vandepitte *et al.* 2014) (Chapter 18).

Chapters 15–19 in the invasion genomics section illustrate some of the different ways in which genomic data can provide insight into the biology of invasive species. Cristescu (Chapter 15) describes recent advances in our ability to reconstruct the often complex demographic history of invasions. The increasing precision with which source populations can be identified and invasion routes determined represents a triumph for invasion biology. Although her focus is on aquatic invasion, Cristescu provides a general primer on the molecular markers, genomic tools, and analytical approaches that can be used to reconstruct invasion histories. She also highlights challenges for the field, such as the efficient detection of early invasions and estimation of the frequency of failed invasions, and offers suggestions on how to overcome these challenges using genomic approaches.

Whereas most studies of invasion success focus on a single invasive species, comparative studies involving multiple invasive lineages offer the possibility of identifying phenotypic or molecular changes that are commonly associated with invasion success (or those that are idiosyncratic). In one of the first comparative genomic studies of this sort, Hodgins and colleagues (Chapter 16) search for changes in protein sequence associated with invasiveness across the plant family Asteraceae. The Asteraceae is the largest family of flowering plants and includes many noxious invaders such as thistles, knapweeds, ragweeds, and dandelions. Analyses of transcriptome assemblies from 35 species, including six major invasive species, identified numerous genes that have been targeted by positive selection. However, there is little evidence of repeatability in positively selected genes across invasive lineages, implying that there are multiple evolutionary pathways to invasion success in the Asteraceae. Interestingly, Hodgins and colleagues failed to find evidence for an increase in the fraction of deleterious mutations in invasive lineages—a theoretical expectation of range expansion (Chapter 13). This suggests that multiple introductions and post-introduction gene flow may mitigate the effects of population bottlenecks often associated with invasions.

For invasions to be successful, introduced species must thrive in the new range. Rapid adaption could facilitate invasion to novel environmental conditions in the introduced range, but a key question is: should such adaptation necessarily involve fitness trade-offs? That is, when alleles are associated with fitness increases in one habitat, are they also associated with fitness declines in a different habitat? Hamilton *et al.* (Chapter 17) explore this question in the model species, *Arabidopsis thaliana*, through genome-wide association analyses of fitness traits in European populations grown in natural sites in Rhode Island, USA. As expected, European populations from locations with the closest climate match to Rhode Island had the highest fitness on average. However, there was surprisingly little evidence for fitness trade-offs: most alleles associated with colonization success in Rhode Island were neutral in other habitats. These results are counterintuitive because globally favorable alleles should sweep to fixation—only alleles exhibiting fitness trade-offs are expected to remain polymorphic. More studies using model species to examine the genetic details associated with colonization success could help to solve this mystery.

One of the classic examples of rapid adaptive evolution in an invasive species, and of biological control going horribly wrong, involves the introduction of cane toads (*Rhinella marina*) to control cane beetles (*Dermolepida albohirtum*) in tropical Australia. The toads quickly spread across tropical and subtropical Australia, with devastating effects on native ecological communities due to predation of nontarget invertebrates, spreading of disease, and toxic effects on naive predators. Rollins *et al.* (Chapter 18) review earlier studies that demonstrate rapid evolution of several traits that have accelerated the dispersal of cane toads, including longer legs. This phenotypic shift during range expansion has

occurred despite evidence of a substantial introduction bottleneck and reduced neutral variation. They then describe new data investigating the genomic basis of dispersal in cane toads. Increased dispersal capability has been accompanied by numerous changes in gene expression, including up-regulation of genes associated with cellular repair and metabolism. The latter result is consistent with increased activity and endurance at the range front relative to the range core.

Chapter 19 by Zhang and colleagues explores the role of epigenetic changes in the evolution of invasive species. Epigenetics refers to an information layer above the DNA level that can affect phenotypic variation via processes such as DNA methylation, histone modification, and small RNAs. Because genetic variation is sometimes limited in invasive populations, it has been suggested that epigenetic variation might compensate for this loss of genetic diversity. Zhang and colleagues show that this may indeed be the case for invasive Japanese knotweed (*Fallopia japonica*), which reproduces clonally through long creeping rhizomes or by rhizome fragments. Using both conventional and methylation-sensitive amplified fragment length polymorphisms (AFLPs) to assay variation, they demonstrate that 83 phenotypically variable collections of Japanese knotweed from central Europe represent a single genetic clone, with 27 distinct epigenotypes. Thus, at least in asexually reproducing lineages, epigenetic changes may contribute to local phenotypic differentiation and possibly also to invasion success.

The unprecedented level of resolution offered by genomic data promises to revolutionize the emerging field of invasion genetics. This includes identifying the geographic origin(s) of invaders, the timing and demography of expansion, and the extent of gene flow (past or present) with other populations or species (Chapter 15). Likewise, many of the key questions about invasion genetics, such as the predominant source(s) of genetic variation underlying rapid adaptive evolution in invasive populations (Chapter 14), the role of chromosomal rearrangements in invasions (Chapter 10), the extent of genetic load in invasive populations (Chapters 13 and 16), and the importance of epigenetic change (Chapter 19) are likely to be resolved at least, in part, by population and evolutionary genomic approaches in the next decade. Perhaps most importantly, genomic tools provide an efficient and comprehensive means for detecting the molecular genetic changes associated with successful invasions (Chapters 16 and 18) and for connecting these changes to the traits responsible for invasion success (Chapter 17).

REFERENCES

Basu C, Halfhill MD, Mueller TC, Stewart CN (2004) Weed genomics: new tools to understand weed biology. *Trends in Plant Science*, **9**, 391–398.

Hodgins KA, Lai Z, Nurkowski K, Huang J, Rieseberg LH (2013) The molecular basis of invasiveness: differences in gene expression of native and introduced common ragweed (*Ambrosia artemisiifolia*) in stressful and benign environments. *Molecular Ecology*, **22**, 2496–2510.

Lockwood BL, Somero GN (2011) Transcriptomic responses to salinity stress in invasive and native blue mussels (genus *Mytilus*). *Molecular Ecology*, **20**, 517–529.

Peng Y, Lai Z, Lane T *et al.* (2014) *De novo* genome assembly of the economically-important weed *Conyza canadensis* using integrated data from multiple sequencing platforms. *Plant Physiology*, **166**, 1241–1254.

Stewart CN, Tranel PJ, Horvath D *et al.* (2009) Evolution of weediness and invasiveness: charting the course for weed genomics. *Weed Science*, **57**, 451–462.

Vandepitte K, De Meyer T, Helsen K *et al.* (2014) Rapid genetic adaptation precedes the spread of an exotic plant species. *Molecular Ecology*, **23**, 2157–2164.

Chapter 15

GENETIC RECONSTRUCTIONS OF INVASION HISTORY

Melania E. Cristescu

Department of Biology, McGill University, Montreal, QC H3A 1B1, Canada

Abstract

A diverse array of molecular markers and constantly evolving analytical approaches have been employed to reconstruct the invasion histories of the most notorious invasions. Detailed information on the source(s) of introduction, invasion route, type of vectors, number of independent introductions and pathways of secondary spread has been corroborated for a large number of biological invasions. In this review, I present the promises and limitations of current techniques while discussing future directions. Broad phylogeographic surveys of native and introduced populations have traced back invasion routes with surprising precision. These approaches often further clarify species boundaries and reveal complex patterns of genetic relationships with noninvasive relatives. Moreover, fine-scale analyses of population genetics or genomics allow deep inferences on the colonization dynamics across invaded ranges and can reveal the extent of gene flow among populations across various geographical scales, major demographic events such as genetic bottlenecks as well as other important evolutionary events such as hybridization with native taxa, inbreeding and selective sweeps. Genetic data have been often corroborated successfully with historical, geographical and ecological data to enable a comprehensive reconstruction of the invasion process. The advent of next-generation sequencing, along with the availability of extensive databases of repository sequences generated by barcoding projects opens the opportunity to broadly monitor biodiversity, to identify early invasions and to quantify failed invasions that would otherwise remain inconspicuous to the human eye.

Previously published as an article in *Molecular Ecology* (2015) 24, 2212–2225, doi: 10.1111/mec.13117

INTRODUCTION

Reconstructing the complex history of biological invasions represents an important step in understanding the invasion process. Species that rapidly expand their distributional ranges far beyond their native ranges (experience dramatic range expansions) or outside their original ecological spaces (experience major habitat transitions) provide, in many ways, large-scale natural experiments and the opportunity to investigate dynamic evolutionary processes in real time (Lee 2002; Sax et al. 2007). Often, a deep understanding of invasion history is required for building predictive models of secondary spread and ecological or economic impact, as well as for implementing sound management measures of prevention and rapid intervention (Kulhanex et al. 2011), such as efficient methods of biological control (Barrett 1992). It is not surprising that over the last few decades, the reconstruction of invasion history has become the foundation of many long-term investigations of biological invasions (Estoup & Guillemaud 2010).

In this review, I evaluate recent progress and future trends in reconstructing the history of invasive species. I contrast natural and human-mediated invasions and focus on well-documented aquatic invasions, particularly zooplankton species that are considered to be dispersalists *par excellence*. I discuss the most common genetic approaches and molecular markers that have been used to determine source populations, invasion routes and vectors, as well as the genetic, evolutionary and ecological changes that invasive species or populations undergo during the process of invasion.

Historical perspectives on reconstructing invasion histories

Early natural historians compiled extensive records of natural and human-mediated colonization events, including major distributional shifts, faunal turnovers and extinctions associated with biological invasions. Many of these events were triggered by intentional introductions or habitat alterations that accompanied the successful expansion of modern humans 'out of Africa' about 70 000–90 000 years ago (Brahic 2012). The early observations on biological invasions were based on detailed historical and comparative biogeographical data and were often focused on particular regions (reviewed in Di Castri 1989; Simberloff 2013).

A century after Darwin's influential work, the birth of invasion biology was prompted by two memorable moments. Elton's (1958) book on *The Ecology of Invasions by Animals and Plants* provided the first comprehensive work on biological invasions. Elton found his inspiration in the early faunal history and in the field of ecology that he was advancing at the time. He described with eloquence the significant faunal and floral turnover associated with the continual destruction of natural dispersal barriers due to increased human trade and travel. Although Elton's work brought biological invasions into the spotlight, the effort of synthesizing the knowledge on biological invasions was firmly advanced by the influential volume on *The Genetics of Colonizing Species* edited by Herbert Baker and Ledyard G. Stebbins in Baker & Stebbins 1965 (Barrett 2015). The broad synthesis inspired by the Asilomar meeting 50 years ago greatly stimulated the research field by placing biological invasions into an evolutionary context and by setting up a conceptual framework for studying the evolution of invasiveness (weediness, persistence, dispersal). *The Genetics of Colonizing Species* covered many aspects on the phenotypic properties that facilitate the invasion process. The properties of the gene pools and the special genetic architecture of invasive species also received close attention. Dobzansky (1965) and Carson (1965) took advantage of the well-understood genetics of *Drosophila* at that time to make one of the first attempts to compare levels of genetic diversity in invasive and non-invasive species. The last fifty years has been marked heavily by the genetic revolution that greatly facilitated the solidification of the field of invasion genetics. The genetic reconstruction of invasion history has been central to many research programmes aimed at understanding the causes and consequences of invasiveness. Determining the geographical origin of introduced populations often provided essential reference points for ecological and evolutionary studies that involved contrasting native and introduced populations (Milne & Abbott 2000; Hierro et al. 2005; Dlugosch & Parker 2008).

What can molecular markers tell us about the invasion process?

The common use of molecular markers in ecological and evolutionary studies has enabled us to reconstruct the evolutionary history of biological invasions, sometime with surprising precision (Handley *et al.* 2011). Molecular approaches have become commonplace and have been used to ask important questions about the invasion process (Sakai *et al.* 2001). What is the taxonomic identity of invasive species? What are their native areas and their sources of introduction? How do these species manage to travel around the world? What are the vectors of primary introduction and vectors of secondary spread? Are we dealing with single or multiple introduction events? How are these serendipitous voyages shaping the genetic landscape of the introduced populations? Are demographic bottlenecks experienced during the invasion process? Are invasive populations depleted of standing genetic variation?

Ultimately, molecular approaches proved to be effective when exploring invasion histories (Estoup & Guillemaud 2010) and opened the door for extensive investigations of the evolutionary and ecological forces that shape invasiveness (Sakai *et al.* 2001; Lee 2002). Although technical challenges are still associated with next-generation sequencing, rapid progress is being made on the use of population genomics approaches and metabarcoding techniques. For example, genotype-by-sequencing (GBS) techniques, such as RAD-seq (Baird *et al.* 2008), that involve the use of large numbers of markers randomly distributed across genomes are being used increasingly for organisms for which few genomic resources presently exist. These methods enable the transition from coarse molecular approaches to dense population genomics surveys. Such high molecular resolution is often necessary to accurately resolve the invasion history of species and identify severe bottleneck events associated with the invasion process. The power of next-generation sequencing will probably be combined with new computational methods and will allow researchers to test complex invasion scenarios (Estoup *et al.* 2010; Fitzpatrick *et al.* 2012). Moreover, metabarcoding approaches open the possibility of applying molecular identification techniques to very complex communities. While this method is commonly used to infer basic biodiversity estimates in healthy or disturbed ecosystems, it also offers the possibility of rapid and relatively inexpensive methods for the detection of invasive species. Such advanced molecular approaches will soon enable the application of 'forensic' methods to invasion biology by allowing the detection of traces of extracellular DNA when invasive species are still at very low abundance and their presence cannot be identify by morphological methods.

CONTRASTING NATURAL AND HUMAN-MEDIATED INVASIONS

Changes in the geographical range of species are recognized to be natural processes (Elton 1958; Lodge 1993). The distribution of species constantly expands and contracts, and these changes can trigger dynamic ecological and evolutionary processes, leading to radiations and extinctions, which ultimately shape the Earth's biodiversity. For example, over geological time, many marine species colonized estuaries, rivers and lakes and eventually colonized the land. It is easy to appreciate that some invasive waves occured abruptly, over ecological time scales. This is in contrast with the major colonizations and recolonizations of particular regions that often involve geological times (e.g. after glaciations; Wilson *et al.* 2009; Estoup & Guillemaud 2010). The temporal and geographical scales that we consider when investigating natural invasions are often vast (Box 1). However, contemporaneous invasions have received much closer investigation than historical invasions. Inherently, the field of invasion biology has a strong anthropogenic focus due to the problem that we are currently facing: the homogenization of biodiversity. Natural barriers to dispersal and gene flow are continually being removed due to accelerated human trade and travel (Mooney & Cleland 2001). Moreover, human activities contribute to extensive worldwide homogenization of the environment through agriculture practices and urbanization (Sax & Brown 2000). By focusing on anthropogenic factors, we recognize humans as the main force facilitating the global hitch-hiking of species (directly or indirectly). In addition, human welfare is often used to evaluate the impact of these invasions. As a consequence of this strong anthropogenic focus, the terminology associated with invasive species is constantly evolving and revised (reviewed by Colautti & MacIsaac 2004). Extensive efforts have been directed towards the implementation of specific and consistent terminology, particularly when focusing on human-mediated invasions. Terms such as colonizing, invasive and weedy have been replaced by nonindigenous, exotic, introduced and naturalized. In this study, I will use the term invasive in a very broad sense to refer to the species that rapidly spread (over ecological times as opposed to geological times), irrespective of their ecological or economic impact. I will discuss species that have an exceptionally high ability to establish populations in a geographical or ecological space not previously occupied. By doing

Box 1 Contrasting the history of natural and human-mediated invasions

Characteristics	Natural Invasion	Human-mediated invasions
Temporal scale and general considerations	Historical or recent invasions that occur naturally without the intervention of humans. The time frame of investigation is large, and natural invasions appear rather episodic when taking into consideration the entire evolutionary history of the species under consideration (Lewontin 1965). Many species experience episodes of rapid and dramatic range expansions (invasions) followed by episodes of stasis or contractions.	Recent invasions that are facilitated by anthropogenic factor: vectors of transport or environmental disturbances. Many vectors of transport allow species to overcome physical barriers. Environmental disturbances eliminate ecological barriers that historically restricted range expansions. Invasion waves are still episodic with respect to the long evolutionary history of species but the evolutionary trajectory of most invasive species is hard to predict.
Origin of invasions and invasion routes	The origins of natural invasions are often inferred based on fossil evidence or genetic data corroborated with geographical evidence. Such inferences are often vague. Repeated introductions from multiple sources are probably to be rare and are seldom documented.	The source(s) of invasion can be identified with relatively higher precision. Multiple introductions appear to be the norm rather than the exception.
Rates of invasions (number of primary introductions per unit of time)	Low rates of invasions due to relatively stable physical and ecological barriers. Higher rates of invasions are expected after periods of massive natural disturbances (e.g. glacial retreat).	Highly accelerated rates of invasion when compared to natural invasion.
Rates of secondary spread	Rates of spread are expected to be higher than during regular natural range expansions. Natural invasions occur much more rapidly (over ecological times) than the major colonizations and recolonizations of particular regions that often involved geological times (e.g. after glaciations; Wilson *et al*. 2009; Estoup & Guillemaud 2010).	Rates of secondary spread are often significantly accelerated by anthropogenic vectors of secondary spread or other anthropogenic disturbances.
Molecular approaches	Sequence-based markers such as mtDNA markers are often employed.	Fast evolving and highly diagnostic microsatellite markers have been used more often than sequence-based markers.
Analytical approaches	Basic population genetics analyses based on the neutral theory of molecular evolution are often applied.	As human-mediated invasions have very restricted timescales, much smaller than the timescale of mutation and genetic drift, analyses based on neutral theory cannot accurately resolve such dynamics (Fitzpatrick *et al*. 2012). As a consequence, the level of genetic differentiation among introduced populations is often dominated by historical contingency rather than the steady-state dynamic (Marisco *et al*. 2011; Fitzpatrick *et al*. 2012).

this, I aim to facilitate the comparison and contrast between natural and human-mediated invasions and to partially depart from the anthropogenic perspective.

NATURAL INVASIONS: LESSONS FROM ZOOPLANKTON SPECIES

Zooplankton organisms can be studied in the context of invasion biology in ways that enable us to contrast natural and human-mediated invasions. This may come as a surprise, as few zooplankton species are considered to be notorious invaders and are on our radar when implementing preventive measures. In fact, zooplankton species have been regarded as cosmopolitan for nearly a century. Their apparent widespread distribution has been linked to their production of resting eggs, which allows zooplankton species to travel long distances and disperse between continents. This view of zooplankton cosmopolitanism has slowly been invalidated (reviewed in Hebert & Cristescu 2002). Detailed morphological and ecological studies have revealed that many species with a broad geographical distribution represent complexes of closely related species that often have restricted, allopatric distributions (Frey 1982; Havel & Shurin 2004).

Detailed molecular data have also revealed an unexpectedly high level of regional endemism and population-level differentiation, suggesting that geographical and/or ecological barriers to dispersal have been effective in maintaining a high level of continental and regional endemism even in species considered to be exceptional dispersalists (Hebert & Wilson 1994; Taylor et al. 1998). For example, detailed phylogeographic surveys of cladoceran species that were assumed to have Holarctic distributions, such as *Polyphemus pediculus* and *Leptodora kindtii* revealed unexpectedly high levels of cryptic endemism (Xu et al. 2009; Millette et al. 2011). Many species of *Daphnia* thought to have geographical distributions that span entire continents are known to have marked phylogeographic structure produced by restricted levels of gene flow among populations. The *Daphnia pulex* group (*sensu* lato) represents a rich species complex of about 12 species with distinct ecology (Adamowicz et al. 2009). The best investigated species of this complex, the North American *D. pulex* (*sensu stricto*) and *D. pulicaria*, are probably to be among the most subdivided species yet documented (Lynch et al. 1999) due to their unusually high level of among-population structure

across small geographical scales, which suggests the existence of strong ecological barriers that restrict gene flow (Cristescu et al. 2012). The quest to identify true cosmopolitan species has also failed when investigating zooplankton groups with smaller body size and higher effective population sizes, such as rotifers (Gómez et al. 2002). Despite the growing evidence of strong regionalism and increased endemism, zooplankton species remain regarded as exceptionally good colonists. When barriers (either physical or ecological) that previously separated biota for millions of years are naturally passed or disrupted, zooplankton species can attain very long-distance colonization events (Adamowicz et al. 2009) or significant ecological transitions (Cristescu et al. 2012). The achieved regional distribution and the restriction of gene flow, despite exceptional dispersal abilities, are often considered to be the result of a strong priority effect shaped by invasion order (Robinson & Dickerson 1987), but also by the rates and timing of invasions (Robinson & Edgemont 1988). Propagule banks (dormant stages of plants and animals that remain viable for decades) are expected to also buffer resident populations against local extinction and to enhance the priority effect on subsequent immigrants (Mergeay et al. 2007). These attributes make zooplankton species particularly suitable for elucidating prolonged invasion histories and contrasting natural and human-mediated invasions.

During the last 50 years, at least five cladoceran species invaded the Great Lakes region of North America alone, suggesting an intercontinental rate of human-mediated dispersal of at least one invasion per decade (Hebert & Cristescu 2002). Using phylogeographic data and genetic divergence between closely allied species of cladocerans, Hebert & Cristescu (2002) estimated that rates of human-mediated invasions are thousands of times higher than natural, intercontinental colonization events. Biological invasion is, of course, a natural process. Many species experience periods of dramatic range expansion. However, the rate and the geographical and ecological scale of human-mediated invasions are incredible today and dramatically different from those of the geological past.

Genetic data also allow us to investigate patterns of very rapid intracontinental range expansions, particularly postglacial recolonization routes. We can compare the genetic structure of populations and/or species that inhabit glaciated regions and experienced strong environmental disturbances, with populations that persisted in more stable, unglaciated regions. For example, demographic analysis using Tajima's D (Tajima 1989), Fu's F_S (Fu 1997) and mismatch distributions (Rogers & Harpending 1992) of lineages that show contrasting distributional patterns allows the detection of non-neutral evolution and rapid population expansion events. Such approaches have been used in conjunction with phylogeographic and haplotype network analyses to identify the centre of origin (equivalent to the source of introduction) of populations or species that inhabit previously glaciated regions or more stable, unglaciated regions. For example, the conventional belief is that many plant and animal species in Europe recolonized large parts of the continent from southern refugia that largely correspond to southern European peninsulas, which were less influenced by glaciations. However, for the predatory cladoceran *Leptodora kindtii*, detailed phylogeographic and network analyses based on mitochondrial markers have revealed a pattern which sharply contrasts with this perception. Southern lineages remained localized in very narrow distributional ranges, sometimes in single habitats, while cryptic refugia situated in the central (Carpathian region) or northern parts of the continent have been responsible for the recolonization of large parts of the European continent and large parts of Eurasia (Millette *et al.* 2011). Similar cryptic, northern refugia, corresponding to the Russian Yaroslave region and Siberia, have been identified for other zooplankton species such as the onychopod cladoceran *Polyphemus pediculus* (Xu *et al.* 2009). Paleoreconstructions of northern Eurasia indicate that large ice-dammed paleolakes covered large parts of the White Sea basin and west Siberia (Mangerud *et al.* 2004). These extensive lakes were formed due to obstruction of the north-flowing rivers whose drainages were diverted towards the Caspian basin. This facilitated the southern dispersal of species trapped among the ice sheets. The persistence of zooplankton species in northern refugia was probably facilitated by their ability to produce diapausing eggs known for their ability to withstand desiccation and freezing and to maintain viability for a prolonged period of time (sometimes hundreds of years). This adaptation could allow zooplankton species to persist in regions in which aquatic habitats underwent more dramatic geological ephemerality. What is particularly impressive is the high 'invasive' ability of those lineages that persisted under such harsh conditions. The phylogeography of *L. kindti*, reconstructed by Millette *et al.* (2011), revealed that sister clades/species within each of the

major geographical regions analysed (North America, Europe and Far East Asia) contain clades that show marked distributional differences. Each of the very narrowly distributed clades situated in previously unglaciated areas of Oregon, Greece, Turkey and Japan has a sister clade characterized by a very widespread distribution that spans entire continents and often shows a clear signature of relatively recent population expansion. The widespread clades arguably contain populations that are more invasive than the clades that were identified in single habitats, which are often remnants of paleolakes. As expected, population genetics analyses conducted on freshwater zooplankton and fishes across both glaciated and unglaciated regions revealed marked genetic differences (Bernatchez & Wilson 1998). While glaciated regions are generally occupied by populations that show a marked signature of recent demographic expansion and generally low genetic diversity (low nucleotide diversity, but high haplotype diversity), unglaciated regions harbour populations with a higher level of intrapopulation genetic diversity (in particular nucleotide diversity due to the presence of divergent haplotypes), larger genetic splits between sister clades and no signature of recent demographic expansions. These dramatic differences in distributional ranges and genetic structure between closely related species that inhabit formally glaciated regions and those found in unglaciated regions are also relevant to invasion biology. These patterns provide support for a hypothesis, proposed by Lee & Gelembiuk (2008), suggesting that the evolutionary history of disturbances might select for life history traits that favour rapid colonization, such as rapid growth, persistence, high somatic plasticity and long-distance dispersal ability.

HUMAN-MEDIATED INVASIONS

Invasion forensics: identifying the source(s) of invasions and reconstructing the invasion routes

When investigating human-mediated invasions, their relatively recent history (centuries and decades vs. millennia) facilitates approaches that mimic forensic techniques, with the goal of reconstructing the most likely invasion scenarios. However, the correct identification of the source population or route of invasion is not always straightforward. The native areas of invasive species can be unknown or very large and difficult to explore exhaustively (Lombaert *et al.* 2011). For many species with a long history of invasion (several decades) and nearly cosmopolitan distribution, the native ranges are an educated guess at best. This is often the case for marine invasive species such as the colonial or solitary tunicates. Carlton (1996) noted that the general approach when working with populations that are not demonstrably native or introduced is still to classify all populations without records of introduction as being 'native' rather than *cryptogenic populations*. For example, Carlton (1996) conservatively estimates that about 100 aquatic species in San Francisco Bay are cryptogenic. This conventional approach of classifying cryptogenic populations as native introduces significant noise to large-scale data sets and obscures the reconstruction of invasion history. Native populations may also be genetically homogenous, due to a recent homogenization effect induced by human activities, high levels of natural gene flow or a short evolutionary history of the lineage under investigation. However, given that genetic homogeneity is never fully achieved, such situations can be resolved by employing very fast evolving marker or genomic approaches that cover a large portion of the genome and a large number of individuals. Moreover, native populations are sometime not completely disconnected from the gene pool that encompasses populations of the invaded ranges. This is particularly common when investigating invasions that start as outbreaks, species that successfully enter new ecological spaces while only slowly expanding beyond the native geographical range. Furthermore, human-mediated invasions often involve complex routes with multiple sources of introduction and repeated introductions (Wilson *et al.* 2009). Such complex invasive scenarios which often achieve network patterns can greatly complicate the reconstruction of the invasion steps.

Despite these general difficulties, the genetic reconstruction of source populations and routes of invasion have proven to be a fruitful endeavour (reviewed in Estoup & Guillemaud 2010). Many studies identified the source introduction(s) and invasion routes with high precision. For example, the study of the Ponto-Caspian invader *Cercopagis pengoi*, the fishhook water-flea (Cristescu *et al.* 2001), demonstrated the usefulness of mitochondrial DNA (mDNA) markers in locating the source of invasion as well as the corresponding invasion corridor. The lagoons of the Black Sea were identified as the source of the Baltic population of *Cercopagis*.

This initial introduction resulted in a severe bottleneck that virtually wiped out the intrapopulation genetic variation in mtDNA. Moreover, the genetic similarity between populations in the Baltic Sea and those in the Great Lakes suggested a subsequent transfer of animals from the Baltic Sea to North America (Cristescu *et al.* 2001). A few other early studies conducted on the southern house mosquito, *Culex quinquefasciatus*, the vector of avian malaria, pointed to the complexity of the invasion process and the importance of identifying invasive populations that act as major sources or hubs of secondary spread (Fonseca *et al.* 2000; Estoup *et al.* 2001). Such transit populations have been termed 'invasive bridgeheads' (Estoup & Guillemaud 2010) and are considered to either foster favourable evolutionary shifts or to be situated in the path of major vectors. These early studies also demonstrated that fine-scale reconstruction of invasion history is possible when the native region harbours significant genetic structure, and both the invasive and native areas are sampled systematically. Moreover, these studies revealed regions and/or habitats that acted as donors for a larger number of species with shared evolutionary history but divergent life history attributes (Ricciardi & MacIsaac 2000; Cristescu *et al.* 2004). A large number of more recent studies combined historical and distributional data with genetic data (genetic patterns observed within and between populations across native and introduced ranges) to reconstruct complex scenarios of invasions for taxa with various life history attributes (brown algae, Voisin *et al.* 2005; crabs, Darling *et al.* 2008; clams, Hoos *et al.* 2010; toads, Estoup *et al.* 2010; insects, Lombaert *et al.* 2010). Collectively, these studies revealed the potential of genetic analyses to reveal detailed patterns of introduction, establishment and spread of invasive species.

Both nuclear and mitochondrial (or chloroplast) markers have been employed successfully with many studies making use of multiple markers with distinct levels of resolution (e.g. organelle markers coupled with nuclear microsatellite markers). The methods commonly used involve building dendrograms (such as UPGMA or neighbour joining, e.g. Saitou & Nei 1987) and parsimony networks often interpreted in a geographical context. Detailed population genetic analyses frequently accompany conclusions derived from interpreting dendrograms and networks. Markers are generally tested for the level of resolution provided, for concordance with Hardy–Weinberg (HW) expectations and for linkage disequilibrium (e.g. nonrandom association of alleles between loci). Several traditional methods are often implemented to determine population structure and demographic history (Box 2). Population assignment methods that allow pairing of introduced populations with their corresponding source population(s) represent an effective tool for identifying population structure. For example, methods of clustering multilocus genotypes into genetically discrete groups such as those implemented in STRUCTURE (Pritchard *et al.* 2000), BAPS (Corander *et al.* 2003) or GENELAND (Guillot *et al.* 2005) are commonly used to infer donor populations and reconstruct invasion routes. More recently, the method based on approximate Bayesian computations (ABC) allows the evaluation of alternative invasion scenarios. ABC uses a model-based inference in a Bayesian setting (Beaumont *et al.* 2002) and provides probabilities (with confidence intervals) for all invasion scenarios considered. The array of population genetics approaches available have proved to be useful in reconstructing invasion routes and inferring the number of independent introductions despite their inherent limitations. However, these well-established methods are not always fully applicable to invasion studies. It is generally accepted that as human-mediated invasions have very restricted timescales, much smaller than the timescale of mutation and genetic drift, analyses based on neutral theory cannot always accurately resolve such dynamics and have to be used cautiously (Fitzpatrick *et al.* 2012).

Founding events

Theory predicts that founding events will drastically deplete genetic variation in invading populations (Nei *et al.* 1975; Roman & Darling 2007). Thus, invasive populations are expected to be less genetically diverse than the populations from which they are derived. However, invasive species persist and adapt to new conditions despite the likely reduction in genetic variation. While a severe reduction in genetic diversity could reduce adaptive potential when such reduction involves quantitative traits, other mechanisms of genomic reconfiguration, the fate of which is driven by genetic drift or a combination of genetic drift and selection, can provide populations with unique adaptive opportunities. Such serendipitous events could involve intraspecific and interspecific admixture which could provide invasive populations with the ability to explore diverse ecological spaces. Moreover, hybridization

followed by fixation of large-scale inversions might also provide reproductive isolation from donor populations or from other introduced populations, generating unexpected barriers to gene flow (Hoffmann & Rieseberg 2008). Chromosomal inversions can spread when such inversions capture locally adapted alleles facilitating range expansions (Prevosti *et al.* 1988; Kirkpatrick & Barrett 2015). Furthermore, prolonged periods of isolation and low population size result in severe inbreeding that could potentially reduce the mutational load of invasive populations. Because inbreeding increases homozygosity, and hence the effectiveness of selection against detrimental alleles that are fully or partially recessive, severe inbreeding can result in a reduction of the frequency of such detrimental alleles and a significant fitness rebound (Crnokrak & Barrett 2002). Phenotypic changes that provide invasive populations the ability to explore divergent niches are also expected to occur due to random sampling of the genetic diversity of source populations (Keller & Taylor 2008).

One of the biggest surprises revealed by genetic studies has been the realization that invasive species are not always experiencing the expected reduction in genetic diversity. Many studies that compared neutral genetic diversity of invasive populations to that of native populations or related, noninvasive species reported comparable levels of genetic variation in native and introduced populations. Moreover, for many of the worldwide marine invaders, such as the European green crab, *Carcinus maenas* (Darling *et al.* 2008), the brown alga *Undaria pinnatifida* (Voisin *et al.* 2005), or the violet tunicate *Botrylloides violaceus* (Bock *et al.* 2011), invasive populations vary broadly in the degree to which genetic diversity is retained in invasive populations. These studies revealed no strong correlation between the genetic diversity retained by an introduced population and its invasive potential (Roman & Darling 2007; Dlugosch & Parker 2008). For example, one of the least diverse populations of *C. maenas* (from the western USA) has been the most successful (nearly 2000 km coastline expansion), while the highly diverse Cape Town population has remained static (Darling *et al.* 2008). The large comparative study conducted by Dlugosch & Parker (2008) on 80 species of animals, plants and fungi revealed that losses of neutral variation are sometime detectable but are certainly not ubiquitous. Fitzpatrick *et al.* (2012) suggests that this is probably explained by the fact that genetic drift is not efficient in removing large amounts of genetic

---

Box 2 Analytical methods employed when reconstructing invasion histories

---

Phylogenies	Phylogenetic approaches (both distance based, e.g. Saitou & Nei 1987; and character based) are often used to infer the taxonomic identity of invasive species by exploring their relationship with the noninvasive relatives. Moreover, these approaches are often used to infer genetic relationships among native and introduced populations and investigate patterns of range expansions on postglacial time scales. Due to the generally short evolutionary time frame covered, such data sets involve a low number of informative characters or short overall genetic distances. Moreover, phylogenetic relationships can be difficult to interpret when the source population(s) is not sampled. These limitations make the reconstruction of true phylogeny difficult (Estoup & Guillemaud 2010).
Population genetics inferences	Population genetics provides a solid mathematical framework for understanding evolutionary processes in natural populations. Markers are generally tested for their level of resolution, concordance with Hardy–Weinberg (HW) expectations and for linkage disequilibrium (i.e. nonrandom association of alleles between loci). Programs such as GENEPOP (Raymond & Rousset 1995), FSTAT (Goudet 1995), ARLEQUIN (Schneider et al. 2000) and MSA (Dieringer & Schlötterer 2003) are used often to investigate marker properties as well as population diversity (proportion of polymorphic loci, allelic richness, hetorozygosity).
	Recent bottlenecks are often detected by testing for reduced allele number and excess of heterozygosity compared to that expected for an equilibrium population. BOTTLENECK (Cornuet & Luikart 1996) is the program commonly used for detecting recent effective population size reductions from allele frequency data. The disadvantage of this method is the limited period of time over which the test is able to show a bottleneck effect, the large bottleneck size, number of loci and sample sizes needed to generate high statistical power.
	A good understanding of invasion history requires often the identification of population structure and migration rates across native and invasive ranges. Assignment tests based on clustering multilocus genotypes into groups represent one of the most robust ways of investigating population structure. The computer program STRUCTURE (Pritchard et al. 2000) is used often to estimate the most likely number of actual populations from the full genotype data set by comparing posterior probabilities across a range of possible population numbers. The programs (BAPS) developed by Corander et al. (2003) and GENELAND developed by Guillot et al. (2005) estimate the number of populations using information about the source populations.
	A fundamental problem in invasion biology is distinguishing between ongoing gene flow (an equilibrium situation) and recent separation of populations with decreased or no gene flow (a nonequilibrium situation). One useful approach is to estimate F_{ST} and gene flow between all pairs of populations and to look for the effect of isolation by distance (Slatkin 1993). Recent geographical isolation is not expected to show an inverse correlation between gene flow and geographical distance (Hutchison & Templeton 1999).
	In general, the methods described take poorly into account the effect of demographic stochasticity or genetic history. Given that stochasticity plays an important role during the invasion process and exhaustive sampling is often not feasible, population genetics results need to be interpreted cautiously.
Using approximate Bayesian computation (ABC) to evaluate distinct invasion scenarios	The model-based method called approximate Bayesian computation (ABC, Beaumont et al. 2002) has been used to make inferences about the invasion routes. Posterior probabilities of different models and/or the posterior distribution of the demographic parameters under a given model are determined by measuring the similarity between the observed data set and a large number of simulated data sets. Several ABC programs such as popABC (Lopes et al. 2009), DIYABC (Cornuet et al. 2008) and ABCtoolbox (Wegmann et al. 2010) provide user-friendly interfaces that allow researchers to test complex invasion scenarios.

Population genomics inferences	NGS technologies coupled with powerful software platforms enable researchers to analyse tens of thousands of genetic markers. Often genomes are subsampled at homologous locations using genotype by sequencing (GBD) or restriction site-associated DNA sequencing (RAD-seq; Davey *et al.* 2011). Several recent softwares such as Stacks (Catchen *et al.* 2013a), SAMtools/BCFtools (Le *et al.* 2009) and the Genome Analysis Toolkit (GATK, McKenna *et al.* 2010) produce core population genomic summary statistics and SNP-by-SNP statistical tests. These tests are often conducted across a reference genome. However, Stacks was developed specifically for projects that do not benefit from a well-characterized reference genome.
Phylogenomics	Phylogenetic studies using genome data (e.g. GBS markers) have the potential to elucidate complex evolutionary histories that are obscured by hybridization events or prolonged periods of gene flow after lineage splitting and incomplete lineage sorting (Catchen *et al.* 2013b). When applied to invasive species, such an approach can elucidate the phylogenetic relationship among native and invasive populations or among invasive lineages and their noninvasive relatives.

variation from populations experiencing expansions soon after introduction. Simulations of isolated populations experiencing sudden decline with or without recovery show that rapid recovery and/or immigration can make severe bottlenecks undetectable (Fitzpatrick *et al.* 2012). Population bottlenecks generally generate a pattern of 'heterozygosity excess' due to a deficiency of rare alleles (Nei *et al.* 1975). Recent bottlenecks are often detected by testing for reduced allele number and excess of heterozygosity compared to that expected for an equilibrium population.

Single or multiple introduction events?

Several recent studies on invasion histories have focused deeper on the role that single or multiple introduction events might play in shaping the genetic diversity of invasive populations (Novak & Mack 2005; Roman & Darling 2007; Dlugosch & Parker 2008). Single introduction events followed by subsequent introductions and bottlenecks are expected to result in a significant reduction of genetic diversity. On the other hand, recurrent gene flow due to repeated introductions is expected to largely mitigate a drastic reduction of genetic diversity (Dlugosch & Parker 2008). Introduced populations are expected to have higher probability of establishment and spread if the propagule pressure is maintained high, due to the high number of individuals being released at a particular

location (Simberloff 2009; Blackburn *et al.* 2015). This theoretical prediction has been confirmed by empirical studies. Multiple introductions are often correlated with successful establishment (Barrett & Husband 1990) because repeated introductions have the combined benefit of maintaining high population size and increased genetic variation. The study conducted by Kolbe *et al.* (2004) on the brown anole revealed eight independent introductions in Florida from across the native range of the invasive lizard. The authors postulate that the infusion of genetic material from distinct native sources resulted in introduced populations that harbour substantially more genetic variation than the source populations. The case of the brown anole is certainly not an isolated case. When investigating invasions with a relatively long history, multiple introductions appear to be the norm rather than the exception (Facon *et al.* 2003; Kelly *et al.* 2006).

GENETIC RECONSTRUCTION OF INVASION HISTORY: WHERE NEXT?

Recent advancements driven by the genomics revolution have enabled rapid, deep and relatively inexpensive characterization of genetic patterns in natural populations, including the relationship between native and introduced populations, and the pathways that link these populations (Dlugosch *et al.* 2013; Puzey & Vallejo-Marín 2014; Box 3). Often, such approaches

---

Box 3 Future directions in reconstructing invasion histories

Using metabarcoding approaches to

1 Infer large-scale patterns of natural or human-mediated connectivity between native and invasive populations or among invasive populations.
2 Implement methods of early detection of invasive species in complex communities or environmental samples. Close surveillance of vulnerable habitats situated near entry points would provide important information on the presence or absence of invasive species,
3 Identify failed invasions. We currently lack robust estimates on the rates of failed invasions. This is mainly due to the difficulties associated with identifying introduced species that subsequently fail to establish. Close surveillance programs based on metabarcoding approaches would allow us to obtain better estimates for rates of failed invasions.

Using invasions with simple and well-known histories to

4 Test the efficiency of basic methods of population genetics and/or population genomics inferences (Fitzpatrick *et al.* 2012).

Using population genomics to

5 Identify major evolutionary events that accompany different stages of the invasion history. For example, genomic studies provide the opportunity to investigate the extent to which demographic, historical and selective processes shape the genomic structure of invasive species (Puzey & Vallejo-Marín 2014). Population genomics approaches can improve the characterization of demographic history by greatly improving the number of loci analysed (Fitzpatrick *et al.* 2012; Peery *et al.* 2012; Catchen *et al.* 2013b).

Using interdisciplinary approaches to

6 Integrate genetic, historical and geographical data to reconstruct the history of the invasion process (Estoup *et al.* 2010; Ascunce *et al.* 2011).
7 Integrate population genetics/genomics within an experimental framework (Ellison *et al.* 2011; Gladieux *et al.* 2015).

---

involve methods of complexity reduction such as transcriptome sequencing (Dlugosch *et al.* 2013; Hodgins *et al.* 2015) or RAD sequencing (Baird *et al.* 2008; Roda *et al.* 2013). However, for studies that benefit from a well-characterized reference genome, whole-genome resequencing is often employed for genotyping SNPs markers (Savolainen *et al.* 2013; Puzey & Vallejo-Marín 2014). Such high-throughput methods that involve genomewide markers provide a global view of the genome and enable deep screening of large numbers of populations. As a consequence, population genomics approaches greatly facilitate the identification of demographic events, the identification of loci that experience selection during the invasion process or other evolutionary events (Stapley *et al.* 2010; Puzey & Vallejo-Marín 2014). It is generally recognized that microsatellite-based bottleneck tests often do not detect bottlenecks in populations known to have experienced dramatic declines (Peery *et al.* 2012). This observation was followed by simulations that confirmed that bottleneck tests have limited statistical power to detect bottlenecks largely as a result of the limited sample sizes

typically used in microsatellite-based studies and poorly estimated values for mutation model parameters. The application of population genomics approach to invasive species is probably to improve the characterization of demographic history by greatly increasing the number of loci analysed (Davey *et al.* 2011; Peery *et al.* 2012).

Several genomics studies indicate that population genomic approaches can provide a solid understanding on the genomic consequences of invasion. By analysing whole-genome sequences of 22 populations of *Mimulus guttatus*, Puzey & Vallejo-Marín (2014) demonstrated that introduced plants in the UK are characterized by a 50% reduction in neutral (synonymous) genetic diversity and suggested a common origin for non-native populations. Hohenlohe *et al.* (2010) used RAD tags data and confirmed the long-standing biogeographical hypothesis that the large panmictic oceanic populations of stickleback have repeatedly colonized freshwater habitats giving rise to phenotypically divergent populations. The genomic signature of both balancing and divergent selection was remarkably consistent across the populations investigated. Moreover, Catchen *et al.* (2013b) used a RAD tags data set to test the hypothesis of recent stickleback introduction into central Oregon, where this species was only recently documented. The authors documented a clear genetic division between coastal and inland populations as well as the role of introgressive hybridization in coastal populations and recent expansion in central Oregon. Although genomic approaches are still in their infancy and the pipelines necessary for analysing genomic data are only emerging, the great opportunity for exploring invasion histories and the underlying evolutionary consequences is overwhelming.

A suite of genetic tools and computational approaches have been developed to sequence and detect organisms from complex environmental samples, including those found at low abundance or which are partially degraded (Box 3). These provide a sensitive approach to monitoring biodiversity (Bik *et al.* 2012; Thomsen *et al.* 2012), detecting invasive species (Lodge *et al.* 2012; Pochon *et al.* 2013; Darling 2014) and inferring large-scale patterns of natural or human-mediated connectivity between populations. The method known as metabarcoding combines next-generation sequencing (NGS) with traditional barcoding techniques (Hebert *et al.* 2003) in order to successfully circumvent the laborious effort otherwise necessary to identify single organisms from complex mixtures (reviewed in Taberlet *et al.* 2012). Many technical limitations have been identified. Several solutions have been proposed to overcome them, and many are under development (Bik *et al.* 2012; Cristescu 2014). These metabarcoding techniques often need to be carefully calibrated against simple biological assemblages of well-known composition in order to assess the most effective methods of recovering biodiversity. The ultimate goal is that each biological or ecological species (reproductively isolated units or distinct units with characteristic ecological requirements) will be represented by a single genetic cluster or OTU. Comprehensive databases comprised of repository sequences would allow researchers to link their OTUs to corresponding Linnean species and access valuable ecological information (Cristescu 2014). Recent studies provide a framework for the application of metabarcoding to monitor biodiversity and for the early detection of invasive species in complex environmental samples (Talbot *et al.* 2014). Moreover, long-term surveillance of vulnerable habitats situated near entry points would provide valuable data on introduced, exotic

species that fail to invade (Box 2). Such approaches can be also employed to quantify propagule pressure associated with the major vectors of introductions or secondary spread. For example, species assemblages commonly transported in ballast water or as hull fouling on transoceanic vessels can be also monitored efficiently. Genetic information on the source of invasions, invasion vectors and routes, the number of introductions, and the pathways of secondary spread can be also used by managers and other decision makers to support conservation projects and other regulations that minimize accidental introductions (Estoup & Guillemaud 2010; Darling 2014).

CONCLUSIONS

Given the complex evolutionary forces and ecological settings involved during the invasion process, the reconstruction of invasion history remains central to studies that aim to understand the mechanisms driving successful invasions. Molecular tools have provided effective ways of tracing the history of both natural and human-mediated invasions. Many attributes of historical and contemporaneous biological invasions are alike, but often the methods (specific molecular markers or analytical approaches) applied to study them need careful consideration. Biological invasions are indeed natural processes; however, their rate of introduction and geographical scale are extraordinarily high today and dramatically different from the recent past. As one of the most successful colonizing species, humans have an inherent desire to understand the fundamental nature of the invasion process and the important historical events that shape invasion trajectories. Biological invasions have provided complex natural experiments that have stimulated deep reflections on the ecological footprint of humanity and our future as a species.

ACKNOWLEDGEMENTS

I would like to thank the organizers for the invitation to participate to the Genetics of Colonizing Species Symposium that inspired this volume. The manuscript benefited from valuable comments provided by the anonymous reviewers the editor, Dr. Loren Rieseberg, and laboratory members. Funding was provided by the Natural Science and Engineering Research Council (NSERC).

REFERENCES

Adamowicz SJ, Petrusek A, Cobourne JK *et al.* (2009) The scale of divergence: a phylogenetic appraisal of intercontinental allopatric speciation in a passively dispersed freshwater zooplankton genus. *Molecular Phylogenetics and Evolution*, **50**, 423–436.

Ascunce MS, Yang C-C, Oakey J *et al.* (2011) Global invasion history of the fire ant *Solenopsis invicta*. *Science*, **331**, 1066–1068.

Baird NA, Etter PD, Atwood TS *et al.* (2008) Rapid SNP discovery and genetic mapping using sequenced RAD markers. *PLoS One*, **3**, e3376.

Baker HG, Stebbins GL (eds.) (1965) *The Genetics of Colonizing Species*. Acadmic Press, New York City, New York.

Barrett SCH (1992) Genetics of weed invasions. In: *Applied Population Biology* (eds Jain SK, Botsford LW), pp. 91–119. Kluwer Academic Publisher, Dordrecht, Netherlands.

Barrett SCH (2015) Foundations of invasion genetics: the Baker and Stebbins legacy. *Molecular Ecology*, **24**, 1927–1941.

Barrett SCH, Husband BC (1990) The genetics of plant migration and colonization. In: *Plant Population Genetics, Breeding, and Genetic Resources* (eds Brown AHD, Clegg MT, Kahler AL, Weir BS), pp. 254–277. Sinauer Associates Inc., Sunderland, Massachusetts.

Beaumont MA, Zhang WY, Balding DJ (2002) Approximate Bayesian computation in population genetics. *Genetics*, **162**, 2025–2035.

Bernatchez L, Wilson CC (1998) Comparative phylogeography of Nearctic and Palearctic fishes. *Molecular Ecology*, **17**, 2563–2565.

Bik HM, Porazinska DL, Creer S *et al.* (2012) Sequencing our way towards understanding global eukaryotic biodiversity. *Trends in Ecology and Evolution*, **27**, 233–243.

Blackburn TM, Lockwood JL, Cassey P (2015) The influence of numbers on invasion success. *Molecular Ecology*, **24**, 1942–1953.

Bock DG, Zhan A, Lejeusne C *et al.* (2011) Looking at both sides of the invasion: patterns of colonization in the violet tunicate *Botrylloides violaceus*. *Molecular Ecology*, **20**, 503–516.

Brahic C (2012) Our true dawn: pinning down human origins. *New Scientist*, **2892**, 34–37.

Carlton JT (1996) Biological invasions and cryptogenic species. *Ecology*, **77**, 1653–1655.

Carson HL (1965) Chromosomal morphism in geographically widespread species of Drosophila. In: *The Genetics of Colonizing Species* (eds Baker HG, Stebbins GL), pp. 503–527. Academic press, New York.

Catchen J, Hohenlohe PA, Bassham S *et al.* (2013a) Stacks: and analysis tool set for population genomics. *Molecular Ecology*, **22**, 3124–3140.

Catchen J, Bassham S, Wilson T *et al.* (2013b) The population structure and recent colonization history of Oregon threespine stickleback determined using RAD-seq. *Molecular Ecology*, **22**, 2864–2883.

Colautti RI, MacIsaac HJ (2004) A neutral terminology to define 'invasive' species. *Diversity and Distributions*, **10**, 135–141.

Corander J, Waldmann P, Sillanpaa MJ (2003) Bayesian analysis of genetic differentiation between populations. *Genetics*, **163**, 367–374.

Cornuet JM, Luikart G (1996) Description and power analysis of two tests for detecting recent population bottlenecks from allele frequency data. *Genetics*, **144**, 2001–2014.

Cornuet JM, Santos F, Beaumont MA *et al.* (2008) Inferring population history with DIY ABC: a user-friendly approach to approximate Bayesian computation. *BMC Bioinformatics*, **24**, 2713–2719.

Cristescu ME (2014) From barcoding single individuals to metabarcoding biological communities: towards an integrative approach to the study of global biodiversity. *Trends in Ecology and Evolution*, **29**, 566–571.

Cristescu ME, Hebert PDN, Witt JDS *et al.* (2001) An invasion history for *Cercopagis pengoi* based on mitochondrial gene sequences. *Limnology and Oceanography*, **46**, 224–229.

Cristescu ME, Witt JDS, Grigorovich IA *et al.* (2004) Dispersal patterns for the Ponto-Caspian amphipod *Echinogammarus ischnus*: invasion waves from the Pleistocene to the present. *Heredity*, **92**, 197–203.

Cristescu ME, Constantin A, Bock DG *et al.* (2012) Speciation with gene flow and the genetics of habitat transitions. *Molecular Ecology*, **21**, 1411–1422.

Crnokrak P, Barrett SCH (2002) Purging the genetic load: a review of the experimental evidence. *Evolution*, **56**, 2347–2358.

Darling JA (2014) Genetic studies of aquatic biological invasions: closing the gap between research and management. *Biological Invasions*, **16**, 1–21.

Darling JA, Bagley MJ, Roman J *et al.* (2008) Genetic patterns across multiple introductions of the globally invasive crab genus *Carcinus*. *Molecular Ecology*, **17**, 4992–5007.

Davey JW, Hohenlohe PA, Etter PD *et al.* (2011) Genome-wide genetic marker discovery and genotyping using next-generation sequencing. *Nature Reviews Genetics*, **12**, 499–510.

Di Castri F (1989) History of biological invasions with special emphasis on the old world. In: *Biological Invasions: A Global Perspective* (Drake JA *et al.*), pp. 1–30. John Wiley & Sons, Oxford.

Dieringer D, Schlötterer C (2003) Microsatellite Analyses (MSA): a platform-independent analysis tool for large microsatellite data sets. *Molecular Ecology Notes*, **3**, 167–169.

Dlugosch KM, Parker M (2008) Founding events in species invasions: genetic variation, adaptive evolution, and the role of multiple introductions. *Molecular Ecology*, **17**, 431–449.

Dlugosch KM, Lai Z, Bonin A *et al.* (2013) Allele identification for transcriptome-based population genomics in the invasive plant *Centaurea solstitialis*. *G3: Genes Genomes Genetics*, **3**, 359–367.

Dobzansky T (1965) "Wild" and "domestic" species of *Drosophila*. In: *The Genetics of Colonizing Species* (eds Baker HG, Stebbins GL), pp. 533–546. Academic press, New York.

Ellison CE, Hall C, Kowbel D *et al.* (2011) Population genomics and local adaptation in wild isolates of a model microbial eukaryote. *Proceedings of the National Academy of Sciences, USA*, **108**, 2831–2836.

Elton CS (1958) *The Ecology of Invasions by Animals and Plants*. Methuen, London.

Estoup A, Guillemaud T (2010) Reconstructing routes of invasion using genetic data: why, how and so what? *Molecular Ecology*, **19**, 4113–4130.

Estoup A, Wilson IJ, Sullivan C *et al.* (2001) Inferring population history from microsatellite and enzyme data in serially introduced cane toads, *Bufo marinus*. *Genetics*, **159**, 1671–1687.

Estoup A, Baird SJE, Ray N *et al.* (2010) Combining genetic, historical and geographical data to reconstruct the dynamics of bioinvasions: application to the cane toad *Bufo marinus*. *Molecular Ecology Resources*, **10**, 886–901.

Facon B, Pointier J-P, Glaubrecht M *et al.* (2003) A molecular phylogeography approach to biological invasions of the New World by parthenogenetic Thiarid snails. *Molecular Ecology*, **12**, 3027–3039.

Fitzpatrick BM, Fordyce JA, Niemiller ML *et al.* (2012) What can DNA tell us about biological invasions? *Biological Invasions*, **14**, 245–253.

Fonseca DM, LaPointe DA, Fleischer RC (2000) Bottlenecks and multiple introductions: population genetics of the vector of avian malaria in Hawaii. *Molecular Ecology*, **9**, 1803–1814.

Frey DG (1982) Questions concerning cosmopolitanism in Cladocera. *Archiv für Hydrobiologie*, **93**, 484–502.

Fu YX (1997) Statistical tests of neutrality of mutations against population growth, hitchhiking and background selection. *Genetics*, **147**, 915–925.

Gladieux P, Feurtey A, Hood ME *et al.* (2015) The population biology of fungal invasions. *Molecular Ecology*, **24**, 1969–1986.

Gómez A, Serra M, Carvalho GR *et al.* (2002) Speciation in ancient cryptic species complexes: evidence from the molecular phylogeny of *Brachionus plicatilis* (Rotifera). *Evolution*, **56**, 1341–1444.

Goudet J (1995) FSTAT (V.1.2): a computer program to estimate F-statistics. *Journal of Heredity*, **86**, 485–486.

Guillot G, Mortier F, Estoup A (2005) GENELAND: a computer package for landscape genetics. *Molecular Ecology Notes*, **5**, 712–715.

Handley LJL, Estoup A, Evans DM *et al.* (2011) Ecological genetics of invasive alien species. *Biological Control*, **56**, 409–428.

Havel JE, Shurin JB (2004) Mechanisms, effects, and scales of dispersal in freshwater zooplankton. *Limnology and Oceanography*, **49**, 1229–1238.

Hebert PDN, Cristescu MEA (2002) Genetic perspective on invasions: the case of Cladocera. *Canadian Journal of Fisheries and Aquatic Sciences*, **59**, 1229–1234.

Hebert PDN, Wilson CC (1994) Provincialism in plankton – endemism and allopatric speciation in Australian *Daphnia*. *Evolution*, **48**, 1333–1349.

Hebert PDN, Cywinska A, Ball SL *et al.* (2003) Biological identification through DNA barcodes. *Proceedings of the Royal Society of London Series B, Biological Sciences*, **270**, 313–321.

Hierro JL, Maron JL, Callaway RM (2005) A biogeographical approach to plant invasions: the importance of studying exotics in their introduced and native range. *Journal of Ecology*, **93**, 5–15.

Hodgins KA, Bock DG, Hahn MA *et al.* (2015) Comparative genomics in the Asteraceae reveals little evidence for parallel evolutionary change in invasive taxa. *Molecular Ecology*, **24**, 2226–2240.

Hoffmann AA, Rieseberg LH (2008) Revisiting the impact of inversions in evolution: from population genetic markers to drivers of adaptive shift and speciation? *Annual Review of Ecology Evolution and Systematics*, **39**, 21–42.

Hohenlohe PA, Bassham S, Etter PD *et al.* (2010) Population genomics of parallel adaptation in threespine stickleback using sequenced RAD tags. *PLoS Genetics*, **6**, DOI:10.1371/journal.pgen.1000862

Hoos PM, Miller AW, Ruiz GM *et al.* (2010) Genetic and historical evidence disagree on the likely sources of the Atlantic amethyst gem clam *Gemma gemma* (Totten, 1834) in California. *Diversity and Distribution*, **16**, 582–592.

Hutchison DW, Templeton AR (1999) Correlation of pairwise genetic and geographic distance measures: inferring the relative influences of gene flow and drift on the distribution of genetic variability. *Evolution*, **53**, 1898–1914.

Keller SR, Taylor DR (2008) History, chance and adaptation during biological invasions: separating stochastic phenotypic evolution from response to selection. *Ecology Letters*, **11**, 852–866.

Kelly DW, Muirhead JR, Heath DD *et al.* (2006) Contrasting patterns in genetic diversity following multiple invasions of fresh and brackish waters. *Molecular Ecology*, **15**, 3641–3653.

Kirkpatrick M, Barrett B (2015) Chromosome inversions, adaptive cassettes, and the evolution of species' ranges. *Molecular Ecology*, **24**, 2046–2055.

Kolbe JJ, Glor RE, Schettino LR *et al.* (2004) Genetic variation increases during biological invasion by a Cuban lizard. *Nature*, **431**, 177–181.

Kulhanex SA, Ricciardi A, Leung B (2011) Is invasion history a useful tool for predicting the impacts of the world's worst aquatic invasive species? *Ecological Applications*, **21**, 189–202.

Le H, Handsaker B, Wysoker A *et al.*, 1000 Genome Project Data Processing Subgroup (2009) The sequence alignment map format and SAMtools. *Bioinformatics (Oxford, England)*, **25**, 2078–2079.

Lee CE (2002) Evolutionary genetics of invasive species. *Trends in Ecology and Evolution*, **17**, 386–391.

Lee CE, Gelembiuk GW (2008) Evolutionary origins of invasive populations. *Evolutionary Applications*, **1**, 427–448.

Lewontin RC (1965) Selection for colonizing ability. In: *The Genetics of Colonizing Species* (eds Baker HG, Stebbins GL), pp. 79–92. Academic press, New York.

Lodge DM (1993) Biological invasions: lessons for Ecology. *Trends in Ecology and Evolution*, **8**, 133–137.

Lodge DM, Turner CR, Jerde CL *et al.* (2012) Conservation in a cup of water: estimating biodiversity and population abundance from environmental DNA. *Molecular Ecology*, **21**, 2555–2558.

Lombaert E, Guillemaud T, Cornuet J-M *et al.* (2010) Bridgehead effect in the worldwide invasion of the biocontrol harlequin ladybird. *PLoS ONE*, **5**, e9743.

Lombaert E, Guillemaud T, Thomas CE *et al.* (2011) Inferring the origin of populations introduced from a genetically structured native range by approximate Bayesian computation: case study of the invasive ladybird *Harmonia axyridis*. *Molecular Ecology*, **20**, 4654–4670.

Lopes JS, Balding D, Beaumont MA (2009) PopABC: a program to infer historical demographic parameters. *Bioinformatics*, **25**, 2747–2749.

Lynch M, Pfrender M, Spitze K *et al.* (1999) The quantitative and molecular genetic architecture of a subdivided species. *Evolution*, **53**, 100–110.

Mangerud J, Jakobsson M, Alexanderson H *et al.* (2004) Ice-dammed lakes and rerouting of the drainage of northern Eurasia during the last glaciation. *Quaternary Science Reviews*, **23**, 1313–1332.

Marisco TD, Wallace LE, Ervin GN *et al.* (2011) Geographic patterns of genetic diversity from the native range of *Cactoblastis cactorum* (Berg) support the documented history of invasion and multiple introductions for invasive populations. *Biological Invasions*, **13**, 857–868.

McKenna A, Hanna M, Banks E *et al.* (2010) The genome analysis toolkit: a MapReduce framework for analyzing next-generation DNA sequencing data. *Genome Research*, **20**, 1297–1303.

Mergeay J, Vanoverbeke J, Verschuren D *et al.* (2007) Extinction, recolonization, and dispersal through time in a planktonic crustacean. *Ecology*, **88**, 3032–3043.

Millette KL, Xu S, Witt JDS *et al.* (2011) Pleistocene-driven diversification in freshwater zooplankton: genetic patterns of refugial isolation and postglacial recolonization in *Leptodora kindtii*. *Limnology and Oceanography*, **56**, 1725–1736.

Milne RI, Abbott RJ (2000) Origin and evolution of invasive naturalized material of *Rhododendron ponticum* L. in the British Isles. *Molecular Ecology*, **9**, 541–556.

Mooney HA, Cleland EE (2001) The evolutionary impact of invasive species. *Proceedings of the National Academy of Sciences, USA*, **98**, 5446–5451.

Nei M, Maruyama T, Chakraborty R (1975) The bottleneck effect and genetic variability in populations. *Evolution*, **29**, 1–10.

Novak SJ, Mack RN (2005) Genetic bottleneck in alien plant species: influences of mating systems and introduction dynamics. In: *Species Invasions: Insights into Ecology and Biogeography* (eds Sax DF, Stachowicz JJ, Gaines SD), pp. 201–228. Sinauer Associates, Sunderland, Massachusetts.

Peery MZ, Kirby R, Reid BN *et al.* (2012) Reliability of genetic bottleneck tests for detecting recent population declines. *Molecular Ecology*, **21**, 3403–3418.

Pochon X, Bott NJ, Smith KF *et al.* (2013) Evaluating detection limits of next-generation sequencing for the surveillance and monitoring of international marine pests. *PLoS One*, **8**, e73935.

Prevosti A, Ribo G, Serra L *et al.* (1988) Colonization of America by *Drosophila subobscura*: experiment in natural populations that supports the adaptive role of chromosomal inversion polymorphism. *Proceedings of the Natural Academy of Sciences, USA*, **85**, 5597–5600.

Pritchard JK, Stephens J, Donnelly P (2000) Inference of population structure using multilocus genotype data. *Genetics*, **155**, 945–959.

Puzey J, Vallejo-Marín M (2014) Genomics of invasion: diversity and selection in introduced populations of monkeyflowers. *Molecular Ecology*, **23**, 4472–4485.

Raymond M, Rousset F (1995) GENEPOP (version 1.2): population genetic software for exact tests and ecumenicism. *Journal of Heredity*, **83**, 239.

Ricciardi A, MacIsaac HJ (2000) Recent mass invasion of the North American Great Lakes by Ponto-Caspian species. *Trends in Ecology & Evolution*, **15**, 62–65.

Robinson JV, Dickerson JE (1987) Does invasion sequence affect community structure? *Ecology*, **68**, 587–595.

Robinson JV, Edgemont MA (1988) An experimental evaluation of the effect of invasion history on community structure. *Ecology*, **69**, 1410–1417.

Roda F, Ambrose L, Walter GM *et al.* (2013) Genomic evidence for the parallel evolution of coastal forms in the *Senecio lautus* complex. *Molecular Ecology*, **22**, 2941–2952.

Rogers AR, Harpending H (1992) Population growth makes waves in the distribution of pairwise genetic differences. *Molecular Biology and Evolution*, **9**, 552–569.

Roman J, Darling JA (2007) Paradox lost: genetic diversity and the success of aquatic invasions. *Trends in Ecology and Evolution*, **22**, 454–464.

Saitou N, Nei M (1987) The neighbor-joining method: a new method for reconstructing phylogenetic trees. *Molecular Biology and Evolution*, **4**, 406–425.

Sakai AK, Allendorf FW, Holt JS *et al.* (2001) The population biology of invasive species. *Annual Review of Ecology and Systematics*, **32**, 305–332.

Savolainen O, Lascoux M, Merila J (2013) Ecological genomics of local adaptation. *Nature Reviews Genetics*, **14**, 807–820.

Sax DF, Brown JH (2000) The paradox of invasion. *Global Ecology and Biogeography*, **9**, 363–371.

Sax DF, Stachowicz JJ, Brown JH *et al.* (2007) Ecological and evolutionary insights from species invasions. *Trends in Ecology and Evolution*, **22**, 465–471.

Schneider S, Roessli D, Excoffier L (2000) *Arlequin Ver. 2.00: A Software For Population Genetic Data Analysis.* Genetics and Biometry Laboratory, University of Geneva, Bern, Switzerland.

Simberloff D (2009) The role of propagule pressure in biological invasions. *Annual Review of Ecology, Evolution and Systematics*, **40**, 81–102.

Simberloff D (2013) *Invasive Species. What Everyone Needs to Know.* Oxford University Press, New York.

Slatkin M (1993) Isolation by distance in equilibrium and non-equilibrium populations. *Evolution*, **47**, 264–279.

Stapley J, Reger J, Feulner PGD *et al.* (2010) Adaptation genomics: the next generation. *Trends in Ecology and Evolution*, **25**, 705–712.

Taberlet P, Coissac E, Pompanon F *et al.* (2012) Towards next-generation biodiversity assessment using DNA metabarcoding. *Molecular Ecology*, **21**, 2045–2050.

Tajima F (1989) Statistical method for testing the neutral mutation hypothesis by DNA polymorphism. *Genetics*, **123**, 585–595.

Talbot JM, Bruns TD, Taylor JW *et al.* (2014) Endemism and functional convergence across the North American soil mycobiome. *Proceedings of the National Academy of Sciences, USA*, **111**, 6341–6346.

Taylor DJ, Finston TL, Hebert PDN (1998) Biogeography of a widespread freshwater crustacean: pseudocongruence and cryptic endemism in the North American *Daphnia laevis* complex. *Evolution*, **52**, 1648–1670.

Thomsen PF, Kielgast J, Iversen LL *et al.* (2012) Monitoring endangered freshwater biodiversity using environmental DNA. *Molecular Ecology*, **21**, 2565–2573.

Voisin M, Engel CR, Viard F (2005) Differential shuffling of native genetic diversity across introduced regions in a brown alga: aquaculture vs. maritime traffic effects. *Proceedings of the National Academy of Sciences, USA*, **102**, 5432–5437.

Wegmann D, Leuenberger C, Neuenschwander S *et al.* (2010) ABCtoolbox: a versatile toolkit for approximate Bayesian computations. *BMC Bioinformatics*, **11**, 116.

Wilson JRU, Dormontt EE, Prentis PJ *et al.* (2009) Something in the way you move: dispersal pathways affect invasion success. *Trends in Ecology and Evolution*, **24**, 136–144.

Xu S, Hebert PDN, Kotov AA *et al.* (2009) The noncosmopolitanism paradigm of freshwater zooplankton: insights from the global phylogeography of the predatory cladoceran *Polyphemus pediculus* (Linnaeus, 1761) (Crustacea, Onychopoda). *Molecular Ecology*, **18**, 5161–5179.

Chapter 16

COMPARATIVE GENOMICS IN THE ASTERACEAE REVEALS LITTLE EVIDENCE FOR PARALLEL EVOLUTIONARY CHANGE IN INVASIVE TAXA

Kathryn A. Hodgins,* Dan G. Bock,† Min A. Hahn,† Sylvia M. Heredia,† Kathryn G. Turner,† and Loren H. Rieseberg†‡

*School of Biological Sciences, Monash University, Clayton, VIC 3800, Australia
†Department of Botany, University of British Columbia, 1316-6270 University Blvd., Vancouver, BC V6T 1Z4, Canada
‡Department of Biology, Indiana University, Bloomington, IN 47405, USA

Abstract
Asteraceae, the largest family of flowering plants, has given rise to many notorious invasive species. Using publicly available transcriptome assemblies from 35 Asteraceae, including six major invasive species, we examined evidence for micro- and macro-evolutionary genomic changes associated with invasion. To detect episodes of positive selection repeated across multiple introductions, we conducted comparisons between native and introduced genotypes from six focal species and identified genes with elevated rates of amino acid change (dN/dS). We then looked for evidence of positive selection at a broader phylogenetic scale across all taxa. As invasive species may experience founder events during colonization and spread, we also looked for evidence of increased genetic load in introduced genotypes. We rarely found evidence for parallel changes in orthologous genes in the intraspecific comparisons, but in some cases we identified changes in members of the same gene family. Using among-species comparisons, we detected positive selection in 0.003–0.69% and 2.4–7.8% of the genes using site and stochastic branch-site models, respectively. These genes had diverse putative functions, including defence response, stress response and herbicide resistance, although there was no clear pattern in the GO terms. There was no indication that introduced genotypes have a higher proportion of deleterious alleles than native genotypes in the six focal species, suggesting multiple introductions and admixture mitigated the impact of drift. Our findings provide little evidence for common genomic responses in invasive taxa of the Asteraceae and hence suggest that multiple evolutionary pathways may lead to adaptation during introduction and spread in these species.

Previously published as an article in *Molecular Ecology* (2015) 24, 2226–2240, doi: 10.1111/mec.13026

INTRODUCTION

Anthropogenic changes have transformed the global landscape and, although many species are suffering from habitat loss and extinction as a result, invasive species have thrived. Invasive species, those that spread outside their natural range and proliferate (Gray & Mack 1986), can have negative impacts on the economy and the environment, which provides considerable incentive to understand the factors that contribute to their success (Stewart et al. 2009). However, the study of invasive species offers opportunities beyond the applied realm. Several decades ago, Baker (1974) proposed that weedy and invasive species were excellent subjects of evolutionary study because of their abundance, ease in growing, documented introduction history and recent evolutionary response to new environments (e.g. herbicide resistance). Since that time, considerable effort has been applied to identify the ecological factors contributing to invasion success (e.g. Catford et al. 2009; Lockwood et al. 2009), but more recently there has been renewed interest in understanding the evolutionary changes associated with invasion (Dlugosch & Parker 2008; Hodgins & Rieseberg 2011; Colautti & Barrett 2013; Turner et al. 2014). Now that genomic information is becoming abundant for a wide array of nonmodel organisms, including invasive species, we can begin to uncover the genomic causes and consequences of biological invasions.

A comparative genomics approach to invasive species research has the potential to identify adaptive genetic changes that are commonly associated with the evolution of invasive species, as well as those that are idiosyncratic. This tactic may also help pinpoint which changes were selected for after introduction, and which have evolved in the native range to predispose certain groups to become problematic invaders. Such information could help elucidate why invasive species are abundant in some lineages but not in others (Kuester et al. 2014). The identification of selected changes in invasive populations with independent origins will provide evidence for common genetic responses to colonization. Similarly, finding adaptive responses in the same functional groups of genes among invaders might reveal shared trade-offs that contribute to invasion success, even if the particular genes involved differ among populations or species (Lai et al. 2008; Mayrose et al. 2011; Guggisberg et al. 2013; Hodgins et al. 2013b).

While positive selection may play an important role in the evolution of invasive species, introduced populations often experience founder events and an increased likelihood of repeated bottlenecks (Barrett & Shore 1989; Dlugosch & Parker 2008). Because of these processes, significant losses of both allelic richness and heterozygosity in introduced populations are common and gains of diversity are infrequent (Dlugosch & Parker 2008). Population expansion during the spread of an invader across a landscape can have important consequences for genetic variation and fitness. As a species increases its range, genetic drift is predicted to be stronger at the leading edge of the expanding wave front, due to low population densities (Slatkin & Excoffier 2012; Peischl et al. 2013). This can have consequences for spatial patterns of neutrally evolving variants (Klopfstein et al. 2006; Slatkin & Excoffier 2012), as well as those under selection (Lehe et al. 2012; Peischl et al. 2013). Genetic drift on the leading edge of range expansions can result in a long-lasting accumulation of deleterious mutations over most of a species range. A recent study revealed that population expansion has left its imprint on the spatial distribution of neutral and deleterious alleles in humans (Peischl et al. 2013). Similarly, organisms that have experienced considerable expansion in range size due to their association with humans, such as weeds, may be expected to bear a significant expansion load, yet this remains to be tested.

The Asteraceae family, home to many of the world's worst weeds and invasive species (Baker 1974), is an ideal system for a comparative genomics approach to invasive species research. For example, 588 species in the Asteraceae are considered globally invasive including two of the 33 plants listed by the IUCN as the 100 World's Worst Invasive Alien Species, and 94 species on the US Federal and state noxious weed list (Chamberlain & Szöcs 2013; EOL 2014; GISD 2014; USDA, NRCS 2014). Furthermore, extensive genomic resources have been developed for this family through the Compositae Genome Project (www.cgpdb.ucdavis.edu). These include transcriptome assemblies from upwards of 40 different species, many of which are crops, feral weeds, crop wild relatives and invasive species (Barker et al. 2008; Lai et al. 2012; Scaglione et al. 2012; Hodgins et al. 2013b), as well as assemblies of introduced and native genotypes from six invasive species and four out groups to the family. Although the traits associated with successful Asteraceae invasive species vary across taxa (Muth & Pigliucci 2006), herbicide resistance (Holt *et al.* 2013), as well as growth–defence/stress tolerance trade-offs are commonly observed in weedy species in this family (Hodgins & Rieseberg 2011; Mayrose *et al.* 2011; Guggisberg *et al.* 2013; Turner *et al.* 2014) including those that are the subject of this study.

Here we implement a comparative genomics approach to examine changes in protein evolution associated with invasion in Asteraceae species at both micro- and macro-evolutionary scales using transcriptome data. First, we asked whether there was evidence of recent parallel evolutionary change among species during invasion by comparing native and introduced genotypes from six focal species. We tested for positive selection in introduced genotypes by identifying genes with elevated nonsynonymous nucleotide substitutions (dN) relative to synonymous substitutions (dS). Following this, we took a broader approach and identified rapidly evolving genes across the family and specifically in weedy lineages. This allowed us to ascertain the genes that may contribute to the propensity of certain groups to become invasive. We then examined changes in genome-wide rates of deleterious mutation to assess whether there was a shift in the genetic load associated with introduction.

MATERIALS AND METHODS

Data set

We used previously published de novo transcriptome assemblies (Barker *et al.* 2008; Lai *et al.* 2012; Scaglione *et al.* 2012; Hodgins *et al.* 2013b) for 39 species, which include 35 Asteraceae and four out group taxa (Table S1, Supporting information; www. cgpdb.ucdavis.edu). We classified each of the species as invasive or noninvasive (see Table S1, Supporting information) using the Encyclopedia of Life invasive species comprehensive list, which was accessed programmatically on August 12, 2014 using the taxize package in R (Chamberlain & Szöcs 2013). For some of the species represented in our data set, multiple transcriptomes were available. We included these samples in our ortholog identification and preferentially selected 454 and Illumina transcriptomes as they were generally more complete than Sanger assemblies. For *Helianthus annuus* (annual sunflower), *Ambrosia trifida* (giant ragweed), *A. artemisiifolia* (common ragweed), *Centaurea diffusa* (diffuse knapweed),

C. solstitialis (yellow starthistle) and *Cirsium arvense* (Canada thistle), de novo assemblies from native and introduced samples were available. From here on, we refer to these six species as our 'focal invasive taxa' (see Appendix S1, Supporting information for species descriptions). The six focal species were used for micro-evolutionary comparisons and the 35 species as well as the out groups were used for the macro-evolutionary comparisons.

Ortholog identification and alignments

For each transcriptome, we removed redundant transcripts, representing alternatively spliced transcripts, alleles or close paralogs, by clustering using Cd-Hit-Est (94% identity, word size = 8 and both strands were compared; Li & Godzik 2006; Fu *et al.* 2012). A single representative sequence from each cluster was retained and used for further analyses. We identified the most likely open reading frames for all orthologs using Transdecoder (option –search_pfam; Haas *et al.* 2013). Open reading frames were translated and annotated through BLASTP to the TAIR 10 database (The Arabidopsis Information Resource; arabidopsis.org). To validate the predicted open reading frames from Transdecoder, we only retained those with a pfam hit or hit to *A. thaliana*. On average, this resulted in 20% of the predicted proteins being removed from the analysis. Confirmation of the predicted proteins reduced the possibility that complete and partial open reading frames were misidentified, reducing a potential source of error, although we acknowledge that we are likely missing important loci that have no homology to proteins in these databases. We conducted an all-against-all BLASTP (e-10). Using these results, we identified orthologs with ORTHAGOGUE version 1.0.2 applying the bit score option and 50% overlap (Li *et al.* 2003). We performed the ortholog identification for all transcriptomes across the entire tree, as well as for subsets of the data used in further analyses. These subsets consisted of (i) species in Asteroideae and Carduoideae, the two subfamilies that contain our focal invasive taxa, (ii) the six focal invasive taxa, and (iii) triplets representing one transcriptome for each introduced and native genotype of our focal invasive taxa, and a third transcriptome from a closely related noninvasive out group (cf. triplet analyses below). For all comparisons, we only used one-to-one orthologs.

We extracted the predicted coding sequences for each transcript and aligned the nucleotide sequences for each orthogroup using Prank (+F option). We used the codon model for the alignments, which is preferred over the amino acid model (Löytynoja & Goldman 2008). We chose Prank because it takes evolutionary relationships into account, outperforming other alignment programs (Löytynoja & Goldman 2008; Fletcher & Yang 2010; Markova-Raina & Petrov 2011; Jordan & Goldman 2012). To automatically remove any sequences resulting in alignment errors, we used Guidance (Penn *et al.* 2010), which generates replicate alignments using a slightly perturbed guide tree with Prank (amino acid model, 30 bootstraps) as the bootstrap aligner, for those alignments with at least four or more sequences. We used a conservative Guidance sequence score cut-off of 0.9 and repeated the Prank alignments. Additionally, to remove potential paralogs, we used a tree-based approach (see Appendix S1, Supporting information).

Species tree construction

We identified the 70 orthogroups with the fewest missing taxa and concatenated the sequences for each species. We visually inspected these data as well as the individual gene trees and removed any sequences representing likely paralogs. We constructed a maximum-likelihood tree in RAXML version 8.0.6 (Stamatakis 2006) using the GTR+G model, which was selected based on JMODELTEST 2.1.4 results (Darriba *et al.* 2012). We did not include gene partitions to prevent overparameterization of the model. The tree was rooted using four out-groups to the Asteraceae family. We used the resulting tree topology for the downstream analyses of positive selection.

Pairwise comparisons of native and introduced transcriptomes

For all orthogroups for which a native and introduced genotype of our focal invasive taxa had one-to-one orthologs, we conducted pairwise comparisons to determine divergence at nonsynonymous and synonymous sites in the coding sequence using PAML version 4.5 (Yang 1997, 2007; see Appendix S1, Supporting information). Using orthogroups represented across multiple native and introduced pairs, we examined

whether patterns of divergence were conserved within and among species by determining whether there was a significant positive correlation in dN/dS ratio (Spearman's rho). Positive correlations would indicate consistency in the strength of selection (purifying and/ or positive) among species.

Site-specific positive selection

We conducted an analysis of site-specific tests of positive selection using PAML 4.5 to identify genes that were rapidly evolving across the family, and within the two subfamilies (Asteroideae and Carduoideae) that contain our focal weedy species. For the family-wide analysis, we examined orthogroups with at least three species present in the Asteroideae, Carduoideae and Cichorioideae subfamilies. For the Asteroideae and Carduoideae analysis, only orthogroups with at least four species in each group were used. We trimmed and unrooted the maximum-likelihood species tree to the species in each alignment, using the R package Ape (Paradis *et al.* 2004). We then applied the site model in CODEML to estimate dN and dS at each codon averaged across all branches in the tree. We tested for sites evolving by positive selection by comparing M1a (nearly neutral) and M2a (positive selection), and M7 (beta) against M8 (beta & ω) (F3X4, and transition/ transversion ratios estimated). Twice the difference in log-likelihood values M7:M8 (2 d.f.) and M1a:M2a (2 d.f.), comparisons were assessed for statistical significance using the χ^2 distribution. Because the M8 model can be influenced by starting parameters, we ran the program using different starting values of ω (0.4 and 1.5). To statistically minimize the false discovery rate (FDR), we compared *P*-values to critical values calculated based on α = 0.05 (p.adjust package in R; Benjamini & Hochberg 1995).

PAML branch and branch-site models

While the site-specific analyses identify genes that are rapidly evolving at specific amino acids across the tree, they do not identify changes in the evolutionary rate that are specific to particular branches. To do this, we implemented PAML's branch and branch-site models for two data subsets. The first consisted of triplets of sequences represented by native and introduced transcriptomes for each of our six focal invasive species and a closely related noninvasive out group. The out groups we used were *Parthenium argentatum* for ragweeds, *Carthamus palaestinus* for C. *diffusa*, C. *solstitialis* and C. *solstitialis*, and *Echinacea angustifolia* for H. *annuus*. In each of these triplets, we marked the introduced genotype as foreground branch (i.e. the branch that is tested for evidence of positive selection). The second data subset consisted of native and introduced genotypes for the six focal invasive taxa. For this analysis, we retained orthogroups containing at least two native-introduced pairs and marked all introduced genotypes as foreground branches.

Branch models allow ω to vary among branches but not among sites and therefore may detect positive selection in specific genes along foreground lineages. We compared the null model, which estimates one ω for all branches, with the alternative model, which estimates one ω for the foreground branches and one ω for the background branches. To identify codons that display evidence of positive selection in specific genes and along foreground branches, we used PAML's branch-site models. We compared the null model, which fixes $\omega_2 = 1$ (Zhang *et al.* 2005) with the alternative model, in which ω_2 is estimated ($\omega_2 \geq 1$). To avoid the detection of local peaks, we ran the branch-site models using different starting ω values (ω = 0.5, 1, 1.5, 2). Other parameters as well as the significance were assessed similarly to the site models, but using one d.f. for likelihood ratio tests (see Appendix S1, Supporting information).

Stochastic branch-site models

We also implemented branch-site models for positive selection using Fitmodel (Guindon *et al.* 2004). The Fitmodel analysis was used for the family-wide data set, as well as for the Asteroideae and Carduoideae subfamilies. We limited our analysis to alignments with six or more sequences due to computational time and because the power to detect selection diminishes when there are few species in the tree (Lu & Guindon 2014). Switching models allow each codon site to change the selective regime and thus be affected by selective pressures at different time points. To test for evidence of positive selection varying down branches, we compared the M1a model (no positive selection) to the M2a model, which included selection. In both cases, switching was allowed to occur. Sites with episodes of positive selection ($P > 0.90$) were detected a

posteriori using a Bayesian approach (Guindon *et al.* 2004). The likelihood ratio test statistic with one d.f. was assumed based on results of simulations (Lu & Guindon 2014). Similarly to the PAML analysis, all putatively selected sites were confirmed by visually inspecting the alignments.

Functional annotation and GO analysis

Using the results of the BLASTP to the TAIR 10 database (cf. above), we selected the top hit for each species assigned to each orthogroup. We were specifically interested in examining positive selection in target-site resistance (TSR) genes for several herbicides (Table S2, Supporting information), due to the repeated evolution of herbicide resistance in Asteraceae weeds (Holt *et al.* 2013). We conducted a BLASTP against the predicted proteins for each species and then identified putative TSR orthogroups using these annotations. We assigned GO terms to each orthogroup based on the GO A. *thaliana* mappings to the top hits and performed a GO enrichment analysis using topGO and the parent–child method (Alexa *et al.* 2006; Grossmann *et al.* 2007; see Appendix S1, Supporting information). To identify broader patterns, we then identified GO slim terms for each gene using the TAIR 10 database and conducted a chi-squared test to compare the number of loci found in each category for the background and selected loci.

Identification of deleterious alleles

Using the pairwise alignments, we wrote custom scripts to identify if SNPs caused amino acid changes between the introduced and native transcriptomes for each of the six focal invasive species. The deleterious effects of amino acid substitutions were then predicted for proteins derived from each gene with Provean (Choi *et al.* 2012). Provean uses homologous sequences identified by PSI-BLAST against protein databases (nr protein database) and identifies changes in sequence similarity of a query sequence to the protein sequence homolog by comparing an alignment-based score before and after the amino acid change to the query sequence. We used the recommended threshold of −2.5 for the identification of deleterious sites using the nr database and a paired *t*-test to determine whether there were differences in the number of deleterious amino acids separating each native and introduced comparison.

We then identified out groups using alignments within each subfamily to determine derived amino acid changes and eliminated any sites where both amino acids were found in the out group species as these were potential paralogs. We applied a Fisher's exact test to examine differences in the proportion of derived deleterious amino acids between the native and introduced transcriptomes for each of the six species and then assessed overall significance using a paired t-test. If multiple comparisons could be conducted within a species, we took the average for each range.

RESULTS

Pairwise comparisons of native and introduced transcriptomes

Our pairwise comparison of dN/dS ratios between native and introduced transcriptomes, performed using 2836–8368 total orthogroups depending on the focal species considered, identified between 0.6% and 2.6% of loci with elevated evolutionary rates (Table S3, Supporting information). All rapidly evolving genes were unique to each species, except one that was shared between *C. arvense* and *A. trifida*, for which no homolog in *A. thaliana* was identified. However, using the all-against-all blasts, we did identify 20 overlapping homolog groups (i.e. where the reciprocal blast hits were identified) among species for rapidly evolving genes (Table S3, Supporting information). For the conserved orthologs identified among the focal species, we found significant correlations between species, except in some comparisons with *A. trifida*, as well as stronger correlations within species, where multiple comparisons could be conducted (Fig. 1). When multiple native and introduced comparisons were possible within each species, we tested all combinations and present average values in Fig. 1.

Among the quickly evolving genes identified, there were 21 GO terms overrepresented for *A. artemisiifolia*, including defence response (four genes), 27 in *A. trifida*, nine in *C. diffusa*, 24 in *C. solstitialis*, seven in *C. arvense* and three terms in *H. annuus* (Table S3, Supporting information; $P < 0.05$). However, the GO analyses of the specific terms did not reveal any significant overrepresentation of terms after correcting for multiple tests for any of the pairwise comparisons. Examination of the GO slim terms identified significant differences in the distribution of terms in *C. solstitialis*, *C. arvense* and *H. annuus* (Table 1).

PAML branch and branch-site models

The final tree topology (Fig. 2) of our species tree generally agreed with the previously published Asteraceae supertree (Funk *et al.* 2005). The branch and branch-site models for the triplet data sets, performed using 939–2590 total orthogroups depending on the focal species considered, revealed evidence of positive selection in <1.28% and 5.75% of orthogroups, respectively, for each focal species (Table S4, Supporting information). For those genes where orthology could be identified, none of the rapidly evolving genes were shared across species. Using the all-against-all blasts, we did identify 24 overlapping homolog groups among two to three species (Table S4, Supporting information). The corresponding TAIR hits and topGO results for these orthogroups are given in Table S4 (Supporting information). The branch model performed for all six focal species, which considered information from 1792 total orthogroups, concomitantly revealed evidence of positive selection in three orthogroups (Table S2 and Table S5, Supporting information). The branch-site model implemented for the same data set, however, did not identify any significant orthogroups.

Site-specific positive selection

For the M1a:M2a comparison, after FDR correction, we found evidence of positive selection in one of the 397 (0.003%; Table S6, Supporting information) orthogroups across the entire family. Similarly, for Carduoideae and Asteroideae, 32 of the 7654 (0.42%) and 16 of the 12,416 (0.13%) orthogroups showed evidence of positive selection, respectively. For the M7:M8 comparison, we found evidence for positive selection in six genes (0.02%) across the entire family, in 53 genes (0.69%) for Carduoideae and in 55 genes (0.44%) for Asteroideae. Most of these genes were also identified using the M1a:M2a models.

For the Carduoideae, we found 10 GO terms overrepresented at $P < 0.05$ including response to biotic stimulus, vegetative phase change, response to UV, response to wounding and response to virus (Table S6, Supporting information). For the Asteroideae, we identified 40 GO terms overrepresented at $P < 0.05$ including GO terms related to response to abiotic stimulus and cell wall biogenesis (Table S6, Supporting information). However, these terms were not significant

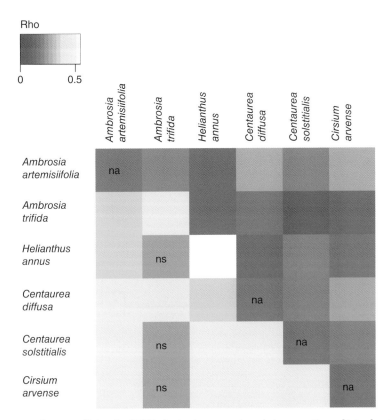

Fig. 1 The Spearman correlation coefficient for dN/dS of pairwise comparisons between native and introduced samples for six species (upper triangle) and their significance (lower triangle). Multiple comparisons within a species were available in some cases (diagonal). $P > 0.05$ (dark grey), $P < 0.05$ (medium grey) and $P < 0.001$ (light grey).

Table 1 Results of chi-squared tests comparing the number of genes in each GO slim category showing evidence of positive selection in pairwise comparisons between native and introduced genotypes relative to those that did not

Species	χ^2	d.f.	P	Top three GO slim categories overrepresented
Ambrosia artemisiifolia	52.59	43	0.15	
Ambrosia trifida	58.23	43	0.06	
Centaurea diffusa	41.95	43	0.52	
Centaurea solstitialis	64.29	43	0.02	Other binding, other enzyme activity, other cellular components, cell wall
Cirsium arvense	70.58	43	0.005	Unknown cellular components, other metabolic processes, transcription factor activity
Helianthus annuus	67.54	43	0.009	Hydrolase activity, response to stress, cell wall

after FDR correction. When comparing the GO slim categories between background and selected genes, the Carduoideae was marginally significantly different ($\chi^2 = 58.63$, d.f. = 43, $P = 0.056$; with *other molecular functions, binding, enzyme activity* and *response to abiotic or biotic stimulus* the most overrepresented terms) and the Asteroideae was not significantly different ($\chi^2 = 46.35$, d.f. = 43, $P = 0.34$).

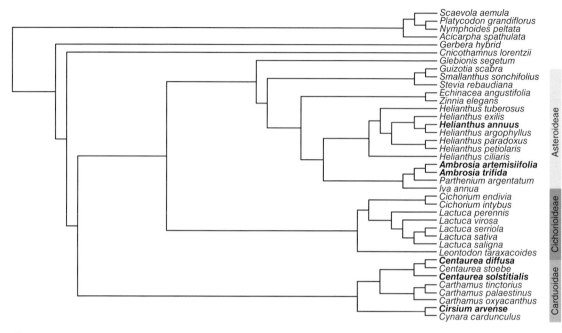

Fig. 2 The tree topology used in the PAML and Fitmodel analysis. Bold font is used to indicate our focal invasive taxa.

Table 2 PAML results for three orthogroups that showed evidence of significant changes in dN/dS ratio (ω) in native vs. introduced genotypes using branch models

Orthogroup	TAIR ID	Description	Significant foreground branches	Likelihood ratio[†]
Ortho group4063	AT5G42990.1	Ubiquitin-conjugating enzyme 18	*A. artemisiifolia, H. annuus*	27.58***
Ortho group5021	AT5G34930.1	Arogenate dehydrogenase	*A. trifida, H. annuus*	17.05*
Ortho group5575	AT3G12120.1	Fatty acid desaturase 2 (FAD2)	*A. trifida, H. annuus*	53.06***

[†]Significance after FDR correction.
*$P < 0.05$;
***$P < 0.001$.

Stochastic branch-site models

Across the entire family, we found evidence for positive selection in 29 of the 397 genes (M1a+S1:M2a+S1; Table S6, Supporting information) after FDR correction (7.8%). All genes possessed at least one site showing evidence of positive selection (Bayesian posterior probability >0.90) in branches leading to at least one introduced species (Table S6, Supporting information; for designations, see Table S1, Supporting information). Within the Carduoideae, we found evidence of positive selection in 55 of the 2259 genes after FDR correction (2.4%). Of those genes, 46 had at least one site showing evidence of positive selection (Bayesian posterior probability >0.90) and 44 of those showed evidence of positive selection in branches leading to at least one introduced species (*Centaurea, Cynara* and *Cirsium*) with two additional loci showing evidence of positive selection in *Carthamus* alone. We identified significant evidence for positive selection in 212 of the 6356 genes in the Asteroideae (3.3%). Of those genes, 83 had at least one site showing evidence of positive selection (Bayesian posterior probability >0.90) and all but three of those showed evidence of positive selection in branches leading to at least one introduced species (*A. artemisiifolia, A. trifida, G. segetum, H. annuus, H. ciliaris, H. petiolaris, H. tuberosus* or *I. annua*).

We then identified the putative function of the significant genes using the BLASTP results to the *A. thaliana* proteins (Table S6, Supporting information). Across the entire family, we found five GO terms overrepresented ($P < 0.05$; Table S6, Supporting information). For the Carduoideae, we found 14 GO terms overrepresented at $P < 0.05$ including response to herbicide (one gene; Table S6, Supporting information). For the Asteroideae, we identified 20 GO terms overrepresented at $P < 0.05$ including GO terms related to response to osmotic stress (seven genes), response to gibberellin-mediated signalling pathway (two genes) and reproductive process (12 genes; Table S6, Supporting information). However, these GO terms were not significant after FDR correction. The analysis of the GO slim terms revealed a marginally significant difference in the proportion of genes in the different categories between the background and selected loci across the family ($\chi^2 = 54.74$, d.f. = 25, $P = 0.10$) with *other enzyme, response to abiotic or biotic stimulus* and *response to stress* as the most overrepresented GO slim terms. There was no significant difference in the Asteroideae ($\chi^2 = 43.18$, d.f. = 43, $P = 0.46$) or Carduoideae ($\chi^2 = 30.5$, d.f. = 43, $P = 0.92$).

Herbicide resistance target genes

We identified filtered orthogroup alignments homologous to several TSR genes (Table S2, Supporting information). Homologs of α-tubulin, β-tubulin, and ALS were found in the Carduoideae and Asteroideae alignments as well as psbA and PPX2L in the Asteroideae alignments. For both the PAML site model M1a:M2a and Fitmodel analysis, we identified ALS as significant in the Asteroideae ($q < 0.05$ after correcting for multiple comparison for the TSR genes only). Comparisons of the sites putatively under selection (amino acids 484 and 598 in the *A. thaliana* protein) were not known amino acids conferring resistance in other species (see www.weedscience.org/Mutations/MutationDisplayAll.aspx), and those known to confer resistance were not represented in our alignments for that gene. Several homologs of α-tubulin and β-tubulin were identified, and both models of site-specific positive selection in PAML were significant for a predicted β-tubulin in the Carduoideae ($q < 0.05$ as significant after correcting for multiple comparison for the TSR genes).

Proportion of deleterious alleles

We found no effect of introduction on the number of deleterious alleles among the six focal species ($t_5 = 0.52$, $P = 0.63$). The same was true when we restricted our analysis to the proportion of deleterious derived alleles in each of the transcriptomes. Individual Fisher's exact tests of the paired samples demonstrated differences in the proportion of derived deleterious alleles for different species: a significantly greater proportion of deleterious variants in the introduced sample from Australia in *H. annuus*, no significant difference in *A. artemisiifolia* and some *A. trifida* comparisons, and significantly lower number of deleterious variants in the introduced range for the remaining comparisons (Fig. 3; Table S7, Supporting information). A paired *t*-test of the per cent difference between the native and introduced transcriptomes revealed no general pattern ($t_5 = -0.82$, $P = 0.45$).

DISCUSSION

Positive selection in introductions

To test whether there were common adaptive changes at the genetic level, we examined evidence for positive selection in introduced genotypes across multiple species. Detecting the genomic signature of selection has an advantage in that no a priori determination of the traits under selection is required. However, common garden studies comparing native and introduced populations of *A. artemisiifolia, C. solstitialis, C. arvense* and *C. diffusa*, as well as comparisons between weedy and wild *H. annuus* have been conducted and striking similarities have emerged (Eriksen *et al.* 2012; Guggisberg *et al.* 2013; Hodgins *et al.* 2013a; Turner *et al.* 2014). In these studies, introduced populations tended to be larger with higher reproductive output and delayed reproduction relative to native populations. Introduced populations of all three species also show evidence of reduced abiotic tolerance and in particular are more susceptible to drought; comparisons of weedy and wild *H. annuus* show a similar pattern (Mayrose *et al.* 2011). In all four species, there is limited evidence for trade-offs of competitive ability with defence, although in most cases the effects of specialist herbivores have not been examined (but see Turner *et al.* 2014). These parallel evolutionary responses to introduction suggest that there may be replicated genetic changes during

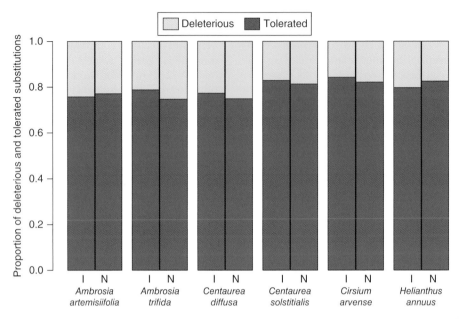

Fig. 3 The proportion of derived deleterious alleles in native (N) and introduced (I) samples across six species identified using Provean. Average values are presented for species with multiple native and introduced samples.

introduction among these species, either in terms of particular genes under selection or the functional groups commonly targeted.

We found little evidence of repeated adaptive change among any of the six focal species investigated in this study. While, in most cases, dN/dS ratios were correlated across species, this appeared to be due to consistent differences in the strength of purifying selection among loci (Fig. 1). Our analyses identified many genes showing evidence of positive selection. However, there was little overlap among species in the identity of positively selected orthologs. For example, only three genes, including one (FAD2) implicated in seed dormancy, as well as seed oil content during sunflower domestication (Linder 2000; Chapman & Burke 2012), have evidence for repeated adaptive evolution using our branch models, but these patterns are only replicated in two species (Table 2). Comparisons of genes inferred to be under positive selection during domestication in this family have found little commonality between any two domestication events using this approach (Kane *et al.* 2011). Notably, this pattern was found despite similar selective pressures and strong artificial selection across thousands of years. Given the shorter timescale and the more diverse ecological conditions likely encountered during invasions relative to an agricultural envi-

ronment, the fact that there is little evidence of repeatability in positively selected genes in our comparisons between native and introduced genotypes is not surprising. Whether the lack of repeatability at the genetic level during invasion will be found across other taxonomic groups remains to be seen. However, given the variable evidence for consistent phenotypic change during invasion (e.g. Bossdorf *et al.* 2005; Colautti *et al.* 2009), and the complex genetic architecture of many of the putative traits under selection (e.g. plant growth and defence response), we predict that our findings are a harbinger of what is to come when comparing multiple species.

We identified evidence that genes in the same family were evolving rapidly across species, suggesting that at least in some cases, similar types of genes were under selection. These genes have diverse functions in *A. thaliana*, including defence response (e.g. AT4G16890, AT3G15850, AT2G25620), abiotic stress response particularly osmotic stress (e.g. AT5G59320) and flavonol biosynthesis (AT5G54160). Although there was limited evidence for the overrepresentation of functional categories in the divergent genes for each species, we found many putatively selected loci important for abiotic and biotic responses. For example, in pairwise comparisons of *A. artemisiifolia*, four genes were annotated as defence response genes, and in *H. annuus*, five genes were annotated as response to stress and a further four were annotated as response to abiotic or biotic stimulus. These GO slim categories were some of the most overrepresented in pairwise comparisons of *C. arvense*. In microarray experiments of *C. arvense* and *A. artemisiifolia* (Guggisberg *et al.* 2013; Hodgins *et al.* 2013a), differences in gene expression between the native and introduced range have also involved stress response genes and, in particular, genes associated with secondary metabolism and detoxification such as the cytochrome P450 gene family. Although differences in expression are not necessarily expected to coincide with changes in coding sequence (e.g. Renaut *et al.* 2012; but see Chapman *et al.* 2013), we identified one differentially expressed gene and putatively selected loci in *A. artemisiifolia*, a UDP-glucosyl transferase (AT1G22340). These genes are involved in metabolism of plant hormones, all major classes of plant secondary metabolites and xenobiotics such as herbicides (Ross *et al.* 2001). Analyses of gene expression between native and introduced populations are underway in the remaining species, and comparisons will offer additional insights into the repeatability of molecular changes during invasion.

The power of our approach relies on our ability to examine large numbers of genes across multiple species to determine commonalities in the signature of selection. However, we lack sufficient sampling within each species to determine whether the observed differences are consistently associated with introductions within each species. Large-scale population genomic sampling for species is underway using a variety of techniques (e.g. transcriptome resequencing, GBS and RAD tags), and these data will be important to determine whether the genes and functional groups that we have identified have likely diverged as a result of selective processes during introduction and spread. Some of the candidate loci could be due to mistakenly identifying close paralogs as orthologs, although all of the orthologs that we identified showed high sequence similarity (Table S3, Supporting information). Other processes may be maintaining variation within and among populations of these species that are not associated with invasion. For example, we might expect that some of the genes identified as highly divergent, such as disease resistance genes, are maintained by balancing selection within the species generally (Bergelson *et al.* 2001). The recent occurrence of the introduc-

tions (100–200 years) and the rapid evolutionary change needed for this approach to be successful, where multiple amino acid changes are often required to identify diverging genes, suggests that standing variation segregating in the initial introductions would likely be the origin of the putatively selected loci (Prentis *et al.* 2008). In contrast, any divergence between the native and introduced ranges that is due to de novo mutations in the introduced range is unlikely to be identified using this approach, given the time needed for new mutations to arise.

Rates of positive selection in the Asteraceae

We found that a small fraction of genes exhibited significant evidence of positive selection using the site models (<1%), while the stochastic branch-site models identified many more candidate loci (2–8%). Site models are mainly able to detect recurrent diversifying selection and lack the capacity to detect episodic adaptive evolution. Our estimates of site-specific positive selection are lower than previous findings (~4% for annual *Helianthus* and *Lactuca* species vs. <1% for each subfamily; Kane *et al.* 2011). This could be in part because of the broader phylogenetic scale that we were examining and suggests that a particular site is rarely the subject of selection over longer evolutionary timescales. In keeping with this, our stochastic branch-site models identified a higher proportion of candidate loci and the proportion was greater in the analysis across the family relative to the subfamilies, as more diverse taxa were added. In some cases, the Bayesian posterior analysis of the stochastic branch-site models did find evidence that the same gene was the target of selection across the entire tree, but the particular sites would often vary among lineages. In other cases, selection was clearly episodic or recurrent down a particular section of the tree.

However, both approaches likely considerably underestimate the actual rate of positive selection for a number of reasons. First, by identifying orthology using a blast-based approach, we are potentially missing rapidly evolving genes that are highly divergent. Moreover, missing or incomplete transcripts are common in some of the assemblies and limited our ability to assess selection across the entire set of species or proteins in many cases. Second, by necessity, we restricted our analysis to one-to-one orthologs, which likely biases our sample towards single-copy genes that can be subject to stronger purifying selection than multicopy gene families (Waterhouse *et al.* 2011). Ortholog identification may have particularly impacted our ability to detect selection at broader phylogenetic scales. Third, we rigorously trimmed and filtered regions that were difficult to align and may harbour diverging sections of the gene. However, conserved regions might be more likely to contain positively selected substitutions (Bazykin & Kondrashov 2012) and removing problematic regions that generate false positives should improve our capacity to detect positive selection (Jordan & Goldman 2012; Privman *et al.* 2012).

Genes under selection in the Asteraceae

Many of the genes that appear to be rapidly evolving in the Asteraceae appear to be likely targets of positive selection in plants in general. Although our gene ontology analysis did not reveal a strong pattern in the types of genes overrepresented in the set of positively selected loci (Table S6, Supporting information), in most cases, the number of selected loci was small which limited the power of the test. In plants, defence genes, in particular leucine-rich repeats of ʀ genes, provide some of the best examples of co-evolutionary 'arms races' resulting in rapid evolutionary change (Bergelson *et al.* 2001). We identified many defence-related genes, particularly in the Carduoideae, including ENHANCED DISEASE SUSCEPTIBILITY 1 (AT3G48090), which impacts salicylic acid (SA) levels and enhances susceptibility to pathogen infection (Falk *et al.* 1999; Nawrath *et al.* 2002), CUCUMOVIRUS MULTIPLICATION 1 (AT4G18040) and BINDING TO TOMV RNA 1 (AT4G37760), both of which are involved in viral resistance (Gao *et al.* 2004; Fujisaki & Ishikawa 2008; Contreras-Paredes *et al.* 2013). Many other stress response genes were identified, including GRX480, a member of the glutaredoxin family that is induced by SA (Ndamukong *et al.* 2007), heat-shock proteins (e.g. AT5G02500 AT3G47650), as well as glutathione S-transferases (e.g. AT2G30860, AT1G78380), involved in abiotic and biotic stress response, including drought stress and herbicide resistance (Rouhier *et al.* 2008; Powles & Yu 2010). In the Asteroideae, abiotic stress response genes predominated, including seven genes related to osmotic stress response, such as LOW EXPRESSION

OF OSMOTICALLY RESPONSIVE GENES 2 (AT2G36530; Ishitani *et al.* 1997) and a glycine-rich RNA-binding protein (AT3G23830), important for cold tolerance and osmotic stress response in *A. thaliana* (Kwak *et al.* 2011). Taken together, these data indicate that genes important for environmental responses appear to be evolving rapidly in this family, which is consistent with previous findings (Kane *et al.* 2011).

Conflicts between genomes can be an important driver of evolutionary rate in plants (e.g. Fujii *et al.* 2011) and this appears to also be the case in our study. We identified candidate genes involved in interactions between the nucleus and either the mitochondrion or the chloroplast including maternal effect embryo arrest proteins (AT3G10110 and AT5G05950; Pagnussat *et al.* 2005). High rates of reproductive protein evolution are also common in both animals and plants (reviewed in Clark *et al.* 2006) and we identified several reproductive genes, particularly in the Asteroideae where 12 were identified. For example, we found several genes involved in male and female gametophyte development (e.g. APK3 AT3G03900, Mob1 AT5G45550 and AT4G30930; Portereiko *et al.* 2006; Mugford *et al.* 2010; Galla *et al.* 2011), which were also localized to the mitochondrial or chloroplast, as well as two putative genes involved in the regulation of flowering time (FCA and CONSTANS-like; Macknight *et al.* 1997; Reeves & Coupland 2001; Hassidim *et al.* 2009).

The evolution of herbicide resistance is increasingly common in weeds (Holt *et al.* 2013) and is known to occur in several species used in this study (e.g. ALS inhibitors in *A. trifida*, *A. artemisiifolia* and *H. annuus*, and resistance to synthetic auxins in *C. arvense* and *C. solstitialis* http://www.weedscience.org/Summary/home.aspx), although the resistance status to various herbicides of the genotypes used in this analysis was not known. There were several genes showing evidence of positive selection that play important roles in detoxification and response to xenobiotics (e.g. glutathione S-transferases, Cytochrome p450s; Powles & Yu 2010; Délye *et al.* 2013) and could be important for herbicide response in this family. However, there are many genes that are known to be the direct targets of commonly applied herbicides (i.e. TSR genes). We tested these for evidence of positive selection and found elevated evolutionary rates in two cases (β-tubulin in Carduoideae and ALS in the Asteroideae). Several ALS-resistant weeds species, such as *H. annuus* and ragweeds, are present in our analysis; however, none of the substitutions in these alignments coincided with known resistance alleles, which suggests either the evolution of novel resistance mutations or a separate selective mechanism driving evolutionary change in these loci. Several key TSR genes (e.g. EPSPS, the target of glyphosate) were not found in our alignments; however, studies are ongoing in ragweeds and other Asteraceae (e.g. *Conyza* spp.) to uncover the genetic basis of herbicide resistance and relative importance of TSR and non-TSR mechanisms in the group.

Genetic load in the introduced range

We predicted that the demographic changes, such as repeated bottlenecks and founder events associated with establishment and spread during invasion, would result in higher levels of genetic load in introduced genotypes. However, our estimates of genetic load across the six focal species did not reveal a general increase in the proportion of deleterious variants in introduced samples. One of the comparisons (native vs. introduced *H. annuus* from Australia) revealed a significantly greater proportion of deleterious alleles in the introduced genotype, suggesting that this population may have been subjected to higher levels of drift. An isozyme study of genetic diversity in Australian *H. annuus* populations, where it was likely introduced as an ornamental, revealed high levels of variation within, but also among populations, as well as an apparent lack of isolation by distance; this pattern is consistent with multiple distinct source populations (Dry & Burdon 1986). However, the study also found relatively high inbreeding coefficients, and the authors suggested this was due to small population sizes, founder events and potentially some degree of selfing. This may also explain why we found evidence of higher genetic load in this particular comparison.

In contrast to *H. annuus*, the introduced samples from the other five focal species were from a portion of the species' range where they are considered invasive and form large population sizes. These comparisons showed statistically equivalent levels of load or even the reverse pattern with a significantly higher proportion of deleterious alleles in the native samples. With the exception of *A. trifida*, where no genetic data are yet available, previous studies of the introduction history of all of these species show evidence of multiple introductions and high levels of genetic variation within

populations (Sun 1997; Marrs *et al.* 2007; Blair & Hufbauer 2010; Chun *et al.* 2010; Gaudeul *et al.* 2011; Guggisberg *et al.* 2012). *Cirsium arvense*, *A. artemisiifolia* and *C. solstitialis* lack substantial isolation by distance in the introduced range, giving further support for repeated long distance dispersal from multiple source populations as the primary driver of genetic variation within the introduced range. Although drift in colonizing species is predicted to increase genetic load, hybridization and population admixture through multiple introductions might limit its severity. Future analyses of the frequency of these deleterious alleles, the degree of admixture across the landscape and the spatial distribution of genetic load within and between the native and introduced range will be essential to more conclusively assess this hypothesis.

Conclusions

Our micro- and macro-evolutionary comparisons revealed that generally there is little overlap in specific genes that appeared to be under positive selection. In our comparisons of native and introduced genotypes across species, there was little commonality in the particular orthologs under selection, despite similarities in phenotypic response during invasion. In some cases, we found similar types of genes under positive selection across two or three introductions, which opens some possibilities of common responses to invasion at the genomic level, but our data suggest that repeated changes in the same gene are unlikely. At the macro-evolutionary scale, we found a similar pattern, where the sites, genes and functional groups showing evidence of positive selection generally differed between subfamilies. These findings point to the idiosyncratic nature of positive selection, where the targets of selection shift across sites, genes and lineages. Moreover, it suggests that selection pressures may be highly variable and outcomes dependent on historical contingency and species-specific genetic constraints, and highlights the possibility of multiple genetic solutions to the challenge of adapting to similar types of environmental change.

ACKNOWLEDGEMENTS

We thank RI Colautti, EBM Drummond and I Mayrose for advice; Z Lai, LO Oliviera, KM Dlugosch and M Barker for data contributions; and Armando Geraldes as well as three anonymous reviewers for their insightful comments. Funding was provided by a Natural Sciences and Engineering Research Council (NSERC) Discovery grant to LHR, NSERC Vanier CGS and Killam Doctoral Fellowship to DGB, and a Swiss National Science Foundation grant to MAH.

REFERENCES

Alexa A, Rahnenführer J, Lengauer T (2006) Improved scoring of functional groups from gene expression data by decorrelating GO graph structure. *Bioinformatics*, **22**, 1600–1607.

Baker HG (1974) The evolution of weeds. *Annual Review of Ecology and Systematics*, **5**, 1–24.

Barker MS, Kane NC, Matvienko M *et al.* (2008) Multiple paleopolyploidizations during the evolution of the Compositae reveal parallel patterns of duplicate gene retention after millions of years. *Molecular Biology and Evolution*, **25**, 2445–2455.

Barrett SCH, Shore JS (1989) Isozyme variation in colonizing plants. In: *Isozymes in Plant Biology* (eds Soltis D, Soltis P), pp. 106–126. Dioscorides Press, Portland, Oregon.

Bazykin GA, Kondrashov AS (2012) Major role of positive selection in the evolution of conservative segments of *Drosophila* proteins. *Proceedings of the Royal Society B*, **279**, 3409–3417.

Benjamini Y, Hochberg Y (1995) Controlling the false discovery rate: a practical and powerful approach to multiple testing. *Journal of the Royal Statistical Society: Series B*, **57**, 289–300.

Bergelson J, Kreitman M, Stahl EA, Tian D (2001) Evolutionary dynamics of plant R-genes. *Science*, **292**, 2281–2285.

Blair AC, Hufbauer RA (2010) Hybridization and invasion: one of North America's most devastating invasive plants shows evidence for a history of interspecific hybridization. *Evolutionary Applications*, **3**, 40–51.

Bossdorf O, Auge H, Lafuma L *et al.* (2005) Phenotypic and genetic differentiation between native and introduced plant populations. *Oecologia*, **144**, 1–11.

Catford JA, Jansson R, Nilsson C (2009) Reducing redundancy in invasion ecology by integrating hypotheses into a single theoretical framework. *Diversity and Distributions*, **15**, 22–40.

Chamberlain SA, Szöcs E (2013) Taxize: taxonomic search and retrieval in R [v2; ref status: indexed, http://f1000r.es/24v]. *F1000Research*, **191**, 1–28.

Chapman MA, Burke JM (2012) Evidence of selection on fatty acid biosynthetic genes during the evolution of cultivated sunflower. *Theoretical and Applied Genetics*, **125**, 897–907.

Chapman MA, Hiscock SJ, Filatov DA (2013) Genomic divergence during speciation driven by adaptation to altitude. *Molecular Biology and Evolution*, **30**, 2553–2567.

Choi Y, Sims GE, Murphy S, Miller JR, Chan AP (2012) Predicting the functional effect of amino acid substitutions and indels. *PLoS ONE*, **7**, e46688.

Chun YJ, Fumanal B, Laitung B, Bretagnolle F (2010) Gene flow and population admixture as the primary post-invasion processes in common ragweed (*Ambrosia artemisiifolia*) populations in France. *The New Phytologist*, **185**, 1100–1107.

Clark NL, Aagaard JE, Swanson WJ (2006) Evolution of reproductive proteins from animals and plants. *Reproduction*, **131**, 11–22.

Colautti RI, Barrett SCH (2013) Rapid adaptation to climate facilitates range expansion of an invasive plant. *Science*, **342**, 364–366.

Colautti RI, Maron JL, Barrett SCH (2009) Common garden comparisons of native and introduced plant populations: latitudinal clines can obscure evolutionary inferences. *Evolutionary Applications*, **2**, 187–199.

Contreras-Paredes CA, Silva-Rosales L, Daròs J-A, Alejandri-Ramírez ND, Dinkova TD (2013) The absence of eukaryotic initiation factor eIF(iso)4E affects the systemic spread of a Tobacco etch virus isolate in *Arabidopsis thaliana*. *Molecular Plant-Microbe Interactions: MPMI*, **26**, 461–470.

Darriba D, Taboada GL, Doallo R, Posada D (2012) jModelTest 2: more models, new heuristics and parallel computing. *Nature Methods*, **9**, 772.

Délye C, Jasieniuk M, Le Corre V (2013) Deciphering the evolution of herbicide resistance in weeds. *Trends in Genetics: TIG*, **29**, 649–658.

Dlugosch KM, Parker IM (2008) Founding events in species invasions: genetic variation, adaptive evolution, and the role of multiple introductions. *Molecular Ecology*, **17**, 431–449.

Dry PJ, Burdon JJ (1986) Genetic structure of natural populations of wild sunflowers (*Helianthus annuus* L.) in Australia. *Australian Journal of Biological Sciences*, **39**, 255–270.

Encyclopedia of Life [EOL] (2014) "Invasive Species Comprehensive List". Available from http://eol.org/collections/55367. Accessed 13 Sept 2014.

Eriksen RL, Desronvil T, Hierro JL, Kesseli R (2012) Morphological differentiation in a common garden experiment among native and non-native specimens of the invasive weed yellow starthistle (*Centaurea solstitialis*). *Biological Invasions*, **14**, 1459–1467.

Falk A, Feys BJ, Frost LN *et al.* (1999) EDS1, an essential component of R gene-mediated disease resistance in *Arabidopsis* has homology to eukaryotic lipases. *Proceedings of the National Academy of Sciences, USA*, **96**, 3292–3297.

Fletcher W, Yang Z (2010) The effect of insertions, deletions, and alignment errors on the branch-site test of positive selection. *Molecular Biology and Evolution*, **27**, 2257–2267.

Fu L, Niu B, Zhu Z, Wu S, Li W (2012) CD-HIT: accelerated for clustering the next-generation sequencing data. *Bioinformatics*, **28**, 3150–3152.

Fujii S, Bond CS, Small ID (2011) Selection patterns on restorer-like genes reveal a conflict between nuclear and mitochondrial genomes throughout angiosperm evolution. *Proceedings of the National Academy of Sciences, USA*, **108**, 1723–1728.

Fujisaki K, Ishikawa M (2008) Identification of an *Arabidopsis thaliana* protein that binds to tomato mosaic virus genomic RNA and inhibits its multiplication. *Virology*, **380**, 402–411.

Funk VA, Bayer RJ, Keeley S et al. (2005) Everywhere but Antarctica: using a supertree to understand the diversity and distribution of the Compositae. *Biologiske Skrifter*, **55**, 343–373.

Galla G, Zenoni S, Marconi G et al. (2011) Sporophytic and gametophytic functions of the cell cycle-associated Mob1 gene in *Arabidopsis thaliana* L. *Gene*, **484**, 1–12.

Gao Z, Johansen E, Eyers S et al. (2004) The potyvirus recessive resistance gene, *sbm1*, identifies a novel role for translation initiation factor eIF4E in cell-to-cell trafficking. *The Plant Journal*, **40**, 376–385.

Gaudeul M, Giraud T, Kiss L, Shykoff JA (2011) Nuclear and chloroplast microsatellites show multiple introductions in the worldwide invasion history of common ragweed, *Ambrosia artemisiifolia*. *PLoS ONE*, **6**, e17658.

Global Invasive Species Database [GISD] (2014) 100 of the world's worst invasive alien species. Available from http://www.issg.org/database/species/search.asp?st=100ss&fr=1&str=&lang=EN. Accessed 13 September 2014.

Gray A, Mack R (1986) Do invading species have definable genetic characteristics? *Philosophical Transactions of the Royal Society B: Biological Sciences*, **314**, 655–674.

Grossmann S, Bauer S, Robinson PN, Vingron M (2007) Improved detection of overrepresentation of Gene-Ontology annotations with parent child analysis. *Bioinformatics*, **23**, 3024–3031.

Guggisberg A, Welk E, Sforza R et al. (2012) Invasion history of North American Canada thistle, *Cirsium arvense*. *Journal of Biogeography*, **39**, 1919–1931.

Guggisberg A, Lai Z, Huang J, Rieseberg LH (2013) Transcriptome divergence between introduced and native populations of Canada thistle, *Cirsium arvense*. *The New Phytologist*, **199**, 595–608.

Guindon S, Rodrigo AG, Dyer KA, Huelsenbeck JP (2004) Modeling the site-specific variation of selection patterns along lineages. *Proceedings of the National Academy of Sciences, USA*, **101**, 12957–12962.

Haas BJ, Papanicolaou A, Yassour M et al. (2013) De novo transcript sequence reconstruction from RNA-seq using the Trinity platform for reference generation and analysis. *Nature Protocols*, **8**, 1494–1512.

Hassidim M, Harir Y, Yakir E, Kron I, Green RM (2009) Over-expression of CONSTANS-LIKE 5 can induce flowering in short-day grown *Arabidopsis*. *Planta*, **230**, 481–491.

Hodgins KA, Rieseberg L (2011) Genetic differentiation in life-history traits of introduced and native common ragweed (*Ambrosia artemisiifolia*) populations. *Journal of Evolutionary Biology*, **24**, 2731–2749.

Hodgins KA, Lai Z, Nurkowski K, Huang J, Rieseberg LH (2013a) The molecular basis of invasiveness: differences in gene expression of native and introduced common ragweed (*Ambrosia artemisiifolia*) in stressful and benign environments. *Molecular Ecology*, **22**, 2496–2510.

Hodgins KA, Lai Z, Oliveira LO et al. (2013b) Genomics of Compositae crops: reference transcriptome assemblies and evidence of hybridization with wild relatives. *Molecular Ecology Resources*, **14**, 166–177.

Holt JS, Welles SR, Silvera K et al. (2013) Taxonomic and life history bias in herbicide resistant weeds: implications for deployment of resistant crops. *PLoS ONE*, **8**, e71916.

Ishitani M, Xiong L, Stevenson B, Zhu JK (1997) Genetic analysis of osmotic and cold stress signal transduction in *Arabidopsis*: interactions and convergence of abscisic acid-dependent and abscisic acid-independent pathways. *The Plant Cell*, **9**, 1935–1949.

Jordan G, Goldman N (2012) The effects of alignment error and alignment filtering on the sitewise detection of positive selection. *Molecular Biology and Evolution*, **29**, 1125–1139.

Kane NC, Barker MS, Zhan SH, Rieseberg LH (2011) Molecular evolution across the Asteraceae: micro- and macroevolutionary processes. *Molecular Biology and Evolution*, **28**, 3225–3235.

Klopfstein S, Currat M, Excoffier L (2006) The fate of mutations surfing on the wave of a range expansion. *Molecular Biology and Evolution*, **23**, 482–490.

Kuester A, Conner JK, Culley T, Baucom RS (2014) How weeds emerge: a taxonomic and trait-based examination using United States data. *The New Phytologist*, **202**, 1055–1068.

Kwak KJ, Park SJ, Han JH et al. (2011) Structural determinants crucial to the RNA chaperone activity of glycine-rich RNA-binding proteins 4 and 7 in *Arabidopsis thaliana* during the cold adaptation process. *Journal of Experimental Botany*, **62**, 4003–4011.

Lai Z, Kane NC, Zou Y, Rieseberg LH (2008) Natural variation in gene expression between wild and weedy populations of *Helianthus annuus*. *Genetics*, **179**, 1881–1890.

Lai Z, Kane NC, Kozik A et al. (2012) Genomics of Compositae weeds: EST libraries, microarrays, and evidence of introgression. *American Journal of Botany*, **99**, 209–218.

Lehe R, Hallatschek O, Peliti L (2012) The rate of beneficial mutations surfing on the wave of a range expansion. *PLoS Computational Biology*, **8**, e1002447.

Li W, Godzik A (2006) Cd-hit: a fast program for clustering and comparing large sets of protein or nucleotide sequences. *Bioinformatics*, **22**, 1658–1659.

Li L, Stoeckert CJ Jr, Roos DS (2003) OrthoMCL: identification of ortholog groups for eukaryotic genomes. *Genome Research*, **13**, 2178–2189.

Linder CR (2000) Adaptive evolution of seed oils in plants: accounting for the biogeographic distribution of saturated and unsaturated fatty acids in seed oils. *The American Naturalist*, **156**, 442–458.

Lockwood JL, Cassey P, Blackburn TM (2009) The more you introduce the more you get: the role of colonization pressure and propagule pressure in invasion ecology. *Diversity and Distributions*, **15**, 904–910.

Löytynoja A, Goldman N (2008) Phylogeny-aware gap placement prevents errors in sequence alignment and evolutionary analysis. *Science*, **320**, 1632–1635.

Lu A, Guindon S (2014) Performance of standard and stochastic branch-site models for detecting positive selection among coding sequences. *Molecular Biology and Evolution*, **31**, 484–495.

Macknight R, Bancroft I, Page T *et al.* (1997) FCA, a gene controlling flowering time in *Arabidopsis*, encodes a protein containing RNA-binding domains. *Cell*, **89**, 737–745.

Markova-Raina P, Petrov D (2011) High sensitivity to aligner and high rate of false positives in the estimates of positive selection in the 12 *Drosophila* genomes. *Genome Research*, **21**, 863–874.

Marrs RA, Sforza R, Hufbauer RA (2007) When invasion increases population genetic structure: a study with *Centaurea diffusa*. *Biological Invasions*, **10**, 561–572.

Mayrose M, Kane NC, Mayrose I, Dlugosch KM, Rieseberg LH (2011) Increased growth in sunflower correlates with reduced defences and altered gene expression in response to biotic and abiotic stress. *Molecular Ecology*, **20**, 4683–4694.

Mugford SG, Matthewman CA, Hill L, Kopriva S (2010) Adenosine-5'-phosphosulfate kinase is essential for *Arabidopsis* viability. *FEBS Letters*, **584**, 119–123.

Muth NZ, Pigliucci M (2006) Traits of invasives reconsidered: phenotypic comparisons of introduced invasive and introduced noninvasive plant species within two closely related clades. *American Journal of Botany*, **93**, 188–196.

Nawrath C, Heck S, Parinthawong N, Métraux J (2002) EDS5, an essential component of salicylic acid–dependent signaling for disease resistance in *Arabidopsis*, is a member of the MATE transporter family. *The Plant Cell*, **14**, 275–286.

Ndamukong I, Abdallat AA, Thurow C *et al.* (2007) SA-inducible *Arabidopsis* glutaredoxin interacts with TGA factors and suppresses JA-responsive PDF1.2 transcription. *The Plant Journal: For Cell and Molecular Biology*, **50**, 128–139.

Pagnussat GC, Yu H-J, Ngo QA *et al.* (2005) Genetic and molecular identification of genes required for female gametophyte development and function in *Arabidopsis*. *Development*, **132**, 603–614.

Paradis E, Claude J, Strimmer K (2004) APE: analyses of phylogenetics and evolution in R language. *Bioinformatics*, **20**, 289–290.

Peischl S, Dupanloup I, Kirkpatrick M, Excoffier L (2013) On the accumulation of deleterious mutations during range expansions. *Molecular Ecology*, **22**, 5972–5982.

Penn O, Privman E, Landan G, Graur D, Pupko T (2010) An alignment confidence score capturing robustness to guide tree uncertainty. *Molecular Biology and Evolution*, **27**, 1759–1767.

Portereiko MF, Sandaklie-Nikolova L, Lloyd A *et al.* (2006) NUCLEAR FUSION DEFECTIVE1 encodes the *Arabidopsis* RPL21M protein and is required for karyogamy during female gametophyte development and fertilization. *Plant Physiology*, **141**, 957–965.

Powles SB, Yu Q (2010) Evolution in action: plants resistant to herbicides. *Annual Review of Plant Biology*, **61**, 317–347.

Prentis PJ, Wilson JRU, Dormontt EE, Richardson DM, Lowe AJ (2008) Adaptive evolution in invasive species. *Trends in Plant Science*, **13**, 288–294.

Privman E, Penn O, Pupko T (2012) Improving the performance of positive selection inference by filtering unreliable alignment regions. *Molecular biology and Evolution*, **29**, 1–5.

Reeves PH, Coupland G (2001) Analysis of flowering time control in *Arabidopsis* by comparison of double and triple mutants. *Plant Physiology*, **126**, 1085–1091.

Renaut S, Grassa CJ, Moyers BT, Kane NC, Rieseberg LH (2012) The population genomics of sunflowers and genomic determinants of protein evolution revealed by RNAseq. *Biology*, **1**, 575–596.

Ross J, Li Y, Lim E, Bowles DJ (2001) Higher plant glycosyltransferases. *Genome Biology*, **2**, reviews 3004.

Rouhier N, Lemaire SD, Jacquot J-P (2008) The role of glutathione in photosynthetic organisms: emerging functions for glutaredoxins and glutathionylation. *Annual Review of Plant Biology*, **59**, 143–166.

Scaglione D, Lanteri S, Acquadro A *et al.* (2012) Large-scale transcriptome characterization and mass discovery of SNPs in globe artichoke and its related taxa. *Plant Biotechnology Journal*, **10**, 956–969.

Slatkin M, Excoffier L (2012) Serial founder effects during range expansion: a spatial analog of genetic drift. *Genetics*, **191**, 171–181.

Stamatakis A (2006) RAxML-VI-HPC: maximum likelihood-based phylogenetic analyses with thousands of taxa and mixed models. *Bioinformatics*, **22**, 2688–2690.

Stewart CN, Tranel PJ, Horvath DP *et al.* (2009) Evolution of weediness and invasiveness: charting the course for weed genomics. *Weed Science*, **57**, 451–462.

Sun M (1997) Population genetic structure of yellow star-thistle (*Centaurea solstitialis*), a colonizing weed in the western United States. *Canadian Journal of Botany*, **75**, 1470–1478.

Turner KG, Hufbauer RA, Rieseberg LH (2014) Rapid evolution of an invasive weed. *The New Phytologist*, **202**, 309–321.

USDA, NRCS (2014) *The PLANTS Database* (http://plants.usda.gov, 13 September 2014). National Plant Data Team, Greensboro, North Carolina.

Waterhouse RM, Zdobnov EM, Kriventseva EV (2011) Correlating traits of gene retention, sequence divergence, duplicability and essentiality in vertebrates, arthropods, and fungi. *Genome Biology and Evolution*, **3**, 75–86.

Yang Z (1997) PAML: a program package for phylogenetic analysis by maximum likelihood. *Computer Applications in the Biosciences: CABIOS*, **13**, 555–556.

Yang Z (2007) PAML 4: phylogenetic analysis by maximum likelihood. *Molecular Biology and Evolution*, **24**, 1586–1591.

Zhang J, Nielsen R, Yang Z (2005) Evaluation of an improved branch-site likelihood method for detecting positive selection at the molecular level. *Molecular Biology and Evolution*, **22**, 2472–2479.

DATA ACCESSIBILITY

Assembly data are publicly available (see Table S1, Supporting information). Code for several of the analyses performed in this study is available as a GitHub repository (https://github.com/kgturner/InvasionSyndicate).

SUPPORTING INFORMATION

Additional supporting information can be found in the online version of the *Molecular Ecology* article.

Table S1 Assemblies used in this study.

Table S2 Known target site resistance genes for several common herbicides that were examined in this study.

Table S3 Loci with dN/dS > 1 for pairwise comparisons between native and introduced transcriptomes, their best hit in the TAIR database, the identity of homologous genes among species, and gene ontology terms (P < 0.05).

Table S4 Significant loci for branch and branch-site models in paml for comparisons between native and introduced genotypes and a closely related outgroup.

Table S5 paml results for three orthogroups that showed evidence of significant changes in dN/dS ratio (ω) in native vs. introduced genotypes.

Table S6 Genes identified as significant from the Fitmodel and paml site models analyses along with their functional descriptions based on BLASTP to TAIR.

Table S7 Total number of amino acid differences predicted to be deleterious using Provean and the total number of amino acid substitutions that are predicted to be derived deleterious alleles.

Table S8 Total number of alignments tested for each analysis.

Appendix S1 Supplementary methods.

Chapter 17

THE ROLE OF CLIMATE ADAPTATION IN COLONIZATION SUCCESS IN *ARABIDOPSIS THALIANA*

Jill A. Hamilton,[*‡] *Miki Okada,*[*] *Tonia Korves,*[†] *and Johanna Schmitt*[*]

[*]Department of Evolution and Ecology, University of California, Davis, CA 95616, USA
[†]Data Analytics Department, The MITRE Corporation, 202 Burlington Rd., Bedford, MA 01730, USA
[‡]Department of Biological Sciences, North Dakota State University, Fargo, ND 58102, USA

Abstract
Understanding the genetic mechanisms that contribute to range expansion and colonization success within novel environments is important for both invasion biology and predicting species-level responses to changing environments. If populations are adapted to local climates across a species' native range, then climate matching may predict which genotypes will successfully establish in novel environments. We examine evidence for climate adaptation and its role in colonization of novel environments in the model species, *Arabidopsis thaliana*. We review phenotypic and genomic evidence for climate adaptation within the native range and describe new analyses of fitness data from European accessions introduced to Rhode Island, USA, in spring and fall plantings. Accessions from climates similar to the Rhode Island site had higher fitness indicating a potential role for climate pre-adaptation in colonization success. A genomewide association study (GWAS), and genotypic mean correlations of fitness across plantings suggest the genetic basis of fitness in Rhode Island differs between spring and autumn cohorts, and from previous fitness measurements in European field sites. In general, these observations suggest a scenario of conditional neutrality for loci contributing to colonization success, although there was evidence of a fitness trade-off between fall plantings in Norwich, UK, and Rhode Island. GWAS suggested that antagonistic pleiotropy at a few specific loci may contribute to this trade-off, but this conclusion depended upon the accessions included in the analysis. Increased genomic information and phenotypic information make *A. thaliana* a model system to test for the genetic basis of colonization success in novel environments.

Previously published as an article in *Molecular Ecology* (2015) 24, 2253–2263, doi: 10.1111/mec.13099

INTRODUCTION

Adaptation to novel environments during colonization is critical for species establishment, persistence and spread during range expansions or introductions to new geographic regions (Keller & Taylor 2008; Franks & Munshi-South 2014). Identifying those mechanisms that contribute to rapid adaptation in colonizing species is therefore essential for predicting biological invasions as well as species responses to climate change. The initial establishment of colonizing prop-agules depends on the fitness of colonizing genotypes in the new environment (Donohue *et al.* 2010; Colautti & Barrett 2013). Successful establishment in novel environments may be facilitated if past selection in the site-of-origin has pre-adapted colonizing genotypes for conditions in the site of introduction (Maron *et al.* 2004; Aitken & Whitlock 2013; Vandepitte *et al.* 2014) and constrained if colonizing genotypes are maladapted (Keller & Taylor 2008). However, other genetic and ecological factors may also determine colonization success, such as phenotypic plasticity (Donohue 2005; Chevin & Lande 2011) or the amount and correlation structure of genetic variation available for evolutionary response. To understand mechanisms of adaptation during colonization requires a combination of phenotypic data on traits and performance in natural conditions, genomic polymorphism data and geographic information about the sites of origin of different genotypes (Colautti & Barrett 2013). Here, we synthesize previous studies and offer new analyses using the model plant species *Arabidopsis thaliana* to show how such data can be combined to explore genetic mechanisms underlying colonization ability, with particular attention to the climate pre-adaptation hypothesis.

ARABIDOPSIS THALIANA: A MODEL COLONIZING SPECIES

The inbreeding annual plant *Arabidopsis thaliana* inhabits a wide range of climates across its native Eurasian range. In the wild, the species occurs in rocky areas and bare ground and is a frequent colonizer of naturally or anthropogenically disturbed habitats (Mitchell-Olds & Schmitt 2006). *A. thaliana* likely underwent major demographic and range expansions associated with the spread of agriculture to Europe, following postglacial recolonization of Eurasia (Sharbel

et al. 2000; Mitchell-Olds & Schmitt 2006; Beck *et al.* 2008; Francois *et al.* 2008). Adaptation to this novel anthropogenic environment is hypothesized to have involved selective sweeps (Toomajian *et al.* 2006). In recent centuries, *A. thaliana* has spread beyond its native range and become naturalized worldwide (Hoffmann 2002; Beck *et al.* 2008), reflecting its success as a colonizing species. This widely naturalized status makes *A. thaliana* a useful system to study the genetic and ecological factors underlying successful naturalization that is essential in the conceptual framework of the introduction, naturalization and invasion continuum (Blackburn *et al.* 2011; Richardson & Pyšek 2012).

Arabidopsis thaliana is also a genetic model species (Meinke *et al.* 1998). A wide range of genetic resources such as knockout mutants, transgenic overexpression lines and recombinant inbred mapping populations have been developed and used to elucidate the architecture of developmental and physiological pathways underlying ecologically important traits. A reference genome was sequenced in 2000 (Initiative 2000) – the first genome sequence for any plant species. An especially important resource is the large number of inbred lines collected across the species' geographic range, many of which are available in stock centres (Li *et al.* 2014) and have been extensively resequenced (Clark *et al.* 2007; Ossowski *et al.* 2008; Weigel & Mott 2009; Cao *et al.* 2011; Weigel 2012; Long *et al.* 2013) and characterized for genetic polymorphisms (Nordborg *et al.* 2005; Atwell *et al.* 2010; Horton *et al.* 2012). Consequently, much is known about the genetic structure of populations.

Eurasian populations of *A. thaliana* display substantial geographic structure as well as admixture, likely shaped by the history of postglacial recolonization from multiple refugia and the westward migration that accompanied the spread of agriculture in Europe (Sharbel *et al.* 2000; Schmuths *et al.* 2004; Mitchell-Olds & Schmitt 2006; Beck *et al.* 2008; Francois *et al.* 2008; Platt *et al.* 2010; Horton *et al.* 2012). Continuous patterns of longitudinal and latitudinal isolation by distance covarying with local climate have contributed to significant population genetic structure within the native range (Stinchcombe *et al.* 2004; Nordborg *et al.* 2005; Samis *et al.* 2008, 2012; Fournier-Level *et al.* 2011; Horton *et al.* 2012). In contrast, weak population structure among North American accessions (Jorgensen & Mauricio 2004; Platt *et al.* 2010; Samis *et al.* 2012) suggests populations in the non-native range result from multiple sources of colonization, likely via human-mediated dispersal. This weak population structure and substantial within population variation within the North American range together reflect the more recent history of the North American populations (Jorgensen & Mauricio 2004). These patterns leave the specific colonization history of the North American populations ambiguous. Although *A. thaliana* is primarily a selfing species (95–99% in the native range and 92% in North America; Bomblies *et al.* 2010; Jorgensen & Mauricio 2004; Pico *et al.* 2008; Platt *et al.* 2010), estimates based on the distribution of heterozygous markers across the genome suggest the outcrossing rate is likely sufficient to generate novel genotypes through recombination where distinct haplotypes are in close proximity (Platt *et al.* 2010). Considerable variation attributable to recombination within populations is maintained despite selfing rates, with 'hotspots' of such variation observed in certain rural populations in Europe likely due to demographic differences between rural and urban areas (Bomblies *et al.* 2010).

Because *A. thaliana* is largely inbreeding, it is possible to replicate the same genotypes in common garden experiments to measure quantitative genetic variation in phenotypic traits of interest. Such studies are increasingly being conducted in the field, providing important information on the genetic basis of ecologically important traits such as germination timing, flowering time, herbivore susceptibility, or fitness under natural conditions (Mauricio 1998; Brachi *et al.* 2013a; Wilczek *et al.* 2014). If genetic polymorphism data are available for the same genotypes, it is then possible to map loci associated with those traits. Several studies have used quantitative trait locus (QTL) mapping in segregating mapping populations to map genomic regions associated with natural variation in specific traits under field conditions (Weinig *et al.* 2002, 2003a,b; Brachi *et al.* 2010; Huang *et al.* 2010; Ågren *et al.* 2013; Fournier-Level *et al.* 2013; Dittmar *et al.* 2014; Oakley *et al.* 2014). The increasing availability of genomic data for many natural accessions is also making it possible to perform candidate gene association tests (Korves *et al.* 2007; Ehrenreich *et al.* 2009; Scarcelli & Kover 2009; Brock *et al.* 2010, 2012; Lovell *et al.* 2013) and genome-wide association studies (GWAS) to identify allelic polymorphisms associated with intraspecific trait variation (Atwell *et al.* 2010; Brachi *et al.* 2010, 2013a; Fournier-Level *et al.* 2011). GWAS allows finer resolution than QTL mapping

approaches, but is complicated by the need to control for *A. thaliana's* strong geographic population structure, which may be confounded with broad-scale geographic gradients in selection on traits and their underlying loci. This population structure can produce false-positive associations for certain traits (Atwell *et al.* 2010); on the other hand controlling for population structure may produce false-negative associations (Brachi *et al.* 2010, 2011). Finally, an advantage of working with natural accessions is that geographic and climatic information from the site of origin can be combined with genomic and phenotypic data to infer historical patterns of local adaptation, as we illustrate below.

TESTING FOR EVIDENCE OF CLIMATE ADAPTATION IN THE NATIVE RANGE

Geographic clines in phenotype and/or genotype are considered strong circumstantial evidence for historical adaptation to environmental gradients (Slatkin 1973; Endler 1977). Common garden experiments with *A. thaliana* accessions have demonstrated significant trait correlations with latitude (Li *et al.* 1998; Stinchcombe *et al.* 2004; Lempe *et al.* 2005), altitude (Pico 2012) and climate-of-origin (Hannah *et al.* 2006; Christman *et al.* 2008; Montesinos-Navarro *et al.* 2011; Zhen *et al.* 2011; Lewandowska-Sabat *et al.* 2012; Zuther *et al.* 2012; Manzano-Piedras *et al.* 2014). Reciprocal transplant experiments reveal strong evidence for local adaptation to climatically extreme sites in Sweden and Italy (Ågren & Schemske 2012, Ågren *et al.* 2013), although these experiments could not formally test whether climate was the causal selective agent. To explicitly test for adaptation to climate requires the provenance testing approach used by foresters (Rehfeldt *et al.* 1999; Aitken *et al.* 2008) in which accessions originating across a species range are grown in multiple common gardens in different climates across the range. If populations have adapted to their local climate, then we expect accessions originating in climates similar to each test site to perform better than accessions from climates dissimilar to the test conditions, as is often observed for growth or survival in forest trees (Rehfeldt *et al.* 1999; Wang *et al.* 2010; Alberto *et al.* 2013; Sork *et al.* 2013). Wilczek *et al.* (2014) performed such a test with *A. thaliana*, planting 241 accessions collected across the native Eurasian range into four common garden sites spanning the native climate range, from a Mediterranean site at the southern range limit in Spain to a subarctic site at the northern range limit in Finland, and from an oceanic environment in England to a continental climate in eastern Germany. Evidence for broad-scale regional adaptation was observed in every site except for Finland; although Nordic accessions had higher relative fitness in Finland than in any other site, they were outperformed by accessions from Germany. Both latitude and climate-of-origin predicted fitness in all of the plantings, supporting the hypothesis of historical adaptation to climate. However, there was also evidence of regional adaptational lag (Aitken *et al.* 2008); in every region, 'immigrant' genotypes originating from lower latitudes and historically warmer climates had higher relative fitness on average than local genotypes. This result suggests that recent climate warming may create a mismatch between local conditions and the historical adaptive optimum and that migration from historically warmer sites may facilitate in situ adaptation to climate change. Additional studies that have looked into regional adaptation identified factors in addition to climate, including soil and competition to be important at the within-region spatial scale (Lewandowska-Sabat *et al.* 2012; Brachi *et al.* 2013b; Manzano-Piedras *et al.* 2014). Consequently, while climatic factors are clearly important to adaptation in a changing climate, long-range introductions may need to consider additional abiotic and biotic factors that may contribute to success in novel environments.

It is also possible to test for the genomic signature of adaptation to climate in *A. thaliana*, taking advantage of rapidly expanding databases of sequence polymorphism. Climate-SNP associations and enrichment analyses for different polymorphic classes within associated SNPs suggest climate explains more variation than expected by chance (Fournier-Level *et al.* 2011; Hancock *et al.* 2011; Lasky *et al.* 2012). Genomic data for natural accessions planted in a range of natural environments have provided the unique opportunity to associate traits important to adaptation in a range of native climates (recently reviewed in Weinig *et al.* 2014). Nonrandom distribution of alleles conferring fitness advantages within the native range point towards the importance of natural selection in shaping genetic variation within the native range (Fournier-Level *et al.* 2011; Hancock *et al.* 2011; Lasky *et al.* 2012). Climate-correlated SNPs appear to be significantly enriched within coding regions for genes important to a number of biological processes, including

energy metabolism and water availability (Hancock *et al.* 2011; Lasky *et al.* 2012).

The geographic extent of these climate-correlated SNPs is attributed to 'hard selective sweeps' resulting in a narrow distribution of SNPs associated with local climates (Hancock *et al.* 2011). Alleles associated with fitness in European sites exhibit different climate distributions than genomic controls, suggesting that they may be involved in adaptation to climate. In some cases, high-fitness alleles exhibit greater climate specialization for certain climatic factors (such as winter temperature), suggesting that they have been favoured in certain climates. For other climatic factors (such as winter precipitation), low-fitness alleles exhibit greater climate specialization, suggesting that while they may be effectively neutral in their native climate, they may be selected against as they become deleterious beyond their native range (Fournier-Level *et al.* 2011).

Where adaptation to local climates has strongly influenced genetic structure, we expect specific climatic factors to influence the distribution of genetic variation in the native range. Lasky *et al.* (2012) suggest early spring and winter temperatures, in addition to geography, explain the majority of SNP variation. This is supported by genomic associations with fitness estimates, as alleles associated with increased fitness appear to be particularly limited by temperature, including temperature within the coolest quarter of the year (Fournier-Level *et al.* 2011). Alleles associated with high survival and fecundity in each of the four European sites were more locally abundant than genomic controls, suggesting a role in local adaptation (Fournier-Level *et al.* 2011). Across-site genetic correlations were weak, suggesting that the genomic basis of local adaptation within the native range is environment-specific corresponding to a pattern of conditional neutrality (Anderson *et al.* 2011, 2013; Fournier-Level *et al.* 2011; Colautti *et al.* 2012; Weinig *et al.* 2014). Teasing apart the influence of climate and geography in the native range is particularly important, as population structure may contribute to spurious associations separate from those associations resulting from natural selection, making predictions into novel environments challenging (Brachi *et al.* 2011; Hancock *et al.* 2011; Korte & Farlow 2013). Continually evolving statistical approaches to genome-wide associations, in combination with complementary tools such as redundancy analysis (RDA), multivariate methods, species distribution modelling, among others (Engelhardt & Stephens 2010;

Ovaskainen *et al.* 2011; Banta *et al.* 2012; Lasky *et al.* 2012; Berg & Coop 2014; Zhou & Stephens 2014) offer a range of tools to incorporate with current genomic resources to expand our understanding of the genomic basis for adaptation in native and novel environments.

TESTING FOR CLIMATE PRE-ADAPTATION

Climate matching is often used to predict whether species will become invasive in the introduced range. Species originating in similar climates to the region of introduction are expected to be pre-adapted to their new environment and therefore more likely to establish and spread (Nunez & Medley 2011; Petitpierre *et al.* 2012; Dainese *et al.* 2014). However, if local climate adaptation has occurred across the native range, then genotypes originating in different parts of the native range may exhibit different degrees of pre-adaptation to the climate in the site of introduction, and therefore differential colonization success (Bossdorf *et al.* 2008; Zenni *et al.* 2014). Multiple introductions are common in invasive species (Sakai *et al.* 2001; Bossdorf *et al.* 2005; Novak & Mack 2005; Dlugosh & Parker 2008; Wilson *et al.* 2009; Estoup & Guillemaud 2010), suggesting that introductions may often include a number of genetic variants. If so, the climate in the site of introduction may selectively favour genotypes pre-adapted to that climate (Bossdorf *et al.* 2008; but see Zenni *et al.* 2014).

Provenance test experiments at sites in the introduced range make it possible to assess the factors contributing to within-species variation in colonization ability (Donohue *et al.* 2005; Samis *et al.* 2012; Gundale *et al.* 2014; Zenni *et al.* 2014). Using this approach, Rutter & Fenster (2007) grew a spring annual cohort of 21 Eurasian *A. thaliana* accessions in a field common garden in Maryland. Latitude and several spring climatic variables in the site of origin predicted fruit production in the common garden, evidence that climate matching can contribute to genotypic performance in introduced populations.

NEW ANALYSES: IS THERE EVIDENCE OF CLIMATE PRE-ADAPTATION IN *A. THALIANA*?

To test for evidence of climate pre-adaptation further, we performed new analyses of data from Korves *et al.* (2007), who performed a larger common garden experiment with Eurasian accessions planted in autumn and spring germination cohorts in a field site in Bristol, Rhode Island, USA (Korves *et al.* 2007). A subset of 169 accessions with genotypic information was used to test for associations of allelic variation within the candidate gene FRIGIDA (FRI) for flowering time and fitness (Korves *et al.* 2007). We used fitness phenotypes from the same accessions in this common garden to test whether climate in the site of origin predicts genotypic performance in Rhode Island. Latitude and longitude for the site of origin were used to obtain climate data for the period 1961–1990 from the International Panel for Climate Change database. Monthly mean values for daily average temperature were calculated from 10-arcminute grids from www.cru.uea.ac.uk (New *et al.* 2002) and imported theses data into ARCVIEW GIS software (ESRI, Redlands, CA).

If climate matching contributed to fitness in the site of introduction, we would expect to observe a significant negative quadratic regression coefficient of fitness in Rhode Island on climate-of-origin, with an 'optimum' climate–of–origin similar to that of the Rhode Island field site. We tested this prediction by fitting polynomial regressions on climate-of-origin of the 169 accessions with fitness phenotypes in fall and spring planting cohorts in Rhode Island using *R 3.0.3* (R Development Core Team 2014). Regressions of fecundity of the spring cohort on mean annual temperature and April temperature of origin provided strong evidence for climate matching (Fig. 1b,d). Accessions originating from climates matching both historical Rhode Island April temperatures and mean April temperature in the year of the experiment (2003) exhibited greater fecundity on average (Fig. 1d). In the autumn cohort, we also observed a significant negative quadratic regression coefficient of survival on mean annual temperature, but the 'optimal' temperature of origin was several degrees cooler than the optimal temperature for the spring cohort and cooler than the Rhode Island historical average (Fig. 1a, *F*-statistic = 9.00, $P < 0.001$). Thus, in the autumn cohort, genotypes from historically cooler climates had a fitness advantage. However, the mean January temperature in Rhode Island during the year of the experiment was cooler than historical thirty-year averages and closely matched the optimal January temperature of origin (Fig. 1c, *F*-statistic = 9.35, $P < 0.001$). This result suggests that climate pre-adaptation to winter temperatures did contribute to over-winter survival in the autumn cohort in Rhode Island and that overwinter

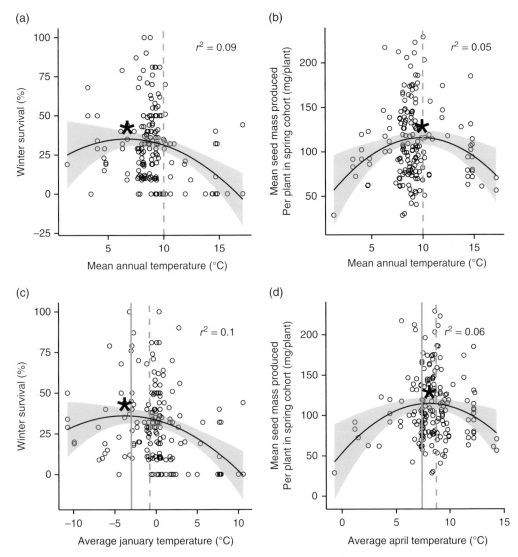

Fig. 1 Quadratic regression of per cent winter survival (a, c) and seed mass per plant (b, d) from the fall and spring planting cohort in Rhode Island versus individual historical mean annual temperature (°C) from the climate-of-origin (a, b) or mean annual January (c) or April (d) temperature (°C) from the climate-of-origin. The vertical dashed line indicates the historical mean annual or monthly temperature at the field site in Rhode Island based on thirty-year averages, and the solid line indicates the average monthly temperature at the field site during the experiment in 2003. The asterisk signifies the inflection point of the quadratic model, indicating the optimum climate-of-origin.

mortality may select against accessions from warmer climates. The difference in optimal temperature of origin between seasonal cohorts suggests that climate matching may favour establishment of accessions from different climates in different seasonal cohorts.

Moreover, interannual temperature variation may favour accessions from different climates in different years. For example, Wilczek *et al.* (2014) observed higher relative fitness among genotypes from warmer locations relative to more local genotypes in a year when average temperatures were warmer than historical averages. While populations on average may be preadapted to novel climates within their native range, climatic variation within the site of introduction may influence the establishment success of particular accessions in a given year or season.

Although climate matching contributed to colonization success in Rhode Island, it did not explain all of the observed variation in fitness. The genotypes with highest fitness in the introduced site originated in climates with temperatures similar to the Rhode Island site, but other genotypes from similar matched climates had much lower fitness (Fig. 1). Similarly, Manzano-Piedras *et al.* (2014) observed effects of climate-of-origin, but also considerable variance in fitness among Iberian *A. thaliana* genotypes originating in similar climates grown in a common garden experiment in Madrid, Spain. Several factors may contribute to this fitness variance. Pre-adaptation to the experimental site could have also involved other environmental factors in the site of origin, such as precipitation, soil, disturbance, herbivores, or pathogens, which may have varied among sites with similar climates. Moreover, nonadaptive processes such as fixation of deleterious alleles by genetic drift in small populations may have constrained the potential for climate pre-adaptation and reduced the fitness of certain source populations. Thus, climate matching may be necessary but not sufficient for colonization success. Better environmental information from the sites of collection (e.g. Brachi *et al.* 2013b) could help to dissect causes of genetic variation in colonization success.

NEW ANALYSES: GENETIC BASIS OF PERFORMANCE IN COLONIZING POPULATIONS

To investigate the genetic basis of fitness in novel environments, we took advantage of fitness estimates of genotypes planted in multiple common garden experiments in the native range described in Fournier-Level *et al.* (2011) with fitness of the same genotypes planted in novel seasonal environments of Rhode Island. In general, genotypic means were not significantly correlated across sites and seasons for both survival and fecundity across seasonal cohorts (Table 1). Fournier-Level *et al.* (2011) observed that accessions differed significantly in fitness within native sites and that local adaptation likely has an environment-specific genetic

Table 1 Spearman rank correlations with genotypic mean survival in the fall (upper) and mean fecundity (lower) in the spring in Rhode Island with sites in the native range of *Arabidopsis*. The correlation between Rhode Island fall survival and Rhode Island spring fecundity was positive. All pairwise comparisons for genotypic means are based on $n = 104$ genotypes

Site	Norwich Fall	Norwich Spring	Halle	Valencia
Genotypic mean correlations for Survival				
Rhode Island Fall	**−0.286**	−0.144	−0.134	0.129
Genotypic mean correlations for Fecundity				
Rhode Island Spring	−0.07	−0.012	0.06	0.02

Bold indicates a significant correlation ($P < 0.05$).

basis across much of the native range of *Arabidopsis thaliana*. Thus, the genetic basis of colonization success in Rhode Island may be different from that within the native range. These observations support the hypothesis of conditional neutrality (Anderson *et al.* 2011, 2013), that is, that alleles that contribute to fitness within select environments may be effectively neutral in different environments. However, an apparent trade-off in fitness between fall plantings in Rhode Island and in Norwich was also observed (Table 1). Genotypic mean estimates of survival were significantly negatively correlated between these two experimental plantings (Table 1). This may suggest that in some cases, genetic trade-offs contribute to colonization success within novel environments, indicating a possible role for antagonistic pleiotropy where there is a common genetic basis (Anderson *et al.* 2011, 2013). Taking advantage of genomewide data available for *Arabidopsis* accessions planted in both environments along with phenotypic data allows us to examine whether pleiotropic loci contribute to fitness in the native and novel environment.

NEW ANALYSES: GENOMEWIDE ASSOCIATIONS IN THE NATIVE AND NOVEL ENVIRONMENTS

Fournier-Level *et al.* (2011) used a genomewide association study (GWAS) to identify individual loci that contributed to survival in the fall planting at Norwich, UK, based on 157 accessions. We used the same methods and 213 248 SNP markers as Fournier-Level *et al.* (2011) to perform a GWAS of survival in the Rhode Island fall cohort for the 104 accessions in that experiment for which genotypic data were available for the same SNP markers. This allowed us to identify loci associated with colonization success in the novel Rhode Island environment and ask whether SNPs strongly associated with fitness in the introduced site overlapped with those associated with fitness in Norwich (Fournier-Level *et al.* 2011). The methods herein replicate those used by Fournier-Level *et al.* (2011); including the use of a mixed model approach implemented in the EMMAX software, a kinship matrix to eliminate confounding due to genetic relatedness, and a minor allele frequency cut-off greater than 0.1 for associated loci.

Comparing the top 100 and 200 survival-associated SNPs in fall plantings from Norwich as identified by Fournier-Level *et al.* (2011) with those associated with survival in Rhode Island, we identified three and four common loci with fitness associations both in the native and novel range, respectively. Individual allelic effects of these loci were in opposing directions, suggesting possible antagonistic pleiotropy (Ågren *et al.* 2013). Loci associated with increased survival within the novel North American environment were associated with reduced fitness in the native range, supporting hypotheses associated with genetic trade-offs. To validate these results, we performed simulations to test the probability of identifying the same top 100 or 200 SNPs in association with fitness in both the native and novel environment. Accessions observed in both Rhode Island and Norwich plantings were shuffled 1000 times while maintaining pairs of fitness values per genotype to produce nonheritable, but correlated traits with the same distribution. As the list of accessions originally used from Fournier-Level was based on 157 accessions as opposed to 104 accessions with genotypic data in Rhode Island, we shuffled the 53 accessions that did not overlap between the two sites separately. The null distribution based on these simulations suggests that identifying common fitness-associated SNPs is not expected by random chance and observing three and four overlapping fitness-associated SNPs is highly significant (Fig. S1a,b, Supporting information). The SNPs, found on chromosomes one (1_23412991), three (3_16596178), four (4_8712143), and five (5_25258712), respectively, span the *Arabidopsis* genome, suggesting that the genetic effect of the individual alleles on fitness is likely independent, although the influence of linkage cannot be ruled out. The genetic variance (R^2) explained by an individual SNP range from 0.13 for each of these SNPs in Norwich and 0.12 to 0.15 in Rhode Island. These results suggest that in some cases, genetic trade-offs may contribute to fitness in novel environments, although as this pattern was only observed in the fall planting cohort, it may suggest colonization success has a season-specific basis. Previous observations have identified winter environmental factors as strongly contributing to fitness (Fournier-Level *et al.* 2011; Lasky *et al.* 2012) and are supported here by our analysis of climate pre-adaptation for winter survival in Rhode Island (Fig. 1). Most deaths in Rhode Island were caused by cold damage, whereas in Norwich, there was high mortality due to pathogen infection. If there is a pleiotropic trade-off between cold tolerance and defence (Rausher 2001; Des Marais & Juenger 2010), then release from pathogens in the novel environment could facilitate colonization by disease-susceptible but cold-tolerant genotypes.

While these results provide some support for the importance of genetic trade-offs during colonization of novel environments, one of the concerns is the possible influence of an unbalanced set of accessions, particularly if GWAS is sensitive to genomic background and choice of genotypes included in the experiment (Huang *et al.* 2012; Wray 2013). We compared the SNPs associated with fitness based on 157 accessions observed in Norwich fall planting cohorts as identified by Fournier-Level *et al.* (2011) with those based on 104 accessions with available genotypic data in the fall planting cohort in Rhode Island (Fig. S2, Supporting information). The lack of phenotypic data in Rhode Island available for those additional genotypes may influence the GWAS results. To test this possibility, we repeated the GWA using only those accessions observed in both environments and generated a null distribution for SNPs in common based on 1000 GWAS simulations, using methods previously described. The null distribution based on these simulations suggests that unlike the unbalanced GWA design, our observations did not deviate from the null distribution when using a balanced set of accessions (Fig. S1c,d, Supporting information). Within the top 100-associated SNPs, no common fitness-associated SNPs were observed, and within the top 200-associated SNPs, there was only one, and simulations indicate the probability of observing at least one SNP in common does not deviate significantly from the null distribution, contrasting with observations from the unbalanced GWA design. These results suggest there is little evidence for a common genetic basis for fitness between the two environments and points towards the sensitivity of the GWAS to genomic background and accessions chosen for the association analysis. Interestingly, the one locus observed in common within the top 200 fitness-associated SNPs using the balanced design was also observed in the unbalanced design. Thus, we cannot discount the fact that the GWA may have had reduced power to detect a common genetic basis observed for the reduced set of accessions. GWAS is sensitive to genomic background and sample choice, pointing towards the importance of careful consideration of the genotypes used to identify the genetic basis for fitness across environments. Another, important caveat is that the loci associated with fitness may vary from year to year Ågren & Schemske (2012); thus, the SNPs associated with colonization success in our study may be a subset of those contributing to successful colonization over time. Ideally, such studies should be replicated in multiple years.

NEXT STEPS: GENOMIC ANALYSIS OF CLIMATE SUITABILITY

Genomewide association studies can identify the top SNPs associated with colonization success, but additional analyses are required to test whether those SNPs are involved in climate adaptation or pre-adaptation. For example, Fournier-Level et al. (2011) used MaxEnt climate models to test whether alleles associated with fitness in the native range occupied specific climate spaces. Similarly, to test for climate pre-adaptation, we can model the climate niche in the native range of SNP alleles associated with colonization success in Rhode Island and ask whether the native climate distribution predicts high climate suitability in the introduced site relative to genomic controls. A converse approach is to identify loci strongly associated with climate and ask whether they predict fitness in a particular site. Hancock et al. (2011) used this approach to predict the fitness of accessions grown in a common garden in Lille, France from the number of 'favourable' climate-associated alleles harboured by each accession. We can use a similar analysis to ask whether strongly climate-associated alleles predict colonization success in the introduced range. Such analyses are currently underway.

LESSONS FROM *A. THALIANA* FOR UNDERSTANDING THE GENETICS OF COLONIZING SPECIES

The model species *A. thaliana* demonstrates how genomic tools and resources can be applied to test hypotheses about the genetic basis of colonization success in introduced populations. The Arabidopsis community can take advantage of a reference genome, extensive genomewide polymorphism data, stock centres providing worldwide collections from diverse climates, and published phenotypic data from multiple common garden experiments across the native and introduced range. These community resources and publicly available data have facilitated integrative studies of the genetic basis of adaptation and increased our ability to predict phenotypes across different environments. We have also learned that genetic resources become much more valuable for studies of adaptation when good ecological and environmental data are available from the sites where accessions were collected (e.g. Brachi et al. 2013b). Recent studies also suggest that incorporating different spatial scales into GWAS may be fruitful (Brachi et al. 2013b) and that understanding geographic variation in adaptation to biotic factors is an important future direction (Zust et al. 2012).

As genomic information becomes more accessible for many other species, it will be increasingly possible to apply similar integrated approaches to understanding the genetic basis of adaptation and colonization of novel environments in other species of interest. Genomic data are most valuable when combined with geographic information and phenotypic data from field experiments, and phenotypic data may soon be the limiting factor for understanding the genetics of adaptation. Recent successes with Arabidopsis show that rapid progress requires the following components: (i) wide geographic collections of accessions spanning the range of environments across the native and introduced ranges of species of interest, with detailed environmental data (climate, soil, biotic factors) for each accession; (ii) characterization of genomic variation for these accessions; and (iii) data on phenotypic variation and fitness for the same accessions, measured in common gardens across environments within both the native and introduced range. Investment in collaborative efforts to develop such resources will produce large rewards in terms of rapid scientific progress. Predicting pre-adaptation to novel environments will help us to understand the evolutionary dynamics of introduced species and may help to inform assisted migration strategies in the face of rapid climate change.

ACKNOWLEDGEMENTS

We are grateful to the organizers of the Symposium on Genetics of Colonizing Species for the opportunity to develop our ideas on this topic. We thank A.M. Wilczek, M.D. Cooper and J. Roe for overseeing the European field experiments discussed here, and A. Fournier-Level, A. Korte and M. Nordborg for GWAS expertise, E. de Moor, J. Plaut, N. Reese and B. Singh assisted with the Rhode Island field experiment. We thank Daniel Runcie, Joe Hereford, Aurore Bontemps, Alejandra Martinez Berdeja, Mark Taylor and Chenoa Wilcox for helpful discussions. T. Korves' affiliation with The MITRE Corporation is provided for identification purposes only and is not intended to convey or imply MITRE's concurrence with, or support for, the positions, opinions or viewpoints expressed by the author. Our work was supported by NSF grants DEB-9976997, EF-045759, IOS-0935589 and DEB-1020111.

REFERENCES

Ågren J, Schemske DW (2012) Reciprocal transplants demonstrate strong adaptive differentiation of the model organism *Arabidopsis thaliana* in its native range. *New Phytologist*, **194**, 1112–1122.

Ågren J, Oakley CG, McKay JK, Lovell JT, Schemske DW (2013) Genetic mapping of adaptation reveals fitness tradeoffs in *Arabidopsis thaliana*. *Proceedings of the National Academy of Sciences, USA*, **110**, 21077–21082.

Aitken SN, Whitlock MC (2013) Assisted gene flow to facilitate local adaptation to climate change. *Annual Review of Ecology, Evolution and Systematics*, **44**, 367–388.

Aitken SN, Yeaman S, Holliday JA, Wang T, Curtis-McLane S (2008) Adaptation, migration or extirpation: climate change outcomes for tree populations. *Evolutionary Applications*, **1**, 95–111.

Alberto F, Aitken S, Alia R *et al.* (2013) Potential for evolutionary responses to climate change – evidence from tree populations. *Global Change Biology*, **19**, 1645–1661.

Anderson JT, Willis JH, Mitchell-Olds T (2011) Evolutionary genetics of plant adaptation. *Trends in Genetics*, **27**, 258–266.

Anderson JT, Lee C, Rushworth CA, Colautti RI, Mitchell-Olds T (2013) Genetic trade-offs and conditional neutrality contribute to local adaptation. *Molecular Ecology*, **22**, 699–708.

Atwell S, Huang Y, Vilhjalmsson B *et al.* (2010) Genome-wide association study of 107 phenotypes in *Arabidopsis thaliana* inbred lines. *Nature*, **465**, 627.

Banta JA, Ehrenreich IM, Gerard S *et al.* (2012) Climate envelope modelling reveals intraspecific relationships among flowering phenology, niche breadth and potential range size in *Arabidopsis thaliana*. *Ecology Letters*, **15**, 769–777.

Beck JB, Schmuths H, Schaal BA (2008) Native range genetic variation in *Arabidopsis thaliana* is strongly geographically structured and reflects Pleistocene glacial dynamics. *Molecular Ecology*, **17**, 902–915.

Berg JJ, Coop G (2014) The population genetic signature of polygenic local adaptation. *PLoS Genetics*, **10**, e1004412.

Blackburn TM, Pyšek P, Bacher S *et al.* (2011) A proposed unified framework for biological invasions. *Trends in Ecology & Evolution*, **26**, 333–339.

Bomblies K, Yant L, Laitinen R *et al.* (2010) Local-scale patterns of genetic variability, outcrossing and spatial structure in natural stands of *Arabidopsis thaliana*. *PloS Genetics*, **6**, e1000890.

Bossdorf O, Auge H, Lafuma L *et al.* (2005) Phenotypic and genetic differentiation between native and introduced plant populations. *Oecologia*, **144**, 1–11.

Bossdorf O, Lipowsky A, Prati D (2008) Selection of prea-dapted populations allowed *Senecio inaequidens* to invade Central Europe. *Diversity and Distributions*, **14**, 676–685.

Brachi B, Faure N, Horton MW *et al.* (2010) Linkage and association mapping of *Arabidopsis thaliana* flowering time in nature. *PloS Genetics*, **6**, e1000940.

Brachi B, Morris GP, Borevitz JO (2011) Genome-wide association studies in plants: the missing heritability is in the field. *Genome Biology*, **12**, 232.

Brachi B, Faure N, Bergelson J, Cuguen J, Roux F (2013a) Genome-wide association mapping of flowering time in *Arabidopsis thaliana* in nature: genetics for underlying components and reaction norms across two successive years. *Acta Botanica Gallica*, **160**, 205–219.

Brachi B, Villoutreix R, Faure N *et al.* (2013b) Investigation of the geographical scale of adaptive phenological variation and its underlying genetics in *Arabidopsis thaliana*. *Molecular Ecology*, **22**, 4222–4240.

Brock M, Maloof JN, Weinig C (2010) Genes underlying quantitative variation in ecologically important traits: PIF4 (PHYTOCHROME INTERACTING FACTOR 4) is associated with variation in internode length, flowering time, and fruit set in *Arabidopsis thaliana*. *Molecular Ecology*, **19**, 1187–1199.

Brock MT, Kover PX, Weinig C (2012) Natural variation in GA1 associates with floral morphology in *Arabidopsis thaliana*. *New Phytologist*, **195**, 58–70.

Cao J, Schneeberger K, Ossowski S *et al.* (2011) Whole-genome sequencing of multiple *Arabidopsis thaliana* populations. *Nature Genetics*, **43**, 956–963.

Chevin L, Lande R (2011) Adaptation to marginal habitats by evolution of increased phenotypic plasticity. *Journal of Evolutionary Biology*, **24**, 1462–1478.

Christman MA, Richards JH, McKay JK *et al.* (2008) Genetic variation in *Arabidopsis thaliana* for night-time leaf conductance. *Plant, Cell and Environment*, **31**, 1170–1178.

Clark RM, Schweikert G, Toomajian C *et al.* (2007) Common sequence polymorphisms shaping genetic diversity in *Arabidopsis thaliana*. *Science*, **317**, 338–342.

Colautti RI, Barrett SCH (2013) Rapid adaptation to climate facilitates range expansion of an invasive plant. *Science*, **342**, 364.

Colautti RI, Lee C, Mitchell-Olds T (2012) Origin, fate and architecture of ecologically relevant genetic variation. *Current Opinion in Plant Biology*, **15**, 199–204.

Dainese M, Kuhn I, Bragazza L (2014) Alien plant species distribution in the European Alps: influence of species' climatic requirements. *Biological Invasions*, **16**, 815–831.

Des Marais D, Juenger TE (2010) Pleiotropy, plasticity, and the volution of plant abiotic stress tolerance. *Annals of the New York Academy of Sciences*, **1206**, 56–79.

Dittmar EL, Oakley CG, Ågren J, Schemske DW (2014) Flowering time QTL in natural populations of *Arabidopsis thaliana* and implications for their adaptive value. *Molecular Ecology*, **23**, 4291–4303.

Dlugosh KM, Parker IM (2008) Founding events in species invasions: genetic variation, adaptive evolution, and the role of multiple introductions. *Molecular Ecology*, **17**, 431–449.

Donohue K (2005) Niche construction through phenological plasticity: life history dynamics and ecological consequences. *New Phytologist*, **166**, 83–92.

Donohue K, Dorn LA, Griffith C *et al.* (2005) The evolutionary ecology of seed germination of *Arabidopsis thaliana*: variation natural selection on germination timing. *Evolution*, **59**, 758–770.

Donohue K, Rubio de Casa R, Burghardt LT, Kovach K, Willis C (2010) Germination, postgermination adaptation and species ecological ranges. *Annual Review of Ecology, Evolution and Systematics*, **41**, 293–319.

Ehrenreich IM, Hanzawa Y, Chou L *et al.* (2009) Candidate gene association mapping of Arabidopsis flowering time. *Genetics*, **183**, 325–335.

Endler JA (1977) *Geographic Variation, Speciation, and Clines. Monographs in Population Biology*, Princeton University Press, Princeton, New Jersey.

Engelhardt BE, Stephens M (2010) Analysis of population structure: a unifying framework and novel methods based on sparse factor analysis. *PloS Genetics*, **6**, e1001117.

Estoup A, Guillemaud T (2010) Reconstructing routes of invasion using genetic data: why, how and so what? *Molecular Ecology*, **19**, 4113–4130.

Fournier-Level A, Korte A, Cooper MD *et al.* (2011) A map of local adaptation in *Arabidopsis thaliana*. *Science*, **334**, 86–89.

Fournier-Level A, Wilczek AM, Cooper MD *et al.* (2013) Paths to selection on life history loci in different natural environments across the native range of *Arabidopsis thaliana*. *Molecular Ecology*, **22**, 3552–3566.

Francois O, Blum MGB, Jakobsson M, Rosenberg NA (2008) Demographic history of European populations of *Arabidopsis thaliana*. *PloS Genetics*, **4**, e1000075.

Franks SJ, Munshi-South J (2014) Go forth, evolve and prosper: the genetic basis of adaptive evolution in invasive species. *Molecular Ecology*, **23**, 2137–2140.

Gundale MJ, Pauchard A, Langdon B *et al.* (2014) Can model species be used to advance the field of invasion ecology? *Biological Invasions*, **16**, 591–607.

Hancock AM, Brachi B, Faure N *et al.* (2011) Adaptation to climate across the *Arabidopsis thaliana* genome. *Science*, **334**, 83–86.

Hannah MA, Wiese D, Freund S *et al.* (2006) Natural genetic variation of freezing tolerance in Arabidopsis. *Plant Physiology*, **142**, 98–112.

Hoffmann MH (2002) Biogeography of *Arabidopsis thaliana* (L.) Heynh. (Brassicaceae). *Journal of Biogeography*, **29**, 125?134.

Horton MW, Hancock AM, Huang Y *et al.* (2012) Genome-wide patterns of genetic variation in worldwide *Arabidopsis thaliana* accessions from the RegMap panel. *Nature Genetics*, **44**, 212–217.

Huang X, Schmitt J, Dorn LA *et al.* (2010) The earliest stages of adaptation in an experimental plant population: strong selection on QTLs for seed dormancy. *Molecular Ecology*, **19**, 1335–1351.

Huang W, Richards S, Carbone M *et al.* (2012) Epistasis dominates the genetic architecture of Drosophila quantitative traits. *Proceedings of the National Academy of Sciences, USA*, **109**, 15553–15559.

Initiative TAG (2000) Analysis of the genome sequence of the flowering plant *Arabidopsis thaliana*. *Nature*, **408**, 796–815.

Jorgensen S, Mauricio R (2004) Neutral genetic variation among wild North American populations of the weedy plant *Arabidopsis thaliana* is not geographically structured. *Molecular Ecology*, **13**, 3403–3413.

Keller SR, Taylor DR (2008) History, chance and adaptation during biological invasion: separating stochastic phenotypic evolution from response to selection. *Ecology Letters*, **11**, 852–866.

Korte A, Farlow A (2013) The advantages and limitations of trait analysis with GWAS: a review. *Plant Methods*, **9**, 29.

Korves TM, Schmid KJ, Caicedo AL *et al.* (2007) Fitness effects associated with the major flowering time gene FRIGIDA in *Arabidopsis thaliana* in the field. *The American Naturalist*, **169**, E141–E157.

Lasky J, Des Marais D, McKay JK *et al.* (2012) Characterizing the genomic variation of *Arabidopsis thaliana*: the roles of geography and climate. *Molecular Ecology*, **21**, 5512–5529.

Lempe J, Balasubramanian S, Sureshkumar S *et al.* (2005) Diversity of flowering responses in wild *Arabidopsis thaliana* strains. *PloS Genetics*, **1**, e6.

Lewandowska-Sabat AM, Fjellheim S, Rognli OA (2012) The continental-oceanic climatic gradient impose clinal variation in vernalization response in *Arabidopsis thaliana*. *Environmental and Experimental Botany*, **78**, 109–116.

Li B, Suzuki J, Hara T (1998) Latitudinal variation in plant size and relative growth rate in *Arabidopsis thaliana*. *Oecologia*, **115**, 293–301.

Li D, Dreher K, Knee E *et al.* (2014) *Arabidopsis Database and Stock Resources.* Springer, New York.

Long Q, Rabanal FW, Meng D *et al.* (2013) Massive genomic variation and strong selection in *Arabidopsis thaliana* lines from Sweden. *Nature Genetics*, **45**, 884–891.

Lovell JT, Juenger TE, Michaels S *et al.* (2013) Pleiotropy of FRIGIDA enhances the potential for multivariate adaptation. *Proceedings of the Royal Society B: Biological Sciences*, **1763**, 1–8.

Manzano-Piedras E, Marcer A, Alonso-Blanco C, Pico FX (2014) Deciphering the adjustment between environment and life history in annuals: lessons from a geographically-explicit approach in *Arabidopsis thaliana*. *PLoS ONE*, **9**, e87836.

Maron JL, Vila M, Bommarco R, Elmendort S, Beardsley P (2004) Rapid evolution of an invasive plant. *Ecological Monographs*, **74**, 261–280.

Mauricio R (1998) Costs of resistance to natural enemies in field populations of the annual plant *Arabidopsis thaliana*. *The American Naturalist*, **151**, 20–28.

Meinke DW, Cherry JM, Dean C, Rounsley SD, Koornneef M (1998) *Arabidopsis thaliana*: a model plant for genome analysis. *Science*, **282**, 662–682.

Mitchell-Olds T, Schmitt J (2006) Genetic mechanisms and evolutionary significance of natural variation in Arabidopsis. *Nature*, **441**, 947–952.

Montesinos-Navarro A, Wig J, Pico FX, Tonsor SJ (2011) *Arabidopsis thaliana* populations show clinal variation in a climatic gradient associated with altitude. *New Phytologist*, **189**, 282–294.

New M, Lister D, Hulme M, Makin I (2002) A high-resolution data set of surface climate over global land areas. *Climate Research*, **21**, 1–25.

Nordborg M, Hu T, Ishino Y *et al.* (2005) The pattern of polymorphism in *Arabidopsis thaliana*. *PloS Biology*, **3**, 3196.

Novak SJ, Mack RN (2005) *Genetic Bottlenecks in Alien Plant Species: Influence of Mating Systems and Introduction Dynamics.* Sinauer Associates, Sunderland, Massachusetts.

Nunez M, Medley KA (2011) Pine invasions: climate predicts invasion success; something else predicts failure. *Diversity and Distributions*, **17**, 703–713.

Oakley CG, Ågren J, Atchison RA, Schemske DW (2014) QTL mapping of freezing tolerance: links to fitness and adaptive trade-offs. *Molecular Ecology*, **23**, 4304–4315.

Ossowski S, Schneeberger K, Clark RM *et al.* (2008) Sequencing of natural strains of *Arabidopsis thaliana* with short reads. *Genome Research*, **18**, 2024–2033.

Ovaskainen O, Karhunen M, Zheng C, Arias J, MerilÄ J (2011) A new method to uncover signatures of divergent and stabilizing selection in quantitative traits. *Genetics*, **189**, 621–632.

Petitpierre B, Kueffer C, Broennimann O *et al.* (2012) Climatic niche shifts are rare among terrestrial plant invaders. *Science*, **335**, 1344–1348.

Pico FX (2012) Demographic fate of *Arabidopsis thaliana* cohorts of autumn- and spring-germinated plants along an altitudinal gradient. *Journal of Ecology*, **100**, 1009–1018.

Pico FX, Mendez-Vigo B, Martinez-Zapater J, Alonso-Blanco C (2008) Natural genetic variation of *Arabidopsis thaliana* is geographically structured in the Iberian Peninsula. *Genetics*, **180**, 1009–1021.

Platt A, Horton MW, Huang Y *et al.* (2010) The scale of population structure in *Arabidopsis thaliana*. *PloS Genetics*, **6**, e1000843.

R Development Core Team (2014) *R: A Language and Environment for Statistical Computer.* Computing RFfS, Vienna, Austria.

Rausher M (2001) Co-evolution and plant resistance to natural enemies. *Nature*, **411**, 857–864.

Rehfeldt GE, Ying CC, Spittlehouse DL, Hamilton DA (1999) Genetic responses to climate in *Pinus contorta*: niche breadth, climate change, and reforestation. *Ecological Monographs*, **69**, 375–407.

Richardson DM, Pyšek P (2012) Naturalization of introduced plants: ecological drivers of biogeographical patterns. *New Phytologist*, **196**, 383–396.

Rutter MT, Fenster CB (2007) Testing for adaptation to climate in *Arabidopsis thaliana*: a calibrated common garden approach. *Annals of Botany*, **99**, 529–536.

Sakai AK, Allendorf FW, Holt JS *et al.* (2001) The population biology of invasive species. *Annual Review of Ecology and Systematics*, **32**, 305–332.

Samis KE, Heath K, Stinchcombe JR (2008) Discordant longitudinal clines in flowering time and phytochrome C in *Arabidopsis thaliana*. *Evolution*, **62**, 2971–2983.

Samis KE, Murren CJ, Bossdorf O *et al.* (2012) Longitudinal trends in climate drive flowering time clines in North American *Arabidopsis thaliana*. *Ecology and Evolution*, **2**, 1162–1180.

Scarcelli N, Kover PX (2009) Standing genetic variation in FRIGIDA mediates experimental evolution of flowering time in Arabidopsis. *Molecular Ecology*, **18**, 2039–2049.

Schmuths H, Hoffmann MH, Bachmann K (2004) Geographic distribution and recombination of genomic fragments on the short arm of chromosome 2 of *Arabidopsis thaliana*. *Plant Biology*, **6**, 128–139.

Sharbel TF, Haubold B, Mitchell-Olds T (2000) Genetic isolation by distance in *Arabidopsis thaliana*: biogeography and postglacial colonization of Europe. *Molecular Ecology*, **9**, 2109–2118.

Slatkin M (1973) Gene flow and selection in a cline. *Genetics*, **75**, 733–756.

Sork VL, Aitken S, Dyer R *et al.* (2013) Putting the landscape into the genomics of forest trees: approaches for understanding local adaptation and population responses to a changing climate. *Tree Genetics & Genomes*, **9**, 901–911.

Stinchcombe JR, Weinig C, Ungerer M *et al.* (2004) A latitudinal cline in flowering time in *Arabidopsis thaliana* modulated by the flowering time gene FRIGIDA. *Proceedings of the National Academy of Sciences, USA*, **101**, 4712–4717.

Toomajian C, Hu T, Aranzana M *et al.* (2006) A nonparametric test reveals selection for rapid flowering in the Arabidopsis genome. *PLoS Biology*, **4**, e137.

Vandepitte K, Meyer T, Helson K *et al.* (2014) Rapid genetic adaptation precedes the spread of an exotic plant species. *Molecular Ecology*, **23**, 2157–2164.

Wang T, O'Neill GA, Aitken S (2010) Integrating environmental and genetic effects to predict responses of tree populations to climate. *Ecological Applications*, **20**, 153–163.

Weigel D (2012) Natural variation in *Arabidopsis*: from molecular genetics to ecological genomics. *Plant Physiology*, **158**, 2–22.

Weigel D, Mott R (2009) The 1001 genomes project for *Arabidopsis thaliana*. *Genome Biology*, **10**, 107.

Weinig C, Ungerer M, Dorn LA *et al.* (2002) Novel loci control variation in reproductive timing in *Arabidopsis thaliana* in natural environments. *Genetics*, **162**, 1875–1884.

Weinig C, Dorn LA, Kane NC *et al.* (2003a) Heterogeneous selection at specific loci in natural environments in *Arabidopsis thaliana*. *Genetics*, **165**, 321–329.

Weinig C, Stinchcombe JR, Schmitt J (2003b) QTL architecture of resistance and tolerance traits in *Arabidopsis thaliana* in natural environments. *Molecular Ecology*, **12**, 1153–1163.

Weinig C, Ewers B, Welch SM (2014) Ecological genomics and process modeling of local adaptation to climate. *Current Opinion in Plant Biology*, **18**, 66–72.

Wilczek AM, Cooper MD, Korves TM, Schmitt J (2014) Lagging adaptation to warming climate in *Arabidopsis thaliana*. *Proceedings of the National Academy of Sciences, USA*, **111**, 7906–7913.

Wilson J, Dormontt EE, Prentis PJ, Lowe A, Richardson DM (2009) Something in the way you move: dispersal pathways affect invasion success. *Trends in Ecology & Evolution*, **24**, 136–144.

Wray GA (2013) Genomics and the evolution of phenotypic traits. *Annual Review of Ecology and Systematics*, **44**, 51–72.

Zenni R, Bailey JK, Simberloff D (2014) Rapid evolution and range expansion of an invasive plant are driven by provenance-environment interactions. *Ecology Letters*, **17**, 727–735.

Zhen Y, Dhakal P, Ungerer M (2011) Fitness benefits and costs of cold acclimation in *Arabidopsis thaliana*. *The American Naturalist*, **178**, 44–52.

Zhou X, Stephens M (2014) Efficient multivariate linear mixed model algorithms for genome-wide association studies. *Nature Methods*, **11**, 407–411.

Zust T, Heichinger C, Grossniklaus U *et al.* (2012) Natural enemies drive geographic variation in plant defenses. *Science*, **338**, 116–119.

Zuther E, Schulz E, Childs L, Hincha D (2012) Clinal variation in the non-acclimated and cold-acclimated freezing tolerance of *Arabidopsis thaliana* accessions. *Plant, Cell and Environment*, **35**, 1860–1878.

DATA ACCESSIBILITY

Genotype data for the SNPs used in this study are available at http://regmap.uchicago.edu.

Phenotypic data for common gardens within the native range are available at the Dryad Digital Repository at http://dx.doi.org/10.5061/dryad.37f9t.

Phenotypic data for common gardens within Rhode Island are available as supplemental material in (Korves *et al.* 2007).

SUPPORTING INFORMATION

Additional supporting information can be found in the online version of the *Molecular Ecology* article.

Fig. S1 Counts for commonly observed SNPs associated with survival in fall plantings of Norwich, UK, and Rhode Island, USA, using an unbalanced (A,B) and balanced (C,D) number of accessions based on 1000 genomewide association simulations where fitness correlations were maintained between environments, but accession ids were shuffled.

Fig. S2 Map of 104 accessions (blue) planted in both fall cohorts at Norwich, UK, and Rhode Island, USA, and 53 additional accessions (red) that were additionally planted only at the European common garden test sites.

Chapter 18

A GENETIC PERSPECTIVE ON RAPID EVOLUTION IN CANE TOADS (*RHINELLA MARINA*)

Lee A. Rollins,* Mark F. Richardson,* and Richard Shine†

*Centre for Integrative Ecology, School of Life and Environmental Sciences, Deakin University, Pigdons Road, Geelong, VIC 3217, Australia
† School of Biological Sciences A08, University of Sydney, Sydney, NSW 2006, Australia

Abstract

The process of biological invasion exposes a species to novel pressures, in terms of both the environments it encounters and the evolutionary consequences of range expansion. Several invaders have been shown to exhibit rapid evolutionary changes in response to those pressures, thus providing robust opportunities to clarify the processes at work during rapid phenotypic transitions. The accelerating pace of invasion of cane toads (*Rhinella marina*) in tropical Australia during its 80-year history has been well characterized at the phenotypic level, including common-garden experiments that demonstrate heritability of several dispersal-relevant traits. Individuals from the invasion front (and their progeny) show distinctive changes in morphology, physiology and behaviour that, in combination, result in far more rapid dispersal than is true of conspecifics from long-colonized areas. The extensive body of work on cane toad ecology enables us to place into context studies of the genetic basis of these traits. Our analyses of differential gene expression from toads from both ends of this invasion-history transect reveal substantial upregulation of many genes, notably those involved in metabolism and cellular repair. Clearly, then, the dramatically rapid phenotypic evolution of cane toads in Australia has been accompanied by substantial shifts in gene expression, suggesting that this system is well suited to investigating the genetic underpinnings of invasiveness.

Previously published as an article in *Molecular Ecology* (2015) 24, 2264–2276, doi: 10.1111/mec.13184

Invasion Genetics: The Baker and Stebbins Legacy, First Edition. Edited by Spencer C. H. Barrett, Robert I. Colautti, Katrina M. Dlugosch, and Loren H. Rieseberg.

INTRODUCTION

At the 1965 *Genetics of Colonizing Species* conference celebrated by this issue of *Molecular Ecology*, several authors noted that invasive species offer excellent research opportunities to investigate fundamental questions in evolutionary biology (Baker & Stebbins 1965). Notably, invaders sometimes evolve rapidly in response to the novel pressures they encounter in their new homes, and the suite of evolutionary processes at work on an expanding range edge includes phenomena that do not apply to populations in spatial equilibrium (Phillips *et al.* 2010a). Plant biologists have taken the lead in investigating the genetic basis of rapid adaptation in colonizing taxa, by exploiting the power of common-garden experiments to tease apart the relative roles of environmentally induced vs. genetically canalized (heritable, inflexibly expressed) factors as underpinnings of spatial variation in phenotypic traits (see Colautti *et al.* 2009). More recently, studies on animal taxa such as house sparrows (*Passer domesticus*) and western bluebirds (*Sialia mexicana*) have added to our understanding of traits important to invasion and suggested an important role for behaviour in invader evolution (Duckworth & Badyaev 2007; Liebl & Martin 2012, 2014). Although amphibians have played little role in the scientific literature on the genetic basis of invasiveness, a system that we are studying in tropical Australia—the invasion of the cane toad, *Rhinella marina* (Fig. 1)—is well suited to such an analysis. The ecology of this system has been studied extensively (more than a decade of research yielding more than 100 articles) and these data will be

essential to our future understanding of whether genetic (and epigenetic) changes have influenced invasiveness of cane toads in Australia. In this study, we review the evidence for rapid evolution of dispersal-enhancing traits in cane toads in the course of their Australian invasion and provide data on the differences in gene expression that may play a major role in underpinning those dramatic phenotypic changes.

THE AUSTRALIAN CANE TOAD INVASION

Cane toad introduction history

Cane toads are large bufonid anurans (typically, 300 g as adults: Lever 2001). Females attain sexual maturity at 6–24 months of age and can produce >30 000 small eggs per clutch. The eggs are laid in a water body (typically, a pond) and externally fertilized by an amplectant male. Eggs hatch within 2 days at high water temperatures, producing small tadpoles (<20 mm long) that grow rapidly (depending upon temperature and food supply). Under ideal conditions, tadpoles can attain metamorphosis in <14 days (Cabrera-Guzmán *et al.* 2011).

Cane toads are native to South and Central America (Zug & Zug 1979), although genetic studies reveal strong divergences within this range (e.g. between *R. marina* east vs. west of the Andes: Slade & Moritz 1998) and hint that multiple taxa may be included within the species currently known as *R. marina*

Fig. 1 Native to forested areas in Central and South America, cane toads have colonized harsh arid landscapes within Australia. The first panel shows an artificial water body in the semi-desert habitat near Longreach in central Queensland, and the second panel shows a high density of adult toads around such a water body at night. Photographs by M. Greenlees. (*See insert for color representation of the figure.*)

(Vallinoto *et al.* 2010). Cane toads were brought to the Caribbean islands (in the hope that the toads would consume insect pests of commercial sugar cane: Lever 2001), and from there to Hawaii in 1932 (Turvey 2009). In 1935, 101 Hawaiian toads were collected and brought to north-eastern Queensland and their progeny were liberated in cane-growing areas across 1200 km of Queensland coastline.

The toads thrived. In the 80 years since they were released in Queensland, cane toads have spread across the Australian tropics all the way to Western Australia and have penetrated into superficially unsuitable areas such as arid deserts and high cool mountains (Florance *et al.* 2011; McCann *et al.* 2014). Toads reached the northern tip of Queensland in 1994 (Burnett 1997) and their southward expansion is decelerating as toads move into cooler climates (Urban *et al.* 2008). Comparisons of dispersal patterns across these range edges can clarify climatic impacts on dispersal. As well as improving our ability to predict future invasion patterns, such studies also could help to anticipate the response of native species to environmental change. Adept at stowaway dispersal, toads continue to be distributed outside their main range in trucks (White & Shine 2009). A satellite population of breeding cane toads exists in the suburbs of Sydney, 400 km south of any other population. Cane toads can achieve high abundances within their invaded range in Australia, especially in the years immediately after they colonize a new site (Freeland 1986; Brown *et al.* 2013). This spectacular success, coupled with the toad's ecological impact on native fauna (Shine 2010), has stimulated extensive research on the cane toad invasion. Arguably, *R. marina* in Australia has become one of the most intensively studied invasive systems worldwide.

Evolution of toads within Australia

In the 80 years since they were introduced to Australia, the toads have experienced two powerful sets of evolutionary forces: those imposed by the novel conditions (both biotic and abiotic) that they have encountered, and those imposed by the process of invasion (continuous range expansion). The result has been substantial phenotypic changes, as revealed by geographic comparisons across the toads' invasion range within Australia (Table 1).

Because toads at the invasion front encounter abundant predators, mortality rates may be high (Brown *et al.* 2013). Perhaps as a result, the relative size of toxin glands is highest in toads at the invasion front and decreases in long-colonized areas (where predators are rare, having been killed or having learned not to eat toads: Phillips & Shine 2005, 2006). However, the contribution of genetic vs. environmental influences to the change in parotoid gland size remains unclear because exposure to stressful conditions (alarm pheromone) during larval life directly induces increased relative size of parotoids in metamorph toads (Hagman *et al.* 2009; Tingley *et al.* 2012).

The same ambiguity occurs with phenotypic shifts seen in toads that have colonized highly arid and cold environments. At the fringe of the desert, toads exhibit modified water balance. Compared to toads from a mesic environment, desert toads have higher rates of evaporative water loss, but also gain water more rapidly and have better locomotor performance when dehydrated (Tingley *et al.* 2012). In dangerously desiccating conditions, adult toads switch to diurnal activity to maintain hydric balance (Webb *et al.* 2014). In such environments, toads may utilize their physiological stress responses to conserve water (Jessop *et al.* 2013). Toads captured in high cold montane regions have lower critical thermal maxima, enabling them to function even at low temperatures, but laboratory trials show that this shift is accomplished through fast-acting phenotypic plasticity, not canalized genetically coded mechanisms (McCann *et al.* 2014). Evolved plasticity may be a key to toad invasion success. Further work is required to tease apart the causal mechanisms underlying such responses.

Cane toads have been expanding their range across Australia for 80 years—that is, for approximately 80 generations. This process of continuous range expansion has subjected the toads to an array of novel evolutionary processes. For example, by definition, toads at the invasion front occur at low density. Evolutionary theory predicts a trade-off between competition and reproductive effort (MacArthur & Wilson 1967). Thus, generation after generation, we might expect selection to favour 'r-selected' life history traits at the invasion front, compared to 'K-selected' traits in long-colonized areas (Phillips *et al.* 2010b). Trials in which offspring from field-collected toads were raised in common-garden conditions have confirmed higher growth rates in larvae from invasion-front parents (Phillips 2009).

Table 1 Phenotypic traits that exhibit significant geographic variation within the geographic range of cane toads in Australia, with an evaluation of the evidence for evolutionary change. Table summarizes traits for which invasion-front ('frontal') toads have been reported to differ from conspecifics in long-colonized areas

Trait	Pattern of divergence	Evidence for		References
		Plasticity	Heritability	
Morphology				
Relative leg length	Frontal toads have longer back legs	No	No	Phillips *et al*. (2006)
Body size	Frontal toads are larger	No	No	Phillips & Shine (2005)
Parotoid gland size	Frontal toads have relatively larger parotoids	Yes	No	Phillips & Shine (2005); Hagman *et al*. (2009)
Spinal arthritis	More common in long-legged frontal toads	Yes	No	Brown *et al*. (2007)
Locomotor performance				
Endurance	Conflicting results (higher in frontal, or no difference)	No	No	Lewellyn *et al*. (2010); Tracy *et al*. (2012)
Immunology				
Investment	Frontal toads show lower metabolic increase	No	No	Llewellyn *et al*. (2012)
Plasticity	Fast-dispersing toads change immune responses	Yes	No	Brown & Shine (2014)
Mechanisms	Frontal toads upregulate some functions, downregulate others	Yes	Yes	Brown *et al*. (2015a)
Parasite burdens				
Lungworms	Frontal toads lack lungworms	No	No	Phillips *et al*. (2010b)
Behaviour				
Daily dispersal rate	Frontal toads disperse much further	No	Yes	Alford *et al*. (2009); Phillips *et al*. (2010c)
Directionality	Frontal toads remain longer in 'dispersal mode'	No	No	Lindström *et al*. (2013)
	Frontal toads follow straighter paths	No	Yes	Brown *et al*. (2014)
	No invasion-related shift in compass orientation	No	No	Brown *et al*. (2015b)

The invasion process also imposes novel pressures on immune function. Invasive species commonly lose native-range parasites and diseases in the course of translocation, or during their spread in the new range (because low host densities at the front impede transmission: Phillips *et al*. 2010b). The invader will encounter many new pathogens, but most may be harmless because of a lack of co-evolutionary history with the invader. These processes might favour lowered overall investment into immunocompetence, or a shift in the specific forms of immune defence. At the same time, the extreme muscular effort associated with rapid dispersal at the invasion front may directly modify components of the immune response and/or increase the risks of inappropriate auto-immune rejection of waste products from extreme activity

(Brown & Shine 2014). Also, reduced energy investment into the immune system might liberate resources to be expended on dispersal, thus facilitating rapid expansion of the front.

As predicted from this latter idea, immune response to a bacterial challenge differed between invasion-front (western) and range-core (eastern) toads raised in a common-garden design (based on elevation in metabolic rate: Llewellyn *et al*. 2012). Also, dispersal actively affects immune responses: radio-tracked toads that moved further in a seven-day trial had an upregulated cell-mediated immune response but a downregulated systemic immune response (Brown & Shine 2014). Studies on common-garden progeny showed the inverse situation, whereby the aspects of immune response upregulated by exercise were downregulated in the (nonexercised) invasion-front toads (Brown *et al.* 2015a). That symmetry suggests an adaptive response, whereby evolution has caused a reduction in the intensity of immune responses most likely to be increased by the exertion of invasion-front animals.

The evolutionary acceleration of rate of dispersal

Expanding populations are subject to selection for increased dispersal rate as well as increased reproductive ability and decreased competitive ability (Burton *et al.* 2010). Consistent with the idea that the invasion process favours an evolved increase in dispersal rate, dispersal-enhancing morphological traits are common in the vanguard of range-expanding populations (Travis & Dytham 2002; Phillips *et al.* 2008; Forsman *et al.* 2011).

The spread of cane toads through Australia has been documented in detail, allowing robust modelling of the determinants of spread rate. Invasion speed has accelerated dramatically, from 10 to 15 km/year after introduction to almost 60 km/year in 2006 (Urban *et al.* 2007; see Figs 2 and 3). That acceleration cannot be attributed to human-assisted translocation nor to environmental features that facilitate toad dispersal (Urban *et al.* 2008). Tellingly, toads that are captured at a range of sites, and radio-tracked at a common site, exhibit the same differentials in dispersal as seen in their home populations: invasion-front toads disperse further and faster than do toads from long-colonized areas (Phillips *et al.* 2010c). Studies on physiological aspects such as endurance have produced conflicting results (Llewelyn *et al.* 2010; Tracy *et al.* 2012). Overall, higher dispersal rates at the invasion front are a characteristic of individual toads, not of environmental conditions.

The shifts in dispersal rate through space are mirrored by changes through time. Radio-tracking of toads at a single site, beginning with the initial arrival of toads, shows substantial shifts in the years after invasion (Lindström *et al.* 2013). For the first two years after toads arrived, they moved long distances almost every night, in consistent directions. As the years passed, the toads spent less time in dispersive mode, moved smaller distances per night and decreased path straightness (Lindström *et al.* 2013; Brown *et al.* 2014).

Were these changes due to the evolution of canalized traits (i.e. 'disperse rapidly, whatever the environ-mental conditions'), or environmentally driven (e.g. 'disperse rapidly only if you encounter low conspecific densities, as occurs at the invasion front')? By raising the offspring of toads from a range of locations at a single site, then radio-tracking their offspring, investigators have shown that the progeny inherit their parents' pattern of dispersal. The offspring of invasion-front toads disperse faster than those of range-core toads, reflecting significant heritability of dispersal rate (Phillips *et al.* 2010c) and path straightness (Brown *et al.* 2014), but not compass directionality (Brown *et al.* 2015b). The evidence for phenotypic evolution in cane toads within their invaded range in Australia is summarized in Table 1.

Under the Darwinian paradigm (including its modern versions), the evolution of traits that enhance rates of dispersal at an expanding invasion front is driven by adaptation. That is, such traits have been fashioned by natural selection, because the ability to stay near the front of the invasion (which requires fast dispersal) enhances individual fitness of toads. That interpretation is plausible, in that toads at the invasion front obtain access to better food supplies (because of the scarcity of conspecific competitors) and thus exhibit higher feeding rates, faster growth and better body condition (Brown *et al.* 2013). However, toads at the front may also experience higher mortality (faster-dispersing animals were more likely to be killed by predators: Phillips *et al.* 2008), experience high rates of spinal arthritis (Brown *et al.* 2007) and reproduce only infrequently (G. P. Brown, personal communication).

Given the lack of direct evidence that accelerated dispersal rate enhances individual fitness, we should consider nonadaptive explanations also. In this case, there is a strong contender: the process of spatial sorting that assembles dispersal-enhancing phenotypes through space rather than time (Shine *et al.* 2011). Toads have been on a marathon footrace across tropical Australia for 80 generations. Each generation, toads that disperse furthest (e.g. move fastest, or more often, or along straighter paths) end up in the invasion vanguard. Thus, genes that code for rapid dispersal become spatially sorted. When breeding occurs at the front, it can only be between fast-moving individuals—because slower animals have been left behind (Shine *et al.* 2011).

The mingling of paternal and maternal genes during sexual reproduction means that by chance, some of the

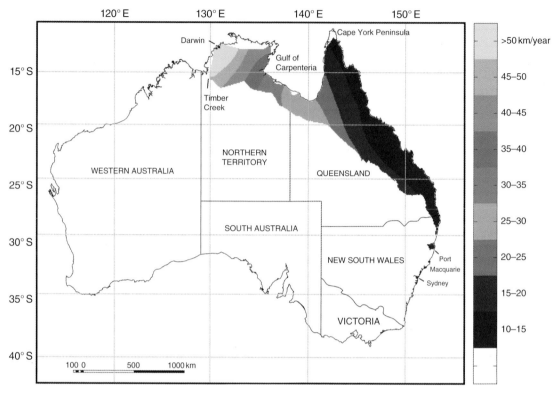

Fig. 2 The geographic spread of cane toads across Australia, showing acceleration of the invasion front through time (Urban *et al.* 2008). Figure by author reprinted with permission from book title published by the University of Chicago Press.

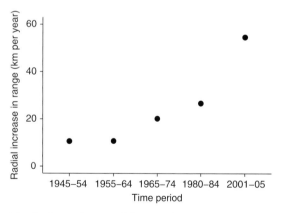

Fig. 3 The acceleration of the rate of geographic spread over time, from Phillips *et al.* (2006).

offspring of these athletic toads will receive 'fast disper-sal' genes from both parents and hence be able to travel even faster than either their mother or their father. Thus, (i) we expect mean dispersal rate to be higher at the range edge (because only fast-moving toads can reach that edge), (ii) the average dispersal rate of the progeny of those toads will be similar to the average dispersal rate of their parents, and hence (iii), some offspring will disperse more rapidly than any individu-als in the preceding (parental) generation (Shine *et al.* 2011). The process is cumulative, producing faster and faster dispersal with every generation. Spatial sorting results in genes for fast dispersal increasingly being colocated within a small subset of individuals (those at the invasion front). Spatial sorting is not a form of natu-ral selection, because it does not require fitness differen-tials. The process of spatial sorting does not change population allele frequencies overall; it simply colocates a subset of alleles (those that code for rapid dispersal) within a limited set of individuals (those at the invasion front). Even without any fitness benefit, traits that accelerate an individual's dispersal inevitably become concentrated at an expanding range edge (Shine *et al.* 2011; Bouin *et al.* 2012). The process of spatial sort-ing occurs in any situation of range expansion—for example, of bacterial cells growing out onto an agar plate, or tumours growing within an organism's body (van Ditmarsch *et al.* 2013; Orlando *et al.* 2013).

Molecular ecology of cane toads

There have been no studies to characterize genes underlying invasiveness in the Australian cane toad population, although previous research has been conducted on the population genetics of this system, all of which is summarized here. A substantial bottleneck occurred during the introduction history of cane toads, as shown by a single mitochondrial haplotype (*ND3* gene) in all individuals sequenced from Hawaii and Australia (Slade & Moritz 1998), little major histocompatibility complex (*MHC*) sequence diversity in Australian populations (Lillie *et al.* 2014), low levels of microsatellite diversity and little evidence for isolation by distance over small spatial scales in this system (Leblois *et al.* 2000). Population comparisons across Queensland indicate low levels of differentiation across sites ($F_{ST} = 0.06$: Estoup *et al.* 2001). Populations spreading southward from the original introduction sites had higher levels of population structure than did those spreading westward (southward transect, $F_{ST} = 0.08$; westward transect, $F_{ST} = 0.01$: Estoup *et al.* 2004). Estoup *et al.* (2004) also reported evidence of isolation by distance along these southern and western transects. Collectively, these results suggest that Australian cane toads are genetically depauperate, despite the overwhelming phenotypic evidence of rapid evolution occurring across the toads' invasive range in Australia (see above). Genomewide comparisons of toads from the core vs. those from the range edge may provide insight into the genetic basis of phenotypic change in this species (Box 1).

FUTURE DIRECTIONS

Biological invasions are of interest not only for their ecological, economic and social impacts but also because they provide superb study systems with which to explore the ways in which living organisms respond to novel challenges. This is especially true for a study species that is readily sampled, maintained, bred and manipulated—such as the cane toad—that will allow us to move beyond simply measuring invasion-associated divergence in phenotypic traits. Building on phenotypic studies and with next-generation

sequencing techniques, we can begin to understand the genetic basis of these traits (e.g. Box 1). The short timescales involved, and the close relatedness of the biological entities being compared, eliminate a vast array of confounding factors that impede most studies in evolutionary biology.

In Australian cane toads, the current evidence regarding phenotypic change mostly involves traits influencing dispersal ability and propensity (i.e. behaviour), and immune function (Table 1). Personality may also be important to invasiveness through its impact on dispersal propensity, which has been shown to be heritable (Doligez *et al.* 2009). Aggressiveness has been linked to dispersal and range expansion in western bluebirds (Duckworth & Badyaev 2007), and these traits are genetically correlated in that system (Duckworth & Kruuk 2009). Range-edge invasive sparrows are more exploratory and more readily eat novel foods than do conspecifics from the core (Liebl & Martin 2012, 2014). Similarly, in assays of exploratory behaviour, dark-eyed juncos from more recently colonized areas were bolder than those from older populations (Atwell *et al.* 2012). Activity scores of invasive birds exposed to novel objects have been associated with polymorphisms in the dopamine receptor D4 (*DRD4*) gene, which is thought to underlie novelty seeking behaviour in vertebrates (Mueller *et al.* 2014). These data suggest that investigating the genetic basis of the role of personality in the Australian toad invasion may be informative, although little is currently known about how personality may influence dispersal in this system.

Epigenetics in invasion

Introduced populations with low genetic diversity sometimes adapt to their new environment, a 'paradox' that has been widely discussed in the invasion literature (Grossniklaus *et al.* 2013). It has been suggested that heritable epigenetic variation may allow invasive species to adapt to novel environments in situations where genetic variation is low (Allendorf & Lundquist 2003; Pérez *et al.* 2006; Rollins *et al.* 2013) and experimental evidence supports this (Pérez *et al.* 2006; Angers *et al.* 2010; Kilvitis *et al.* 2014). Further, the process of invasion itself may increase epigenetic variation through increased contact with environmental stressors (Jablonka 2013) and exposure to population bottlenecks, often seen in introductions (Rapp & Wendel 2005).

Box 1 Genetic basis of dispersal in cane toads

The wealth of ecological data on dispersal-related traits makes this a logical starting point to investigate the genetic basis of invasiveness in this system. Here, we discuss the approach we have taken to identify candidate genes that may be important to dispersal. The speed of the advancing cane toad invasion front in tropical Australia (>50 km/year, Phillips *et al.* 2006; see Fig. 3) predicts that individuals on the leading edge of the invasion represent successful long-distance dispersers (or their offspring). Genetic differences between core and edge populations may clarify which genes underpin dispersal ability, central to a species' ability to invade. This system is particularly amenable to such an approach; the low levels of neutral genetic diversity found in Australian cane toads suggest that signals of selection on dispersal ability are unlikely to be confounded by those from demographic history.

We hypothesized that genes involved in dispersal ability will be expressed in muscle tissue and will include genes for metabolic enzymes. To test this prediction, we have characterized the cane toad muscle transcriptome using adult females collected from the wild (full methods are described in Appendix S1, on-line version of the *Molecular Ecology* article). For the construction of the muscle transcriptome, we used RNA extracted from one sample from each of two localities at the range core (Innisfail and Rossville, Queensland) and two at the range edge (El Questro and Purnululu National Park, Western Australia). We identified 57 580 transcripts after filtering, 21 533 of which were annotated based on

sequence homology (Appendix S1, Tables S2 and S3, on-line version of the *Molecular Ecology* article). Differential expression analysis was conducted between Australian toads from the range core in the east and the range edge in the west ($N = 20$ females, Table S1, on-line version of the *Molecular Ecology* article). We identified 479 transcripts upregulated and 142 downregulated on the range edge (Appendix S1, on-line version of the *Molecular Ecology* article). Below, we consider 1) whether metabolic genes are differentially expressed between regions; 2) the known function of the most differentially expressed genes; and 3) which gene ontology terms are enriched in our differential expression data and how these may relate to dispersal ability.

Expression of metabolic genes

First, we investigated whether genes coding for metabolic enzymes involved in glycolysis and the tricarboxylic acid (TCA) cycle were differentially regulated. Of all enzymes relevant to glycolysis, the pyruvate kinase gene (*PK*) was the only gene differentially expressed (three isoforms downregulated on the range edge). *PK* catalyses the final step in glycolysis and regulates the relative requirements for energy production (when *PK* is upregulated) vs. the need for glycolytic precursors used for nucleic acid and lipid synthesis, which are necessary for cell division (when *PK* is downregulated) (Gupta & Bamezai 2010). Conversely, L-threonine 3-dehydrogenase (*TDH*) was strongly upregulated on the range front

Box 1, Table 1 Genes showing the largest difference in expression between cane toads sampled from the range core and range edge. Transcript identifier, gene name (and abbreviation), *P*-value (indicates significance of expression difference), log-fold change and the number of other differentially expressed isoforms are given. All of these genes are upregulated at the range edge

Transcript ID	Gene	*P*-value	Foldchange (log)	# Other D.E. isoforms
c25693_g8_i1	Krueppel-like factor 9 (*KLF9*)	4.93E-10	3.74	0
c26559_g4_i4	L-threonine 3-dehydrogenase (*TDH*)	7.15E-10	4.14	0
c26683_g2_i2	Proteasome activator complex subunit 4 (*PSME4, PA200*)	7.59E-09	2.72	3
c27340_g6_i4	Cyclin-L2 (*CCNL2*)	1.14E-08	2.31	2
c23773_g1_i4	Ubiquitin conjugation factor E4 B (*UBE4B*)	1.38E-08	3.56	4
c26121_g4_i1	Sestrin-1 (*SESN1*)	1.55E-08	4.57	4
c26435_g1_i2	Sodium- and chloride-dependent taurine transporter (*SLC6A6, TauT*)	1.84E-08	4.02	3
c26220_g1_i2	Krueppel-like factor 15 (*KLF15*)	2.00E-08	3.40	1
c24732_g2_i1	Ras-GEF domain-containing family member 1B (*RASGEF1G*)	2.21E-08	2.34	1
c27349_g2_i1	Protein FAM134B (*FAM134B, JK1*)	4.28E-08	3.40	3

(Table 1). *TDH* is necessary for threonine catabolism (Chen *et al*. 1995); inhibition of *TDH* results in suppression of acetyl-CoA, which is essential to the TCA cycle. Additionally, *TDH* may play a role in epigenetics through its association with acetyl-CoA, which appears to be involved in histone acetylation (Kaelin & McKnight 2013).

Several nuclear-encoded mitochondrial matrix enzymes involved in metabolism of amino acids were upregulated on the range edge. Delta-1-pyrroline-5-carboxylate dehydrogenase (*AL4A1*) is involved in glutamate metabolism and functions in the pathway connecting the urea and TCA cycles (Geraghty *et al*. 1998). Proline dehydrogenase 1 (*PRODH1*) catalyses the first step in the catabolism of proline to glutamate (Phang *et al*. 2008), and two isoforms of *PRODH1* were upregulated on the range edge. *PRODH1* plays an important role in the regulation of apoptosis, and DNA damage induces expression of this gene (Natarajan & Becker 2012). *PRODH1* also alleviates nutritional stress via the production of ATP and may protect against oxidative stress (Natarajan & Becker 2012). Glycine dehydrogenase (*GLDC*) catalyses glycine catabolism, and upregulation of this gene promotes glycolysis (Zhang *et al*. 2012). Two isoforms of *GLDC* were upregulated on the range edge. Branched-chain-amino-acid aminotransferase (*BCAT2*) catalyses the first step of catabolism of valine, leucine and isoleucine, which is regulated by Krueppel-like factor 15 (*KLF15*, discussed below, Table 1). *BCAT2* and *KFL15* are involved in the maintenance of muscle mass (Shimizu *et al*. 2011) and both are upregulated on the range edge. Also involved in branched-chain amino acid catabolism, 2-oxoisovalerate dehydrogenase subunit beta (*BCKDHB, ODBB*) was similarly upregulated on the range edge. *KLF15* also regulates the expression of phosphoenolpyruvate carboxykinase (*PEPCK-M, PCKGM*), which catalyses the conversion of oxaloacetate to phosphoenolpyruvate in glyconeogenesis. *PEPCK* is upregulated under acute mitochondrial stress in skeletal muscle tissue (Ost *et al*. 2015), and two isoforms of *PEPCK* are upregulated in toads from the range edge. In contrast to the many mitochondrial matrix enzymes upregulated on the range edge, probable 4-hydroxy-2-oxoglutarate aldolase (*HOGA1*) was the only one of these enzymes downregulated. *HOGA1* is involved in hydroxyproline metabolism and mutations in this gene are associated with hyperoxaluria leading to renal disease (Riedel *et al*. 2012) and oxalate arthritis (Lorenz *et al*. 2013).

Functions of differentially expressed genes

Our second approach to analysing differentially expressed genes was to examine those genes having the most significantly different expression levels between individuals on the range core vs. range edge. All of the genes having the highest levels of differential expression (Table 1) were upregulated in the west at the range edge. These genes are primarily involved in energetics, response to oxidative stress, immune function and cellular replication. *KLF15* expression is increased by fasting and, as discussed above, affects *BCAT2* and *PEKCK* expression (Jeong *et al*. 2012). *KFL15* is also involved in lipid metabolism, and expression of *KLF15* is induced by exercise and is vital to muscle endurance (Haldar *et al*. 2012). Krueppel-like factor 9 (*KLF9*) encodes a transcription factor involved in multiple cellular processes, including adipogenesis (Wu & Wang 2013). *KLF9* is upregulated by oxidative stress, which can result from many environmental insults including exposure to environmental toxins, UV radiation, heat shock and infection (Ermak & Davies 2002) as well as from exercise (Sanchis-Gomar 2013). Sodium- and chloride-dependent taurine transporter (*SLC6A6,TauT*) appears to have a protective role against muscle atrophy (Uozumi *et al*. 2006), and knockout of this gene can decrease exercise capacity by >80% (Warskulat *et al*. 2004). FAM134B (*JK1*) expression is positively correlated with lipid accumulation (Yuan *et al*. 2014) and is upregulated in response to heat stress (Goto *et al*. 2011). Sestrin 1 (*SESN1*) responds to oxidative stress and DNA damage (Lee *et al*. 2010). Similarly, proteasome activator complex subunit 4 (*PSME4,PA200*) is thought to be involved in DNA repair in response to oxidative stress (Pickering & Davies 2012) and, more recently, it has been suggested that *PSME4* may be involved in MHC class I antigen presentation (Cascio 2014). Ras-GEF domain-containing family member IB (*RASGEF1B*) is expressed during immune response and parasite infection (Andrade *et al*. 2010). Upregulation of cyclin L2 (*CCNL2*) induces apoptosis (Rao *et al*. 2014), which may be a response to oxidative stress. Ubiquitin conjugation factor E4 B (*UBE4B*) is important to normal progression of mitosis and is involved in the degradation of proteins (Koegl *et al*. 1999). We have identified five differentially expressed isoforms of UBE4B in our data (all upregulated on the range edge).

Enrichment of gene ontology terms

As a third approach to understanding differential expression patterns, we used enrichment analysis to identify whether functional categories (gene ontology (GO) terms for biological processes) associated with an a priori suite of genes (those from the reference set used for assembly) contain more differentially expressed genes between regions than expected by chance. The results of this analysis are shown in Table 2 and Fig. 1. The most overrepresented term was 'response to starvation', and this term was only found in genes upregulated on the range edge. The terms 'response to UV' and 'muscle atrophy' were also overrepresented. These results, in combination with our analysis of the most differentially expressed genes (Table 1), suggest that toads on the range edge experience greater environmental stressors than do those in the east. Most of the remaining terms are associated with energy production (e.g. catabolism, nutrient transport). In support of this, we have shown that metabolic enzymes are overwhelmingly upregulated at the invasion front and that many of the most highly differentially expressed genes (Table 1) involve energy production. GO terms associated with genes that were downregulated at the front are primarily regulatory processes.

Why do toads on the range edge exhibit a stronger response to environmental stressors (upregulated cellular repair)? We cannot separate cause and effect: Is this difference because conditions are less hospitable for toads on the range edge (i.e. hotter and drier than the native range of the toad: Tingley et al. 2014), or because these animals experience difficulty coping with the same levels of environmental stress as at the core (because of biological trade-offs involved in accelerated dispersal rates)? Alternatively, are range-edge individuals better able than range-core individuals to respond to the same level of environmental stress? This question could be answered using a common-garden experiment to separate environmental and genetic effects. Arthritis is more prevalent in range-edge toads (Brown et al. 2007), hinting that these results stem from endogenous origins.

Upregulation of genes involved in metabolism indicates increased activity levels in toads on the range front as compared to those from the range core. That inference is strongly supported by radiotelemetry (Brown & Shine 2014). Further, upregulation of *KLF15* is related to increased endurance (Haldar et al. 2012); this gene is strongly upregulated on the invasion front, where toads have been shown to have greater endurance (Llewelyn et al. 2010; but see Tracy et al. 2012). Such genes may underpin the dispersal success of range-edge individuals, and these candidate genes should be considered in a common-garden experiment, as above, to disentangle cause and effect.

Box 1, Table 2 Gene ontology (GO) terms overrepresented in differentially expressed genes. GO term identifier, description of biological process, $\log_{10}$ P-value (indicates significance of overrepresentation) and uniqueness (index showing the inverse of how similar a GO term is to all other terms)

Term_ID	Description	P-value ($\log_{10}$)	Uniqueness
GO:0042594	Response to starvation	−3.4001	0.848
GO:0019216	Regulation of lipid metabolic process	−3.7645	0.729
GO:0009411	Response to UV	−3.7878	0.881
GO:0060213	Positive regulation of nuclear-transcribed mRNA poly(A) tail shortening	−4.0128	0.699
GO:0019470	4-hydroxyproline catabolic process	−4.2676	0.838
GO:0034641	Cellular nitrogen compound metabolic process	−4.2865	0.917
GO:0060356	Leucine import	−4.4191	0.812
GO:0015827	Tryptophan transport	−4.4191	0.807
GO:0033198	Response to ATP	**−4.767**	0.872
GO:0019433	Triglyceride catabolic process	−5.056	0.744
GO:0015734	Taurine transport	−5.5513	0.816
GO:0014889	Muscle atrophy	−5.7721	0.736
GO:0051924	Regulation of calcium ion transport	**−6.1397**	0.579
GO:0050790	Regulation of catalytic activity	**−6.7033**	0.737
GO:0042264	Peptidyl-aspartic acid hydroxylation	**−7.5969***	0.921
GO:0060021	Palate development	−8.6421*	0.803

Bold P-values represent terms overrepresented in downregulated genes.
*Contains genes that are both up- and downregulated on the range edge (font indicates whether more terms are overrepresented in up- vs. downregulated genes on the range edge).

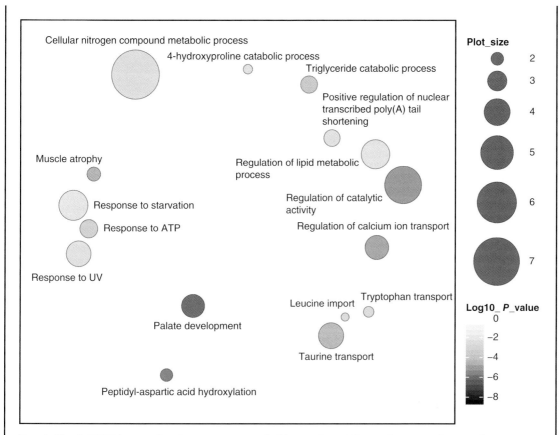

Box 1, Fig. 1 REVIGO scatterplot of GO categories enriched for significant differential expression between the range core and range edge. Benjamini–Hochberg corrected *P* < 0.05. GO categories are clustered by semantic similarity (simRel). Darker circles represent higher enrichment, whereas circle size denotes frequency and breadth of the GO categories.

Two of the ten most highly differentially expressed genes appear to play a role in immune function and parasite resistance (*PSME4* and *RASGEF1B*). As discussed above, immune function of toads on the range edge differs to that in the range core; cell-mediated immune response was upregulated in toads that moved further (Brown & are clustered by semantic similarity (simRel). Darker circles represent higher enrichment, whereas circle size denotes frequency and breadth of the GO categories.Shine 2014) and downregulated in the offspring of invasion-front toads (Brown *et al.* 2015a). Differences in parasite ecology between the edge and core (Kelehear *et al.* 2012) suggest that toads from these regions may respond differently to parasite infection. Our differential expression results suggest that further investigation of these genes with respect to immune function and parasite resistance is warranted.

Our understanding of the role of epigenetics in invasion is poor, especially in animal systems, but certainly warrants further investigation. Leading research on epigenetics in animal invasions, Schrey *et al.* (2012) found a relationship between levels of DNA methylation and age of introduction in invasive house sparrows. Further, a significant negative relationship was shown between genetic and epigenetic (DNA methylation) diversity in this invasion, suggesting that epigenetic variation may compensate for loss of genetic variation (Liebl & Martin 2012).

Epigenetic studies remain technically challenging for most species and, in nonmodel systems, are primarily focused on changes to DNA methylation. While methylation-sensitive amplified fragment length polymorphism (MS-AFLP) is inexpensive and can be conducted in systems lacking reference genomes, the usefulness of this approach is limited because loci identified as differentially methylated are anonymous. Modification of reduced representation genome sequencing using highly multiplexable methods (such as restriction-associated digest sequencing, 'RAD-seq'; or genotyping by sequencing, 'GBS') for methylation analysis has been discussed (Kilvitis *et al.* 2014) and is currently under development (T. P. van Gurp & K. J. F. Verhoeven, personal communication). If candidate genes have been identified, sodium bisulphite sequencing can be used to investigate methylation of single loci. All of these approaches only investigate methylation status; other forms of epigenetic modification (e.g. histone modification) remain challenging in taxa lacking reference genomes.

CONCLUSION

Understanding how traits evolve and how they affect invasiveness is best achieved by a comprehensive approach investigating the outcomes of evolution based on combining phenotypic and genetic data. Cane toads offer an excellent system to investigate invasiveness because of the wealth of ecological information available and the many phenotypic traits that differ across the invasive range in Australia. The genetic basis of these traits has been understudied until recently, but the results presented here suggest that metabolic and immune function genes may be important to a species' ability to invade. Studies of natural populations coupled with manipulative common-garden experiments in this system may provide a clearer understanding of how genetic and epigenetic changes facilitate invasion.

ACKNOWLEDGEMENTS

We thank the organizers of *Invasion Genetics: The Baker and Stebbens Legacy* for inviting this contribution and for useful comments from anonymous reviewers. We thank Joachim Ehlenz and Serena Lam for assistance in the field and Melanie Elphick for assistance with manuscript preparation. Next-generation sequencing was conducted by Macrogen, Inc. Our cane toad research is funded by the Australian Research Council to RS and through a fellowship and funding to LAR from Deakin University.

REFERENCES

Alford RA, Brown GP, Schwarzkopf L, Phillips BL, Shine R (2009) Comparisons through time and space suggest rapid evolution of dispersal behaviour in an invasive species. *Wildlife Research*, **36**, 23–28.

Allendorf FW, Lundquist LL (2003) Introduction: population biology, evolution, and control of invasive species. *Conservation Biology*, **17**, 24–30.

Andrade W, Silva A, Alves VS *et al.* (2010) Early endosome localization and activity of RasGEF1b, a toll-like receptor-inducible Ras guanine-nucleotide exchange factor. *Genes and Immunity*, **11**, 447–457.

Angers B, Castonguay E, Massicotte R (2010) Environmentally induced phenotypes and DNA methylation: how to deal with unpredictable conditions until the next generation and after. *Molecular Ecology*, **19**, 1283–1295.

Atwell JW, Cardoso GC, Whittaker DJ *et al.* (2012) Boldness behavior and stress physiology in a novel urban environment suggest rapid correlated evolutionary adaptation. *Behavioral Ecology*, **23**, 960–969.

Baker HG, Stebbins GL (1965) *The Genetics of Colonizing Species.* Academic Press, New York, NY.

Bouin E, Calvez V, Meunier N *et al.* (2012) Invasion fronts with variable motility: phenotype selection, spatial sorting and wave acceleration. *Comptes Rendus Mathematique*, **350**, 761–766.

Brown GP, Shine R (2014) Immune response varies with rate of dispersal in invasive cane toads (*Rhinella marina*). *PLoS One*, **9**, e99734.

Brown GP, Shilton C, Phillips BL, Shine R (2007) Invasion, stress, and spinal arthritis in cane toads. *Proceedings of the National Academy of Sciences of the USA*, **104**, 17698–17700.

Brown GP, Kelehear C, Shine R (2013) The early toad gets the worm: cane toads at an invasion front benefit from higher prey availability. *Journal of Animal Ecology*, **82**, 854–862.

Brown GP, Phillips BL, Shine R (2014) The straight and narrow path: the evolution of straight-line dispersal at a cane toad invasion front. *Proc. R. Soc. B*, **281**, 20141385.

Brown GP, Phillips BL, Dubey S, Shine R (2015a) Invader immunology: invasion history alters immune system function in cane toads (*Rhinella marina*) in tropical Australia. *Ecology Letters*, **18**, 57–65.

Brown GP, Phillips BL, Shine R (2015b) Directional dispersal has not evolved during the cane toad invasion. *Functional Ecology*, **29**, 830–838.

Burnett S (1997) Colonizing cane toads cause population declines in native predators: reliable anecdotal information and management implications. *Pacific Conservation Biology*, **3**, 65.

Burton OJ, Phillips BL, Travis JMJ (2010) Trade-offs and the evolution of life-histories during range expansion. *Ecology Letters*, **13**, 1210–1220.

Cabrera-Guzmán E, Crossland M, Shine R (2011) Can we use the tadpoles of Australian frogs to reduce recruitment of invasive cane toads? *Journal of Applied Ecology*, **48**, 462–470.

Cascio P (2014) PA28αβ: the enigmatic magic ring of the proteasome? *Biomolecules*, **4**, 566–584.

Chen Y-W, Dekker EE, Somerville RL (1995) Functional analysis of *E. coli* threonine dehydrogenase by means of mutant isolation and characterization. *Biochimica et Biophysica Acta*, **1253**, 208–214.

Colautti RI, Maron JL, Barrett SCH (2009) Common garden comparisons of native and introduced plant populations: latitudinal clines can obscure evolutionary inferences. *Evolutionary Applications*, **2**, 187–199.

van Ditmarsch D, Boyle KE, Sakhtah H *et al.* (2013) Convergent evolution of hyperswarming leads to impaired biofilm formation in pathogenic bacteria. *Cell Reports*, **4**, 697–708.

Doligez B, Gustafsson L, Part T (2009) 'Heritability' of dispersal propensity in a patchy population. *Proc. R. Soc. B*, **276**, 2829–2836.

Duckworth RA, Badyaev AV (2007) Coupling of dispersal and aggression facilitates the rapid range expansion of a passerine bird. *Proceedings of the National Academy of Sciences of the USA*, **104**, 15017–15022.

Duckworth RA, Kruuk LEB (2009) Evolution of genetic integration between dispersal and colonization ability in a bird. *Evolution*, **63**, 968–977.

Ermak G, Davies KJ (2002) Calcium and oxidative stress: from cell signaling to cell death. *Molecular Immunology*, **38**, 713–721.

Estoup A, Wilson IJ, Sullivan C, Cornuet J-M, Moritz C (2001) Inferring population history from microsatellite and enzyme data in serially introduced cane toads, *Bufo marinus*. *Genetics*, **159**, 1671–1687.

Estoup A, Beaumont M, Sennedot F, Moritz C, Cornuet J-M (2004) Genetic analysis of complex demographic scenarios: spatially expanding populations of the cane toad, *Bufo marinus*. *Evolution*, **58**, 2021–2036.

Florance D, Webb JK, Dempster T *et al.* (2011) Excluding access to invasion hubs can contain the spread of an invasive vertebrate. *Proc. R. Soc. B*, **278**, 2900–2908.

Forsman A, Merila J, Ebenhard T (2011) Phenotypic evolution of dispersal-enhancing traits in insular voles. *Proc. R. Soc. B*, **278**, 225–232.

Freeland W (1986) Populations of cane toad, *Bufo marinus*, in relation to time since colonization. *Wildlife Research*, **13**, 321–329.

Geraghty MT, Vaughn D, Nicholson A *et al.* (1998) Mutations in the Δ1-pyrroline 5-carboxylate dehydrogenase gene cause type II hyperprolinemia. *Human Molecular Genetics*, **7**, 1411–1415.

Goto K, Oda H, Kondo H *et al.* (2011) Responses of muscle mass, strength and gene transcripts to long-term heat stress in healthy human subjects. *European Journal of Applied Physiology*, **111**, 17–27.

Grossniklaus U, Kelly WG, Ferguson-Smith AC, Pembrey M, Lindquist S (2013) Transgenerational epigenetic inheritance: how important is it? *Nature Reviews Genetics*, **14**, 228–235.

Gupta V, Bamezai RNK (2010) Human pyruvate kinase M2: a multifunctional protein. *Protein Science*, **19**, 2031–2044.

Hagman M, Hayes R, Capon R, Shine R (2009) Alarm cues experienced by cane toad tadpoles affect post metamorphic morphology and chemical defences. *Functional Ecology*, **23**, 126–132.

Haldar SM, Jeyaraj D, Anand P *et al.* (2012) Kruppel-like factor 15 regulates skeletal muscle lipid flux and exercise adaptation. *Proceedings of the National Academy of Sciences of the USA*, **109**, 6739–6744.

Jablonka E (2013) Epigenetic inheritance and plasticity: the responsive germline. *Progress in Biophysics and Molecular Biology*, **111**, 99–107.

Jeong JY, Jeoung NH, Park K-G, Lee I-K (2012) Transcriptional regulation of pyruvate dehydrogenase kinase. *Diabetes & Metabolism Journal*, **36**, 328–335.

Jessop TS, Letnic M, Webb JK, Dempster T (2013) Adrenocortical stress responses influence an invasive vertebrate's fitness in an extreme environment. *Proc. R. Soc. B*, **280**, 20131444.

Kaelin WG Jr, McKnight SL (2013) Influence of metabolism on epigenetics and disease. *Cell*, **153**, 56–69.

Kelehear C, Brown GP, Shine R (2012) Rapid evolution of parasite life history traits on an expanding range-edge. *Ecology Letters*, **15**, 329–337.

Kilvitis H, Alvarez M, Foust C *et al.* (2014) Ecological epigenetics. In: *Ecological Genomics* (eds Landry CR, Aubin-Horth N), pp. 191–210. Springer, the Netherlands.

Koegl M, Hoppe T, Schlenker S *et al.* (1999) A novel ubiquitination factor, E4, is involved in multiubiquitin chain assembly. *Cell*, **96**, 635–644.

Leblois R, Rousset F, Tikel D, Moritz C, Estoup A (2000) Absence of evidence for isolation by distance in an expanding cane toad (*Bufo marinus*) population: an individual-based analysis of microsatellite genotypes. *Molecular Ecology*, **9**, 1905–1909.

Lee JH, Bodmer R, Bier E, Karin M (2010) Sestrins at the crossroad between stress and aging. *Aging*, **2**, 369.

Lever C (2001) *The Cane Toad. The History and Ecology of Successful Colonist*. Westbury Academic and Scientific Publishing, Otley, UK.

Liebl AL, Martin LB (2012) Exploratory behaviour and stressor hyper-responsiveness facilitate range expansion of an introduced songbird. *Proc. R. Soc. B*, **279**, 4375–4381.

Liebl AL, Martin LB (2014) Living on the edge: range edge birds consume novel foods sooner than established ones. *Behavioral Ecology*, **25**, 1089–1096.

Lillie M, Shine R, Belov K (2014) Characterisation of major histocompatibility complex class I in the Australian cane toad, *Rhinella marina*. *PLoS One*, **9**, e102824.

Lindström T, Brown GP, Sisson SA, Phillips BL, Shine R (2013) Rapid shifts in dispersal behavior on an expanding range edge. *Proceedings of the National Academy of Sciences of the USA*, **110**, 13452–13456.

Llewellyn D, Thompson MB, Brown GP, Phillips BL, Shine R (2012) Reduced investment in immune function in invasion-front populations of the cane toad (*Rhinella marina*) in Australia. *Biological Invasions*, **14**, 999–1008.

Llewelyn J, Phillips BL, Alford RA, Schwarzkopf L, Shine R (2010) Locomotor performance in an invasive species: cane toads from the invasion front have greater endurance, but not speed, compared to conspecifics from a long-colonised area. *Oecologia*, **162**, 343–348.

Lorenz EC, Michet CJ, Milliner DS, Lieske JC (2013) Update on oxalate crystal disease. *Current Rheumatology Reports*, **15**, 1–9.

MacArthur RH, Wilson EO (1967) *The Theory of Island Biogeography*. Princeton University Press, Princeton, NJ.

McCann S, Greenlees MJ, Newell D, Shine R (2014) Rapid acclimation to cold allows the cane toad to invade montane areas within its Australian range. *Functional Ecology*, **28**, 1166–1174.

Mueller J, Edelaar P, Carrete M *et al.* (2014) Behaviour-related DRD4 polymorphisms in invasive bird populations. *Molecular Ecology*, **23**, 2876–2885.

Natarajan SK, Becker DF (2012) Role of apoptosis-inducing factor, proline dehydrogenase, and NADPH oxidase in apoptosis and oxidative stress. *Cell Health and Cytoskeleton*, **2012**, 11.

Orlando PA, Gatenby RA, Brown JS (2013) Tumor evolution in space: the effects of competition colonization tradeoffs on tumor invasion dynamics. *Frontiers in Oncology*, **3**, 45.

Ost M, Keipert S, van Schothorst EM *et al.* (2015) Muscle mitohormesis promotes cellular survival via serine/glycine pathway flux. *The FASEB Journal*, **29**, 1314–1328.

Pérez J, Nirchio M, Alfonsi C, Muñoz C (2006) The biology of invasions: the genetic adaptation paradox. *Biological Invasions*, **8**, 1115–1121.

Phang JM, Donald SP, Pandhare J, Liu Y (2008) The metabolism of proline, a stress substrate, modulates carcinogenic pathways. *Amino Acids*, **35**, 681–690.

Phillips BL (2009) The evolution of growth rates on an expanding range edge. *Biology Letters*, **5**, 802–804.

Phillips BL, Shine R (2005) The morphology, and hence impact, of an invasive species (the cane toad, *Bufo marinus*): changes with time since colonisation. *Animal Conservation*, **8**, 407–413.

Phillips BL, Shine R (2006) Allometry and selection in a novel predator–prey system: Australian snakes and the invading cane toad. *Oikos*, **112**, 122–130.

Phillips BL, Brown GP, Webb JK, Shine R (2006) Invasion and the evolution of speed in toads. *Nature*, **439**, 803.

Phillips BL, Brown GP, Travis JM, Shine R (2008) Reid's paradox revisited: the evolution of dispersal kernels during range expansion. *The American Naturalist*, **172**, S34–S48.

Phillips BL, Brown GP, Shine R (2010a) Evolutionarily accelerated invasions: the rate of dispersal evolves upwards during the range advance of cane toads. *Journal of Evolutionary Biology*, **23**, 2595–2601.

Phillips BL, Brown GP, Shine R (2010b) Life-history evolution in range-shifting populations. *Ecology*, **91**, 1617–1627.

Phillips BL, Kelehear C, Pizzatto L *et al.* (2010c) Parasites and pathogens lag behind their host during periods of host range advance. *Ecology*, **91**, 872–881.

Pickering AM, Davies KJ (2012) Differential roles of proteasome and immunoproteasome regulators Pa28αβ, Pa28γ and Pa200 in the degradation of oxidized proteins. *Archives of Biochemistry and Biophysics*, **523**, 181–190.

Rao N, Song F, Jhamb D *et al.* (2014) Proteomic analysis of fibroblastema formation in regenerating hind limbs of *Xenopus laevis* froglets and comparison to axolotl. *BMC Developmental Biology*, **14**, 32.

Rapp RA, Wendel JF (2005) Epigenetics and plant evolution. *New Phytologist*, **168**, 81–91.

Riedel TJ, Knight J, Murray MS *et al.* (2012) 4-Hydroxy-2-oxoglutarate aldolase inactivity in primary hyperoxaluria type 3 and glyoxylate reductase inhibition. *Biochimica et Biophysica Acta*, **1822**, 1544–1552.

Rollins LA, Moles AT, Lam S *et al.* (2013) High genetic diversity is not essential for successful introduction. *Ecology and Evolution*, **3**, 4501–4517.

Sanchis-Gomar F (2013) Sestrins: novel antioxidant and AMPK-modulating functions regulated by exercise? *Journal of Cellular Physiology*, **228**, 1647–1650.

Schrey AW, Coon CAC, Grispo MT *et al.* (2012) Epigenetic variation may compensate for decreased genetic variation with introductions: a case study using house sparrows (*Passer domesticus*) on two continents. *Genetics Research International*, **2012**, 1–7.

Shimizu N, Yoshikawa N, Ito N *et al.* (2011) Crosstalk between glucocorticoid receptor and nutritional sensor mTOR in skeletal muscle. *Cell Metabolism*, **13**, 170–182.

Shine R (2010) The ecological impact of invasive cane toads (*Bufo marinus*) in Australia. *The Quarterly Review of Biology*, **85**, 253–291.

Shine R, Brown GP, Phillips BL (2011) An evolutionary process that assembles phenotypes through space rather than through time. *Proceedings of the National Academy of Sciences of the USA*, **108**, 5708–5711.

Slade RW, Moritz C (1998) Phylogeography of *Bufo marinus* from its natural and introduced ranges. *Proc. R. Soc. B*, **265**, 769.

Tingley R, Greenlees MJ, Shine R (2012) Hydric balance and locomotor performance of an anuran (*Rhinella marina*) invading the Australian arid zone. *Oikos*, **121**, 1959–1965.

Tingley R, Vallinoto M, Sequeira F, Kearney MR (2014) Realized niche shift during a global biological invasion. *Proceedings of the National Academy of Sciences of the USA*, **111**, 10233–10238.

Tracy CR, Christian KA, Baldwin J, Phillips BL (2012) Cane toads lack physiological enhancements for dispersal at the invasive front in Northern Australia. *Biology Open*, **1**, 37–42.

Travis JM, Dytham C (2002) Dispersal evolution during invasions. *Evolutionary Ecology Research*, **4**, 1119–1129.

Turvey N (2009) A toad's tale. *Hot Topics from the Tropics*, **1**, 1–10.

Uozumi Y, Ito T, Takahashi K *et al.* (2006) Myogenic induction of taurine transporter prevents dexamethasone-induced muscle atrophy. In: *Taurine 6* (eds Oja S, Saransaari P), pp. 265–270. Springer, USA.

Urban MC, Phillips BL, Skelly DK, Shine R (2007) The cane toad's (*Chaunus [Bufo] marinus*) increasing ability to invade Australia is revealed by a dynamically updated range model. *Proc. R. Soc. B*, **274**, 1413–1419.

Urban MC, Phillips BL, Skelly DK, Shine R (2008) A toad more traveled: the heterogeneous invasion dynamics of cane toads in Australia. *The American Naturalist*, **171**, E134–E148.

Vallinoto M, Sequeira F, Sodré D *et al.* (2010) Phylogeny and biogeography of the *Rhinella marina* species complex (Amphibia, Bufonidae) revisited: implications for Neotropical diversification hypotheses. *Zoologica Scripta*, **39**, 128–140.

Warskulat U, Flögel U, Jacoby C *et al.* (2004) Taurine transporter knockout depletes muscle taurine levels and results in severe skeletal muscle impairment but leaves cardiac function uncompromised. *The FASEB Journal*, **18**, 577–579.

Webb JK, Letnic M, Jessop TS, Dempster T (2014) Behavioural flexibility allows an invasive vertebrate to survive. *Biology Letters*, **10**, 20131014.

White A, Shine R (2009) The extra-limital spread of an invasive species via 'stowaway' dispersal: toad to nowhere? *Animal Conservation*, **12**, 38–45.

Wu Z, Wang S (2013) Role of kruppel-like transcription factors in adipogenesis. *Developmental Biology*, **373**, 235–243.

Yuan Z, Song D, Wang Y (2014) The novel gene pFAM134B positively regulates fat deposition in the subcutaneous fat of *Sus scrofa*. *Biochemical and Biophysical Research Communications*, **454**, 554–559.

Zhang WC, Shyh-Chang N, Yang H *et al.* (2012) Glycine decarboxylase activity drives non-small cell lung cancer tumor-initiating cells and tumorigenesis. *Cell*, **148**, 259–272.

Zug GR, Zug PB (1979) *The Marine Toad, Bufo marinus: A Natural History Resume of Native Populations*. Smithsonian Institution Press, Washington, DC.

DATA ACCESSIBILITY

Raw RNA-Seq data are available on NCBI SRA (under BioProject PRJNA277985). Final de novo transcriptome assembly, gene expression estimates for all individuals and differential expression between individuals are available on Dryad doi:10.5061/dryad.m5298. Transcriptome annotations, assembly statistics and sequence homology may be found in the on-line version of the *Molecular Ecology* article.

SUPPORTING INFORMATION

Additional supporting information can be found in the online version of the *Molecular Ecology* article.

Table S1 Next-generation sequencing information for individual samples.

Table S2 Summary statistics of de novo transcriptome assembly for *Rhinella marina* muscle tissue.

Table S3 Overview of functional annotation results.

Table S4 Annotation metatable.

Table S5 Orthologous transcripts identified between muscle tissue from this study and liver tissue for Nourisson *et al.* 2014.

Fig. S1 ERCC technical and diagnostic plots produced by the *erccdashboard*. Each sample type contained $n = 10$ biological replicates.

Fig. S2 MA plot of differentially expressed genes identified between the range-core and range-edge.

Appendix S1 Methods.

Chapter 19

EPIGENETICS OF COLONIZING SPECIES? A STUDY OF JAPANESE KNOTWEED IN CENTRAL EUROPE

Yuan-Ye Zhang,* Madalin Parepa,*† Markus Fischer,* and Oliver Bossdorf *†

* Institute of Plant Sciences, University of Bern, Altenbergrain 21, CH-3013 Bern, Switzerland
† Institute of Evolution and Ecology, University of Tübingen, Auf der Morgenstelle 5, D-72076 Tübingen, Germany

Abstract

Some of the world's most successful invasive plants have spread across large geographic areas while retaining little or no genetic diversity. Because of this lack of heritable variation, evolutionary hypotheses are usually not invoked when attempting to explain the success of these species. However, heritable trait variation within and among invasive populations could also be created through epigenetic or other nongenetic processes, particularly in clonal invaders where somatic changes can potentially persist indefinitely. We tested this possibility in a collection of 83 genetically identical clones of the invasive Japanese knotweed *Fallopia japonica* collected across Central Europe and propagated in a common environment for several years. Using regular as well as methylation-sensitive amplified fragment length polymorphism (AFLP) markers, we found that all clones indeed belonged to the same genotype but to 27 different epigenotypes. The different knotweed clones were also phenotypically differentiated. Path analysis indicated that among-clone phenotypic variation is partly correlated with epigenetic variation, as well as both directly and indirectly (through epigenetic variation) with climates of origin. Our results thus suggest a potential role of epigenetic variation in the geographic spread and invasion success of Japanese knotweed. More generally, our study highlights the need to incorporate large-scale epigenetic screens in studies of genetically uniform clonal invaders.

Invasion Genetics: The Baker and Stebbins Legacy, First Edition. Edited by Spencer C. H. Barrett, Robert I. Colautti, Katrina M. Dlugosch, and Loren H. Rieseberg.
© 2017 John Wiley & Sons, Ltd. Published 2017 by John Wiley & Sons, Ltd.

INTRODUCTION

Some of the world's most successful invasive plants are almost or entirely genetically uniform in their introduced range. Examples include the "Bermuda buttercup" (*Oxalis pes-caprae*; Ornduff 1987), Japanese knotweed (*Fallopia japonica*; Hollingsworth & Bailey 2000), alligator weed (*Alternanthera philoxeroides*; Geng *et al.* 2007), the hawkweed *Hieracium aurantiacum* (Loomis & Fishman 2009), the invasive grass *Pennisetum setaceum* (Le Roux *et al.* 2007), and the water hyacinth (*Eichhornia crassipes*; Zhang *et al.* 2010). Researchers attempting to explain the huge success of these species usually argue that they must possess some inherent ecological advantage—for example, a particularly large degree of phenotypic plasticity ("general-purpose genotype"; Baker 1965; Loomis & Fishman 2009; Oplaat & Verhoeven 2015), superior means of spreading with human help (e.g., Gravuer *et al.* 2008; Wilson *et al.* 2009), or preadaptation to disturbed habitats (Prinzing *et al.* 2002)—that is large enough to outweigh the disadvantages of genetic uniformity. Evolutionary hypotheses for invasion success (Bossdorf *et al.* 2005; Barrett *et al.* 2008; Dlugosch & Parker 2008; Prentis *et al.* 2008) are usually not invoked in these species, because these require the presence of heritable variation in the introduced range. A possibility that so far has not yet been considered is that these genetically almost or entirely uniform invaders harbor heritable variation created through epigenetic or other nongenetic mechanisms of inheritance.

Epigenetic mechanisms such as DNA methylation or histone modifications can create heritable variation in plant phenotypes, and thus maintain evolutionary potential even in the absence of DNA sequence variation (e.g., Johannes *et al.* 2009; Zhang *et al.* 2013). Studies in natural plant populations usually find that there is extensive epigenetic variation within and among natural populations, and that this epigenetic variation often exceeds and only partly correlates with DNA sequence variation (e.g., Vaughn *et al.* 2007; Richards *et al.* 2012; Schmitz *et al.* 2013; Medrano *et al.* 2014; Schulz *et al.* 2014). One possible origin for natural epigenetic variation is spontaneous epimutation, which occurs at higher frequencies than mutations of DNA sequence (Verhoeven *et al.* 2010; van der Graaf *et al.* 2015). Invasive populations that went through an extreme genetic bottleneck may thus accumulate novel epigenetic variation much more quickly

than DNA sequence variation, and it is therefore possible that they are more variable at the epigenetic than at the genetic level.

The second possible origin of natural epigenetic variation is inheritance of environmentally induced epigenetic changes. Recent studies have repeatedly demonstrated that environmental differences can induce epigenetic changes, which may in turn become inherited to offspring (e.g., Verhoeven *et al.* 2010; Kou *et al.* 2011; Bilichak *et al.* 2012; Rasmann *et al.* 2012). Although there is currently much debate about the true importance of such environmentally induced transgenerational effects (Pecinka & Scheid 2012; Heard & Martienssen 2014), much of this debate is about whether epigenetic changes can be passed on to sexual offspring (i.e., whether they can be meiotically inherited). An aspect, which so far has received surprisingly little attention, is that many plants reproduce vegetatively, and that for many species this is the main mode of reproduction. For vegetative reproduction, no germline needs to be passed on, and epigenetic changes may potentially persist forever through mitotic inheritance. Hence, clonal plants should generally exhibit the greatest potential for environmentally driven epigenetic differentiation (Latzel & Klimešová 2010; Verhoeven & Preite 2013; Douhovnikoff & Dodd 2015). As all of the aforementioned genetically uniform invaders are clonal plants, it is an intriguing possibility that epigenetically based habitat adaptation—through epimutation and selection, or through persistent environmentally induced epigenetic changes—is a key mechanism explaining their huge success despite their genetic uniformity.

It is important to note that besides epigenetic changes, further mechanisms can cause transgenerational phenotypic effects, including simple nutritional effects, the transmission of defense chemicals, hormones or other signaling molecules, and the vertical transmission of endophytic mutualists. All of these mechanisms are encompassed by the broader concepts of maternal environmental effects or transgenerational plasticity (Roach & Wulff 1987; Rossiter 1996; Herman & Sultan 2011). Many previous studies have demonstrated such transgenerational plasticity and its adaptive significance at the level of plant phenotype (e.g., Galloway & Etterson 2007; Whittle *et al.* 2009; Latzel *et al.* 2014). However, these efforts usually did not address the underlying physiological or epigenetic mechanisms. In any case, as for epigenetic mechanisms, clonal plants—including genetically uniform invaders—have great potential also for other, non-epigenetic transgenerational effects.

One of the best-known cases of a genetically uniform plant invader is Japanese knotweed (*Fallopia japonica*), where a single clone has spread aggressively through a broad range of habitats in temperate Europe and North America (Beerling *et al.* 1994; Bailey & Conolly 2000; Grimsby *et al.* 2007; Gerber *et al.* 2008; Bailey *et al.* 2009). A previous study of Japanese knotweed found that invasive populations in the United States are indeed epigenetically differentiated between different habitats in the field (Richards *et al.* 2012). Here, we took advantage of an extensive live collection of invasive Japanese knotweed clones from different origins across Central Europe to test for heritable phenotypic and epigenetic variation in the species. Specifically, we asked the following questions: (1) Do genetically uniform populations of Japanese knotweed harbor significant epigenetic variation? (2) If yes, is this epigenetic variation accompanied by significant variation in phenotype? (3) What are the relationships between environmental, epigenetic, and phenotypic variation?

MATERIALS AND METHODS

Study system

Japanese knotweed (*F. japonica* (Houtt.) Ronse Decr. var. *japonica*) is a perennial member of the Polygonaceae native to Japan, Korea, and China. It is a tall and vigorous forb with a gynodioecious breeding system and the ability to reproduce and spread vegetatively through an extensive rhizome network (Smith *et al.* 2007). During the nineteenth century, the species was introduced to Europe and North America as an ornamental and forage plant (Bailey & Conolly 2000). In its introduced range, *F. japonica* invades ruderal habitats and river banks, where it often forms dense monospecific stands, decreases native biodiversity (Gerber *et al.* 2008; Aguilera *et al.* 2010), and alters nutrient cycles (Pyšek 2009). Previous research suggests that multiple factors are involved in the dominance and invasion success of *F. japonica*, its sister species *Fallopia sachalinensis*, and their hybrid *Fallopia × bohemica*: invasive knotweeds not only spread extremely rapidly through clonal growth (Bímová *et al.* 2003; Pyšek *et al.* 2003) but also appear to suppress native plants through allelopathy and other soil-mediated effects (Siemens & Blossey 2007; Murrell *et al.* 2011; Parepa *et al.* 2013b), and they possess a

superior ability to rapidly take up nutrients and take advantage of nutrient pulses (Parepa *et al.* 2013a). The ecological impacts, together with substantial structural damage in urban areas and high removal costs make Japanese knotweed one of the most problematic plant invaders of temperate regions worldwide (Lowe *et al.* 2000). In its invasive European range, *F. japonica* appears to be genetically uniform and is represented by only a single female clone (Hollingsworth & Bailey 2000; Mandák *et al.* 2005; Krebs *et al.* 2010).

In 2005, fresh rhizomes were collected in 83 invasive populations of *F. japonica* across seven different regions in Central Europe, spanning a broad geographic and climatic range from populations near Kiel, close to the North Sea in Northern Germany, to populations in the Ticino region in Southern Switzerland (Krebs *et al.* 2010; Fig. 1a,b). In each local population, one individual clone was collected. All clones had originally been genotyped using random amplified polymorphic DNA (RAPD) markers and, together with leaf morphological characteristics and flow cytometry estimates of ploidy levels, verified to be genetically uniform *F. japonica* var. *japonica* (Krebs *et al.* 2010).

Rhizome cuttings from these invasive populations were used to establish a live collection at the Botanical Garden of the University of Marburg, Germany, where all clones were grown under the same environmental conditions for 2 years. In 2007, the collection was moved to the University of Bern, where we re-planted rhizome cuttings from all clones into 121 pots with a 1:1 mixture of sand and fresh field soil (RICOTER Erdaufbereitung AG, Aarberg, Switzerland).

Molecular analyses

In early 2010, we collected fresh and intact leaves from each of the 83 clones in our live collection and dried them on silica gel. From each sample, we extracted total genomic DNA using the Qiagen DNeasy Plant Mini kit (Qiagen Inc., Valencia, CA). We used AFLP markers to reexamine genotypic variation among our samples, following a modified AFLP protocol as in Zhang *et al.* (2010). To test for epigenetic variation among samples, we analyzed their DNA methylation profiles using the methylation-sensitive amplified polymorphism (MSAP) method (Reyna-López *et al.* 1997; Cervera *et al.* 2002). MSAP is a modification of the AFLP method where the frequent cutter *MseI* is replaced by either of the two methylation-sensitive

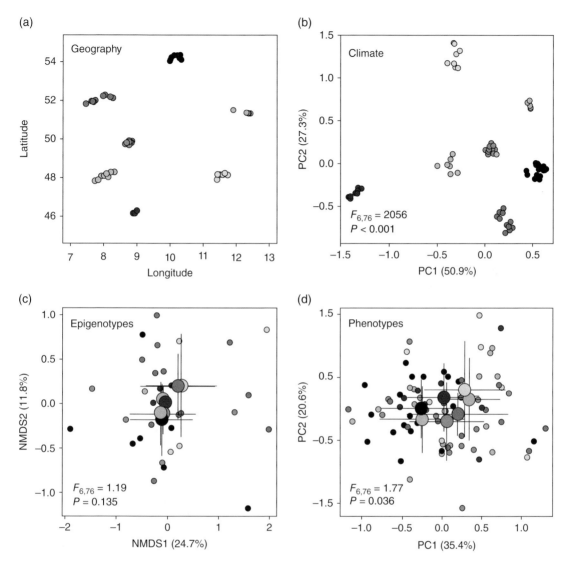

Fig. 1 Geographic (a), climatic (b), epigenetic (c), and phenotypic (d) distances among Japanese knotweed clones. Small symbols represent 83 different Central European origins of invasive Japanese knotweed (*Fallopia japonica*), with different colors for each of the seven geographic regions (see Table 1). The large symbols in the last two panels represent the regional means (±SD). Climate and phenotype plots are based on principal component analysis (PCA) of nine bioclimatic variables and seven phenotypic traits, respectively, whereas the epigenetic plot represents a non-metric multidimensional scaling (NMDS) analysis of 19 polymorphic methylation-sensitive amplified polymorphism (MSAP) markers. The test statistics are the results of multivariate tests for regional differentiation (see section "Materials and Methods" for details). (*See insert for color representation of the figure.*)

restriction enzymes *Msp*I and *Hpa*II. *Msp*I and *Hpa*II both recognize CCGG cutting site, but *Msp*I is blocked when the outer C is methylated, while *Hpa*II is blocked when either of the Cs is fully methylated. To assess the fragment sizes of AFLP and MSAP markers, we used an

ABI 3730 Genetic Analyzer (Applied Biosystems, Waltham, MA) and then determined AFLP and MSAP profiles using Genemapper 3.7 (Applied Biosystems, Waltham, MA). We scored 285 AFLP markers with four primer combinations (E-ACT/M-CAG, E-ACA/M-CAC,

E-ACG/M-CAA, and E-AGC/M-CAC), and 324 MSAP markers with four primer combinations (E-ACG/H-AAT, E-ACG/H-CAT, E-AGC/H-AAT, and E-AAC/H-CAC) with either *Msp*I or *Hpa*II digestion.

To assess error rates in AFLP and MSAP fingerprinting, we repeated both methods on 23 samples (Bonin *et al.* 2004; Pompanon *et al.* 2005). Error rates were 0.02, 0.03, and 0.31% per sample and locus for AFLP, MSAP with *Msp*I, and MSAP with *Hpa*II, respectively, resulting in an average of 0.5, 0.1, and 1.0 errors per multilocus fingerprint. Compared to previous studies, these error rates were rather low, indicating that our peak scoring and (epi-)genotyping were reasonably robust. The higher error rate of the *Hpa*II restriction enzyme likely resulted from the greater levels of polymorphism detected with this enzyme than with *Msp*I (see later text). In molecular marker studies, error rates often increase with the number of polymorphic loci studied (Arnaud-Haond *et al.* 2007), probably because a larger number of loci more likely includes highly polymorphic—and more error-prone—individual loci.

We assigned the 83 knotweed clones to genotypes and epigenotypes, based on their pair-wise sample mismatch distances for AFLP markers, MSAP (*Msp*I), and MSAP (*Hpa*II) markers, respectively, using GenoDive (Meirmans & van Tienderen 2004). Since for all of these markers scoring errors cannot be ruled out and may create a peak at lower genetic or epigenetic distances (Meirmans & van Tienderen 2004; Arnaud-Haond *et al.* 2007), we used a mismatch threshold for assigning two samples to different genotypes or epigenotypes. Based on the estimated 1.0 errors per multilocus epigenotype for *Hpa*II, we allowed one mismatch between samples for this restriction enzyme. As this was already the lowest possible threshold, we used the same for AFLP and *Msp*I data, even though these markers had lower error rates. Thus, samples were generally assigned to different genotypes or epigenotypes if they differed in at least two AFLP or MSAP markers.

To characterize epigenetic diversity across samples, as well as within and among the different regions, we calculated the total numbers of polymorphic MSAP loci for the *Msp*I cutter, the *Hpa*II cutter, or both combined, separately for each region, or across all samples. Finally, we explored the epigenetic similarity of different origins using non-metric multidimensional scaling (NMDS; *vegan* package in R; Oksanen *et al.* 2015) of multi-locus MSAP profiles, using only the 19 loci with levels of polymorphism above 0.05. To test for regional differentiation, we used the *adonis* function in *vegan*, which hierarchically partitions the variance in distance matrices, in our case a Euclidean matrix based on the first and second NMDS axes.

Phenotypic variation

To test for phenotypic variation among the 83 knotweed clones, we set up a 2-year common garden experiment in spring 2008. We planted rhizome cuttings of 8 cm length, each with two intact nodes, into 61 pots filled with a 1 : 1 mixture of sand and fresh field soil, at 5 cm below the soil surface. Since there was some variation in the thickness of the planted rhizomes, we recorded the diameter of each. We planted five replicate pots for each clone. The 415 pots were arranged in five blocks in an experimental garden at the University of Bern, with one replicate of each clone in each block, and complete randomization within blocks. In the fall of 2008, we measured the leaf chlorophyll content of each clone, using a chlorophyll meter (SPAD-502, Konika Minolta, Osaka, Japan). We then cut all aboveground biomass, determined the total leaf area of each plant using a leaf area meter (LI-3100, Li-Cor, Lincoln, NE), dried and weighed leaves and remaining biomass, and determined the specific leaf area (SLA) of each plant as the ratio of total leaf area to total leaf dry weight, as well as its total dry aboveground biomass. In the spring of 2009, all clones resprouted from the rhizomes. To avoid nutrient depletion, we added slow-release fertilizer to each pot once. In the fall of 2009, we measured chlorophyll content and SLA, as described earlier, and then harvested the entire plants, including roots and rhizomes, which were carefully washed and separated. All plant parts were dried and weighed.

We analyzed the phenotype data with analysis of covariance, with initial size as a covariate, and experimental block, region of origin, and clone nested within region as categorical factors. As initial size, we used the rhizome volume, calculated from length and diameter measurements. We did these analyses for seven phenotypic traits: SLA in 2008 and 2009, leaf chlorophyll content in 2008 and 2009, total aboveground dry biomass in 2008, dry rhizome biomass in 2009, and total belowground (rhizome + root) biomass in 2009.

To explore overall phenotypic similarities among different origins, we conducted a principal component analysis (PCA; *vegan* package in R) using all seven phenotypic traits, and we then analyzed the Euclidean distance matrix based on the first and second PC with the *adonis* function in R to test for regional differentiation in multivariate phenotypes.

Relationships between environmental, epigenetic, and phenotypic variation

To quantify climatic variation, we used long-term (1950–2000) data from the WorldClim database (www.worldclim.org). Specifically, we used nine bioclimatic variables that we considered to be most important to describe mean climate as well as climatic variability: annual mean temperature, total annual precipitation, the minima and maxima of temperature and precipitation (temperature of the coldest/hottest month and precipitation of the driest/wettest month), seasonality (=standard deviation) of temperature and precipitation, and isothermality (=temperature mean diurnal range/annual temperature range). To explore regional climate differences, we conducted a PCA and tested for regional differences as with the phenotypic data, using a Euclidean distance matrix based on the first two climate PCs.

To examine the relationships between environmental, epigenetic, and phenotypic variation, we used structural equation modeling (SEM), which related the multivariate MSAP and phenotypic data with climatic data from the origins of the knotweed clones. We used the *lavaan* package in R (Rosseel 2012) to test for the effects of climate (nine variables) and epigenetic variation (first two NMDS axes) on phenotypic variation (seven traits), also allowing indirect effects of climate on phenotype through epigenetic variation. We started with a full model where each climate variable had a direct effect on each phenotypic trait as well as the two epigenetic variables, and also each of the two epigenetic variables affected each phenotypic trait. We then removed climate variables that did not show any significant correlation with epigenotype or phenotype, as well as phenotypic traits that were not significantly affected by any of the climate or epigenetic variables, to obtain the most parsimonious structural equation model.

RESULTS

Molecular variation

We found significant epigenetic variation, but only very little genetic variation among the 83 knotweed clones. A small fraction of AFLP markers (12 out of

285, 4.2%) were polymorphic, with mostly one, and rarely two or three mismatches between samples. GenoDive assigned all samples to one clone, that is, for samples that differed by two or three loci all one-mismatch intermediates existed, so that all were assigned to the same clone. MSAP markers, in contrast, were more variable, particularly when the restriction enzyme *Hpa*II was used. With the *Msp*I cutter, 7% of the MSAP loci (24 out of 324) were polymorphic, but with *Hpa*II there were 25% polymorphic loci (81 out of 324; Table 1). With one mismatch allowed between samples, the 83 samples were assigned to 6 or 27 unique epigenotypes using the *Msp*I and *Hpa*II cutter, respectively. Because of the low levels of polymorphism in the *Msp*I data, adding the *Msp*I data to the *Hpa*II data hardly altered the total the number of polymorphic loci or epigenotypes identified, and did not affect the conclusions drawn. Therefore, we only present the MSAP analyses based on the *Hpa*II cutter. Out of the 27 epigenotypes identified, one epigenotype accounted for about 2/3 of our samples. This epigenotype was present in all of the seven regions, and dominant (>50% of the samples) in five of them, whereas most of the other epigenotypes were rare and occurred in only one or two regions, with one or two origins per region (Table 1). The analysis of the NMDS-based distance matrix did not find significant regional differentiation in multilocus MSAP epigenotypes (Fig. 1c).

Phenotypic variation

Many of the phenotypic traits measured were strongly affected by the spatial block in our experimental garden and, to a lesser degree, by the initial sizes of the planted rhizomes (Table 2). In several of the traits, we found significant differences between geographic regions, or among origins (=clones) within regions (Table 2). There were significant region effects on SLA in 2008, the chlorophyll content in 2009, and in the final total belowground biomass. For instance, plants from the regions of Kiel and Freiburg consistently produced more belowground biomass, but had lower SLA, than the plants from Leipzig and München (Fig. 2). These findings for individual phenotypic traits were corroborated by the analysis of multivariate phenotypic distances, which also indicated significant regional differentiation in phenotype (Fig. 1d). Besides regional differences, we also found significant

Table 1 Results of MSAP marker analysis, using the *Hpa*II restriction enzyme, that quantify DNA methylation variation in common-garden offspring of invasive Japanese knotweed from 83 Central European origins

Region	Number of samples[*]	Polymorphic loci	% of loci polymorphic	Number of epigenotypes	Epigenotypes[†]
Kiel	20	42	13%	10	e1(10×), e2, e3(2×), e4, e5, e6, e7, e8, e9, e10
Osnabrück	16	37	11%	8	e1(9×), e11, e12, e13, e14, e15, e16, e17
Leipzig	7	8	2%	2	e1(6×), e18
Darmstadt	13	23	7%	5	e1(9×), e2, e19, e20, e21
München	8	34	10%	6	e1(3×), e22, e23, e24, e25, e26
Freiburg	10	7	2%	1	e1(10×)
Ticino	9	5	2%	3	e1(7×), e4, e27
Total	83	81	25%	27	

[*]Within a region of origin, each sample represents a geographically distinct location.
[†]Different epigenotypes are denoted by e1–e27. If an epigenotype was found multiple times within a region, this is indicated in brackets.

Table 2 Analysis of variance of phenotypic variation in common-garden offspring from 83 origins of the invasive Japanese knotweed

Phenotypic traits	Block (d.f. = 4)		Initial size (d.f. = 1)		Region (d.f. = 6)		Origin (region) (d.f. = 76)	
	F-ratio	P-value	F-ratio	P-value	F-ratio	P-value	F-ratio	P-value
SLA (2008)	184.27	<0.001	24.40	<0.000	2.35	**0.031**	1.42	**0.020**
SLA (2009)	11.56	**0.001**	1.79	0.182	0.52	0.795	0.93	0.651
Chlorophyll (2008)	180.10	<0.001	4.75	**0.030**	0.50	0.808	0.96	0.563
Chlorophyll (2009)	151.93	<0.001	1.06	**0.304**	2.42	**0.027**	0.94	0.619
Aboveground mass (2008)	44.50	<0.000	17.41	<0.000	0.82	0.555	0.92	0.668
Rhizome mass (2009)	0.04	0.839	37.29	<0.000	1.76	0.107	1.40	**0.024**
Total belowground mass (2009)	1.29	0.257	0.46	0.499	2.57	**0.019**	1.12	0.247

The plants originated from seven different regions in Central Europe. Initial size is the estimated volume of the planted rhizomes. Significant *P*-values are in bold.
d.f., degrees of freedom and the residual d.f. = 323; SLA, specific leaf area.

differences among origins within regions for SLA 2008 and final rhizome mass (Table 2).

Relationships between environmental, epigenetic, and phenotypic variation

The most parsimonious SEM linking environmental, epigenetic, and phenotypic variation contained only five of the nine tested climate variables and five of the seven measured phenotypic traits (Fig. 3). The fit of the model was good (RMSEA = 0.000, *P* = 0.680).

Interestingly, variables describing climate means were dropped from the model, and those maintained in the model were all related to climate variability and extremes, indicating that these may be most relevant for explaining variability of Japanese knotweed. The model indicated that phenotypic variation among clones of Japanese knotweed is related to both environmental and epigenetic variation, with two groups of traits: (1) variation in SLA was significantly related to climate of origin but especially to the epigenetic variation among clones. Some of these strong epigenetic influences appeared to be climate-related, whereas

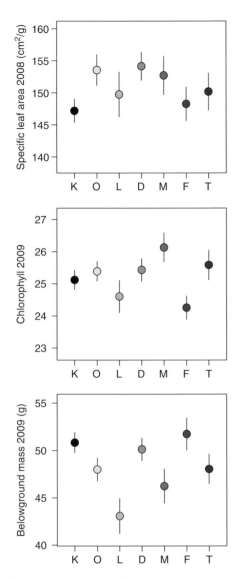

Fig. 2 Regional variation in phenotype among invasive Japanese knotweed clones. The data (means ± SD) are from common-garden progeny of 83 Central European origins of *Fallopia japonica* distributed across seven geographic regions: K = Kiel, O = Osnabrück, L = Leipzig, D = Darmstadt, M = München, F = Freiburg, and T = Ticino. The colors are as in Fig. 1. (*See insert for color representation of the figure.*)

others were independent of climate; (2) variation in above- and belowground biomass traits was significantly correlated with several of the climate variables, but not with epigenetic variation. The strongest

patterns were found for belowground biomass, which was significantly related to four of the five climate variables, in particular precipitation variability.

DISCUSSION

Some invasive plants appear to colonize large ranges without any genetic diversity. This is surprising, given the importance of genetic diversity for adaptation and thus, presumably, range expansion. Here, we find that genetically uniform clones of Japanese knotweed, one of the world's most successful plant invaders (Lowe *et al.* 2000), harbor substantial epigenetic and phenotypic variation, and that this variation is associated with climate of origin and thus possibly involved in habitat adaptation.

Our AFLP analysis of the 83 clones of Japanese knotweed indicated that all of them belonged to the same genetic clone, or were at least nearly identical, which is consistent with the results of several previous molecular studies of European populations (Hollingsworth & Bailey 2000; Mandák *et al.* 2005; Krebs *et al.* 2010). In contrast, MSAP markers detected substantial DNA methylation variation and assigned our 83 samples to 27 different epigenotypes. Although we cannot entirely rule out DNA sequence variation in our samples—rare somatic mutations cannot be distinguished from scoring errors by the AFLP method (Douhovnikoff & Dodd 2003; Bonin *et al.* 2004; Meirmans & van Tienderen 2004)—our results indicate that epigenetic variation is at least an order of magnitude larger than genetic variation among Central European origins of Japanese knotweed. As all Japanese knotweed in Europe is descended from a single clone (Bailey & Conolly 2000), there are two possible explanations: (1) spontaneous epimutations have accumulated much faster than sequence mutations during the more than 150 years since the species was introduced, or (2) environmental differences in the source populations induced epigenetic changes that have persisted through several years of vegetative propagation in a common environment.

If the observed epigenetic variation is the result of spontaneous epimutation, it could be largely neutral, without any functional and ecological significance. DNA methylation changes are expected to have the greatest effects when they occur close to transposable elements or in the regulatory regions of genes, but they are thought to be of little functional significance in the

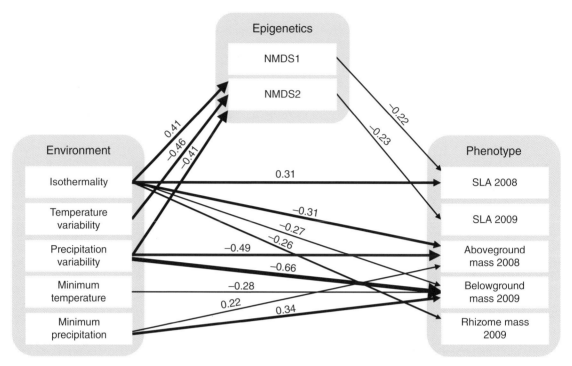

Fig. 3 Relationships between environmental, epigenetic, and phenotypic variation among Japanese knotweed clones. The data describe 83 Central European origins of the invasive Japanese knotweed (*Fallopia japonica*). The epigenetic variables are the first two axes from a non-metric multidimensional (NMDS) analysis of 19 polymorphic MSAP (=DNA methylation) markers. The phenotyping of different clones was after several years of propagation in a common environment. The arrows indicate significant correlations between variables, with arrow thickness scaled to the effect size (values on arrows).

coding region of genes (Mirouze *et al.* 2009; Becker *et al.* 2011; van der Graaf *et al.* 2015). If epigenetic variation is largely due to epimutation, we should find many private epigenotypes that occur only in single locations. This is indeed what we find. Many of the identified epigenotypes are rare, which suggests that even if part of the epigenetic variation in Japanese knotweed is nonrandom, epimutation rates may be high and add enough "noise" for clones to be assigned to different epigenotypes even if they share a common epigenetic response to the environment.

If, in contrast, the observed epigenetic variation is the result of natural selection or environmental induction, then it should be accompanied by systematic variation in phenotype. In our common garden experiment, we found significant variation in several phenotypic traits among the different clones representing different populations and regions of origin. Besides differences in belowground biomass and rhizome production, the

clones significantly differed in two key traits of plant ecological strategy: leaf chlorophyll content and SLA. Leaf chlorophyll is closely related to leaf nitrogen, photosynthetic capacity, and nutrient allocation strategy, whereas SLA is thought to capture one of the main general axes of plant strategy and life history variation (Westoby *et al.* 2002; Wright *et al.* 2004), with high SLA related to fast growth and competitive ability in nutrient-rich environments, and low SLA reflecting a slower growth strategy in less favorable environments. SLA has frequently been found to be positively correlated with plant invasiveness (e.g., Grotkopp *et al.* 2002; Hamilton *et al.* 2005; Grotkopp & Rejmánek 2007).

In summary, the 83 investigated genetically identical knotweed clones were not only epigenetically variable, but they also showed significant variation in several key phenotypic traits related to habitat adaptation. It is important to stress that we measured both epigenetic and phenotypic variation in a common garden, so the observed variation did not reflect phenotypic plasticity, but it was stable variation that persisted for several years and across several vegetative generations, despite identical environmental conditions.

To further corroborate the ecological significance of the observed epigenetic and phenotypic variation, we used a structural equation model that linked the two kinds of variation with the climatic variation of the 83 origins. We found that with regard to their epigenetic and climatic correlates, the studied phenotypic traits fell into three groups: (1) SLA was strongly correlated with epigenetic variation and to a lesser degree with climatic variation. Our data thus support the idea that for this important plant strategy trait different epigenetically based phenotypes of knotweed may have been selected in different areas of the invasive European range. (2) Biomass traits were significantly correlated with climate of origin, but not with epigenotype. Here, the persistent phenotypic differences may also reflect some aspect of habitat adaptation, for instance, greater root production in resource-limited habitats (Poorter & Nagel 2000). The underlying mechanisms may either be non-epigenetic such as long-term nutritional or physiological effects (Herman & Sultan 2011; Latzel et al. 2014), or the low-resolution MSAP method did not capture the relevant epigenetic variation. (3) Leaf chlorophyll content was neither correlated with epigenetic variation nor with any of the climatic variables, despite significant regional differentiation detected in the analyses of variance. Here, our analysis most likely did not include any of the environmental factors that are relevant for this functional trait.

Altogether, the SEM analysis supports the idea that at least part of the observed phenotypic variation is functionally related to epigenetic variation, and that the observed epigenetic variation cannot entirely result from epimutation. Both kinds of variation thus appear to be ecologically relevant and ultimately driven by environmental differences among habitats of origin. A more thorough understanding of these functional relationships would require (a) more detailed small-scale environmental information from all source populations, in particular about light, soil, and nutrient conditions, and (b) higher-resolution epigenomic data. Both were beyond the scope of our study, and in particular the latter is currently still not feasible in a polyploid species with a large genome such as Japanese knotweed.

There is one previous study on epigenetic variation in invasive knotweed. Richards et al. (2012) also used

AFLP and MSAP markers to analyze 16 invasive Fallopia populations on Long Island, NY. In contrast to our study, however, their samples contained different knotweed species and hybrids, including multiple genotypes of F. japonica. Consistent with our study, they found that epigenetic variation greatly exceeded genetic variation, and that plants with the same AFLP haplotypes differed at the level of DNA methylation. Although the same authors also showed that genetically identical plants from different populations strongly differed in phenotype (Richards et al. 2008), they did not link epigenetic and phenotypic variation, and their study design did not allow testing for environmental correlates. We are aware of only one other epigenetic study of an invasive plant: Gao et al. (2010) analyzed epigenetic variation, again using MSAP, in three (almost) genetically uniform populations of the invasive alligator weed (A. philoxeroides) in China. They, too, found that epigenetic variation was much greater than DNA sequence variation, and that populations from different origins maintained part of their epigenetic variation in a common environment.

Our study demonstrates that genetically uniform clones of Japanese knotweed are epigenetically and phenotypically variable across Central Europe, and that this variation is related to the environment of origin. A plausible interpretation is that Japanese knotweed, despite its genetic uniformity, has adapted to different habitats through epigenetic or other nongenetic means. It is an intriguing possibility that similar results might be found in other successful clonal invaders, and we clearly need more studies that thoroughly address these questions across a larger range of clonal invaders.

Epigenetic variation is generally thought to be more dynamic than genetic variation, and theoretical models have shown that epigenetic inheritance may be adaptive particularly in changing environments (e.g., Geoghegan & Spencer 2012; Klironomos et al. 2013). If epigenetically based adaptation is possible, it might play a particularly important role in biological invasions, and it might be "game-changing" for genetically uniform clonal invaders. Clearly, we need to study not only the genetics, but also the epigenetics of colonizing species.

ACKNOWLEDGMENTS

We are grateful to Christine Krebs for providing the collection of knotweed clones; to Carole Adolf, Andreas Burri, Moritz Joest, Sergio Menjivar, Pius Winiger, and

Silvia Zingg for their help with setup, maintenance, and harvest of the experiment; and to Santiago Soliveres for his help with the SEM analysis. This work was financially supported by the Chinese Scholarship Council (scholarship to YYZ), the Hans-Sigrist-Stiftung (fellowship to OB), and by the Swiss National Science Foundation (grant no. 31EE30-131171 to OB and MF). Molecular analyses were carried out at the Genetic Diversity Center at ETH Zürich.

REFERENCES

Aguilera AG, Alpert P, Dukes JS, Harrington R (2010) Impacts of the invasive plant *Fallopia japonica* (Houtt.) on plant communities and ecosystem processes. *Biological Invasions*, **12**, 1243–1252.

Arnaud-Haond S, Duarte CM, Alberto F, Serrão EA (2007) Standardizing methods to address clonality in population studies. *Molecular Ecology*, **16**, 5115–5139.

Bailey JP, Bímová K, Mandák B (2009) Asexual spread versus sexual reproduction and evolution in Japanese knotweed *s.l.* sets the stage for the "Battle of the Clones". *Biological Invasions*, **11**, 1189–1203.

Bailey JP, Conolly AP (2000) Prize—winners to pariahs—a history of Japanese knotweed *s.l.* (Polygonaceae) in the British Isles. *Watsonia*, **23**, 93–110.

Baker HG (1965) Characteristics and modes of origin of weeds. In: *The Genetics of Colonizing Species* (eds. Baker HG, Stebbins GL), pp. 147–172. Academic Press, New York.

Barrett SCH, Colautti RI, Eckert CG (2008) Plant reproductive systems and evolution during biological invasion. *Molecular Ecology*, **17**, 373–383.

Becker C, Hagmann J, Müller J *et al.* (2011) Spontaneous epigenetic variation in the *Arabidopsis thaliana* methylome. *Nature*, **480**, 245–249.

Beerling DJ, Bailey JP, Conolly AP (1994) *Fallopia japonica* (Houtt.) Ronse Decraene (*Reynoutria japonica* houtt, *Polygonum cuspidatum* Sieb. & Zucc.). *Journal of Ecology*, **82**, 959–979.

Bilichak A, Ilnystkyy Y, Hollunder J, Kovalchuk I (2012) The progeny of *Arabidopsis thaliana* plants exposed to salt exhibit changes in DNA methylation, histone modifications and gene expression. *PLoS One*, **7**, e30515.

Bímová K, Mandák B, Pyšek P (2003) Experimental study of vegetative regeneration in four invasive *Reynoutria* taxa (Polygonaceae). *Plant Ecology*, **166**, 1–11.

Bonin A, Bellemain E, Eidesen PB *et al.* (2004) How to track and assess genotyping errors in population genetics studies. *Molecular Ecology*, **13**, 3261–3273.

Bossdorf O, Auge H, Lafuma L *et al.* (2005) Phenotypic and genetic differentiation between native and introduced plant populations. *Oecologia*, **144**, 1–11.

Cervera MT, Ruiz-García L, Martínez-Zapater JM (2002) Analysis of DNA methylation in *Arabidopsis thaliana* based on methylation-sensitive AFLP markers. *Molecular Genetics and Genomics*, **268**, 543–552.

Dlugosch KM, Parker IM (2008) Founding events in species invasions: genetic variation, adaptive evolution, and the role of multiple introductions. *Molecular Ecology*, **17**, 431–449.

Douhovnikoff V, Dodd RS (2003) Intra-clonal variation and a similarity threshold for identification of clones: application to *Salix exigua* using AFLP molecular markers. *Theoretical and Applied Genetics*, **106**, 1307–1315.

Douhovnikoff V, Dodd RS (2015) Epigenetics: a potential mechanism for clonal plant success. *Plant Ecology*, **216**, 227–233.

Galloway LF, Etterson JR (2007) Transgenerational plasticity is adaptive in the wild. *Science*, **318**, 1134–1136.

Gao L, Geng Y, Li B, Chen J, Yang J (2010) Genome-wide DNA methylation alterations of *Alternanthera philoxeroides* in natural and manipulated habitats: implications for epigenetic regulation of rapid responses to environmental fluctuation and phenotypic variation. *Plant, Cell & Environment*, **33**, 1820–1827.

Geng YP, Pan XY, Xu CY *et al.* (2007) Phenotypic plasticity rather than locally adapted ecotypes allows the invasive alligator weed to colonize a wide range of habitats. *Biological Invasions*, **9**, 245–256.

Geoghegan JL, Spencer HG (2012) Population-epigenetic models of selection. *Theoretical Population Biology*, **81**, 232–242.

Gerber E, Krebs C, Murrell C *et al.* (2008) Exotic invasive knotweeds (*Fallopia* spp.) negatively affect native plant and invertebrate assemblages in European riparian habitats. *Biological Conservation*, **141**, 646–654.

Gravuer K, Sullivan JJ, Williams PA, Duncan RP (2008) Strong human association with plant invasion success for *Trifolium* introductions to New Zealand. *Proceedings of the National Academy of Sciences of the USA*, **105**, 6344–6349.

Grimsby JL, Tsirelson D, Gammon MA, Kesseli R (2007) Genetic diversity and clonal vs. sexual reproduction in *Fallopia* spp. (Polygonaceae). *American Journal of Botany*, **94**, 957–964.

Grotkopp E, Rejmánek M (2007) High seedling relative growth rate and specific leaf area are traits of invasive species: phylogenetically independent contrasts of woody angiosperms. *American Journal of Botany*, **94**, 526–532.

Grotkopp E, Rejmánek M, Rost TL (2002) Toward a causal explanation of plant invasiveness: seedling growth and life-history strategies of 29 pine (*Pinus*) species. *American Naturalist*, **159**, 396–419.

Hamilton MA, Murray BR, Cadotte MW *et al.* (2005) Life-history correlates of plant invasiveness at regional and continental scales. *Ecology Letters*, **8**, 1066–1074.

Heard E, Martienssen RA (2014) Transgenerational epigenetic inheritance: myths and mechanisms. *Cell*, **157**, 95–109.

Herman JJ, Sultan SE (2011) Adaptive transgenerational plasticity in plants: case studies, mechanisms, and implications for natural populations. *Frontiers in Plant Science*, **2**, 102.

Hollingsworth ML, Bailey JP (2000) Evidence for massive clonal growth in the invasive weed *Fallopia japonica* (Japanese knotweed). *Botanical Journal of the Linnean Society*, **133**, 463–472.

Johannes F, Porcher E, Teixeira FK *et al.* (2009) Assessing the impact of transgenerational epigenetic variation on complex traits. *PLoS Genetics*, **5**, e1000530.

Klironomos FD, Berg J, Collins S (2013) How epigenetic mutations can affect genetic evolution: model and mechanism. *BioEssays*, **35**, 571–578.

Kou HP, Li Y, Song XX *et al.* (2011) Heritable alteration in DNA methylation induced by nitrogen-deficiency stress accompanies enhanced tolerance by progenies to the stress in rice (*Oryza sativa* L.). *Journal of Plant Physiology*, **168**, 1685–1693.

Krebs C, Mahy G, Matthies D *et al.* (2010) Taxa distribution and RAPD markers indicate different origin and regional differentiation of hybrids in the invasive *Fallopia* complex in central-western Europe. *Plant Biology*, **12**, 215–223.

Latzel V, Janeček Š, Doležal J, Klimešová J, Bossdorf O (2014) Adaptive transgenerational plasticity in the perennial *Plantago lanceolata*. *Oikos*, **123**, 41–46.

Latzel V, Klimešová J (2010) Transgenerational plasticity in clonal plants. *Evolutionary Ecology*, **24**, 1537–1543.

Le Roux JJ, Wieczorek AM, Wright MG, Tran CT (2007) Super-genotype: global monoclonality defies the odds of nature. *PLoS One*, **2**, e590.

Loomis ES, Fishman L (2009) A continent-wide clone: population genetic variation of the invasive plant *Hieracium aurantiacum* (orange hawkweed; Asteraceae) in North America. *International Journal of Plant Sciences*, **170**, 759–765.

Lowe S, Browne M, Boudjelas S, De Poorter M (2000) *100 of the World's Worst Invasive Alien Species: A Selection from the Global Invasive Species Database*. Invasive Species Specialist Group, Auckland, New Zealand.

Mandák B, Bimová K, Pyšek P, Štěpánek J, Plačková I (2005) Isoenzyme diversity in *Reynoutria* (Polygonaceae) taxa: escape from sterility by hybridization. *Plant Systematics and Evolution*, **253**, 219–230.

Medrano M, Herrera CM, Bazaga P (2014) Epigenetic variation predicts regional and local intraspecific functional diversity in a perennial herb. *Molecular Ecology*, **23**, 4926–4938.

Meirmans PG, van Tienderen PH (2004) GENOTYPE and GENODIVE: two programs for the analysis of genetic diversity of asexual organisms. *Molecular Ecology Notes*, **4**, 792–794.

Mirouze M, Reinders J, Bucher E *et al.* (2009) Selective epigenetic control of retrotransposition in *Arabidopsis*. *Nature*, **461**, 427–430.

Murrell C, Gerber E, Krebs C *et al.* (2011) Invasive knotweed affects native plants through allelopathy. *American Journal of Botany*, **98**, 38–43.

Oksanen J, Blanchet FG, Kindt R *et al.* (2015) vegan: Community Ecology Package. R package version 2.3-0. http://CRAN.R-project.org/package=vegan, accessed February 29, 2016.

Oplaat C, Verhoeven KJF (2015) Range expansion in asexual dandelions: selection for general-purpose genotypes? *Journal of Ecology*, **103**, 261–268.

Ornduff R (1987) Reproductive systems and chromosome races of *Oxalis pes-caprae* L. and their bearing on the genesis of a noxious weed. *Annals of the Missouri Botanical Garden*, **74**, 79–84.

Parepa M, Fischer M, Bossdorf O (2013a) Environmental variability promotes plant invasion. *Nature Communications*, **4**, 1604.

Parepa M, Schaffner U, Bossdorf O (2013b) Help from under ground: soil biota facilitate knotweed invasion. *Ecosphere*, **4**, 31.

Pecinka A, Scheid OM (2012) Stress-induced chromatin changes: a critical view on their heritability. *Plant and Cell Physiology*, **53**, 801–808.

Pompanon F, Bonin A, Bellemain E, Taberlet P (2005) Genotyping errors: causes, consequences and solutions. *Nature Reviews Genetics*, **6**, 847–859.

Poorter H, Nagel O (2000) The role of biomass allocation in the growth response of plants to different levels of light, CO_2, nutrients and water: a quantitative review. *Australian Journal of Plant Physiology*, **27**, 595–607.

Prentis PJ, Wilson JRU, Dormontt EE, Richardson DM, Lowe AJ (2008) Adaptive evolution in invasive species. *Trends in Plant Science*, **13**, 288–294.

Prinzing A, Durka W, Klotz S, Brandl R (2002) Which species become aliens? *Evolutionary Ecology Research*, **4**, 385–405.

Pyšek P (2009) *Fallopia japonica* (Houtt.) Ronse Decr., Japanese knotweed (Polygonaceae, Magnoliophyta). In: *Handbook of Alien Species in Europe* (ed. DAISIE), p. 384. Springer, Berlin.

Pyšek P, Brock JH, Bímová K *et al.* (2003) Vegetative regeneration in invasive *Reynoutria* (Polygonaceae) taxa: the determinant of invasibility at the genotype level. *American Journal of Botany*, **90**, 1487–1495.

Rasmann S, De Vos M, Casteel CL *et al.* (2012) Herbivory in the previous generation primes plants for enhanced insect resistance. *Plant Physiology*, **158**, 854–863.

Reyna-López GE, Simpson J, Ruiz-Herrera J (1997) Differences in DNA methylation patterns are detectable during the dimorphic transition of fungi by amplification of restriction polymorphisms. *Molecular and General Genetics*, **253**, 703–710.

Richards CL, Schrey AW, Pigliucci M (2012) Invasion of diverse habitats by few Japanese knotweed genotypes is correlated with epigenetic differentiation. *Ecology Letters*, **15**, 1016–1025.

Richards CL, Walls RL, Bailey JP *et al.* (2008) Plasticity in salt tolerance traits allows for invasion of novel habitat by Japanese knotweed *s.l.* (*Fallopia japonica* and *F. bohemica*, Polygonaceae). *American Journal of Botany*, **95**, 931–942.

Roach DA, Wulff RD (1987) Maternal effects in plants. *Annual Review of Ecology and Systematics*, **18**, 209–235.

Rosseel Y (2012) lavaan: an R package for structural equation modeling. *Journal of Statistical Software*, **48**, 1–36.

Rossiter MC (1996) Incidence and consequences of inherited environmental effects. *Annual Review of Ecology and Systematics*, **27**, 451–476.

Schmitz RJ, Schultz MD, Urich MA *et al.* (2013) Patterns of population epigenomic diversity. *Nature*, **495**, 193–198.

Schulz B, Eckstein RL, Durka W (2014) Epigenetic variation reflects dynamic habitat conditions in a rare floodplain herb. *Molecular Ecology*, **23**, 3523–3537.

Siemens TJ, Blossey B (2007) An evaluation of mechanisms preventing growth and survival of two native species in invasive bohemian knotweed (*Fallopia* × *bohemica*, Polygonaceae). *American Journal of Botany*, **94**, 776–783.

Smith JMD, Ward JP, Child LE, Owen MR (2007) A simulation model of rhizome networks for *Fallopia japonica* (Japanese knotweed) in the United Kingdom. *Ecological Modelling*, **200**, 421–432.

van der Graaf A, Wardenaar R, Neumann DA *et al.* (2015) Rate, spectrum, and evolutionary dynamics of spontaneous epimutations. *Proceedings of the National Academy of Sciences of the USA*, **112**, 6676–6681.

Vaughn MW, Tanurdžić M, Lippman Z *et al.* (2007) Epigenetic natural variation in *Arabidopsis thaliana*. *PLoS Biology*, **5**, 1617–1629.

Verhoeven KJF, Jansen JJ, van Dijk PJ, Biere A (2010) Stress-induced DNA methylation changes and their heritability in asexual dandelions. *New Phytologist*, **185**, 1108–1118.

Verhoeven KJF, Preite V (2013) Epigenetic variation in asexually reproducing organisms. *Evolution*, **68**, 644–655.

Westoby M, Falster DS, Moles AT, Vesk PA, Wright IJ (2002) Plant ecological strategies: some leading dimensions of variation between species. *Annual Review of Ecology and Systematics*, **33**, 125–159.

Whittle CA, Otto SP, Johnston MO, Krochko JE (2009) Adaptive epigenetic memory of ancestral temperature regime in *Arabidopsis thaliana*. *Botany*, **87**, 650–657.

Wilson JR, Dormontt EE, Prentis PJ, Lowe AJ, Richardson DM (2009) Something in the way you move: dispersal pathways affect invasion success. *Trends in Ecology & Evolution*, **24**, 136–144.

Wright IJ, Reich PB, Westoby M *et al.* (2004) The worldwide leaf economics spectrum. *Nature*, **428**, 821–827.

Zhang YY, Fischer M, Colot V, Bossdorf O (2013) Epigenetic variation creates potential for evolution of plant phenotypic plasticity. *New Phytologist*, **197**, 314–322.

Zhang YY, Zhang DY, Barrett SCH (2010) Genetic uniformity characterizes the invasive spread of water hyacinth (*Eichhornia crassipes*), a clonal aquatic plant. *Molecular Ecology*, **19**, 1774–1786.

DISCUSSION

MELANIA CRISTESCU

TIM WRIGHT — I was interested in your finding of cryptic northern refugia and I am curious whether they were predicted, or whether it makes post hoc sense? Is there something about the assumptions of the interpretation of the haplotype networks that perhaps the place where the most common haplotype is found is not necessarily the place of origin of the expansions?

MELANIA CRISTESCU — I think it is not unlikely to imagine northern refugia in northern parts of the European continent, especially because zooplankton species can remain viable as resting eggs for long periods of time. We found some geological evidence that there were environments that maintain freshwater organisms in the northern parts of Europe.

KATHRYN HODGINS

ROBERT COLAUTTI — With regard to your point about increasing sample sizes, are you referring to more species or more samples of native and introduced populations within a species?

KATHRYN HODGINS — Well, both. If you want to look for genomic differences between native and introduced portions of the range of a species, you need to have good geographical sampling from within each range. But to look for general patterns across species in a family such as Asteraceae, you need data from many species.

RUSSELL LANDE — It may be that the reason you are having difficulties finding patterns among invasive members of the Asteraceae is not because you do not have enough data, because as you pointed out you have quite a lot of data on many species. But there is a lot of heterogeneity in these species. Some of them are old invasions, whereas others are new invasions; some have colonized similar habitats to their native ranges, whereas others have colonized novel habitats. Maybe if you can distinguish these, you can begin to sort things out. You could also look directly at phenotypes and the degree to which local adaption is occurring, which is a missing component of purely genomic approaches.

KATHRYN HODGINS — Yes, I agree. I think having more data could help answer some of these nuances, especially pulling out aspects of the history and ecology of genomic variation.

Invasion Genetics: The Baker and Stebbins Legacy, First Edition. Edited by Spencer C. H. Barrett, Robert I. Colautti, Katrina M. Dlugosch, and Loren H. Rieseberg.
© 2017 John Wiley & Sons, Ltd. Published 2017 by John Wiley & Sons, Ltd.

JOHANNA SCHMITT

JENNIFER LAU – I was wondering about the pattern of immigrants from the warmer south doing better than the locals in high latitude environments. You explained this in terms of evolutionary lag, but another idea that comes from forestry provenance trials is the idea that extremely cold years can be really important at high latitudes. Do you have ideas for differentiating between these two hypotheses? Do you really think it is evolutionary lags, or do you think it could be cold years?

JOHANNA SCHMITT – I agree. The thinking is that maybe the important selective pressures are the really extreme years, and so what we are seeing is that immigrant genotypes did well in a good year, but would have been wiped out in a bad year, and I don't know of a good way to rule that out. I will say that if you look at weather data over the past 20 years, extremely cold years are getting rarer. So my actual hunch is that it has something to do with the germination phase, which we are not able to look at. We know there is a latitudinal cline in dormancy with genotypes from high latitudes having much lower dormancy because they have to germinate right away in the short window for growth. We had a field experiment that Renee Petipas did in Finland—marking and following individuals—and she found there was selection against late germinants. Liana Burghardt, formerly of Kathleen Donohue's lab, has done some modeling showing that if you have even a small amount of dormancy, populations from northern climates will stay in the seed bank for years before they germinate. So if we had been able to look through the entire life cycle, we probably would not be seeing much of a lag. That said, I think in the phase we *did* look at there is some evidence for an evolutionary lag. The other possibility is that the northern populations have a lot of deleterious alleles because of surfing and things like that. But on the other hand, Nordic genotypes do much better at the Finnish site than at the other sites, so I think there is some evidence for local adaptation.

DOUGLAS GILL – A topic that has been absent from our conversations for the past 2 days has been the temporal aspect of recovery with old genotypes being reincorporated into contemporary populations from ancient seed banks. Is there evidence for geographical variation in seed banking in *Arabidopsis thaliana*? Does the species have seed banks?

JOHANNA SCHMITT – There is no ecological evidence of large-scale geographical variation in seed banks, although seed banks do occur in the north, for example, in Sweden and Norway. To my knowledge, nobody has systematically looked for clines in seed banks, although people are doing this a bit in Spain, but not in a way that we can compare over large geographical regions in Europe. We do know that there is a latitudinal cline in seed dormancy, and Liana Burghardt's hydrothermal seed population model of emergence (L. Burghardt *et al.*, 2015, *American Naturalist* **185**: 212–227) would suggest that there may be latitudinal clines in seed banking. On the other hand, there is counter-gradient variation in dormancy that might prevent this from happening. This general phenomenon is something that should also apply to other species.

JAMES RODGER – There has been renewed interest in whether selfing is an evolutionary dead end, particularly in plants (e.g., B. Igic and J.W. Busch, 2013, *New Phytologist* **198**: 386–397). There is plenty of evidence that selfing species are often at the tips of phylogenies, and apparently you do not get old, diverse, highly selfing groups. Yet we have plenty of evidence for local adaptation in selfing species. I would like to hear perspectives on whether people think selfing is indeed an evolutionary dead end. And if so, why?

JOHANNA SCHMITT – Working with recombinant inbred lines in a species like *Arabidopsis thaliana*, which is a highly selfing species with a small amount of outcrossing, I have thought a lot about the potential for epistatic selection. We observe a lot of epistasis in certain traits, and following low levels of outcrossing there is likely to be the generation of novel combinations, and then subsequent selfing allows lineage sorting with new adaptive combinations potentially rising to high frequency. So that might be an example of how selfing may promote rapid adaptation that is not dependent on additive genetic variation.

LOREN RIESEBERG – I think that selfers can adapt; however, because of the accumulation of genetic load I think what happens is that new selfing species originate and replace the old selfers, preventing long branches in selfing groups. However, if selfing species were only competing with themselves, they may go on for a very long time and they would adapt just fine. The problem is that new selfing species are being formed that are better than old selfers, and it is the competition between them that causes the rapid turnover that creates short lineages rather than any inability of selfers to adapt.

MARK KIRKPATRICK – I am fascinated by the research program of looking at how the fitness of alleles change in space. But I am wondering about a possible complication. The single nucleotide polymorphisms (SNPs) that you have identified are likely often not the actual cause of adaptive changes, and in some populations those SNPs may have recombined away from the allele that is actually causing the fitness effects. So rather than analyzing your data SNP by SNP, could you gain more power to capture the actual causative allele by instead looking at haplotypes of several SNPs that are close by on the chromosome?

JOHANNA SCHMITT – You can do this, and we did look for linkage blocks involving sets of tightly linked SNPs to see if there was evidence of sweeps in some of these. We also used geographical associations to identify the top candidates to examine. So you are right. The other point is that now with re-sequencing of these genotypes, we may be able to get a little more fine-grained. Part of the problem is figuring out how to prune so that you do not have redundant information, and then trying to find functional variants.

LOREN RIESEBERG – For plants, in general, there is often a very tight association with soil type. In sunflowers, for example, it really is key to know exactly what the soil characteristics are where populations are growing—what the nitrogen and water content is and so on. It strikes me that with all this vast amount of *Arabidopsis* data, it would be really powerful if information about the soil where the samples were collected was available. Are people getting this?

JOHANNA SCHMITT – That is a really excellent point. And remember there was a lot of scatter around all those climate lines, so clearly there is something else going on, and I am sure that it is partly edaphic. I think a lot of it is also to do with pathogens. For the early collections, there is very little ecological information, so all we can do is to try and find geologic maps and pinpoint where collections were made. This is something I have not wanted to attempt, but it is something that could be done.

LOREN RIESEBERG – Now with the advent of genome-wide association study (GWA mapping), it is important that when collections are made from the field we get

as much information from the site as possible. And so with our sunflower work, we are going back and recollecting many populations to get ecological information about soils, pathogens, and other plants and animals in the area. I think you may have to try and get the *Arabidopsis* community to go back and recollect everything.

JOHANNA SCHMITT – Recollecting in these old sites is what we need to do. There have been some really great new collections from the Iberian Peninsula by Carlos Alonso-Blanco who has got good ecological information. And some collections in Kazakhstan and places like that. But we also need to go back to France and Germany, and Kirsten Bomblies has done that around Tübingen. So the *Arabidopsis* community is starting to do this now.

LEE ANN ROLLINS

DOUGLAS GILL – What is the current evidence of ecological impact of cane toads on mid- to long-term residents of Australia?

LEE ANN ROLLINS – When cane toads first moved through the "top-end" of Australia, there was a great deal of concern that many native animals would be eliminated. Richard Shine is probably the best person to talk about this, but from what I understand the evidence suggests that not all animals that were expected to die have died. Whether or not that is due to resistance or behavioral changes or some combination is an area that people are looking into.

RICHARD SHINE (UNIVERSITY OF SYDNEY) – Basically what happens is that 99.9% of the large predatory lizards and some crocodiles and marsupials are killed, which is catastrophic. But they are not actually extinct, and these species in general seem to recover in areas where toads have not been present for 30–40 years. The end result is that the things these predators usually eat become more abundant, and so most species of wildlife actually become more common.

RUSSELL LANDE – I found the part of your talk on evidence for selective dispersal particularly interesting. I want to bring up an historical example as a prelude to asking you a question. There are early ideas about evolution involving selective dispersal reinforcing a process started by natural selection in R.A. Fisher's 1930 book (*The Genetical Theory of Natural Selection*).

This involves the reduction of eyes in cavefish, and he points out that after this is initiated by natural selection, individuals that can see the light better will leave the cave. And so this speeds up the evolution of loss of eyes or reduction of eyes. So in cane toads, is it a purely nonselective process that you were describing, or is it interacting with some selective advantage related to dispersal along the advancing front?

LEE ANN ROLLINS – I do not think we can say there is no evidence for selection and there is evidence for spatial sorting, and the cane toads do appear to be less fit. But it has not been explicitly tested, although it would be very surprising to me if there is no natural selection occurring.

JOHN PANNELL – Comparisons are often made between populations at the range edge and those at the core, but what about populations just back from the range edge where they may have reached carrying capacity and where there may not be selection on dispersal traits. Is there anything known about whether increased dispersal is then gradually lost?

LEE ANN ROLLINS – It certainly seems to be that not very far behind the expansion front things do change back to some degree, but we have not looked at what is driving that. There is very little genetic or genomic work done on the system, so we cannot really answer that question with what we know right now.

SPENCER BARRETT – What is motivating your current effort to look at epigenetic variation in the cane toad? I am ignorant of your system, and perhaps you have already investigated the quantitative genetics and heritability of ecologically relevant traits, but if not, why jump into epigenetics? Do you know that populations are depauperate in quantitative genetic variability, say compared with Hawaii? That appeared to be why you set up looking at epigenetics as this may possibly provide an additional source of variability.

LEE ANN ROLLINS – Ben Phillips has done a fair amount of work on the heritability of many of the traits that we have studied in this system.

GENERAL QUESTIONS TO PANEL

TROY DAY – One of the themes at this meeting is that we do not have examples of failed invasions to help identify traits that determine success. But we do have some information about the emergence of infectious disease in humans. For example, in flu there are some recorded instances of avian flu introduction into humans that do not actually adapt and cause an invasion. And people are starting to investigate at the molecular level why it might be that certain genotypes of flu are not able to colonize humans. So what are your thoughts on whether we can use information on specific systems like that to understand these processes more generally? In another example, there has been a lot of work looking for things like H.G. Baker's list of "ideal" traits in weeds, but instead concerning risk factors in the emergence of diseases. The common thing that comes out is that certain taxonomic groups, such as protozoa and viruses, are a greater risk than other taxonomic groups, and within viruses RNA viruses are more risky than DNA viruses. I wonder to what extent these kinds of conclusions within the context of infectious diseases are informative more generally.

STEPHAN PEISCHL – I think we should pay more attention to failed invasions and infectious diseases in humans. For example, I think cholera outbreaks often come from the same source, so you have a wave spreading through humans but it does not establish. The outbreaks are always from India, I think, so somehow the virus seems to fail to adapt to other conditions. I think that could be seen as a failed invasion, and it would be interesting to study.

MELANIA CRISTESCU – This is a complex problem because when considering failed invasions, we have to think about at which stage we consider them to have failed. There are multiple stages in the invasion process; so are they a failed invasion because they failed to be introduced, or because they failed to establish? New molecular techniques such as meta-barcoding give us the tools to identify species that are introduced into new ecosystems and many will fail to establish. In future, we are going to have large data sets that will allow us the possibility to answer these questions.

SPENCER BARRETT – Mark van Kleunen in his presentation discussed data from introductions to botanical gardens where some species have become invasive whereas others have not. Indeed, in the invasive plant literature, there have been several studies of this type. There is also the influential paper by Marcel Rejmánek and David Richardson (*Ecology*, 1996, **77**: 1655–1661) that looked at the introduction of pine trees (*Pinus*) to parts of the world to identify which species have become invasive and which have not. From this, they were able to identify several key traits (e.g., seed mass, short juvenile period, and short interval between large seed crops) that were associated with invasiveness. So the botanical literature has information that can address this kind of question.

RAFAEL ZENNI (UNIVERSITY OF TENNESSEE) – We published a paper (R. Zenni and M.A. Nunez, 2013, *Oikos*, **122**: 801–815) with more than a hundred cases of failed invasions trying to address patterns and mechanisms causing failed invasion using data from the biocontrol literature. There are many examples of biocontrol agents introduced all over the place under various conditions, and there is also a world global catalogue of introduction attempts of biocontrol agents. Introductions have been tracked to see which were successful and which were not, as well as the extent to which success was based on population size, climate matching, density, and the frequency of host plant populations. So these are sources that people can look at to address the issue of failed invasions.

REFERENCES

Burghardt L, Metcalf CJ, Wilczek AM, Schmitt J, Donohue K (2015) Modeling the influence of genetic and environmental variation on the expression of plant life cycles across landscapes. *American Naturalist*, **185**, 212–227.

Igic B, Busch JW (2013) Is self-fertilization an evolutionary dead end? *New Phytologist*, **198**, 386–397.

Rejmánek M, Richardson D (1996) What attributes make some plant species more invasive? *Ecology*, **77**, 1655–1661.

Zenni R, Nunez MA (2013) The elephant in the room: the role of failed invasions in understanding invasion biology. *Oikos*, **122**, 801–815.

Chapter 20

WHAT WE STILL DON'T KNOW ABOUT INVASION GENETICS

Dan G. Bock,* Celine Caseys,* Roger D. Cousens,†
Min A. Hahn,* Sylvia M. Heredia,* Sariel Hübner,*
Kathryn G. Turner,* Kenneth D. Whitney,‡
and Loren H. Rieseberg*§

*Department of Botany, University of British Columbia, 1316-6270
University Blvd., Vancouver, BC V6T 1Z4, Canada
†School of BioSciences, The University of Melbourne, Melbourne, VIC 3010, Australia
‡Department of Biology, University of New Mexico, Albuquerque, NM 87131-0001, USA
§Department of Biology, Indiana University, Bloomington, IN 47405, USA

Abstract

Publication of *The Genetics of Colonizing Species* in 1965 launched the field of invasion genetics and highlighted the value of biological invasions as natural ecological and evolutionary experiments. Here, we review the past 50 years of invasion genetics to assess what we have learned and what we still don't know, focusing on the genetic changes associated with invasive lineages and the evolutionary processes driving these changes. We also suggest potential studies to address still-unanswered questions. We now know, for example, that rapid adaptation of invaders is common and generally not limited by genetic variation. On the other hand, and contrary to prevailing opinion 50 years ago, the balance of evidence indicates that population bottlenecks and genetic drift typically have negative effects on invasion success, despite their potential to increase additive genetic variation and the frequency of peak shifts. Numerous unknowns remain, such as the sources of genetic variation, the role of so-called expansion load and the relative importance of propagule pressure vs. genetic diversity for successful establishment. While many such unknowns can be resolved by genomic studies, other questions may require manipulative experiments in model organisms. Such studies complement classical reciprocal transplant and field-based selection experiments, which are needed to link trait variation with components of fitness and population growth rates. We conclude by discussing the potential for studies of invasion genetics to reveal the limits to evolution and to stimulate the development of practical strategies to either minimize or maximize evolutionary responses to environmental change.

Previously published as an article in *Molecular Ecology* (2015) 24, 2277–2297, doi: 10.1111/mec.13032

INTRODUCTION

Ecologists and evolutionary biologists have a love–hate relationship with invasive species, defined here as widespread nonindigenous species. Although we dislike the harm they cause to the economy and environment, we appreciate their attributes as study organisms. They are easy to propagate and often have short generation times and small genomes (at least in plants). In addition, they typically produce very large numbers of offspring and frequently have the capacity for selfing or asexual reproduction, which can facilitate experimentation. Most importantly, at least from a scientific perspective, they represent natural ecological and evolutionary experiments unfolding in a recent historical time frame, thereby providing a window on ecological and evolutionary processes. This aspect is especially valuable to evolutionary biologists, who often are limited to making indirect inferences about evolutionary processes from DNA sequences, museum samples or from brief snapshots of evolution in contemporary populations.

Evolutionary studies of invasive lineages have generated two main kinds of information. First, as alluded to above, they have yielded valuable insights into evolutionary processes, especially with respect to the speed of adaptation and to the role of population bottlenecks in evolution. Second, these studies have informed us regarding the features of the invading organisms themselves and the evolutionary processes and the genetic changes that underlie these features (Lee 2002; Handley *et al.* 2011). Here, we focus on this second kind of information, as our review is about the evolutionary genetics of invasive lineages. However, the inferences we make about the genetics of invaders are broadly relevant to understanding how organisms successfully colonize new environments, regardless of whether they conform to any particular definition of an invasive species.

We first discuss genetic and genomic variation in invasive lineages. We ask about the sources of genetic variation, the roles of intra- and interspecific hybridization in invasions and whether certain kinds of genomic changes might serve as stimuli for invasiveness (see Glossary for terms relevant to invasiveness). We then examine key evolutionary processes, exploring the roles of genetic drift and pre- and post-introduction adaptation in successful invasions. We consider phenomena associated with the adaptive evolution of invasive lineages, including invasion 'lag phases', evolutionary trade-offs and phenotypic plasticity. Lastly, we describe what is known about the architecture of genetic changes associated with successful invasions and evaluate different approaches for identifying these changes. In keeping with the theme of this volume, we consider these topics in the context of what was known by the contributors to the Baker and Stebbins (1965) volume while identifying what we still don't know about each issue. Where possible, we suggest experiments or other kinds of studies that have the potential to address still-unanswered questions.

GENETIC VARIATION

Sources of variation

What is the primary source of genetic variation employed by natural selection during the evolution of invasive lineages? This question, which relates mainly to post-introduction adaptation, was a topic of discussion by the Baker and Stebbins (1965) contributors and remains unsolved. However, analyses of new genomic data sets are beginning to yield answers.

While some successful invaders arrive well-suited to their new environments, the success of others appears to depend on rapid local adaptation. Adaptation relies on two main sources of variation: pre-existing standing genetic variation and new beneficial mutations. Adaptation from standing genetic variation is generally faster and more predictable because standing variants typically have higher initial frequencies, which increases both the probability and speed of their fixation (Barrett & Schluter 2008; Prentis *et al.* 2008). In contrast, adaptation from new mutations is slowed by the waiting period for them to occur and reach fixation, which could be critical to the fate of the invasion. Despite the greater efficiency of adaptation from standing genetic variation compared with that from new mutations, Baker and Stebbins (1965) contributors gave greater credence to the latter (Dobzhansky 1965; Mayr 1965), especially as an explanation for the lag phase (Box 1). A third source of variation, which represents a distinct kind of standing variation, is the introgression of alleles from other species (Hedrick 2013). This process was deemed likely by the Baker and Stebbins (1965) contributors (see Interspecific hybridization, below).

The relative contributions of these sources of variation to the adaptation of invasive lineages are not obvious. Due to intra- and interspecific admixture, invasive lineages often harbour significant levels of standing variation (see below). On the other hand, Fisher's geometric model of adaptation (Fisher 1930) implies that new mutations are more likely to be beneficial in a population that is far from its adaptive optimum, which is likely for a new invader. Moreover, mutation accumulation experiments frequently find evidence of beneficial mutations (Heilbron *et al.* 2014). In practice, the rapid evolution of invasive lineages may involve more than one type of variation, and analytical approaches may fail to distinguish between them, especially when multiple colonizations have occurred or when selection is weak (Hermisson & Pennings 2005).

However, there are features of invasions, as well as new techniques, that may permit the different sources of variation to be determined, at least under some circumstances and for some loci. Most importantly, many invasions are recent, and source regions can often be identified, so one can ask whether variants under selection in the invaded range are present in relevant native populations. Also, by examining herbarium or museum specimens, it may be possible to pinpoint the source of variation and assess how allele frequencies have changed over time. Such an approach was recently

Box 1 Lag phases

Biologists have long noted that, on occasion, the rate of spread of an invasive species accelerates after a long period of quiescence, more rapidly than would be expected on the basis of a standard population model (Fig. 1). The lag phase, the period from introduction to acceleration, can in some instances be over a century in length. Anecdotes of lags abound, although it is rarely possible to estimate their length with any accu-

racy. The data are simply too poor: while the species is in its lag phase, recording intensity is inevitably extremely low and occurrences will be overlooked, while search effort may be increased as the species becomes of concern, creating bias (Cousens & Mortimer 1995). Even so, many invasion ecologists now appear to regard lags as the norm, rather than the exception.

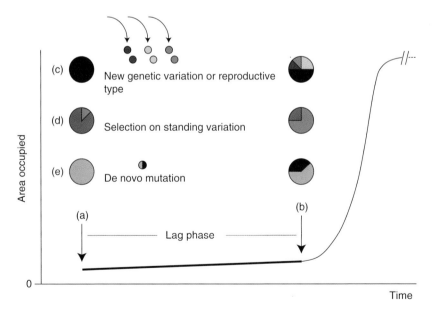

Box 1, Fig. 1 Example of the lag phase and potential genetic causes, showing hypothetical patterns of genetic variation at a single locus at (a) initial establishment and before (b) the onset of accelerated expansion. (c) shows an increase in genetic variation following immigration of new genotypes or new sexes. (d) illustrates selection on standing variation. (e) represents the origin of a *de novo* adaptive mutation. (*See insert for color representation of the figure.*)

There have been many plausible explanations given for lag phases (e.g. Crooks & Soulé 1999), although it is only rarely possible to ascribe a cause to a particular instance. Potential genetic causes include the following: evolution from standing genetic variation or new mutations resulting in increased local adaptation or dispersal capability; the introduction of genotypes that are either more fit or that allow novel, fitter gene combinations to be generated (Kolbe *et al.* 2004); and the introduction of the opposite sex where previously only one sex was present, thus allowing sexual reproduction to occur. Such genetic mechanisms were in the forefront of the thinking of the Baker and Stebbins (1965) contributors, in particular the need for sufficient time to elapse after introduction before a species became adapted to the novel conditions. This thinking persists, even without specific reference to a lag, with many researchers commenting routinely on the requirement for invaders to adapt to novel conditions. In some instances, however, the environment may not be so different from the native range, and plasticity may be sufficient for the invader to spread rapidly. Adaptation will no doubt occur in these situations, but more as a means of fine-tuning than overcoming a major fitness hurdle, and the change in rate of spread may be hard to discern.

There are also several possible nongenetic causes for lags. These include the following: sudden/rapid change in an environmental factor (e.g. land management, introduction of a symbiotic species, loss of a predator); overcoming an Allee effect (Aikio *et al.* 2010); eventual dispersal into another region of more suitable habitat (e.g. across a barrier around the initial introduction site); reaching an area where a more effective dispersal vector is available (Ridley 1930); and a threshold age for physiological maturity (Wangen & Webster 2006). An apparent lag may occur where early occurrence records in fact represent repeated failed establishments, but these are later followed by an event in which the species successfully establishes and then spreads.

It is tempting to argue that we need to better establish the mechanism behind particular instances of lag phase. Do genetic or nongenetic processes predominate and, if so, which ones? However, we face the challenge that we cannot monitor all invasions in fine detail in their early stages—there are simply too many and the areas too extensive—and those having extensive lag phases are only identified well after the lag has ended. Although herbarium or museum specimens may allow us to retrospectively search for genetic changes (e.g. Vandepitte *et al.* 2014), demographic and detailed population data typically will not be available through which to rule out the alternative mechanisms.

employed by Vandepitte *et al.* (2014) to show that the genetic changes underlying flowering time adaptation in colonizing populations of the Pyrenean rocket arose from standing variation. This study and others also illustrate how genome scans for footprints of selection, differentiation and hybridization permit detection of candidate genes and genomic regions that are associated with invasiveness, and when compared with native populations, reveal the likely source(s) of adaptive genetic changes (Prentis *et al.* 2008; Scascitelli *et al.* 2010; Tollenaere *et al.* 2013; Brown *et al.* 2014). Given wider implementation of these approaches over the coming decade, this major question is likely to be resolved soon.

Multiple introductions, genetic diversity and intraspecific admixture

Modern molecular techniques permit reconstruction of the phylogeographic histories of invaders, revealing invasion routes and putative source populations (e.g. Muirhead *et al.* 2008; Gray *et al.* 2014; Martin *et al.* 2014; Zhang *et al.* 2014)—information that typically was not available to the Baker and Stebbins (1965) contributors. Such studies indicate that many successful invasions are associated with multiple introductions and subsequent mixing (Bossdorf *et al.* 2005; Dlugosch & Parker 2008; Simberloff 2009). In part, this association may be driven by a correlation between number of introductions and propagule pressure (Simberloff 2009), because increased propagule pressure is known to aid founding populations in overcoming stochastic processes that would otherwise lead to extinction. However, increased propagule pressure is also likely to be associated with increased genetic diversity as a larger fraction of the native range alleles are likely to be sampled. Genetic diversity can have independent effects on colonization/invasion success through two sets of mechanisms, distinguished by timescale.

On shorter timescales, increased genetic diversity may increase colonization success through predominantly ecological mechanisms analogous to the species diversity effects seen in biodiversity-ecosystem functioning studies. These include selection effects, in which high-diversity founder populations have increased probabilities of containing (and becoming dominated by) a genotype with high invasive potential, and complementarity effects, in which either facilitation between genotypes or trait differences among genotypes lead to improved performance of mixtures over monocultures (Crawford & Whitney 2010; Forsman 2014). Forsman's recent (2014) meta-analysis found a significant positive effect of genetic diversity on measures of colonization success in plants and animals. This result should perhaps be interpreted with caution, as the meta-analysis did not correct for phylogenetic nonindependence and further, included several studies suffering from pseudo-replicated genetic diversity treatments. Further progress in this area will involve experimentally decoupling the purely numeric component of propagule pressure effects from the genetic diversity component and determining why their relative importance can vary in the field (e.g. Erfmeier et al. 2013).

On longer timescales, intraspecific genetic admixture may benefit invaders via the same set of evolutionary mechanisms proposed to benefit interspecific hybrids (Rius & Darling 2014). These include (i) an increase in genetic variation, providing a larger pool of raw material for adaptive evolution (Anderson 1949; Anderson & Stebbins 1954); (ii) the creation of novel or transgressive phenotypes through previously unexplored allele and gene combinations (Stebbins 1969; Lavergne & Molofsky 2007); (iii) heterosis, particularly when stabilized by nonsexual forms of reproduction (Baker 1965); and (iv) the masking or purging of deleterious mutations, which may reduce potentially negative effects of genetic bottlenecks and inbreeding (Ellstrand & Schierenbeck 2000; Keller & Waller 2002). Many observational and a few experimental studies indicate that admixture can contribute to invasion success (e.g. Kolbe et al. 2004; Wolfe et al. 2007; Keller & Taylor 2010; Verhoeven et al. 2011). For example, Keller & Taylor (2010) found that the level of genetic admixture in the invasive plant Silene vulgaris was associated with increased fecundity and thus may contribute to its success in the invaded range. However, in all cases, it is not possible to determine whether the invasion would have been successful without admixture. More experimental manipulations that directly test the effect of admixture on colonization success are needed (Rius & Darling 2014).

While genetic admixture and invasion appear to be linked, the specific roles of the four above-mentioned

evolutionary mechanisms are less clear. In general, increased genetic diversity and the creation of novel genotypes should have long-lasting effects by enhancing the adaptive potential of a population. This may be most beneficial to invaders experiencing novel environmental conditions, whereas in native environments, genetic admixture may result in the loss of local adaptation (Verhoeven et al. 2011). In contrast, the beneficial effects of heterosis are thought to be transitory in sexual populations and thus mainly important to establishment (Rius & Darling 2014). However, this depends on the genetic basis of heterosis (Hochholdinger & Hoecker 2007; Lippman & Zamir 2007). Genetic models for the evolution of heterosis include the following: dominance (enhanced performance due to the masking of deleterious recessive alleles from one parent by dominant alleles from the other parent), overdominance (enhanced performance due to beneficial interactions of alleles from different lineages at a single locus) and epistasis (enhanced performance due to beneficial interactions between loci from different lineages). Heterosis due to dominance (considered most common) and epistasis can be fixed by selection, resulting in the long-term preservation of heterotic effects and purging of genetic load.

Interspecific hybridization

As alluded to previously, the Baker and Stebbins (1965) contributors were well aware of the hypothesis that interspecific hybridization could act as an evolutionary stimulus, perhaps triggering colonizing and invasive behaviour. Several contributors had or were about to publish seminal papers on the issue (Heiser 1951; Anderson & Stebbins 1954; Stebbins 1959; Lewontin & Birch 1966; Panetsos & Baker 1968), and reference to the idea permeated the presentations and discussions. However, it was not obvious at the time how prevalent the process was, with E. O. Wilson asking whether 'introgression commonly results in a considerable increase in the fitness of a species' (Baker & Stebbins (1965) p. 213).

We have a much better handle on the issue today. Ellstrand & Schierenbeck (2000) published lists of species that were both hybrid derived and invasive, describing potential cases among a broad taxonomic array of plant invaders. Importantly, investigators also began to experimentally test performance of hybrids vs. parents in many systems, allowing cases of neutral or incidental hybridization in already-invasive taxa to be distinguished from cases where hybridization is a causal driver of increased invasiveness. Recently, these studies have been compiled in a systematic review and meta-analysis (Hovick & Whitney 2014) focusing on studies in which hybridization has been putatively associated with colonizing behaviour, and in which the performance of hybrids vs. their parental species has been experimentally tested. Meta-analyses of fecundity, survival and size (as proxies for population growth rate, λ) determined that wild hybrids are typically larger and more fecund than their parental species, while not differing in survival. Further, hybrid fecundity generally increases with generation, suggesting that natural selection can play an important role in shaping hybrid performance (and thus invasiveness) over time. However, these results are driven by tests in plants and further work is needed to understand patterns in animals and fungi. Also, hybridization and polyploidy are confounded in this and earlier studies, so more work is needed to understand both their independent and their synergistic effects.

Substantial progress has also been made in the identification of the genomic regions/alleles potentially involved in introgression events in colonizing or invasive species. Specific genes or quantitative trait loci (QTL) have been identified that control introgressing traits such as inflorescence morphology affecting pollination in *Senecio vulgaris* (Kim *et al.* 2008; Chapman & Abbott 2010), resistance to anticoagulant poisons in the Western European house mouse (Song *et al.* 2011), several fitness, ecophysiological, architectural and phenological traits in *Helianthus annuus texanus* (Whitney *et al.* 2015) and fitness traits in crop–wild *Lactuca* hybrids (Hartman *et al.* 2013). Other studies have used molecular signatures of selection or geographical clines in marker frequencies to identify introgressing genomic regions, without identification of the phenotypic traits affected (e.g. *Ambystoma*, Fitzpatrick *et al.* 2009; *Tamarix*, Gaskin & Kazmer 2009). In all of these examples, introgression is associated with increased adaptation, although in most cases, the link between the introgression of specific alleles and increased invasiveness has not been made. Nevertheless, the fact that specific alleles are frequently found that increased adaptation in the recipient species could be interpreted as support for the 'novel phenotypes' mechanism discussed in the previous section, while not ruling out contributions from other mechanisms.

It remains unclear, however, *why* hybridization sometimes results in increased colonization success and sometimes does not. One approach is to view the problem as a genotype × genotype × environment (G×G×E) interaction, where the interactions between the alleles provided by the donor species, the genome of the recipient species and the environment in which the hybrids are located determine whether the overall outcome is an increase, decrease or no change in λ. This approach emphasizes the extremely contingent nature of the process. However, it may be that even if each case is not individually predictable, the process across many cases is predictable at a statistical level. To our knowledge, there have not yet been attempts to evaluate such predictability. We suggest three hypotheses, one previously articulated and two perhaps new: (i) parental species separated by intermediate genetic distances might give rise to more successful hybrids than will less or more divergent parents (Ellstrand & Schierenbeck 2000; see also Stelkens & Seehausen 2009); (ii) the greater the adaptive fit of the donor parent to the environment to be colonized, the higher the chance that beneficial alleles are available via hybridization, and the more successful the hybrid; and (iii) the higher the frequency of hybridization between two parents, the greater the chance of 'hitting the G×G×E jackpot' and thus the greater the probability that a successful hybrid will arise.

Genomic variation

Punctuated changes in the structure and organization of the genome may also contribute to the evolution of invasiveness. Three types of genomic variation—namely polyploidy, genome size variation and chromosomal rearrangements—have been considered in this context.

Polyploidy

The first attempts to address the role of polyploidy in the evolution of invasiveness date from the early part of the twentieth century. These efforts relied on assessments of the frequency of polyploids among invasive species (e.g. Muntzing 1935; Gustafsson 1948) and were generally idiosyncratic, providing inconclusive answers. The views of the Baker and Stebbins (1965) contributors were mixed as well, with Mulligan (1965) writing that 'there is no evidence that polyploid weeds are particularly favoured for the colonization of newly available areas.' In contrast, Ehrendorfer (1965) listed polyploidy as one of the characteristics of good colonizers and provided examples of polyploid species that are considerably more widespread than their diploid progenitors.

Estimates of the frequency of polyploids among invasive species have expanded in both taxonomic scope and accuracy during the past two decades, with the implementation of methods such as flow cytometry, which allow ploidal levels to be identified *en masse*, and with the establishment of electronic databases of genome size and chromosome numbers. With these advances came more frequent reports that polyploids are over-represented among invasive species in regional floras (e.g. Verlaque *et al.* 2002; Pandit *et al.* 2006). In broad taxonomic surveys as well, polyploidy was found to be associated with invasiveness. Pandit *et al.* (2011), for example, compiled data from 81 invasive species and 2356 of their congeners and showed that being invasive is 20% more likely for polyploid species than for closely related diploid species.

However, the factors driving these patterns remain unclear. While studies comparing closely related diploid and polyploid species in their native and introduced ranges (e.g. Hahn *et al.* 2012a) can generate strong hypotheses, experimental work is needed to link the effects of polyploidization to invasiveness. These effects include genetic and epigenetic changes such as the masking of deleterious alleles, fixed heterozygosity and epigenetic remodelling, as well as morphological/physiological changes such as increased body size, altered drought tolerance and altered phenology (Soltis & Soltis 2000; te Beest *et al.* 2012). For allopolyploid invaders, some of these effects may be due to hybridization (e.g. Hegarty *et al.* 2011). Therefore, the most valuable studies will be those that experimentally decouple the effects of hybridization from polyploidy through comparisons of the invasiveness of both diploid parental species with both diploid and polyploid hybrids.

Even if differences in invasive potential are detected between ploidal levels, studies using natural polyploids are likely to overestimate the contribution of genome doubling, because of the confounding effects of genetic differences that accumulate after polyploid formation. To address this issue, experiments should be performed that use not only field-collected samples of varying ploidal levels, but also artificially obtained neo-polyploids. This approach has been used recently by Ramsey (2011) to show that genome duplication as well as post-polyploidization evolution facilitated adaptation of hexaploid cytotypes of the noninvasive wild yarrow (*Achillea borealis*) to Mediterranean habitats on the Pacific coast of North America.

Genome size

Inherently linked to ploidy, genome size has also been proposed to contribute to the evolution of invasiveness, albeit in the opposing direction (te Beest *et al.* 2012). Small genomes have been associated with traits such as short generation time, which may facilitate reproductive success under ephemeral conditions, or small seeds, which may enhance reproductive output and dispersal ability (Knight *et al.* 2005; te Beest *et al.* 2012). Support for this prediction has so far mainly come from broad surveys of the distribution of genome size values among invasive species (e.g. Kubesova *et al.* 2010; Pandit *et al.* 2014).

Evidence from specific systems, which may provide clues to the underlying traits and mechanisms, has, by comparison, been much more difficult to find. One possible exception is the study by Lavergne *et al.* (2010). The authors estimated genome sizes as well as rates of vegetative growth under glasshouse conditions for native (European) and invasive (North American) diploid genotypes of reed canary grass (*Phalaris arundinacea*). Patterns were in the expected direction: invasive genotypes had lower average genome sizes and displayed higher early growth rates than native genotypes. However, many previous reports of intraspecific variation in plant genome size have been discounted due to methodological issues (Greilhuber 1988; Price *et al.* 2000), and it is not clear whether the *Phalaris* study avoided these issues. Future work is required to confirm the genome size variation, elucidate the mechanisms linking genome size and growth rate and to make the connection between early growth rate and propensity to invade in the field.

Chromosomal rearrangements

In *The Genetics of Colonizing Species*, Carson (1965) and Dobzhansky (1965) make a distinction between 'flexible' and 'rigid' chromosomal inversions and discuss their dynamics in populations of cosmopolitan *Drosophila* species. 'Flexible' polymorphisms are shown to vary in frequency along environmental gradients and are hypothesized to contribute to local adaptation. 'Rigid' polymorphisms are shown to maintain unchanged frequencies in drosophilid populations and are proposed to result from heterozygote advantage. Under currently accepted models for the spread of chromosomal inversions (Hoffmann & Rieseberg 2008), 'flexible' polymorphisms can occur if inversions bring together alleles that are locally adapted, with or without epistasis. Similarly, 'rigid' polymorphisms can arise via overdominance, when inverted and noninverted arrangements carry different deleterious alleles (Hoffmann & Rieseberg 2008).

More recent work has provided some experimental support that inversion polymorphisms contribute to adaptation during biological invasions. For example, Prevosti *et al.* (1988) calculated correlations between chromosomal rearrangement frequencies and latitude for populations of *Drosophila subobscura* established along the Pacific coasts of North and South America. Striking similarities were observed along the two latitudinal clines, providing strong indication that chromosomal inversions are adaptive.

Few other studies have established a link between invasion success and inversion polymorphisms (although see Kirkpatrick & Barrett 2015). Even less is known about the potential contributions of other kinds of rearrangements to invasions. Future research should therefore aim to identify the role of chromosomal rearrangements in invasion potential in other systems. Moreover, the genes responsible for the associations between invasion success and inversion polymorphisms are unknown. One approach to address this is to use genome scans to identify targets of spatially variable selection within inversions (e.g. Fabian *et al.* 2012). The success of this approach will depend, however, on whether some recombination has occurred within the inverted region, breaking-up linkage disequilibrium away from the inversion breakpoints.

Epigenetic variation and invasion

The Baker and Stebbins (1965) contributors were aware of epigenetic variation: Waddington had previously coined the term, defining it as 'the branch of biology that studies the causal interactions between genes and their products, which bring the phenotype into being' (Waddington 1942). However, epigenetics did not feature in the symposium discussions, in part because it was largely a theoretical concept. We now know that epigenetic phenomena provide an information layer above the DNA sequence level and can

contribute to variation in gene expression and phenotype via multiple molecular mechanisms including DNA methylation, histone modifications, small RNAs and noncoding RNA (Kinoshita & Jacobsen 2012). Moreover, some epigenetic modifications are elicited by environmental factors and can be transmitted across generations (Verhoeven *et al.* 2010; Dowen *et al.* 2012).

Because invaders often exhibit reduced genetic variation in their new range (Dlugosch & Parker 2008), there has been interest in whether epigenetic variation could ameliorate this apparent handicap. For example, epigenetic diversity appears to compensate for the loss of genetic diversity and inbreeding in recently introduced Kenyan house sparrows (Liebl *et al.* 2013). In Japanese knotweed, successful invasion of diverse habitats was correlated with epigenetic differentiation in response to new and dynamic microclimate conditions (Richards *et al.* 2012). Experimental studies have shown that epigenetic modifications can be induced by specific abiotic and biotic stresses (Verhoeven *et al.* 2010; Dowen *et al.* 2012) and contribute to increased population biomass (Latzel *et al.* 2013). These results are consistent with a possible role for epigenetic variation in invasive species via adaptive phenotypic plasticity and by compensating for losses in genetic variability. However, the adaptive significance of epigenetic variation remains largely unknown, especially in the context of plant invasions. Field and common garden experiments are needed to differentiate between plastic and heritable epigenetic variation and to link this variation to specific phenotypes and to fitness (Richards *et al.* 2010).

EVOLUTIONARY PROCESSES

Genetic drift and invasion

Newly introduced populations often experience a genetic bottleneck, which can have potentially important consequences for their evolution and ultimate fate. The relationship between bottlenecks and variation in Mendelian traits such as molecular markers is well understood theoretically (Wright 1931; Dlugosch & Parker 2008), leading to the following predictions: (i) the loss of Mendelian variation via drift should correlate with both the severity and length of the bottleneck (Wright 1931); (ii) bottlenecks should cause greater reductions in allelic richness than in expected heterozygosity (Nei *et al.* 1975); and (iii) large shifts in allele frequencies are likely, especially for rare alleles that survive the bottleneck (Peischl *et al.* 2013). All three predictions have been validated by empirical studies of species invasions (Dlugosch & Parker 2008; Uller & Leimu 2011; Tsuchida *et al.* 2014). Such losses of diversity and/or drift-induced changes in allele frequencies have the potential to impede adaptive evolution, at least to the extent that Mendelian traits affect fitness. As discussed earlier, however, multiple introductions often restore lost diversity, not infrequently resulting in higher levels of diversity than in the native range because of admixture from genetically different source populations (Dlugosch & Parker 2008; Uller & Leimu 2011).

The effects of genetic bottlenecks on quantitative genetic variation, which is generally believed to underlie the majority of fitness related traits (Falconer & Mackay 1996), are less severe. As pointed out by Lewontin (1965, p. 481):

> If there is colonization by a single fertilized female... one-half of the additive, three-quarters of all the dominance variation, and a large amount of the epistatic variation are present in the offspring.

Moreover, theoretical and empirical studies indicate that population bottlenecks can convert dominance (Robertson 1952; Cockerham & Tachida 1988) and epistatic variance (Goodnight 1988; Whitlock *et al.* 1993; Cheverud & Routman 1995) to additive variance (Bryant *et al.* 1986). While conversion of the former is restricted to cases where the initial frequency of the recessive allele is low in the source population, conversion of epistatic variance is less restricted (Whitlock *et al.* 1993). Thus, it is perhaps unsurprising that comparisons of phenotypic and/or quantitative genetic variation in source and introduced populations have generally failed to find differences in variance (reviewed in Dlugosch & Parker (2008), although see Simberloff *et al.* (2000) and Van Heerwaarden *et al.* (2008) for examples of reduced and increased quantitative genetic variation, respectively). Ample evidence of rapid post-introduction adaptive differentiation (see below) further implies that genetic variation is generally not limiting in invaders.

The Baker and Stebbins (1965) contributors were enthusiastic about the possibility that bottlenecks associated with colonization might enable invaders to reach a new adaptive peak through a process put forward by Wright (1931). These ideas have gained some support from studies demonstrating gains in additive genetic variation following bottlenecks (e.g. Bryant *et al.* 1986), and from recent theory indicating that range expansions increase the frequency of peak shifts (Burton & Travis 2008). Nonetheless, we are unaware of examples where such gains have contributed to Wrightian peak shifts in introduced populations (Van Heerwaarden *et al.* 2008).

Another possible consequence of population bottlenecks and subsequent population expansion is the accumulation of deleterious mutations, which could limit invasion success (Peischl *et al.* 2013). Simulations indicate that extreme drift is created at the wave front of expanding populations because population density is low and growth rate is high (Edmonds *et al.* 2004). New and standing mutations at the wave front can 'surf' to high frequency whether they are neutral or deleterious (Klopfstein *et al.* 2006), creating what has been termed 'expansion load' (Peischl *et al.* 2013). Expansion load can reduce fitness over much of the newly expanded range and persist for thousands of generations.

The extent of expansion load in invading lineages is largely unknown, although an excess of deleterious mutations has been reported in non-African human populations (Peischl *et al.* 2013). Likewise, we are unaware of examples of colonization failure due to expansion load or of the operation of compensating mechanisms such as admixture (see above) or Allee effects (Glossary), which might limit its severity. Methods now exist for detection of deleterious mutations from genomic scan data (e.g. Adzhubei *et al.* 2010), so the extent of expansion load in invading lineages can be estimated (see Hodgins *et al.* 2015 and Peischl & Excoffier 2015). Linking load to failed invasions will be more challenging because of the confounding effects of other demographic and genetic factors associated with invasion success.

Pre-introduction adaptation

For a species to establish in a new location, its intrinsic rate of increase must be positive. This condition will be more likely if there is a close match between native and recipient environments. In other words, species should be pre-adapted to at least some novel geographical locations. Even in cases of a partial environmental match, adaptive phenotypic plasticity may be sufficient for the colonizer to survive and reproduce. Some indication of invasive potential may come from the breadth of the native range: a wide realized niche will increase the possibility that (if introduced) at least one genotype will be suited to a set of novel conditions. Researchers attempting to understand the degree of matching of environments in native and recipient regions, however, face a number of pitfalls associated with extrapolation (Mesgaran *et al*. 2014).

Perhaps because predicting the environmental match between organism and location is both case specific and data intensive, there has been a long-standing tradition of instead searching for universal, pre-adapted traits that are associated with invasive behaviour. For plants, Baker's list of traits (Baker 1965) that together would result in the 'ideal weed' is a touchstone, emphasizing a capacity for asexual reproduction, high fecundity, rapid growth to maturity, phenotypic plasticity and broad environmental tolerance. Similar lists have since been suggested for animals (e.g. Kolar & Lodge 2001). Many of the traits correspond to the concept of 'r-selected' species and are likely to result in high rates of increase.

Tests of the idea that invasive species come pre-equipped with particular traits have had mixed success. Many authors have observed that 'Baker traits' were both present in some noninvasive species and absent in some invasive species (Perrins *et al*. 1992; Mack 1996). More recently, there has been some success in distinguishing traits of invasive vs. noninvasive species (Rejmánek & Richardson 1996; Pyšek & Richardson 2007; van Kleunen *et al*. 2010). For example, Rejmánek & Richardson (1996) used multivariate techniques to identify short juvenile periods, a short interval between large seed crops, and small seed mass as good predictors of increased invasiveness in *Pinus*. A meta-analysis (van Kleunen *et al*. 2010) found evidence for higher values of performance traits in invasive vs. noninvasive plants, although we point out that samples of the former often derive from the invaded range, so trait values for invasives (and thus effect sizes) do not necessarily reflect pre-adaptation and could instead reflect post-introduction adaptation. However, these successes have been balanced with other cases in which invasiveness was not correlated with biological traits (Caley & Kuhnert 2006), and have been followed by scepticism as to whether the pursuit of predictive traits is worthwhile (Thompson & Davis 2011; Moles *et al*. 2012).

The future of the pre-adaptation paradigm is unclear. Certainly, it appears that further comparative studies of traits in highly invasive vs. noninvasive taxa will be carried out. A new approach involves breaking the invasion process into stages to isolate the traits that matter at each stage, with perhaps a higher chance of identifying consistent trait differences between invasive and noninvasive species (van Kleunen *et al*. 2015). It is also likely that others will continue to argue that other factors, for example post-introduction adaptation and the environmental context in which a colonizing species finds itself (e.g. when predators and pathogens have been left behind, aka 'enemy release'), are more explanatory than pre-existing traits in determining invasiveness.

Post-introduction adaptation

The success of many biological invasions may depend on the capacity of invasive species to adapt to novel environmental conditions. Such post-introduction adaptation was considered by a number of the Baker and Stebbins (1965) contributors: although the evidence provided was often indirect, rapid evolution of life history, reproductive and dispersal traits in plants (Ehrendorfer 1965; Harper 1965) and *Drosophila* (Dobzhansky 1965) was considered important to the colonization process (Lewontin 1965). Since then, observational and experimental studies have documented adaptive changes in invasive relative to native populations (reviewed in Dlugosch & Parker 2008; Prentis *et al*. 2008; Whitney & Gabler 2008; Felker-Quinn *et al*. 2013). Rates of evolution can be quite rapid, with many examples occurring in <50 years (Whitney & Gabler 2008). This makes sense, as environmental differences between native and invaded ranges should generate strong selective pressures. Indeed, rates of adaptive phenotypic change may be higher in human disturbed environments than in undisturbed contexts (Hendry *et al*. 2008), and a survey of herbarium specimens across 150 years in Australia showed significantly more morphological changes in introduced species than in Australian natives (Buswell *et al*. 2011).

Increased growth rate or reproductive capacity is frequently reported from field observations in the invaded range (Elton 1958; Crawley 1987; Thebaud & Simberloff 2001; Parker *et al*. 2013; Pandit *et al*. 2014; see also discussion of the EICA hypothesis in Box 2) and increasingly from common garden experiments

Box 2 Evolutionary trade-offs and invasion

Increased performance in competitive ability, size and fecundity of invasive populations relative to their native conspecifics has been addressed by multiple hypotheses, several of which invoke evolutionary trade-offs between self-defence, growth and reproduction (Fig. 1). Underlying all trade-off hypotheses is the assumption that organisms are unable to be both highly competitive or have high reproductive output and be highly tolerant of stressful conditions (Grime 1977). Many of the trade-offs described in the context of invasive species are based on modifications of interspecific interactions in the introduced range, such as release from natural enemies.

The most studied hypothesis in invasive plant species is the evolution of increased competitive ability (EICA) hypothesis, which posits that selection will favour genotypes with reduced allocation to herbivore defence and increased allocation to growth, reproductive output or competitive ability in the absence of herbivores characteristic of the native range (Blossey & Notzold 1995). Although increased performance in invasive individuals relative to natives is often observed in common garden experiments, shifts in defences are less common (Kumschick *et al.* 2013). As EICA is only supported if increases in plant growth are linked to

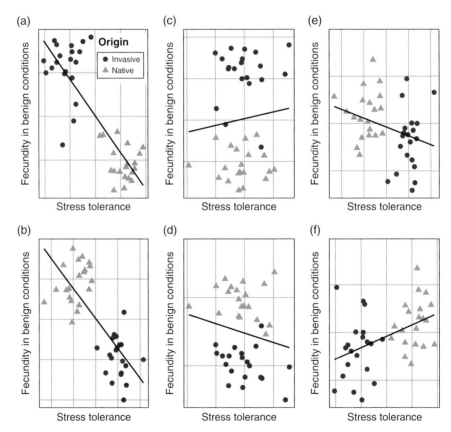

Box 2, Fig. 1 Simulated population means suggestive of evolution in the invaded range in a two-trait comparison. (a) and (b) represent trade-offs in resource allocation for increased fecundity (a) or increased stress tolerance (b) in the invaded range. (c–f) represent the result of introduction bottlenecks for either increased (d) or decreased (e) fecundity, or increased (e) or decreased (f) tolerance, but without correlated change in the other trait.

decreases in defence, evidence for EICA is equivocal (Felker-Quinn *et al.* 2013).

Trade-offs in invasive plant species may not be limited in response to herbivore defence only, but may include tolerance to stressful abiotic conditions (Bossdorf *et al.* 2005; He *et al.* 2010; Turner *et al.* 2014). Such trade-offs have been demonstrated in both natural and invasion contexts. For example, tolerance to serpentine soils in serpentine sunflower or drought in common ragweed comes at the expense of competitive ability and growth rate (Sambatti & Rice 2007; Hodgins & Rieseberg 2011). These studies and others suggest that invasive individuals may evolve a lower tolerance to biotic or abiotic stress to increase competitive ability, vigour and/or fecundity and therefore will perform relatively poorly under stressful conditions (Hodgins & Rieseberg 2011; Lachmuth *et al.* 2011; Kumschick *et al.* 2013; Turner *et al.* 2014). However, detecting a trade-off that has occurred can be complex and can depend on testing performance under the correct stressor. Trade-offs can occur in multiple directions (Fig. 1) and may include dispersal or competitive ability rather than a simple two-way relationship between performance and defence or tolerance (e.g. Burton *et al.* 2010). To thoroughly investigate potential trade-offs in invasive plant species, future studies should consider defence responses to specialist and generalist herbivores separately (Joshi & Vrieling 2005), differentiate constitutive and induced resistance (Kempel *et al.* 2011), assess the level of resistance in different plant tissue types (young and old, above and below ground; Alba *et al.* 2012) and account for confounding abiotic factors, latitudinal origins and climate effects (Colautti *et al.* 2009; Felker-Quinn *et al.* 2013; Rypel 2014). Furthermore, strategies favoured by selection may change over time, between different phases of an invasion (Dietz & Edwards 2006) or depending on the habitats invaded (Lachmuth *et al.* 2011).

(reviewed in Felker-Quinn *et al.* 2013). This improved fecundity could contribute to rapid spread and population growth in the invaded range. However, many experimental studies are limited in scope, thus limiting the generality of their conclusions. Observing the phenotypes of a single generation of native and invasive populations in a common environment may be insufficient to demonstrate adaptation to a novel habitat. Differences between native and invasive populations caused by maternal environmental effects need to be taken into account, as do the effects of latitudinal or environmental clines (Colautti *et al.* 2009).

Some of the best evidence supporting post-introduction adaptation and contemporary evolution in invasive species comes from recapitulation of clinal variation of species in their native ranges. Latitudinal clines in morphological wing traits, chromosome inversion frequencies, genetic variation and physiological resistance in invasive *Drosophila* populations have evolved to parallel native-range clines (Gilchrist *et al.* 2001, 2004, 2008; Hoffmann *et al.* 2002). Likewise, thermal adaptations in body size between a specialist and a generalist invasive *Drosophila* species showed positive corresponding variation between altitudinal and latitudinal gradients (Folguera *et al.* 2008). Finally, similar altitudinal clines in growth and reproductive traits were found for native and invasive populations of Asteraceae plants (Alexander *et al.* 2009).

Although the prevalence of post-introduction adaptation is well established, at least two critical areas remain to be investigated. First, post-introduction evolutionary change complicates risk-assessment schemes which aim to quantify the invasion potential of individual taxa (Whitney & Gabler 2008; Box 3). Such schemes would benefit from the development of better metrics of genetic variation, hybridization propensity and other features associated with adaptive potential. Second, and most importantly, it remains unclear how important post-arrival adaptation is to invasion outcomes. In other words, does evolutionary change ever tip the scales from a failed to a successful invasion? Or does such change simply accelerate the rate (or impact) of an invasion that would have been successful anyway? Answering these questions could close a major chapter in our understanding of the relative importance of evolutionary vs. ecological factors in invasions.

Phenotypic and developmental plasticity

As an alternative to coping with novel environmental conditions through local adaptation (above), a successful invader might employ generalist strategies that produce high performance under a wide range of conditions. In his classic study, Baker (1965) introduced the term 'general purpose genotype' to describe a genotype that possesses broad environmental tolerance and should be frequently found in weeds. Here, we consider the general purpose genotype in the broader framework of phenotypic and developmental plasticity.

In the context of invasions, we are most interested in plastic responses that confer fitness advantages to invaders. Richards *et al.* (2006) outlined three possible scenarios by which an invader may benefit from phenotypic plasticity: (i) plastic responses in morphological and physiological traits that permit fitness to be maintained across different stressful or unfavourable environments (i.e. fitness homeostasis), a 'Jack-of-all-trades' strategy; (ii) an invader may increase its fitness under favourable conditions (i.e. opportunism), a 'Master-of-some' strategy; or (iii) a combination of the first two strategies (i.e. 'Jack-and-Master'), which permits both fitness homeostasis and opportunism.

Many studies have compared phenotypic plasticity of invasive vs. noninvasive species and populations (Rice & Mack 1991; Sexton *et al.* 2002), and in some weedy plants, plasticity has also been shown to be adaptive (Funk 2008; Hahn *et al.* 2012b). However, recent meta-analyses (Davidson *et al.* 2011; Palacio-López & Gianoli 2011) came to conflicting conclusions concerning whether plasticity is generally important in invasions, a result that might be due to the transient evolution of plasticity itself (Lande 2009; Sultan *et al.* 2013). Thus, future meta-analyses should consider time since introduction in the interpretation of such data sets.

Theory predicts that frequent fluctuations of the environment will select for phenotypic and developmental plasticity (Meyers *et al.* 2005), while infrequent fluctuations favour local adaptation. Possibly, anthropogenic disturbance in the ancestral range may select for plastic genotypes that are then pre-adapted for invasion elsewhere (Hufbauer *et al.* 2012). Adaptation to extreme environmental changes, such as at the start of a biological invasion, can also favour rapid evolutionary increase of plasticity (Lande 2009, 2015). After an initial benefit from plasticity, however, an invader may lose its ability to express different phenotypes in favour of the expression of a locally adapted fixed phenotype through genetic assimilation (Pigliucci & Murren 2003; Lande 2015). The latter prediction assumes that there are costs and constraints associated with the maintenance of the genetic and physiological machinery required for plasticity (van Kleunen & Fischer 2005). The transient

nature of plasticity was recognized by the Baker and Stebbins (1965) contributors, who considered the future of general purpose genotypes to be 'rather dim' (Mayr 1965, p. 171). There were also suggestions that plasticity was associated with autogamy, apomixis, vegetative reproduction, hybridization and polyploidy (Baker 1965). While heterosis (whether expressed in diploid or polyploid hybrids) is known to stabilize fitness across environments (Schlichting 1986; Lippman & Zamir 2007), evidence for the other proposed associations remains sparse (te Beest *et al.* 2012; Hahn *et al.* 2012b).

Conceptual and theoretical work during the latter half of the 20th century predicted that two main kinds of genetic mechanisms would be responsible for phenotypic plasticity: (i) loci with environmentally sensitive alleles and (ii) regulatory loci that modify gene expression levels across different environments (Via *et al.* 1995). Since then, numerous 'plasticity genes' have been cloned and characterized and the molecular genetic mechanisms underlying plasticity are more diverse than previously surmised (reviewed in Des Marais *et al.* 2013; Pierik *et al.* 2013). While the majority of loci are environmentally sensitive loci such as photoreceptors or regulatory loci, other kinds of genes are involved as well. There is also evidence that epigenetic modifications such as DNA methylation and chromatin modification play a role in adaptive plasticity, in some cases by providing a type of epigenetic memory that enables accurate prediction and response to future conditions (Bastow *et al.* 2004). A future focus should be to link such mechanisms to evolutionary changes in plasticity that have accompanied biological invasions.

Genetic architecture and invasion

Understanding the genetic and molecular mechanisms that underlie the formation of invasive genotypes has been a central goal of invasion genetics, yet knowledge on the topic remains limited. The few currently available examples indicate—in agreement with theoretical expectations—that invasiveness is often underpinned by a small number of genes. Moreover, rapid evolution in invasive taxa does not appear to be mutation limited (above). Below, we discuss the genetic architecture of invasiveness in the framework of two general approaches, top-down (or forward) genetics and bottom-up (or reverse) genetics.

Box 3 Adaptation in biological control agents

Post-introduction adaptation is of particular interest in introduced biological control agents (Roderick *et al.* 2012) because it can lead to unwanted host shifts or other nontarget interactions in the introduced range. Even without evolutionary change, predicting the ecological impact of biological controls is difficult (Louda *et al.* 2005). The Baker and Stebbins (1965) contributors shared these concerns and discussed the likelihood of host shifts and post-introduction adaptation in biological control species (DeBach

1965; Wilson 1965). Collaborative work between geneticists and biological control workers was encouraged, with the goal of identifying adaptive genetic changes in introduced biological control populations.

Recent data confirm that biological control agents can evolve post-introduction. A well-known example is the cane toad, which was introduced into Australia 70 years ago to control populations of the cane beetle. Unfortunately, cane toads eat essentially

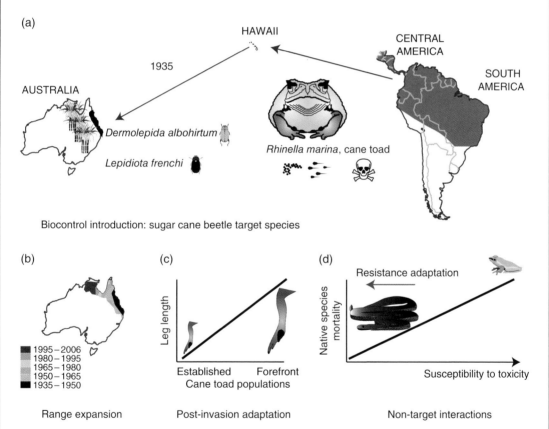

Box 3, Fig. 1 The cane toad (*Rhinella marina*) was introduced (a) to control cane beetles in sugar cane fields in northeastern Queensland in 1935 after successful use in Hawaii. Since then, cane toads have (b) expanded across tropical and subtropical Australia (Urban *et al.* 2008), increasing their rate of spread through, among others (Phillips & Shine 2006; Shine 2010), (c) evolution of longer legs (Phillips *et al.* 2006). Cane toads eat a wide variety of nontarget invertebrates reducing their population sizes. (d) Most native predators have declined as well due to lethal toxic ingestion of the toads, tadpoles and/or eggs with one known exception: the Australian black snake that has evolved physiological resistance to cane toad toxins (Phillips & Shine 2006; Shine 2010). (*See insert for color representation of the figure.*)

anything that moves, spread disease, are toxic to naïve predators and have evolved longer legs—factors that have contributed to the species' rapid spread and devastating ecological impact (Fig. 1; Roderick *et al.* 2012; Rollins & Shine 2015). Phylogenetic studies of host shifts, for example interkingdom host jumps and changes in habitat preference, nutrition mode and ecological role in the fungal genus *Trichoderma* (Chaverri & Samuels 2013), provide an additional cautionary note. Such results should provide pause to the biological control community, as host shifts and other nontarget impacts may become more likely with evolutionary change.

While host shifts remain challenging to predict, modelling techniques using food networks successfully predicted host shifts from native herbivores to nonnative plant species in central Europe (Pearse & Altermatt 2013). Also, genetic improvement of biological control agents through artificial selection (Roderick *et al.* 2012) has the potential to sharpen target specificity and impact, reduce nontarget effects and possibly slow evolutionary responses to new environments.

The top-down approach

The top-down approach starts with knowledge on the phenotypic traits that vary between invasive and noninvasive genotypes, or that have been targets of selection during the evolution of invasiveness. The task then becomes to identify loci that underlie those traits. This can be achieved through candidate gene analyses and through genomewide association or quantitative trait locus (QTL) mapping.

In some cases, dissecting the genetic basis of invasiveness can be relatively straightforward, if a list of candidate genes known to affect the phenotypes under investigation is available. Some of the best-known invasiveness genes come from studies in this category. One example comes from studies of the fire ant (*Solenopsis invicta*), in which multi-queened introduced populations are more ecologically destructive and show less aggression to conspecifics than single-queened native populations (Porter & Savignano 1990). Krieger & Ross (2002) were able to identify *Gp-9*, a gene that encodes an odorant-binding protein, as the locus underlying polymorphism in this social behaviour in *S. invicta*. Another example is the dopamine receptor D4 gene, which is associated with novelty seeking and activity behaviour in introduced populations of yellow-crowned bishops (Mueller *et al.* 2014).

More often than not, no information is available on the likely genetic underpinnings of invasiveness. In this case, efforts have been directed towards finding associations between genetic markers and phenotypes of interest in pools of unrelated individuals, or in experimental populations derived from crosses between parents that show extreme trait values.

This latter approach, known as QTL-mapping, has been used with some success in weed genomics (Basu *et al.* 2004). In allopolyploid invasive Johnson grass

(*Sorghum halepense*), Paterson *et al.* (1995) used crosses between the two species progenitors to understand the genetic basis of rhizomatousness, a weediness trait in this system. A small number of QTLs, most of which show additive or dominant gene action, were identified. More recently, Whitney *et al.* (2015) investigated loci involved in adaptive introgression associated with range expansion in the natural hybrid sunflower *H. annuus texanus*. Three donated QTLs were found that increased components of male and female fitness in the recipient species, likely as pleiotropic effects of phenological and architectural trait QTLs that colocalized with the fitness QTLs.

The bottom-up approach

The bottom-up approach does not require prior knowledge on traits that contribute to the propensity to invade. Instead, this strategy involves searching for changes in gene expression or allele frequency between pools of native and invasive genotypes, and making inferences about the traits involved based on knowledge of gene function.

Transcriptome analyses use microarrays or direct sequencing of RNA to identify genes that are differentially expressed in native and invasive genotypes. Lockwood & Somero (2011), for example, investigated the transcriptional response to low-salinity stress in two species of blue mussels (genus *Mytilus*). One of these, *M. galloprovincialis*, is invasive and has spread along the Pacific coast of California except areas North of Bodega Bay. This area is characterized by lower salinity and is still dominated by the native species *M. trossulus*. The authors performed a microarray analysis of *M. galloprovincialis* and *M. trossulus* individuals grown under benign conditions as well as those simulating abrupt decreases of salinity. Results revealed that most differentially expressed genes in response to salt stress are shared between the two species. Thus, either a small number of genes limit the spread of the invader, or most species-specific differences in tolerance to osmotic stress are mediated downstream of transcription (Lockwood & Somero 2011).

Similar studies have been performed for invasive plants. Hodgins *et al.* (2013), for example, examined differential gene expression between native and invasive genotypes of common ragweed (*Ambrosia artemisiifolia*) across 45 062 unigenes. In this case as well, a small fraction of the genes were differentially expressed between native and invasive samples. The functional categories over-represented among the differentially expressed genes were also in agreement with results from a common garden experiment in this system (Hodgins & Rieseberg 2011) and highlighted genes involved in oxidoreductase activity, response to blue light, as well as abiotic and biotic stress response, as strong candidates for invasiveness genes in this system.

At the genome level, bottom-up approaches rely on finding the signature of positive selection, which can include regions that show high levels of genetic differentiation or shifts in the site frequency spectrum of mutations. Puzey & Vallejo-Marín (2014), for example, performed one such genome scan analysis to detect the signature of positive selection during the invasion of monkeyflowers (*Mimulus guttatus*) in the UK. While a specific target of selection was not identified, genes located in swept regions were shown to be associated with flowering time, as well as biotic and abiotic stress (Puzey & Vallejo-Marín 2014). Moreover, two of these regions were positioned near or at a chromosomal inversion polymorphism associated with a number of morphological and life history differences in monkeyflowers (Puzey & Vallejo-Marín 2014).

In another recent example, Vandepitte *et al.* (2014) investigated the genetic basis of adaptation following the 1824 introduction of the Pyrenean rocket (*Sisymbrium austriacum* subsp. *chrysanthum*) in Belgium using native, contemporary invasive samples and herbarium specimens collected in the introduced area. Six genes involved in flowering were identified as outliers of genetic differentiation and experienced allele frequency changes over the course of the invasion process.

A concern with the bottom-up approach is false positives, which can arise due to nonequilibrium demographic histories (Lotterhos & Whitlock 2014), as well to genomic heterogeneity in mutation and recombination rates (Renaut *et al.* 2014). These issues can be especially problematic in invaders, as generally little is known about their genomes. Also, as previously discussed, populations at the invasion front undergo extreme drift, allowing neutral and deleterious alleles to surf to high frequency, mimicking the signature of selection. Further, the loci identified as 'invasion loci' remain hypotheses until further work confirms that they control actual invasiveness in the field.

The small number of studies investigating the genetic architecture of invasiveness currently precludes the making of many generalizations. It is unclear, for example, whether and how often the genetic architecture of invasiveness traits differs from that of other traits differentiating natural populations or species. For example, are recessive QTLs more frequently established in invasive populations? Theory predicts that the probability of fixation for advantageous mutations is higher if they are dominant (Haldane's sieve; Turner 1977). Because of frequent bottlenecks, this process might be less effective in invasive populations. Also, the extent to which evolution re-uses the same genes or genomic regions during the evolution of invasiveness remains unclear.

CONCLUSIONS

We have learned a great deal about invasion genetics since the *Genetics of Colonizing Species* was published 50 years ago. Thanks in part to the widespread application of molecular marker techniques, we have elucidated the geographical origin(s) of many invaders, as well as their invasion routes. We have discovered that invaders are surprisingly variable genetically and that their variability depends in large part on whether they result from single or multiple introductions. Strong evidence has accumulated in favour of a positive role for intraspecific admixture, hybridization and polyploidy in invasion success. On the other hand, the balance of evidence indicates that population bottlenecks and genetic drift likely have negative or no effects on invasion success, despite the potential for gains in additive genetic variation or increases in the frequency of peak shifts. We understand the environmental conditions favouring the evolution of phenotypic and developmental plasticity and have cloned and functionally characterized genes underlying plasticity and invasiveness. Most importantly, we now know that rapid adaptation of invaders is common and generally does not appear to be limited by genetic variation.

In addition to the things we think we know, certain hypotheses appear to be gaining support, while others are falling out of favour. For example, both theoretical and empirical evidence suggests that natural selection in invaders relies mainly on standing genetic variation. Likewise, there is increasing support for evolutionary trade-offs between abiotic stress tolerance and growth and reproduction, but support for similar trade-offs involving resistance to biotic stress appears to be declining (Box 2).

There also are numerous things that we don't know, which we have highlighted throughout this review and below, including:

• the relative roles of the numeric and genetic diversity components of propagule pressure in successful invasions;
• why hybridization sometimes results in increased colonization success and sometimes does not;
• whether chromosomal rearrangements, epigenetic modifications and shifts in genome size are important contributors to invasion success;
• whether the accumulation of deleterious mutations limits invasions and/or if compensatory mechanisms reduce the severity of expansion load;
• what traits or trait combinations, if any, best predict invasion success;
• why some invaders exhibit strong local adaptation and others do not;
• the generality and main cause of the lag phase;
• whether phenotypic plasticity evolves in a predictable way during the course of an invasion;
• which of the different strategies by which an invader may benefit from adaptive plasticity are most frequent;
• whether the genetic architecture of invasiveness traits differs from that of other traits differentiating natural populations or species;
• the extent of gene re-use during the evolution of invaders.

We also have suggested experiments or approaches to answer these questions. For example, many unknowns relating to sources of genetic variation, chromosomal rearrangements, genetic load, genetic architecture and gene re-use will fall to the power of evolutionary genomic approaches, perhaps within the next decade. Other questions will be more challenging, but manipulative experiments in model organisms have potential for decoupling the numeric and genetic diversity components of propagule pressure, assessing the potential role of epigenetic modification in the colonization of new habitats, and testing theoretical predictions regarding the temporal evolution of phenotypic plasticity. Of course, such experiments complement rather than replace classical reciprocal transplant and field-based selection experiments, which are required to connect trait variation with components of fitness and population growth rates. There also will continue to be a place for comparative studies, especially those that consider context, such as phylogenetic relationships, latitude, environment and the stage of the invasion, when making such inferences.

The Baker and Stebbins (1965) contributors were keenly aware that findings on the genetics of colonizing species, while fascinating in their own right, were important because they ramified throughout evolutionary biology. Evolutionary genetic studies of invasions tell us how species are likely to respond evolutionarily to changes in their environments, whether these changes come about through range expansions or occur *in situ*. We wish to know: What strength of selection (and over how long a period) can populations withstand? How much can they change their phenotype? How is their future evolutionary potential affected by past evolutionary change? More generally, what are the limits to evolution? These questions have become increasingly important as organisms must adapt to a changing world or face extirpation. Evolutionary genetic studies of invasive species have given us hope for the future by demonstrating multiple strategies by which organisms successfully respond to new environments, including rapid evolutionary change. Over the next 50 years, we expect studies of invasion genetics to reveal the limits to evolution (Blows & McGuigan 2015; Day 2015), as well as practical strategies to either minimize evolutionary change (such as in biological control agents) or maximize evolutionary potential (such as in native species facing environmental challenges), depending on the desired outcome.

ACKNOWLEDGEMENTS

We thank Wiley-Blackwell for supporting the Asilomar Conference on Invasion Genetics that stimulated this review. Research support was provided by a Natural Sciences and Engineering Research Council of Canada (NSERC) Discovery grant to LHR, NSERC Vanier CGS and Killam Doctoral Fellowships to DGB, Swiss

National Science Foundation postdoctoral fellowships to CC and MAH, and US National Science Foundation grant DEB 1257965 to KDW and LRH.

REFERENCES

Adzhubei IA, Schmidt S, Peshkin L et al. (2010) A method and server for predicting damaging missense mutations. *Nature Methods*, **7**, 248–249.

Aikio S, Duncan RP, Hulme PE (2010) Lag-phases in alien plant invasions: separating the facts from the artefacts. *Oikos*, **119**, 370–378.

Alba C, Bowers MD, Hufbauer R (2012) Combining optimal defense theory and the evolutionary dilemma model to refine predictions regarding plant invasion. *Ecology*, **93**, 1912–1921.

Alexander JM, Edwards PJ, Poll M, Parks CG, Dietz H (2009) Establishment of parallel altitudinal clines in traits of native and introduced forbs. *Ecology*, **90**, 612–622.

Anderson E (1949) *Introgressive Hybridization*. Chapman & Hall, London.

Anderson E, Stebbins GL (1954) Hybridization as an evolutionary stimulus. *Evolution*, **8**, 378–388.

Baker HG (1965) Characteristics and modes of origin of weeds. In: *The Genetics of Colonizing Species* (eds Baker HG, Stebbins GL), pp. 147–168. Academic Press, New York.

Baker HG, Stebbins GL (1965) *The Genetics of Colonizing Species*. Academic Press, New York, New York.

Barrett RDH, Schluter D (2008) Adaptation from standing genetic variation. *Trends in Ecology & Evolution*, **23**, 38–44.

Bastow R, Mylne JS, Lister C et al. (2004) Vernalization requires epigenetic silencing of FLC by histone methylation. *Nature*, **427**, 164–167.

Basu C, Halfhill MD, Mueller TC, Stewart CN (2004) Weed genomics: new tools to understand weed biology. *Trends in Plant Science*, **9**, 391–398.

te Beest M, Le Roux JJ, Richardson DM et al. (2012) The more the better? The role of polyploidy in facilitating plant invasions. *Annals of Botany*, **109**, 19–45.

Blossey B, Notzold R (1995) Evolution of increased competitive ability in invasive nonindigenous plants: a hypothesis. *Journal of Ecology*, **83**, 887–889.

Blows M, McGuigan K (2015) The distribution of genetic variance across phenotypic space and the response to selection. *Molecular Ecology*, **24**, 2056–2072.

Bossdorf O, Auge H, Lafuma L et al. (2005) Phenotypic and genetic differentiation between native and introduced plant populations. *Oecologia*, **144**, 1–11.

Brown AMV, Huynh LY, Bolender CM, Nelson KG, McCutcheon JP (2014) Population genomics of a symbiont in the early stages of a pest invasion. *Molecular Ecology*, **23**, 1516–1530.

Bryant EH, Mccommas SA, Combs LM (1986) The effect of an experimental bottleneck upon quantitative genetic-variation in the housefly. *Genetics*, **114**, 1191–1211.

Burton OJ, Travis JMJ (2008) The frequency of fitness peak shifts is increased at expanding range margins due to mutation surfing. *Genetics*, **179**, 941–950.

Burton OJ, Phillips BL, Travis JMJ (2010) Trade-offs and the evolution of life-histories during range expansion. *Ecology Letters*, **13**, 1210–1220.

Buswell JM, Moles AT, Hartley S (2011) Is rapid evolution common in introduced plant species? *Journal of Ecology*, **99**, 214–224.

Caley P, Kuhnert PM (2006) Application and evaluation of classification trees for screening unwanted plants. *Austral Ecology*, **31**, 647–655.

Carson HL (1965) Chromosomal morphism in geographically widespread species of *Drosophila*. In: *The Genetics of Colonizing Species* (eds Baker H, Stebbins G). Academic Press, New York.

Chapman MA, Abbott RJ (2010) Introgression of fitness genes across a ploidy barrier. *New Phytologist*, **186**, 63–71.

Chaverri P, Samuels GJ (2013) Evolution of habitat preference and nutrition mode in a cosmopolitan fungal genus with evidence of interkingdom host jumps and major shifts in Ecology. *Evolution*, **67**, 2823–2837.

Cheverud JM, Routman EJ (1995) Epistasis and its contribution to genetic variance-components. *Genetics*, **139**, 1455–1461.

Cockerham CC, Tachida H (1988) Permanency of response to selection for quantitative characters in finite populations. *Proceedings of the National Academy of Sciences, USA*, **85**, 1563–1565.

Colautti RI, Maron JL, Barrett SCH (2009) Common garden comparisons of native and introduced plant populations: latitudinal clines can obscure evolutionary inferences. *Evolutionary Applications*, **2**, 187–199.

Cousens R, Mortimer M (1995) *Dynamics of Weed Populations*. Cambridge University Press, Cambridge.

Crawford KM, Whitney KD (2010) Population genetic diversity influences colonization success. *Molecular Ecology*, **19**, 1253–1263.

Crawley MJ (1987) What makes a community invasible? In: *Colonization, Succession, and Stability: the 26th Symposium of the British Ecological Society Held Jointly with the Linnean Society of London*, pp. 429–453. Blackwell Scientific Publications, Oxford.

Crooks JA, Soulé ME (1999) Lag times in population explosions of invasive species: causes and implications. In: *Invasive Species and Biodiversity Management* (eds Sandlund O, Schei P, Viken A), pp. 103–125. Kluwer Academic Publishers, Dordrecht, the Netherlands.

Davidson AM, Jennions M, Nicotra AB (2011) Do invasive species show higher phenotypic plasticity than native species and, if so, is it adaptive? A meta-analysis. *Ecology Letters*, **14**, 419–431.

Day T (2015) Information entropy as a measure of genetic diversity and evolvability in colonization. *Molecular Ecology*, **24**, 2073–2083.

DeBach P (1965) Some biological and ecological phenomena associated with colonizing entomophagous insects. In: *The Genetics of Colonizing Species* (eds Baker HG, Stebbins GL), pp. 287–303. Academic Press, New York.

Des Marais LD, Hernandez KM, Juenger TE (2013) Genotype-by-environment interaction and plasticity: exploring genomic responses of plants to the abiotic environment. *Annual Review of Ecology, Evolution, and Systematics*, **44**, 5–29.

Dietz H, Edwards PJ (2006) Recognition that causal processes change during plant invasion helps explain conflicts in evidence. *Ecology*, **87**, 1359–1367.

Dlugosch KM, Parker IM (2008) Founding events in species invasions: genetic variation, adaptive evolution, and the role of multiple introductions. *Molecular Ecology*, **17**, 431–449.

Dobzhansky T (1965) "Wild" and "domestic" species of *Drosophila*. In: *The Genetics of Colonizing Species* (eds Baker HG, Stebbins GL), pp. 533–546. Academic Press, New York.

Dowen RH, Pelizzola M, Schmitz RJ et al. (2012) Widespread dynamic DNA methylation in response to biotic stress. *Proceedings of the National Academy of Sciences, USA*, **109**, E2183–E2191.

Edmonds CA, Lillie AS, Cavalli-Sforza LL (2004) Mutations arising in the wave front of an expanding population. *Proceedings of the National Academy of Sciences, USA*, **101**, 975–979.

Ehrendorfer F (1965) Dispersal mechanisms, genetic systems, and colonizing abilities in some flowering plant families. In: *The Genetics of Colonizing Species* (eds Baker HG, Stebbins GL), pp. 331–351. Academic Press, New York.

Ellstrand NC, Schierenbeck KA (2000) Hybridization as a stimulus for the evolution of invasiveness in plants? *Proceedings of the National Academy of Sciences, USA*, **97**, 7043–7050.

Elton CS (1958) *The Ecology of Invasions by Animals and Plants*. Methuen & Company, Ltd, London.

Erfmeier A, Hantsch L, Bruelheide H (2013) The role of propagule pressure, genetic diversity and microsite availability for *Senecio vernalis* invasion. *PLoS One*, **8**, e57029.

Fabian DK, Kapun M, Nolte V et al. (2012) Genome-wide patterns of latitudinal differentiation among populations of *Drosophila melanogaster* from North America. *Molecular Ecology*, **21**, 4748–4769.

Falconer DS, Mackay TFC (1996) *Introduction to Quantitative Genetics*, 4th edn. Pearson education, Essex.

Felker-Quinn E, Schweitzer JA, Bailey JK (2013) Meta-analysis reveals evolution in invasive plant species but little support for Evolution of Increased Competitive Ability (EICA). *Ecology and Evolution*, **3**, 739–751.

Fisher RA (1930) *The Genetical Theory of Natural Selection*. Clarendon Press, Oxford.

Fitzpatrick BM, Johnson JR, Kump DK et al. (2009) Rapid fixation of non-native alleles revealed by genome-wide SNP analysis of hybrid tiger salamanders. *Bmc Evolutionary Biology*, **9**, 176.

Folguera G, Ceballos S, Spezzi L, Fanara JJ, Hasson E (2008) Clinal variation in developmental time and viability, and the response to thermal treatments in two species of *Drosophila*. *Biological Journal of the Linnean Society*, **95**, 233–245.

Forsman A (2014) Effects of genotypic and phenotypic variation on establishment are important for conservation, invasion, and infection biology. *Proceedings of the National Academy of Sciences, USA*, **111**, 302–307.

Funk JL (2008) Differences in plasticity between invasive and native plants from a low resource environment. *Journal of Ecology*, **96**, 1162–1173.

Gaskin JF, Kazmer DJ (2009) Introgression between invasive saltcedars (*Tamarix chinensis* and *T. ramosissima*) in the USA. *Biological Invasions*, **11**, 1121–1130.

Gilchrist GW, Huey RB, Serra L (2001) Rapid evolution of wing size clines in *Drosophila subobscura*. *Genetica*, **112**, 273–286.

Gilchrist GW, Huey RB, Balanya J, Pascual M, Serra L (2004) A time series of evolution in action: a latitudinal cline in wing size in South American *Drosophila subobscura*. *Evolution*, **58**, 768–780.

Gilchrist GW, Jeffers LM, West B et al. (2008) Clinal patterns of desiccation and starvation resistance in ancestral and invading populations of *Drosophila subobscura*. *Evolutionary Applications*, **1**, 513–523.

Goodnight CJ (1988) Epistasis and the effect of founder events on the additive genetic variance. *Evolution*, **42**, 441–454.

Gray MM, Wegmann D, Haasl RJ et al. (2014) Demographic history of a recent invasion of house mice on the isolated Island of Gough. *Molecular Ecology*, **23**, 1923–1939.

Greilhuber J (1988) Self-tanning—a new and important source of stoichiometric error in cytophotometric determination of nuclear DNA content in plants. *Plant Systematics and Evolution*, **158**, 87–96.

Grime JP (1977) Evidence for existence of three primary strategies in plants and its relevance to ecological and evolutionary theory. *American Naturalist*, **111**, 1169–1194.

Gustafsson A (1948) Polyploidy, life-form and vegetative reproduction. *Hereditas*, **34**, 1–22.

Hahn MA, Buckley YM, Müller-Schärer H (2012a) Increased population growth rate in invasive polyploid *Centaurea stoebe* in a common garden. *Ecology Letters*, **15**, 947–954.

Hahn MA, van Kleunen M, Müller-Schärer H (2012b) Increased phenotypic plasticity to climate may have boosted the invasion success of polyploid *Centaurea stoebe*. *PLoS One*, **7**, e50284.

Handley LJL, Estoup A, Evans DM et al. (2011) Ecological genetics of invasive alien species. *Biocontrol*, **56**, 409?428.

Harper JL (1965) Establishment, aggression, and cohabitation in weedy species. In: *The Genetics of Colonizing Species* (eds Baker HG, Stebbins GL), pp. 243–265. Academic Press, New York.

Hartman Y, Uwimana B, Hooftman DAP *et al.* (2013) Genomic and environmental selection patterns in two distinct lettuce crop-wild hybrid crosses. *Evolutionary Applications*, **6**, 569–584.

He W-M, Thelen GC, Ridenour WM, Callaway RM (2010) Is there a risk to living large? Large size correlates with reduced growth when stressed for knapweed populations. *Biological Invasions*, **12**, 3591–3598.

Hedrick PW (2013) Adaptive introgression in animals: examples and comparison to new mutation and standing variation as sources of adaptive variation. *Molecular Ecology*, **22**, 4606–4618.

Hegarty MJ, Batstone T, Barker GL *et al.* (2011) Nonadditive changes to cytosine methylation as a consequence of hybridization and genome duplication in *Senecio* (Asteraceae). *Molecular Ecology*, **20**, 105–113.

Heilbron K, Toll-Riera M, Kojadinovic M, MacLean RC (2014) Fitness is strongly influenced by rare mutations of large effect in a microbial mutation accumulation experiment. *Genetics*, **197**, 981–990.

Heiser Jr CB (1951) Hybridization in the annual sunflowers: *Helianthus annuus* × *H. debilis* var. *cucumerifolius*. *Evolution*, **5**, 42–51.

Hendry AP, Farrugia TJ, Kinnison MT (2008) Human influences on rates of phenotypic change in wild animal populations. *Molecular Ecology*, **17**, 20–29.

Hermisson J, Pennings PS (2005) Soft sweeps: molecular population genetics of adaptation from standing genetic variation. *Genetics*, **169**, 2335–2352.

Hochholdinger F, Hoecker N (2007) Towards the molecular basis of heterosis. *Trends in Plant Science*, **12**, 427–432.

Hodgins KA, Rieseberg LH (2011) Genetic differentiation in life-history traits of introduced and native common ragweed (*Ambrosia artemisiifolia*) populations. *Journal of Evolutionary Biology*, **24**, 2731–2749.

Hodgins KA, Lai Z, Nurkowski K, Huang J, Rieseberg LH (2013) The molecular basis of invasiveness: differences in gene expression of native and introduced common ragweed (*Ambrosia artemisiifolia*) in stressful and benign environments. *Molecular Ecology*, **22**, 2496–2510.

Hodgins KS, Bock DG, Hahn MA *et al.* (2015) Comparative genomics in the Asteraceae reveals little evidence for parallel evolutionary change in invasive taxa. *Molecular Ecology*, **24**, 2226–2240.

Hoffmann AA, Rieseberg LH (2008) Revisiting the impact of inversions in evolution: from population genetic markers to drivers of adaptive shifts and speciation? *Annual Review of Ecology, Evolution, and Systematics*, **39**, 21–42.

Hoffmann AA, Anderson A, Hallas R (2002) Opposing clines for high and low temperature resistance in *Drosophila melanogaster*. *Ecology Letters*, **5**, 614–618.

Hovick SM, Whitney KD (2014) Hybridisation is associated with increased fecundity and size in invasive taxa: meta-analytic support for the hybridisation-invasion hypothesis. *Ecology Letters*, **17**, 1464–1477.

Hufbauer RA, Facon B, Ravigne V *et al.* (2012) Anthropogenically induced adaptation to invade (AIAI): contemporary adaptation to human-altered habitats within the native range can promote invasions. *Evolutionary Applications*, **5**, 89–101.

Joshi J, Vrieling K (2005) The enemy release and EICA hypothesis revisited: incorporating the fundamental difference between specialist and generalist herbivores. *Ecology Letters*, **8**, 704–714.

Keller SR, Taylor DR (2010) Genomic admixture increases fitness during a biological invasion. *Journal of Evolutionary Biology*, **23**, 1720–1731.

Keller LF, Waller DM (2002) Inbreeding effects in wild populations. *Trends in Ecology & Evolution*, **17**, 230–241.

Kempel A, Schadler M, Chrobock T, Fischer M, van Kleunen M (2011) Tradeoffs associated with constitutive and induced plant resistance against herbivory. *Proceedings of the National Academy of Sciences, USA*, **108**, 5685–5689.

Kim M, Cui ML, Cubas P *et al.* (2008) Regulatory genes control a key morphological and ecological trait transferred between species. *Science*, **322**, 1116–1119.

Kinoshita T, Jacobsen SE (2012) Opening the door to epigenetics in PCP. *Plant and Cell Physiology*, **53**, 763–765.

Kirkpatrick M, Barrett B (2015) Chromosome inversions, adaptive cassettes, and the evolution of species' ranges. *Molecular Ecology*, **24**, 2046–2055.

van Kleunen M, Fischer M (2005) Constraints on the evolution of adaptive phenotypic plasticity in plants. *New Phytologist*, **166**, 49–60.

van Kleunen M, Weber E, Fischer M (2010) A meta-analysis of trait differences between invasive and non-invasive plant species. *Ecology Letters*, **13**, 235–245.

van Kleunen M, Dawson W, Maurel N (2015) Characteristics of successful alien plants. *Molecular Ecology*, **24**, 1954–1968.

Klopfstein S, Currat M, Excoffier L (2006) The fate of mutations surfing on the wave of a range expansion. *Molecular Biology and Evolution*, **23**, 482–490.

Knight CA, Molinari NA, Petrov DA (2005) The large genome constraint hypothesis: evolution, ecology and phenotype. *Annals of Botany*, **95**, 177–190.

Kolar CS, Lodge DM (2001) Progress in invasion biology: predicting invaders. *Trends in Ecology & Evolution*, **16**, 199–204.

Kolbe JJ, Glor RE, Schettino LRG *et al.* (2004) Genetic variation increases during biological invasion by a Cuban lizard. *Nature*, **431**, 177–181.

Krieger MJB, Ross KG (2002) Identification of a major gene regulating complex social behavior. *Science*, **295**, 328–332.

Kubesova M, Moravcova L, Suda J, Jarosik V, Pyšek P (2010) Naturalized plants have smaller genomes than their non-invading relatives: a flow cytometric analysis of the Czech alien flora. *Preslia*, **82**, 81–96.

Kumschick S, Hufbauer RA, Alba C, Blumenthal DM (2013) Evolution of fast-growing and more resistant phenotypes in introduced common mullein (*Verbascum thapsus*). *Journal of Ecology*, **101**, 378–387.

Lachmuth S, Durka W, Schurr FM (2011) Differentiation of reproductive and competitive ability in the invaded range of *Senecio inaequidens*: the role of genetic Allee effects, adaptive and nonadaptive evolution. *New Phytologist*, **192**, 529–541.

Lande R (2009) Adaptation to an extraordinary environment by evolution of phenotypic plasticity and genetic assimilation. *Journal of Evolutionary Biology*, **22**, 1435–1446.

Lande R (2015) Evolution of phenotypic plasticity in colonizing species. *Molecular Ecology*, **24**, 2038–2045.

Latzel V, Allan E, Silveira AB *et al.* (2013) Epigenetic diversity increases the productivity and stability of plant populations. *Nature Communications*, **4**, 2875.

Lavergne S, Molofsky J (2007) Increased genetic variation and evolutionary potential drive the success of an invasive grass. *Proceedings of the National Academy of Sciences, USA*, **104**, 3883–3888.

Lavergne S, Muenke NJ, Molofsky J (2010) Genome size reduction can trigger rapid phenotypic evolution in invasive plants. *Annals of Botany*, **105**, 109–116.

Lee CE (2002) Evolutionary genetics of invasive species. *Trends in Ecology & Evolution*, **17**, 386–391.

Lewontin R (1965) Selection for colonizing ability. In: *The Genetics of Colonizing Species* (eds Baker HG, Stebbins GL), pp. 77–94. Academic Press, New York.

Lewontin R, Birch L (1966) Hybridization as a source of variation for adaptation to new environments. *Evolution*, **20**, 315–336.

Liebl AL, Schrey AW, Richards CL, Martin LB (2013) Patterns of DNA methylation throughout a range expansion of an introduced songbird. *Integrative and Comparative Biology*, **53**, 351–358.

Lippman ZB, Zamir D (2007) Heterosis: revisiting the magic. *Trends in Genetics*, **23**, 60–66.

Lockwood BL, Somero GN (2011) Transcriptomic responses to salinity stress in invasive and native blue mussels (genus *Mytilus*). *Molecular Ecology*, **20**, 517–529.

Lockwood JL, Cassey P, Blackburn TM (2009) The more you introduce the more you get: the role of colonization pressure and propagule pressure in invasion ecology. *Diversity and Distributions*, **15**, 904–910.

Lotterhos KE, Whitlock MC (2014) Evaluation of demographic history and neutral parameterization on the performance of FST outlier tests. *Molecular Ecology*, **23**, 2178–2192.

Louda SM, Rand TA, Russell FL, Arnett AE (2005) Assessment of ecological risks in weed biocontrol: input from retrospective ecological analyses. *Biological Control*, **35**, 253–264.

Mack RN (1996) Predicting the identity and fate of plant invaders: emergent and emerging approaches. *Biological Conservation*, **78**, 107–121.

Martin MD, Zimmer EA, Olsen MT *et al.* (2014) Herbarium specimens reveal a historical shift in phylogeographic structure of common ragweed during native range disturbance. *Molecular Ecology*, **23**, 1701–1716.

Mayr E (1965) The nature of colonizations in birds. In: *The Genetics of Colonizing Species* (eds Baker HG, Stebbins GL), pp. 29–43. Academic Press, New York.

Mesgaran MB, Cousens RD, Webber BL (2014) Here be dragons: a tool for quantifying novelty due to covariate range and correlation change when projecting species distribution models. *Diversity and Distributions*, **20**, 1147–1159.

Meyers LA, Ancel FD, Lachmann M (2005) Evolution of genetic potential. *PLoS Computational Biology*, **1**, 236–243.

Moles AT, Flores-Moreno H, Bonser SP *et al.* (2012) Invasions: the trail behind, the path ahead, and a test of a disturbing idea. *Journal of Ecology*, **100**, 116–127.

Mueller JC, Edelaar P, Carrete M *et al.* (2014) Behaviour-related DRD4 polymorphisms in invasive bird populations. *Molecular Ecology*, **23**, 2876–2885.

Muirhead JR, Gray DK, Kelly DW *et al.* (2008) Identifying the source of species invasions: sampling intensity vs. genetic diversity. *Molecular Ecology*, **17**, 1020–1035.

Mulligan G (1965) Recent colonization by herbaceous plants in Canada. In: *The Genetics of Colonizing Species* (eds Baker HG, Stebbins GL), pp. 127–143. Academic Press, New York.

Muntzing A (1935) The evolutionary significance of autopolyploidy. *Hereditas*, **21**, 263–378.

Nei M, Maruyama T, Chakraborty R (1975) The bottleneck effect and genetic variability in populations. *Evolution*, **29**, 1–10.

Palacio-López K, Gianoli E (2011) Invasive plants do not display greater phenotypic plasticity than their native or noninvasive counterparts: a meta-analysis. *Oikos*, **120**, 1393–1401.

Pandit MK, Tan HTW, Bisht MS (2006) Polyploidy in invasive plant species of Singapore. *Botanical Journal of the Linnean Society*, **151**, 395–403.

Pandit MK, Pocock MJO, Kunin WE (2011) Ploidy influences rarity and invasiveness in plants. *Journal of Ecology*, **99**, 1108–1115.

Pandit MK, White SM, Pocock MJO (2014) The contrasting effects of genome size, chromosome number and ploidy level on plant invasiveness: a global analysis. *New Phytologist*, **203**, 697–703.

Panetsos CA, Baker HG (1968) Origin of variation in wild *Raphanus Sativus* (Cruciferae) in California. *Genetica*, **38**, 243–274.

Parker JD, Torchin ME, Hufbauer RA *et al.* (2013) Do invasive species perform better in their new ranges? *Ecology*, **94**, 985–994.

Paterson AH, Schertz KF, Lin YR, Liu SC, Chang YL (1995) The weediness of wild plants—molecular analysis of genes influencing dispersal and persistence of johnsongrass, *Sorghum Halepense* (L) Pers. *Proceedings of the National Academy of Sciences, USA*, **92**, 6127–6131.

Pearse IS, Altermatt F (2013) Predicting novel trophic interactions in a non-native world. *Ecology Letters*, **16**, 1088–1094.

Peischl S, Excoffier L (2015) Expansion load: recessive mutations and the role of standing genetic variation. *Molecular Ecology*, **24**, 2084–2094.

Peischl S, Dupanloup I, Kirkpatrick M, Excoffier L (2013) On the accumulation of deleterious mutations during range expansions. *Molecular Ecology*, **22**, 5972–5982.

Perrins J, Williamson M, Fitter A (1992) Do annual weeds have predictable characters. *Acta Oecologica – International Journal of Ecology*, **13**, 517–533.

Phillips BL, Shine R (2006) An invasive species induces rapid adaptive change in a native predator: cane toads and black snakes in Australia. *Proceedings of the Royal Society B-Biological Sciences*, **273**, 1545–1550.

Phillips BL, Brown GP, Webb JK, Shine R (2006) Invasion and the evolution of speed in toads. *Nature*, **439**, 803.

Pierik R, Mommer L, Voesenek LACJ (2013) Molecular mechanisms of plant competition: neighbour detection and response strategies. *Functional Ecology*, **27**, 841–853.

Pigliucci M, Murren CJ (2003) Perspective: genetic assimilation and a possible evolutionary paradox: can macroevolution sometimes be so fast as to pass us by? *Evolution*, **57**, 1455–1464.

Porter SD, Savignano DA (1990) Invasion of polygyne fire ants decimates native ants and disrupts arthropod community. *Ecology*, **71**, 2095–2106.

Prentis PJ, Wilson JRU, Dormontt EE, Richardson DM, Lowe AJ (2008) Adaptive evolution in invasive species. *Trends in Plant Science*, **13**, 288–294.

Prevosti A, Ribo G, Serra L et al. (1988) Colonization of America by *Drosophila subobscura*: experiment in natural populations that supports the adaptive role of chromosomal-inversion polymorphism. *Proceedings of the National Academy of Sciences, USA*, **85**, 5597–5600.

Price HJ, Hodnett G, Johnston JS (2000) Sunflower (*Helianthus annuus*) leaves contain compounds that reduce nuclear propidium iodide fluorescence. *Annals of Botany*, **86**, 929–934.

Puzey J, Vallejo-Marín M (2014) Genomics of invasion: diversity and selection in introduced populations of monkeyflowers (*Mimulus guttatus*). *Molecular Ecology*, **23**, 4472–4485.

Pyšek P, Richardson DM (2007) Traits associated with invasiveness in alien plants: where do we stand? In: *Biological Invasions* (ed. Nentwig W), pp. 97–125. Springer, Heidelberg.

Ramsey J (2011) Polyploidy and ecological adaptation in wild yarrow. *Proceedings of the National Academy of Sciences, USA*, **108**, 7096–7101.

Rejmánek M, Richardson DM (1996) What attributes make some plant species more invasive? *Ecology*, **77**, 1655–1661.

Renaut S, Owens GL, Rieseberg LH (2014) Shared selective pressure and local genomic landscape lead to repeatable patterns of genomic divergence in sunflowers. *Molecular Ecology*, **23**, 311–324.

Rice KJ, Mack RN (1991) Ecological genetics of *Bromus tectorum*. II. Intraspecific variation in phenotypic plasticity. *Oecologia*, **88**, 84–90.

Richards CL, Bossdorf O, Muth NZ, Gurevitch J, Pigliucci M (2006) Jack of all trades, master of some? On the role of phenotypic plasticity in plant invasions. *Ecology Letters*, **9**, 981–993.

Richards CL, Bossdorf O, Verhoeven KJF (2010) Understanding natural epigenetic variation. *New Phytologist*, **187**, 562–564.

Richards CL, Schrey AW, Pigliucci M (2012) Invasion of diverse habitats by few Japanese knotweed genotypes is correlated with epigenetic differentiation. *Ecology Letters*, **15**, 1016–1025.

Ridley HN (1930) *Dispersal of Plants Throughout the World*. L. Reeve & Co, Ashford.

Rius M, Darling JA (2014) How important is intraspecific genetic admixture to the success of colonising populations? *Trends in Ecology & Evolution*, **29**, 233–242.

Robertson A (1952) The effect of inbreeding on the variation due to recessive genes. *Genetics*, **37**, 188–207.

Roderick GK, Hufbauer R, Navajas M (2012) Evolution and biological control. *Evolutionary Applications*, **5**, 419–423.

Rollins LA, Shine R (2015) A genetic perspective on rapid evolution in cane toads (*Rhinella marina*). *Molecular Ecology*, **24**, 2264–2276.

Rypel AL (2014) Do invasive freshwater fish species grow better when they are invasive? *Oikos*, **123**, 279–289.

Sambatti JBM, Rice KJ (2007) Functional ecology of ecotypic differentiation in the Californian serpentine sunflower (*Helianthus exilis*). *New Phytologist*, **175**, 107–119.

Scascitelli M, Whitney KD, Randell RA et al. (2010) Genome scan of hybridizing sunflowers from Texas (*Helianthus annuus* and *H. debilis*) reveals asymmetric patterns of introgression and small islands of genomic differentiation. *Molecular Ecology*, **19**, 521–541.

Schlichting CD (1986) The evolution of phenotypic plasticity in plants. *Annual Review of Ecology and Systematics*, **17**, 667–693.

Sexton JP, McKay JK, Sala A (2002) Plasticity and genetic diversity may allow saltcedar to invade cold climates in North America. *Ecological Applications*, **12**, 1652–1660.

Shine R (2010) The ecological impact of onvasive cane toads (*Bufo Marinus*) in Australia. *Quarterly Review of Biology*, **85**, 253–291.

Simberloff D (2009) The role of propagule pressure in biological invasions. *Annual Review of Ecology Evolution and Systematics*, **40**, 81–102.

Simberloff D, Dayan T, Jones C, Ogura G (2000) Character displacement and release in the small Indian mongoose, *Herpestes javanicus*. *Ecology*, **81**, 2086–2099.

Soltis PS, Soltis DE (2000) The role of genetic and genomic attributes in the success of polyploids. *Proceedings of the National Academy of Sciences, USA*, **97**, 7051–7057.

Song Y, Endepols S, Klemann N *et al.* (2011) Adaptive introgression of anticoagulant rodent poison resistance by hybridization between old world mice. *Current Biology*, **21**, 1296–1301.

Stebbins GL (1959) The role of hybridization in evolution. *Proceedings of the American Philosophical Society*, **103**, 231–251.

Stebbins GL (1969) The significance of hybridization for plant taxonomy and evolution. *Taxon*, **18**, 26–35.

Stelkens R, Seehausen O (2009) Genetic distance between species predicts novel trait expression in their hybrids. *Evolution*, **63**, 884–897.

Sultan SE, Horgan-Kobelski T, Nichols LM, Riggs CE, Waples RK (2013) A resurrection study reveals rapid adaptive evolution within populations of an invasive plant. *Evolutionary Applications*, **6**, 266–278.

Thebaud C, Simberloff D (2001) Are plants really larger in their introduced ranges? *American Naturalist*, **157**, 231–236.

Thompson K, Davis MA (2011) Why research on traits of invasive plants tells us very little. *Trends in Ecology & Evolution*, **26**, 155–156.

Tollenaere C, Jacquet S, Ivanova S *et al.* (2013) Beyond an AFLP genome scan towards the identification of immune genes involved in plague resistance in *Rattus rattus* from Madagascar. *Molecular Ecology*, **22**, 354–367.

Tsuchida K, Kudo K, Ishiguro N (2014) Genetic structure of an introduced paper wasp, *Polistes chinensis antennalis* (Hymenoptera, Vespidae) in New Zealand. *Molecular Ecology*, **23**, 4018–4034.

Turner JR (1977) Butterfly mimicry: the genetical evolution of an adaptation. *Evolutionary Biology*, **10**, 163–206.

Turner KG, Hufbauer RA, Rieseberg LH (2014) Rapid evolution of an invasive weed. *New Phytologist*, **202**, 309–321.

Uller T, Leimu R (2011) Founder events predict changes in genetic diversity during human-mediated range expansions. *Global Change Biology*, **17**, 3478–3485.

Urban MC, Phillips BL, Skelly DK, Shine R (2008) A toad more traveled: the heterogeneous invasion dynamics of cane toads in Australia. *American Naturalist*, **171**, E134–E148.

Van Heerwaarden B, Willi Y, Kristensen TN, Hoffmann AA (2008) Population bottlenecks increase additive genetic variance but do not break a selection limit in rain forest *Drosophila*. *Genetics*, **179**, 2135–2146.

Vandepitte K, De Meyer T, Helsen K *et al.* (2014) Rapid genetic adaptation precedes the spread of an exotic plant species. *Molecular Ecology*, **23**, 2157–2164.

Verhoeven KJF, Jansen JJ, van Dijk PJ, Biere A (2010) Stress-induced DNA methylation changes and their heritability in asexual dandelions. *New Phytologist*, **185**, 1108–1118.

Verhoeven KJF, Macel M, Wolfe LM, Biere A (2011) Population admixture, biological invasions and the balance between local adaptation and inbreeding depression. *Proceedings of the Royal Society B-Biological Sciences*, **278**, 2–8.

Verlaque R, Aboucaya A, Fridlender A (2002) Invasive alien flora of France: ecology, life-forms and polyploidy. *Botanica Helvetica*, **112**, 121–136.

Via S, Gomulkiewicz R, Dejong G *et al.* (1995) Adaptive phenotypic plasticity—consensus and controversy. *Trends in Ecology & Evolution*, **10**, 212–217.

Waddington CH (1942) The epigenotype. *Endeavour*, **1**, 18–20.

Wangen SR, Webster CR (2006) Potential for multiple lag phases during biotic invasions: reconstructing an invasion of the exotic tree *Acer platanoides*. *Journal of Applied Ecology*, **43**, 258–268.

Whitlock MC, Phillips PC, Wade MJ (1993) Gene interaction affects the additive genetic variance in subdivided populations with migration and extinction. *Evolution*, **47**, 1758–1769.

Whitney KD, Gabler CA (2008) Rapid evolution in introduced species','invasive traits' and recipient communities: challenges for predicting invasive potential. *Diversity and Distributions*, **14**, 569–580.

Whitney KD, Broman KW, Kane NC *et al.* (2015) Quantitative trait locus mapping identifies candidate alleles involved in adaptive introgression and range expansion in a wild sunflower. *Molecular Ecology*, **24**, 2194–2211.

Wilson F (1965) Biological control and the genetics of colonizing species. In: *The Genetics of Colonizing Species* (eds Baker HG, Stebbins GL), pp. 307–330. Academic Press, New York.

Wolfe LM, Blair AC, Penna BM (2007) Does intraspecific hybridization contribute to the evolution of invasiveness?: an experimental test. *Biological Invasions*, **9**, 515–521.

Wright S (1931) Evolution in Mendelian populations. *Genetics*, **16**, 0097–0159.

Zhang B, Edwards O, Kang L, Fuller S (2014) A multi-genome analysis approach enables tracking of the invasion of a single Russian wheat aphid (*Diuraphis noxia*) clone throughout the New World. *Molecular Ecology*, **23**, 1940–1951.

GLOSSARY

Adaptive introgression	The movement of fitness-increasing alleles from one species to another via hybridization and backcrossing.
Additive genetic variance	The proportion of genetic variance in a phenotypic trait that is due to the additive or main effect of alleles.
Adaptive peak	A high point on an adaptive landscape (the surface of a three-dimensional graph that is used to visualize the relationship between genotypes or phenotypes and fitness).
Allee effect	The phenomenon by which the per capita rate of increase decreases, or reaches 0 or negative values, in populations for which conspecifics are not numerous enough.
Dominance genetic variance	The proportion of genetic variance in a phenotypic trait due to dominant gene action, which is an interaction between alleles at the same gene locus.
Epigenetic variation	Functionally relevant variation in the genome that does not involve modifications in the underlying DNA sequence, such as DNA methylation, histone modifications or noncoding RNA.
Epistasis	The condition by which two or more independently inherited genes interact to control a phenotype.
Epistatic (interaction) genetic variance	The proportion of genetic variance in a phenotypic trait that is due to interactions between alleles at two or more gene loci.
Expansion load	Gradual accumulation of deleterious mutations during range expansion, which occurs because of increased genetic drift at the leading edge of the expansion front.
Heterosis	Phenotypic superiority of a hybrid over its parents, due to increased levels of heterozygosity.
Invasiveness	The ability of a species to become widespread when introduced to locations outside its natural geographical range.
Lag phase	The time between initial introduction and subsequent rapid population growth.
Linkage disequilibrium	Nonrandom association of alleles at two or more loci. Note that while such nonrandom associations most commonly result from genetic linkage, they can also arise due to selection or extreme drift.
Local adaptation	Enhanced fitness of local populations compared to nonlocal populations, driven by spatial variation in selection pressures.
Noninvasive species	A nonindigenous species that, contrary to an invasive species, does not achieve widespread distribution in its new environment.
Outbreeding depression	The reduction in fitness for offspring resulting from crosses between individuals of different populations.
Overdominance	The condition by which the heterozygote produces a phenotype more extreme than that of either homozygote.
Phenotypic plasticity	The ability of a genotype to produce different phenotypes in response to environmental variation. This plasticity may be adaptive, maladaptive or neutral.
Pleiotropy	The condition by which a gene affects more than one phenotypic character.
Pre-introduction adaptation	A situation in which an invader is already well adapted to the conditions in its introduced range, typically because of a close match between the native and introduced environments.

(continued)

Propagule pressure	The total number of individuals introduced at a given location, which is the product of the number of introduction events (propagule number) and mean the number of individuals introduced per event (propagule size) (Lockwood *et al.* 2009).
r selection	Selection for increased rates of reproduction, associated with a reduced investment per capita offspring.
Realized niche	The niche an organism occupies in an environment as a result of factors such as competition for resources, which constrain the acquisition of the fundamental (or potential) niche.
Selective sweep	Rapid increase in the frequency of an allele and nearby linked neutral variants under strong positive selection for the allele.
Transgressive Segregation	The formation of extreme phenotypes (relative to those of the parental lines) in segregating hybrid populations.

INDEX

Bold numbers refer to figures and tables.

Invasion Genetics: The Baker and Stebbins Legacy, First Edition. Edited by Spencer C. H. Barrett, Robert I. Colautti,
Katrina M. Dlugosch, and Loren H. Rieseberg.
© 2017 John Wiley & Sons, Ltd. Published 2017 by John Wiley & Sons, Ltd.